W0227375

PLANETARY NEBULAE

NGC 1501, a planetary nebula exhibiting an axis of symmetry, many condensations, and sharp outer edges. Photograph in the light of Hα by Dr R. Minkowski, 200-inch telescope, Mount Wilson and Palomar Observatories photograph. Scale 1″.6/mm. *See also Minkowski's paper, below, p. 456.*

INTERNATIONAL ASTRONOMICAL UNION
UNION ASTRONOMIQUE INTERNATIONALE

SYMPOSIUM No. 34

HELD IN TATRANSKÁ LOMNICA, CZECHOSLOVAKIA
4–8 SEPTEMBER 1967

PLANETARY NEBULAE

EDITED BY

D. E. OSTERBROCK AND C. R. O'DELL

WITH THE EDITORIAL ASSISTANCE OF

E. F. SWAN

D. REIDEL PUBLISHING COMPANY

DORDRECHT-HOLLAND

1968

Published on behalf of
the International Astronomical Union
by
D. Reidel Publishing Company, Dordrecht, Holland

Softcover reprint of the hardcover 1st edition 1968

ISBN-13:978-94-010-3475-3 e-ISBN-13:978-94-010-3473-9
DOI: 10.1007/978-94-010-3473-9

PREFACE

This written account of the Symposium on Planetary Nebulae was prepared from manuscripts submitted by the participants. Nearly every paper that was presented at the meeting is reproduced here, in either complete or abbreviated form. The discussions have been somewhat shortened and rearranged, but we have tried to preserve the essential points and the general tenor of the exchanges. Participants who spoke in the discussion were asked immediately for written remarks, which were then edited, reproduced, and circulated at the meeting by the highly effective local Secretariat organized by Dr Perek. In addition, notes of the discussion taken by Mrs Edith F. Swan and by the undersigned were used.

We wish to thank all the authors for their unusually good cooperation. We are especially grateful to Dr Minkowski, who kindly provided many excellent reproductions of Mount Wilson and Palomar photographs, mostly taken by himself, of various planetary nebulae. We are particularly indebted to Mrs Swan, who attended the Symposium, made notes on the papers and discussions as they occurred, and did much of the checking and editing of the manuscripts. In addition, we are very grateful to Mrs Evelyn Seaver, who also did much of the checking, editing, and retyping of manuscripts, and to Dr B.L. Webster, Miss Rebecca Todd, Mr Joseph Tapscott, and Mr Dennis Schatz, who provided excellent assistance in the preparation of this volume.

D. E. OSTERBROCK
C. R. O'DELL

INTRODUCTION

Some few years ago, interest in planetary nebulae was mainly restricted to the study of certain special problems, such as the development of observational techniques and the study of physical processes in the nebulae. More recent work has led to major advances in our understanding of the planetary nebula phenomenon, and of its significance as a stage in the evolution of the stars. It therefore seemed that an IAU Symposium, bringing together workers concerned with all aspects of these problems, would be very opportune.

The Symposium was held in the Hotel Morava at Tatranská Lomnica, Czechoslovakia, September 4–8, 1967. The Organizing Committee consisted of: M.J. Seaton (Chairman), L.H. Aller, Y. Andrillat, K.H. Böhm, D.E. Osterbrock, L. Perek, S.B. Pikelner, and B.A. Vorontsov-Velyaminov. All local arrangements were the responsibility of Dr L. Perek and Engineer V. Rajsky. The participants at the Symposium were very appreciative of the excellent hospitality provided by their Czechoslovak hosts. Special mention should be made of the hospitality provided by Dr L. Pajdušáková, of the Astronomical Institute of the Slovak Academy of Sciences at Tatranská Lomnica, and of the help provided by members of her staff, and of members of the staff of the Ondřejov Observatory, in the material organization of the Symposium.

Some financial assistance was provided by the Executive Committee of the International Astronomical Union, and was used to help certain participants with travel expenses.

M. J. SEATON

TABLE OF CONTENTS

SESSION III – PHYSICAL PROCESSES

SESSION IV – STRUCTURE AND DYNAMICS

SESSION V – CENTRAL STARS

SESSION VI – ORIGIN AND EVOLUTION

SESSION VII – CONCLUDING DISCUSSION

LIST OF PARTICIPANTS

G. Abell, University of California, Los Angeles, U.S.A.
L. H. Aller, University of California, Los Angeles, U.S.A.
H. Andrillat, Faculté des Sciences, Montpellier, France
Y. Andrillat, Faculté des Sciences, Montpellier, France
P. Andrle, Astronomický ústav ČSAV, Czechoslovakia
V. P. Arhipova, Sternberg State Astronomical Institute, U.S.S.R.
A. Baglin, Institut d'Astrophysique, France
R. Bajcar, Astronomický ústav SAV, Czechoslovakia
F. Bertola, Osservatorio Astronomico, Padova, Italy
M. Blaha, Astronomický ústav ČSAV, Czechoslovakia
K. H. Böhm, Universität Heidelberg, GFR
E. Böhm-Vitense, Universität Heidelberg, GFR
A. A. Boyarchuk, Crimean Astrophysical Observatory, U.S.S.R.
J. H. Cahn, University College London, England
E. R. Capriotti, Perkins Observatory, U.S.A.
C. Chevalier, Institut d'Astrophysique, France
M. Chopinet, Observatoire de l'Université de Bordeaux, France

---

1. Shao
2. Khromov
3. Wares
4. M. Liller
5. Mathews
6. Vaughan
7. Malaise
8. W. Liller
9. Grant
10. Cahn
11. Bertola
12. Evans
13. Pecker
14. Goldsworthy
15. Westerlund
16. Chopinet
17. McCarroll
18. Gershberg
19. Gebbie
20. Feast
21. Arhipova
22. Boyarchuk
23[1]. Dokuchaeva
23. Garstang
24. Kostjakova
25. Hummer
26. Ringuelet-Kaswalder
27. Flower
28. Blaha
29. I. Pronik
30. Rose
31. V. Pronik
32. Hekela
33. Kahn
34. Faulkner
35. Woltjer
36. Kohoutek
37. Sobouti
38. Koelbloed
39. Deinzer
40. Koubský
41. E. Swan
42. Böhm
43. Williams
44. G. Swan
45. Mayer
46. Miller
47. Capriotti
48. O'Dell
49. Bretz
50. Woyk
51. Osterbrock
52. Münch
53. Böhm-Vitense
54. Lambrecht
55. Weidemann
56. Horák
57. Menzel
58. Abell
59. Veteśnik
60. Chevalier
61. Onderlička
62. Lang
63. Bajcár
64. Temesvary
65. Tremko
66. Minkowski
67. Pottasch
68. Miyamoto
69. Perinotto
70. Thompson
71. Pik-Sin The
72. Elsmore
73. Davies
74. Salpeter
75. Terzian
76. Underhill
77. Savedoff
78. Gurzadian
79. Reeves
80. Aller
81. Perek
82. Van Horn
83. M. Seaton
84. Menon
85. J. Seaton
86. Sheglov
87. Thomasson

J. G. Davies, Nuffield Radio Astronomy Laboratories, England
W. Deinzer, Universität Heidelberg, GFR
O. D. Dokuchaeva, Sternberg State Astronomical Institute, U.S.S.R.
M. V. Dolidze, Abastumani Observatory, U.S.S.R.
F. Dossin, European Southern Observatory, Hamburg, GFR
B. Elsmore, Cavendish Laboratory, England
D. S. Evans, Royal Observatory, South Africa
D. J. Faulkner, Mt. Stromlo Observatory, Australia
M. W. Feast, Radcliffe Observatory, South Africa
D. R. Flower, University College London, England
R. H. Garstang, Joint Institute for Laboratory Astrophysics, U.S.A.
K. B. Gebbie, University College London, England
R. E. Gershberg, Crimean Astrophysical Observatory, U.S.S.R.
F. A. Goldsworthy, The University, Leeds, England
I. Grant, Atlas Computer Laboratory, England
G. A. Gurzadian, Burakan Astrophysical Observatory, U.S.S.R.
J. Hekela, Astronomický ústav ČSAV, Czechoslovakia
T. Horák, Svatopluka Čecha, Brno, Czechoslovakia
L. Houziaux, Université de Liège, Belgium
D. C. Hummer, Joint Institute for Laboratory Astrophysics, U.S.A.
M. A. Kaftan-Kassim, State University of New York at Albany, U.S.A.
F. D. Kahn, The University, Manchester, England
G. S. Khromov, Sternberg State Astronomical Institute, U.S.S.R.
R. Kippenhahn, Universitäts-Sternwarte Göttingen, GFR
D. Koelbloed, Astronomical Institute, Amsterdam, The Netherlands
L. Kohoutek, Astronomický ústav ČSAV, Czechoslovakia
E. B. Kostjakova, Sternberg State Astronomical Institute, U.S.S.R.
P. Koubský, Astronomický ústav ČSAV, Czechoslovakia
H. Lambrecht, Universitäts-Sternwarte, Jena, GDR
K. Lang, Astronomicky ústav University J. E. Purkyně, Czechoslovakia
M. H. Liller, Harvard College Observatory, U.S.A.
W. Liller, Harvard College Observatory, U.S.A.
D. Malaise, Université de Liège, Belgique
W. G. Mathews, University of California, San Diego, U.S.A.
P. Mayer, Astronomický ústav Karlovy University, Czechoslovakia
R. McCarroll, Observatoire de Paris, France
T. K. Menon, National Radio Astronomy Observatory, U.S.A.
D. H. Menzel, Harvard Observatory and Smithsonian Astrophysical Observatory, U.S.A.
J. S. Miller, University of California, Santa Cruz, U.S.A.
R. L. Minkowski, University of California, Berkeley, U.S.A.

S. Miyamoto, University of Kyoto, Japan
G. Münch, Mt. Wilson and Palomar Observatories, U.S.A.
C. R. O'Dell, Yerkes Observatory, U.S.A.
D. E. Osterbrock, University of Wisconsin, U.S.A.
B. Onderlička, Astronomický ústav University J. E. Purkyně, Czechoslovakia
L. Perek, Astronomický ústav ČSAV, Czechoslovakia
M. Perinotto, Osservatorio Astrofisico, Asiago, Italy
S. R. Pottash, Sterrekundig Laboratorium Kapteyn, Groningen, The Netherlands
I. N. Pronik, Crimean Astrophysical Observatory, U.S.S.R.
V. I. Pronik, Crimean Astrophysical Observatory, U.S.S.R.
H. Reeves, Institut de Physique Nucléaire, France
A. Ringuelet-Kaswalder, Universidad Nacional de la Plata, Argentina
W. K. Rose, Princeton University, U.S.A.
E. E. Salpeter, Cornell University, U.S.A.
M. P. Savedoff, University of Rochester, U.S.A.
M. J. Seaton, University College London, England
C.-Y. Shao, Harvard College Observatory and Smithsonian Astrophysical Observatory, U.S.A.
Y. Sobouti, Pahlavi University, Iran
E. Swan, University of Wisconsin, U.S.A.
P. V. Sheglov, Sternberg State Astronomical Institute, U.S.S.R.
Y. Terzian, Arecibo Ionospheric Observatory, U.S.A.
Pik-Sin The, Bosscha Observatory, Indonesia
P. Thomasson, University of Manchester, England
A. R. Thompson, Stanford University, U.S.A.
J. Tremko, Astronomický ústav SAV, Czechoslovakia
E. V. Turchaninova, University of Kiev, U.S.S.R.
A. B. Underhill, Sterrewacht Utrecht, The Netherlands
H. M. Van Horn, University of Rochester, U.S.A.
C. Vauclair, Paris, France
A. H. Vaughan Jr., Mt. Wilson and Palomar Observatories, U.S.A.
M. Vetešník, Astronomický ústav University J. E. Purkyně, Czechoslovakia
G. W. Wares, Air Force Cambridge Research Laboratories, U.S.A.
V. Weidemann, Universität, Kiel, GFR
B. E. Westerlund, University of Arizona, U.S.A.
R. E. Williams, University of Arizona, U.S.A.
Ch. Wimel-Pecker, Institut d'Astrophysique, France
L. Woltjer, Columbia University, U.S.A.
E. Woyk (Chvojková, E.), Astronomický ústav ČSAV, Czechoslovakia
J. Zverko, Astronomicky ústav SAV, Czechoslovakia

AN INTRODUCTORY REVIEW

M.J. SEATON
(University College London, England)

> The actors are at hand; and, by their show,
> You shall know all that you are like to know.
> (Shakespeare, *A Midsummer Night's Dream*)

The Organising Committee of this Symposium agreed that we should begin with a prologue and end with an epilogue. The first task of a prologue is to introduce the actors: we have The Eskimo, The Owl, The Helix and The Dumb-Bell (Figure 1) and many more. The *dramatis personae* of all known galactic planetary nebulae has been prepared for us by Dr Perek and Dr Kohoutek, and we are all much indebted to them for their labours. Planetaries in the Magellanic Clouds will be introduced by Dr Westerlund.

Planetary nebulae are easily recognised, they can be observed to great distances and they belong to an interesting intermediate population type. The galactic distribution of planetaries, which will be discussed by Perek, is of great importance for studies of galactic structure; thus of all objects which can be observed in the visible spectrum, the planetaries provide the best means of determining the direction to the galactic centre. The problem of determining distances is one of great importance and of considerable difficulty. It is closely related to the problem of determining the masses of the nebulae.

Consider a uniform spherically symmetric nebula of radius R, ionized mass M_i and surface brightness S in a hydrogen line. It may be shown that

$$R = KS^{-1/5},$$

where K is proportional to $M_i^{2/5}$. The distance is $r = R/\theta$, where θ is the angular radius. Aller and Minkowski first pointed out that the distance could be calculated if a value of M_i is assumed. If a nebula is optically thin for ionizing radiation the ionized mass M_i is equal to the total nebular mass M. Shklovsky showed that it was plausible to assume that planetaries will become optically thin at some later stage in their evolution, and further assumed that the process of mass ejection is complete at this stage, so that M remains constant, and that the same value of M can be used for all nebulae. Using these ideas, he was able to draw some very interesting conclusions concerning the evolution of the nebulae and of their central stars. Further studies

Osterbrock and O'Dell (eds.), Planetary Nebulae, 1–5. © *I.A.U.*

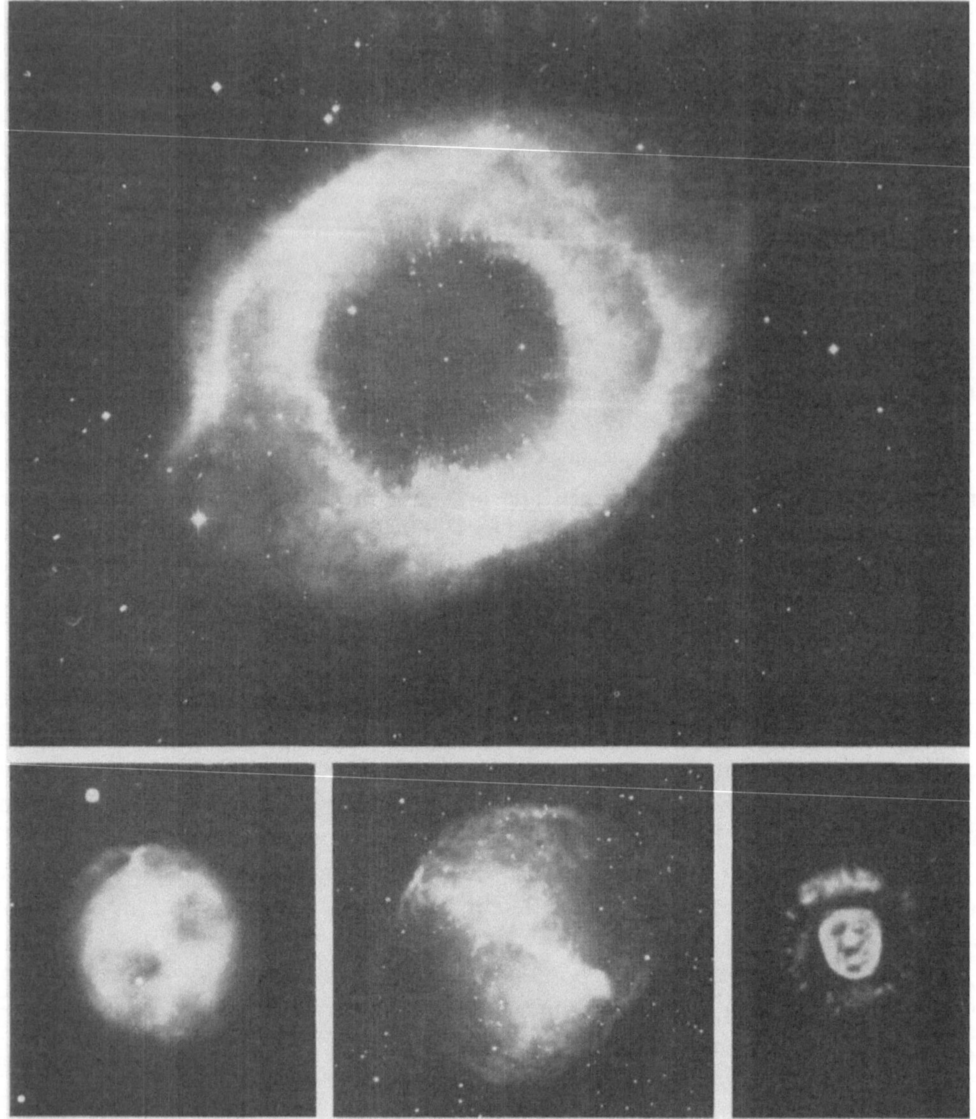

Fig. 1. *Top, NGC 7293, the Helix Nebula. Lower left, NGC 3587, the Owl Nebula. Lower middle, NGC 6853, the Dumb-Bell Nebula. Lower right, NGC 2392, the Eskimo Nebula. All Mount Wilson and Palomar Observatory photographs.*

were made by O'Dell, who calculated $N(R)$, the number of planetaries per unit volume per unit radius interval. This quantity is of importance for discussions of the origin and final evolution of the nebulae. For optically thin nebulae the expansion velocity $\dot{R}$ should be equal to the expansion velocity V measured using the Doppler effect.

Since V is practically constant it is to be expected that $N(R)$ should be a constant, independent of R, for optically thin nebulae. Using Šklovsky's method, O'Dell found $N(R)$ to be constant within a certain range of values of R, and subsequent work by Harman and Seaton showed that this was precisely the range for which the nebulae could be expected to be optically thin. It therefore appears that Šklovsky's method can be relied upon for the determination of the distances of optically thin nebulae. The big question which remains is that of accurately calibrating the distance scale, that is, of determining the constant K. Different estimates of K differ by factors of 2 or more, and the absolute value of $N(R)$, which is proportional to K^{-4}, is therefore very uncertain. I hope that this question of calibrating the distance scale will be discussed in some detail during the Symposium.

Our second Session is concerned with observations of the spectra of planetaries. The very first paper on this subject was written by Huggins in 1864. Having completed extensive studies of laboratory, solar and stellar spectra, Huggins explains why he turned his attention to the nebulae:

"It became therefore an object of great importance, in reference to our knowledge of the visible universe, to ascertain whether this similarity of plan observable among the stars, and uniting them with our sun into one great group, extended to the distinct and remarkable class of bodies known as nebulae. ... Some of the most enigmatical of these wondrous objects are those which present in the telescope small round or slightly oval discs. For this reason they were placed by Sir William Herschel in a class by themselves under the name of Planetary Nebulae."

The stellar spectra observed by Huggins were characterized by continua and by dark absorption lines. He describes his first observation of a planetary nebula, NGC 6543:

"On August 29th, 1864, I directed the telescope armed with the spectrum apparatus to this nebula. At first I suspected some derangement of the instrument had taken place; for no spectrum was seen, but only a short line of light perpendicular to the direction of dispersion. I then found that the light of this nebula, unlike any other extra-terrestrial light which had been subjected by me to prismatic analysis, was not composed of light of different refrangibilities, and therefore could not form a spectrum. A great part of the light from this nebula is monochromatic."

On using a smaller slit width, Huggins was able to observe a number of lines and to identify the Hβ line of hydrogen. He was not able to identify the stronger lines but he speculated that they might "indicate a physical difference in the atoms".

It was not until 1928 that the nebular lines were identified by Bowen as due to forbidden transitions in ions of oxygen, nitrogen and other common elements. I have always felt that Bowen's achievement was remarkable in that, in addition to making the identifications, he gave a very clear discussion of the basic physical processes taking place. At about the time that this work was done, pioneering studies of physical processes in nebulae were begun by Zanstra and by Menzel. With great physical insight they were able to develop methods for the determination of star temperatures, at a time when a quantitative theory of the hydrogen recombination spectrum had not yet been worked out. Observational data required for the application of these methods were first obtained by Zanstra. A series of papers by Menzel and his collaborators, on Physical Processes in Gaseous Nebulae, has made a contribution to our subject of fundamental importance.

A great deal of effort is still being devoted to work on nebular photometry, and I would particularly mention the very extensive observations made by Vorontsov-Velyaminov and his collaborators at the Sternberg Astronomical Institute, and the accurate spectrophometry of individual objects made by Aller and his collaborators. Flux measurements for a considerable number of planetaries have now been made at radio wavelengths, and the conclusions which can be drawn from them will be discussed during the Symposium. Within the near future we may also expect to have observations in the far infrared and in the ultraviolet, and two of our participants have had the courage to accept invitations to make predictions about the results which may be obtained.

Interest in planetary nebulae has led to many pioneering investigations in atomic physics, plasma physics and radiative transfer theory, and the results obtained have been important for many other branches of astrophysics and laboratory physics. In discussing these questions of basic physical processes, in the third session, we should remember that our main concern at this Symposium should be in the astrophysical significance of the results which have been obtained. Better observations and more detailed quantitative studies of physical processes enable us to determine much better chemical composition of the nebulae. We must consider not only the reliability of the results, but also their significance; one may ask, for example, whether the material ejected to form a planetary is contaminated by material which has undergone nuclear reactions during the evolution of the central star.

Our 4th Session will be concerned with questions relating to the structures and dynamics of the nebulae. These may provide important clues to understanding the processes which lead to mass ejection and formation of planetaries. We may also expect that there will be a great deal of interest in the possible importance of magnetic fields. New observations have been made, using interferometric and high-dispersion techniques, which provide a great deal of new information on velocity fields in nebulae.

What can be deduced from the spectra of the central stars? Why do some of these stars have emission lines? To what extent can one justify the conventional assumptions made in model atmosphere calculations? How reliable are attempts to determine temperatures and luminosities of the central stars, and to deduce from observations the evolutionary tracks on the Hertzsprung-Russell diagram? These are some of the questions with which we shall be concerned in the 5th Session.

It has long been recognised that the nebulae evolve in a time of order 10^4 years. This time is obtained on dividing a typical nebular radius by a typical expansion velocity. The work of Šklovsky and of O'Dell has shown that the central stars also undergo considerable evolution in times of this order, and it has recently been recognised that it is possible to observe directly the variabilities in individual nebulae and central stars, within time intervals of order 10 years. Questions of the origin and evolution of the nebulae and of their central stars will be discussed in Session 6. By now it seems to be well established that planetaries evolve towards the white dwarfs, but the nature and the time-scale of this evolution must be discussed, and we must consider whether all white dwarfs are formed from planetaries, or whether planetaries produce only some special group of white dwarfs.

One of the major outstanding problems which we have to consider is that of the origin of planetary nebulae. Many different topics discussed during the Symposium may give important clues to the solution of this problem.

Our epilogue will be spoken by a chorus, led by Dr Osterbrock. Doubtless there will be some discordant notes, but I hope that there will be some harmony as well. At the very end, we may have to quote once again from Shakespeare, from the epilogue to *King Henry the Eighth*:

> 'Tis ten to one this play can never please
> All that are here: some come to take their ease,
> And sleep an act or two; but those, we fear,
> W'have frighted with our trumpets ...

In calling for our play to begin, may I ask that all our speakers should let their trumpets sound loud and clear.

Session I

BASIC DATA

PLANETARY NEBULAE AS A PART OF THE GALAXY

L. Perek
(Astronomický ústav ČSAV, Praha, Czechoslovakia)

1. Introduction

Planetary nebulae form one of the most important subsystems of the Galaxy. If we knew more about this subsystem, an important gap in our ideas about the galactic structure would be filled. Very briefly, we are facing the following situation:

Population I, which contributes to the total mass of the Galaxy by hardly more than 7%, can be tracked almost over the entire Galaxy thanks to the radio observations of neutral hydrogen at 21 cm wavelength. Other data, for a wide solar neighbourhood, follow from observations of Cepheids and early-type stars.

Population II, at the other extreme, is more massive and may contain up to $\frac{1}{4}$ of the total mass of the Galaxy. RR Lyrae stars and globular clusters are bright enough to allow observations far beyond the galactic centre. Data for density distribution and kinematics of this population are available.

The most massive is the disk population with its $\frac{2}{3}$ of the total mass. It is, however, not as well observed as the other populations are. Only a few typical categories of objects are known to belong to the disk population, and out of these the stars of the galactic nucleus and weak-line stars are not observed or easily recognized at large distances; novae are few, and the short-period RR Lyrae stars have small amplitudes and are thus not easy to discover. The last item on the list of disk-population objects is planetary nebulae.

Planetary nebulae, though not extremely luminous, are observable beyond, or at least up to, the galactic centre. They can be systematically detected and are thus the most promising trackers of the disk population. If we knew, e.g., the density of planetary nebulae close to the centre of the Galaxy, we would have an independent check on the mass of the disk population contained in the central bulge, and thus also on the rotation curve derived from the radio observations of neutral hydrogen. Another important problem is the dependence of velocity dispersion on the position in the Galaxy. Its knowledge would help in constructing a dynamical model of the Galaxy representing the velocity distribution as well as the space densities.

2. Discoveries

There are basically two methods for discovering planetaries. If a planetary nebula

Osterbrock and O'Dell (eds.), Planetary Nebulae, 9–22. © I.A.U.

is discovered by one of the methods, it is the confirming evidence of the other method which makes the classification reliable.

The more straightforward method is to recognize planetaries according to the form. It has been in use since the concept of a 'planetary nebula' has come into being towards the end of the 18th century and is still up to date. Only the telescopes have changed. It was William and John Herschel and their contemporaries who first discovered planetary nebulae with this method. One of the most recent applications was to the Palomar Atlas. Several discoveries were made on the paper prints by Krasnogorskaja, Vorontsov-Velyaminov and others, but it was Abell who surveyed the original plates and found 86 new planetary nebulae. A complete survey of the paper prints by Kohoutek yielded 31 more discoveries. The original plates made it possible to discover planetary nebulae with a surface brightness as low as 25 magnitudes per square second of arc ($=16\overset{m}{.}5$ per circle 1′). On the other hand, a planetary nebula can be recognized on a direct plate only if it shows a disk. The two smallest of Abell's planetaries have dimensions of $13 \times 13''$ and $17 \times 15''$, two have diameters of 20″, and all the others are larger. But among the 1036 known planetary nebulae at least 40% are smaller than 13″. These small objects, which are very frequent, especially in the direction of the centre, escape detection with this method even if they are fairly bright. Also over-exposed nebulae in the range 20″ to 40″ might be mistaken for stars on Schmidt camera plates because the only difference is the absence of the diffraction cross.

The second method, based on the emission spectra, came into practice between 1880 and 1910 and some 50 'gaseous nebulae' were discovered at Harvard. In recent applications of the method plates are taken through an objective prism and the planetary nebula is recognized by its emission lines and by the absence of a continuum. In the crowded regions of the Milky Way stellar spectra overlap all too frequently and some information is lost. A red filter is therefore sometimes used to shorten stellar spectra and thus to minimize the overlaps. The emission spectra of faint nebulae are in this case, however, restricted to the Hα line only and the classification is not as reliable as if more lines are observed.

There are several factors that influence the completeness of the survey. One of them is the intensity of the Hα line. An analysis by Henize (1967) of his survey of the Southern hemisphere showed that the incompleteness of one search was 0·23 and 0·03 for Hα intensities 1 and 2 respectively. For medium intense and strong Hα lines the search was practically complete. The repetition of the search reduced the incompleteness to 0·05 and 0·001 respectively.

The effect can also work in reverse, i.e. objects are included in the survey which are not planetary nebulae. Henize (*loc. cit.*) also investigated the purity of his survey. The purity is very high if the Hα image can be resolved, or if there is no continuum at all. The presence of the forbidden lines N1, N2, is a most helpful criterion. If even a faint continuum is present the situation becomes complicated. It is impossible to reject all these objects because even some bright planetary nebulae exhibit a faint continuum. If the Hα line is at least widened, or if it is sharp *and* the forbidden lines show up, the purity is of the order of 0·8. Great line strength, if combined with a faint continuum, is not sufficient for classifying the object as a planetary nebula, the purity being only about 0·2.

The two methods for discovering planetary nebulae complement each other. Large faint planetaries are discovered on the direct plates while small bright objects appear on the spectral plates.

Table 1 lists the telescopes which were used for systematic surveys and Figure 1 shows the areas of the surveys. The survey with the Metcalf telescope by Minkowski in the Northern hemisphere, extended by Henize to the South, is the most comprehensive and homogeneous one, although there is a difference of about $0^{m}.5$, the Southern survey reaching fainter magnitudes.

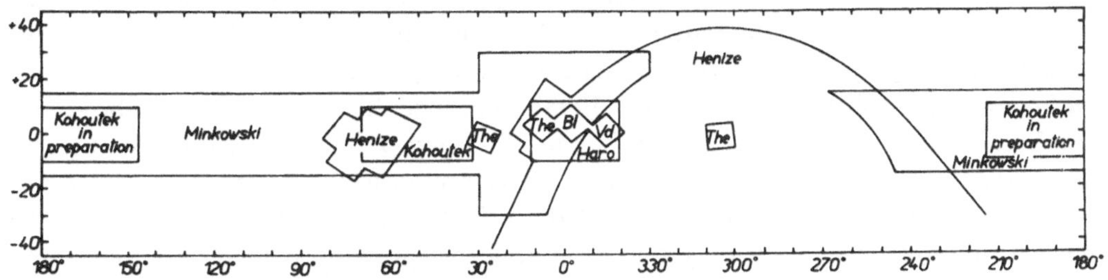

FIG. 1. *Areas of surveys for planetary nebulae.*

Table 1

Telescopes used for surveys of planetary nebulae

Telescope Type	Diameter Focus (cm)	Objective prism Angle	Dispersion (Å/mm)	Author and approx. number of discoveries	
Metcalf	25	15°	450 at Hα	Minkowski	200
Refractor	132			Henize	150
				survey of	470
Bima Sakti	51	6°	312 at Hγ	The	64
Schmidt	126			Blanco	30
				Vandervoort	9
Tonantzintla	66	4°	~ 300 at Hγ	Haro	120
Schmidt	231			Peimbert, Batiz, Costero	24
				Perek	30
Hamburg	80	4°	570 at Hγ	Kohoutek	109
Schmidt	240				
Abastumani	70	8°	180 at Hγ	Apriamašvili	14
Maksutov	210				
Palomar	120	Direct plates		Abell	86
Schmidt	300			Kohoutek	31

No meaningful data can be given for the limiting magnitude of the individual surveys. The limiting magnitude of a typical Schmidt camera of 60 to 70 cm diameter is about 17^m, but this refers to stars. The emission spectra, the non-stellar images, the small difference between the brightness of the background and the surface brightness of some nebulae, and other effects do not allow the definition of a simple limiting magnitude. It is evident, of course, that larger instruments reach fainter objects and thus it appears highly important to complete the survey of the Northern hemisphere with the Hamburg Schmidt and to cover the Southern sky with an equally powerful telescope.

It is interesting to follow the increase of the number of known planetary nebulae with time. The first nebula was discovered and classified by Darquier in 1779. The number increased to more than 60 in the following 80 years. After the appearance of the NGC and the two Index Catalogues, Curtis (1918) listed 102 planetary nebulae. Vorontsov-Velyaminov and Parenago (1931) compiled a list of 131 objects in their study of photographic magnitudes. The catalogue appended to the *Gaseous Nebulae and Novae* by Vorontsov-Velyaminov (1948) contained 288 planetaries and the second catalogue published by the same author in 1962 had 592 entries. Minkowski (1965) mentioned a total of 672 known in 1962. The preliminary catalogue by Perek and Kohoutek (1963) listed 704 nebulae and the printed edition (1967) lists 1036 objects known in 1964. Since then only two discoveries of possible planetary nebulae were published by Kazarjan (1966).

Original observations and measurements of planetary nebulae have been published in a large number of papers and they cannot be mentioned here. We can only refer to the literature given in the *Catalogue of Galactic Planetary Nebulae* by Perek and Kohoutek (1967). A large collection of photographs is contained in the classical paper by Curtis (1918), and a complete set of photographs, intended primarily as finding charts, appeared in Perek and Kohoutek (1967). Among the best photographs available are those taken by Minkowski (1964) with the 200-inch (508-cm) Palomar telescope.

3. Distances

The problem of determining distances of planetary nebulae stands at the base of all further studies. Some nebulae have a unique property which yields the distance of that single object. In this class belong the trigonometric parallax of NGC 7293, the distance of NGC 246 computed from the companion of the central star, the distance of the planetary nebula in the globular cluster M 15, etc. These distances are very important for the individual objects, but unless the methods can be applied to a large group, they can serve as zero points only. The distance scale itself cannot depend on more or less unique circumstances, such as exceptional brightness, size, or closeness, but must be based on physical properties common to all planetary nebulae.

Such properties are in the first place:

Angular dimensions, known for	60% of all planetaries
Proper motions	5%
Angular expansion	< 1%
Radial velocities	33%
Integral or wide-range magnitudes	>50%
Monochromatic magnitudes or surface brightnesses in absolute units	25%

and more refined,

Relative intensities of lines	27%
Central star observations	19%.

The low percentage of *proper motions* can be somewhat increased due to new discoveries but the increase will be slight at best and the mean parallax of $0\overset{''}{.}00079 \pm 11$ derived by Parenago (1946) can hardly be much improved. *Integral* or *wide-range magnitudes* are known for a large number of objects but the information content is low because the ratios of intensities of individual lines change from one planetary nebula to the other, and thus it is unknown, without other data, which lines enter the integral magnitude with what amount.

Radial velocities can be measured for many more objects. Strictly speaking, they can be used for distance determinations for objects moving in circular orbits only. The dispersion of velocities diminishes the accuracy of distance determinations. It is tolerable for Population-I objects which have small dispersions. Planetary nebulae, however, have dispersions of 25–45 km/sec, depending on direction, and the distances derived from radial velocities are subject to considerable errors. Besides, it is the kinematics of planetary nebulae which we wish to determine from radial velocities and we cannot have both kinematics as well as distances at the same time.

Angular expansion which, compared with the radial expansion, yields a very neat method of distance determination, is limited to very close objects. So are *central star observations*. Of the 19%, most are included in the simple statement that a central star has been observed. Detailed observations, however, are rare. Intensities of spectral lines of central stars are known for 1% of the planetary nebulae, and a description or classification of the spectrum is available for only 6% more.

Angular dimensions are known for many nebulae and the number and accuracy of observations can be increased with relatively little effort. Short exposure times are sufficient to register even rather faint planetary nebulae. Good seeing and a long focus are, however, mandatory to show the tiny disks of a very few seconds of arc. *Monochromatic magnitudes* or surface brightnesses have been measured for a considerable number of planetary nebulae and there is no obstacle in principle to extending the measurements to a vast majority of objects. *Relative intensities* of lines require repeating photometric measurements at other wavelengths unless the lines are so close as to require lengthy spectral treatment.

Methods of distance determinations based on angular dimensions and monochromatic magnitudes at a very small number of well-spaced wavelengths have the best chance to give a distance scale of the system of planetary nebulae.

Distance scales were discussed in many papers. We refer to Minkowski (1964, 1965) and to Seaton (1966). Here, we state briefly that the scale of optically thick nebulae can be set up if a mean absolute magnitude is introduced. The methods differ in either neglecting the spread in absolute magnitudes or in respecting it by introducing a term in $(m_s - m_n)$. We quote some representative formulae

Berman (1937):

$$\log r = 0{\cdot}2(H\text{–}A) - \log d + 0{\cdot}064\,(m_s - m_n) + \text{const.}$$

Vorontsov-Velyaminov (1934):

$$\log r = 0{\cdot}2(H\text{–}A) - \log d + \text{const.},$$

where H is the surface brightness, A the correction for extinction presented here in a uniform way in order to make the formulae comparable, d the diameter in seconds of arc, r the distance, and m_s, m_n the magnitudes of the central star and nebula respectively.

Optically thin nebulae, where the whole nebula is observed, require the knowledge of, or an assumption about, the mass. If the deviations from the mean mass are neglected, the corresponding term vanishes into the zero point of the scale. Some representative formulae follow:

Shklovsky (1956):

$$\log r = 0{\cdot}08\,(H\text{–}A) - \log d + \text{const.}$$

Abell (1966):

$$\log r = 0{\cdot}08\; m_{pr} - 0{\cdot}2 \log v + \text{const.}$$

O'Dell (1962):

$$\log r = -\,0{\cdot}2 \log F\;(H\beta) - 0{\cdot}6 \log\; d + \text{const.}$$

Kohoutek (1960, 1961):

$$\log r = 0{\cdot}13\,(H\text{–}A) - \log d + \text{const.}$$

It is interesting to note that this second group of methods is much less sensitive to the interstellar extinction. Compare the coefficient 0·08 with 0·2 of the previously discussed methods! The formula by Shklovsky goes back to Ambarcumjan (1939) and was first used by Minkowski and Aller (1954) for the determination of the mass of the Owl Nebula. Shklovsky established a distance scale with its help. The formulae by

Abell and O'Dell, with the integrated photored magnitude m_{pr}, the volume of the nebula v in cubic seconds of arc, the flux in absolute units $F(H\beta)$, are modifications of the Shklovsky formula. Kohoutek tried to derive masses by assuming that the difference between the absolute magnitude of the nebula and the absolute bolometric magnitude of the star remains constant during its evolution.

In any case, it is necessary to know if the nebula is optically thick or thin. Making the wrong assumption results in a distance which is too large. Minkowski (1964) therefore determined distances of planetary nebulae with sufficiently good photometric data according to both the methods for optically thick and for optically thin nebulae, and the comparison of results gave an indication of which method was the correct one. This seems to be the best proceeding at the present for faint and small planetary nebulae.

Seaton (1966) proposed a method for distance determination using electron densities deduced from relative intensities of forbidden lines. Further the surface brightness at $H\beta$ and the angular dimensions were needed. Spectrophotometric measurements are indispensable for this method because the [N II] lines cannot be separated photometrically from the $H\alpha$ line. Therefore this method can be applied only to bright, well-studied planetary nebulae.

Gershberg (1962) brought forward the fact that practically all planetary nebulae are transparent at centimeter wavelengths and he modified Shklovsky's method by using radio fluxes instead of surface brightnesses in the optical range. His distance determinations are limited to six objects. The radio fluxes can be used directly in Shklovsky's formula for nebulae optically thin both in the Lyman continuum and in the radio range.

An attempt to determine the best distance scale was made by Pskovskij (1959). He compared the distance scales of Vorontsov-Velyaminov, Berman, and Shklovsky by grouping planetaries according to larger (smaller) distances by one scale or other. Then he computed the Oort constant A for these groups and concluded that the best scale would have the least variation in A. The comparison, which spoke in favour of the Shklovsky distance scale, of course carries some weight but two important points should be considered. Although mostly NGC objects were used by Pskovskij, the distances of planetary nebulae are too large to warrant the constancy of A. Second, the value of A itself is dependent on the systematic velocity of the subsystem. Planetary nebulae need not yield the same value of A as Population I does.

4. Galactic Distribution

The distribution of planetary nebulae on the sky (Perek and Kohoutek, 1967, Fig. 2) gives at once the impression of a flattened galaxy seen edge-on with the obscuring layer in the plane of symmetry. This particular distribution is still more conspicuous in the frequencies plotted over galactic longitudes (*loc. cit.*, Fig. 1). The distribution is not quite symmetric, a part of the asymmetry being due to surveys with large Schmidt cameras. There is also an asymmetry present close to the centre (*loc. cit.*, Fig. 3). The deficiency in the South sets in approximately at declination $-35°$ which is about the limit of the Northern observations. It is, however, not easy to reconcile this explanation with the fact that the central region, both North and South of declination $-35°$, has been covered with the same telescopes. Another source of asymmetry may come from the irregular interstellar extinction.

The asymmetry in galactic latitude (*loc. cit.*, Fig. 5) is entirely caused by interstellar extinction. It is prominent near the centre and absent far from that direction.

The very high peak of the frequencies at the centre suggests at first sight that the nebulae seen in that direction are also at the distance of the galactic centre. It is interesting to see how this impression compares with numbers of planetary nebulae observed in other directions.

The comparison is based on the direction towards the centre and on a direction 65° away from the centre. In the last-mentioned direction the distance from the centre varies only between 9 and 11 kpc for the first 10 kpc from the Sun. The average numbers of planetary nebulae per square degree in the individual regions are shown in Table 2. The regions were selected between latitudes 2° and 5° so as to avoid the heavy extinction in the galactic plane. They coincide with the regions of maximum density of planetary nebulae in the direction to the centre.

Table 2

Average observed numbers of planetary nebulae per square degree

$l^{II} \diagdown b^{II}$	$+2°$ to $+5°$	$-2°$ to $-5°$
355° to 0°	3·0	2·5
0° to 5°	0·7	2·8
60° to 70°	0·23	0·30
290° to 300°	0·17	0·13

To compute the numbers of planetary nebulae per square degree, we need first the density near the Sun and second the variation of the density with distance.

If the density near the Sun is computed from a large volume, the incompleteness and the steep fall-off of the density with z might lead to a wrong estimate. On the other hand, a too small volume with a very small number of nebulae inside is subject to random fluctuations. Figure 2 shows the densities (numbers of nebulae per kpc^3) inside spheres of radius r. Each curve bears the name of the method of distance determination. Shklovsky's method leads to about 200 while other methods give about 30 as the extrapolated intersection of the curve with the axis of ordinates. Numerical results of Minkowski's method are not available but his graph of the distribution in the galactic plane contains 49 and 13 planetary nebulae inside squares of 2 and 1 kpc

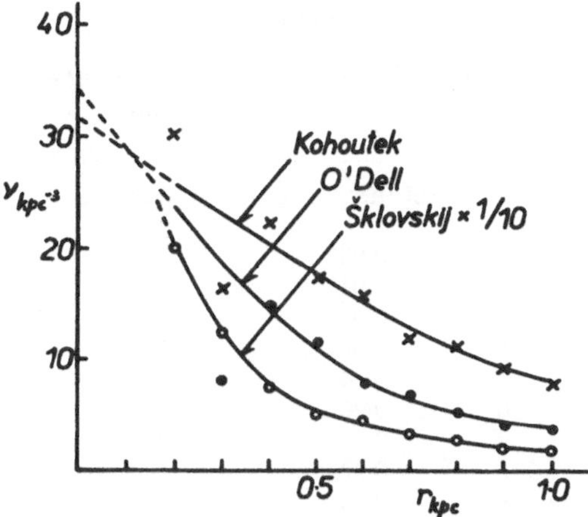

FIG. 2. *Space density of planetary nebulae in the solar vicinity inside spheres of radius r.*

respectively. With respect to a mean distance from the galactic plane of 0·3 to 0·5 kpc, we arrive at values between 26 and 40. It may be concluded that in the solar vicinity there are, as a rough estimate, 30 planetary nebulae per kpc^3.

Assuming, for the sake of simplicity, an exponential density law

$$\nu = \nu_c \exp(-pR - qz),$$

the number $N(r)$ of planetary nebulae in 1 square degree up to a distance r from the Sun is

$$N(r) = \frac{\nu_0 k}{m^3}\left[(x^2 - 2x + 2)\,e^x - 2\right],$$

where

$$x = rm,$$

$$m = \frac{\mathrm{d}\ln\nu}{\mathrm{d}r},$$

$$k = 0{\cdot}01745^2,$$

and ν_0 is the density at the Sun. The logarithmic density gradient m along the radius vector r is readily deduced from the gradients in the principal directions. Let us take values which are characteristic for the disk population

$$\frac{\mathrm{d}\log\nu}{\mathrm{d}R} = -\,0{\cdot}20; \quad \frac{\mathrm{d}\log\nu}{\mathrm{d}z} = -\,2{\cdot}00.$$

These values are close to those quoted by Plaut (1965) for novae which belong to the same population. At the latitude of 3°.5 (centre of the investigated regions), the z-gradient enters with a coefficient of sin 3°.5. The R-gradient enters with the full

Table 3

Computed numbers of planetary nebulae per square degree up to the distance r from the Sun

r_{kpc}	Direction 65° off the centre	Direction to the centre
1	0·003	0·003
2	0·016	0·03
3	0·045	0·12
4	0·087	0·34
5	0·14	0·75
6	0·20	1·54
7	0·26	2·8
8	0·32	4·8
9	0·38	7·9
10	0·44	15

amount in the direction towards the centre and can be neglected in the direction 65° off the centre. Thus we arrive at the values in Table 3.

Let us assume for the moment that we see all planetary nebulae up to the distance *r*. The observed numbers correspond to the computed numbers at 5 to 8 kpc from the Sun in the direction 65° off the centre. In the direction to the centre the observed and computed numbers tally at about 7 kpc from the Sun.

If we do not see all planetaries up to a certain distance, the ratio of the numbers may still be used. The only condition is that the numbers are affected in the same way in both directions. This is fulfilled if the interstellar extinction does not differ appreciably in the first two kiloparsecs in the two directions. The line of sight at that distance reaches a height of 120 pc above the galactic plane and most of the extinction occurs in the layer below. The ratio of the observed numbers is between 10 and 13 and this range of values is attained by the computed numbers not far from 7 kpc from the Sun.

We conclude that a disk population observed to only about 7 kpc from the Sun would show much the same distribution on the sky as the planetary nebulae do. Thus the apparent concentration of planetary nebulae to the direction of the centre cannot be considered a proof that the majority of these nebulae are really close to the centre. Some other evidence is needed to prove that *some* planetary nebulae lie beyond 7 kpc from the Sun. This supporting evidence comes from the radial velocities.

The distribution of planetary nebulae in the galactic plane is illustrated by a graph in Minkowski's (1964) paper. The distribution reaches to 3 or 4 kpc from the Sun, only occasional planetaries lying at larger distances. This is caused by the lack of adequate photometric data for distant objects. An analogous figure by Perek (1963), with distances computed according to Kohoutek's formula from photometric estimates, reaches further from the Sun but fails to show an important concentration at the centre. This is due to the fact that planetary nebulae at the centre have mostly very small or even stellar images, and that their distances could not be determined.

5. Radial Velocities

One hundred radial velocities have been known since the pioneer work of Campbell and Moore in 1918. Very few velocities were added in the following 40 years, until Minkowski and Mayall substantially enlarged the observing material in the late 'fifties by measuring 142 and 134 velocities respectively. Smaller programs or individual velocities were contributed by many astronomers interested in the field.

The total of 348 known radial velocities makes the system of planetary nebulae one of the best studied. It compares more than favourably with the 70 radial velocities of globular clusters, with the 160 velocities of RR Lyrae stars and even with the sample of about 300 velocities of Mira-types stars.

It is quite pleasant to note that there is no part of the Milky Way utterly devoid of radial-velocity measurements. The region of the galactic centre has the largest number of determinations. The numbers drop rapidly toward the Southern Milky Way and somewhat more slowly towards the North. Extensive programs can be set up even today for measuring radial velocities in both hemispheres, and these would serve a very good purpose as will be seen from the following discussion.

The distribution of radial velocities of planetary nebulae in galactic longitudes (galactic latitude below 20°) is quite remarkable (Figure 3). Although the range of velocities near the centre exceeds 500 km/sec, compared to 100–150 km/sec at other directions, the overall picture shows some regularity.

We note at once the straggler M 1–67 at $l^{\mathrm{II}}=50°$, $V=+215$ km/sec, which is the known planetary nebula around Merrill's star. This star is possibly in a hyperbolic, or at least in a very eccentric, galactic orbit (Perek, 1956). There is hardly any doubt

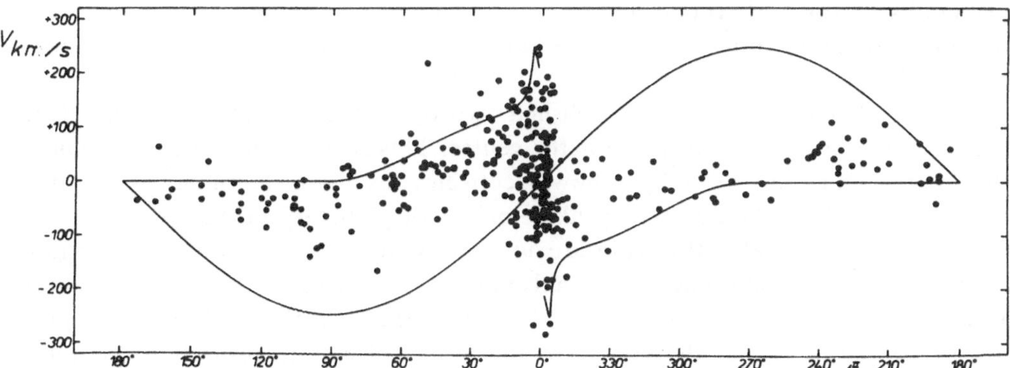

FIG. 3. *Radial velocities plotted against galactic longitudes for nebulae below 20° latitude. The curves limit the permitted area of radial velocities of objects moving in circular orbits.*

about the reality of the large positive value of the radial velocity. It was measured by Merrill in 1938, by Wilson in 1946, by Minkowski (nebula) in 1957, and by Bertola in 1964. All measurements give large values for the star as well as for the nebula.

Other planetaries which may be expected not to conform to the overall picture appear in Figure 4, which shows objects above 20° galactic latitude. There is Ps 1, at

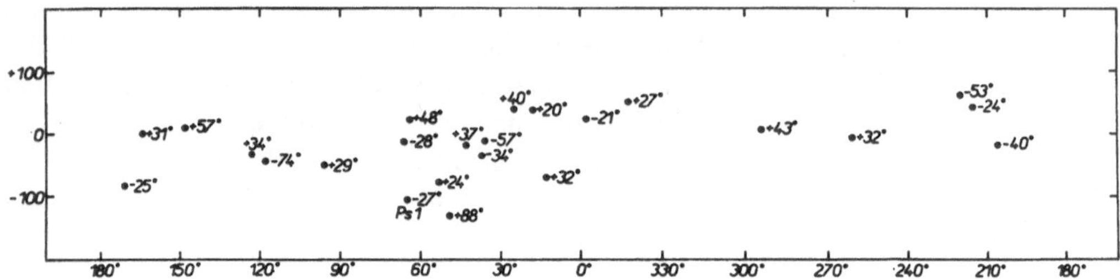

FIG. 4. *Radial velocities of planetary nebulae above 20° latitude with latitude shown.*

$l^{\mathrm{II}} = 65°$, $V = -143$ km/sec, the planetary nebula in the globular cluster M 15. Its kinematics are those of the globular cluster itself. Another is H 4-1, $b^{\mathrm{II}} = +88°$, $V = -133$ km/sec, situated close to the galactic pole.

It was shown by Minkowski (1965) that radial velocities of objects moving in circular orbits should lie inside a permitted area limited by two curves. The first is the sine curve of the reflected motion of the Sun around the centre and the second curve gives the extreme radial velocities with respect to the Sun. The extreme velocities occur in the interval $270° < l^{\mathrm{II}} < 90°$, at points subtending a right angle over the base centre–Sun, and are equal to the circular velocities at those points plotted from the sine curve. The extreme velocity is zero in the interval $90° < l^{\mathrm{II}} < 270°$.

Minkowski (1965) used Schmidt's model (1956) which represents the hydrogen 21-cm observations. He found that too many velocities near the centre lie outside the permitted area and concluded that the kinematics of planetary nebulae could not be explained in terms of circular orbits.

In 1960, Rougoor and Oort published hydrogen observations from the central regions between 100 and 600 pc from the centre. This curve, adjusted to a distance of the Sun from the centre of 10 kpc, shows a conspicuous hump. This hump is prominent in Figure 3 and fits the observed peak velocities of planetary nebulae.

Further we note that, if the orbits are not strictly circular, the limits of the permitted area may be exceeded by amounts roughly equal to the velocity dispersion. The data on velocity dispersions are rather scarce. Delhaye (1965) quotes the results by Wirtz giving 45, 35, 20 km/sec for the dispersions in the R, θ, z directions respectively. Parenago (1946) gives 29 km/sec for the z-dispersion. We find from 23 planetaries above 20° latitude an average projection into the z-axis of 24 km/sec, which is equivalent to a dispersion of 30 km/sec, in agreement with Parenago. Thus the limits of the permitted area may be exceeded by amounts between 35 and 45 km/sec. These values being root-mean-square, correspond to 28 and 36 km/sec average excess velocities respectively. We find from Figure 3 that the limits – with the exception of the sine curve between $355° < l^{II} < 5°$ – are exceeded by 48 nebulae and that the average excess velocity is 24 km/sec, well within the expected range. The average excess velocity in the remaining small part is 72 km/sec, i.e. it is larger by a factor of 3 than anywhere else! This is well illustrated in Figure 5, which shows the central part of Figure 3 with the scale of abscissa blown up.

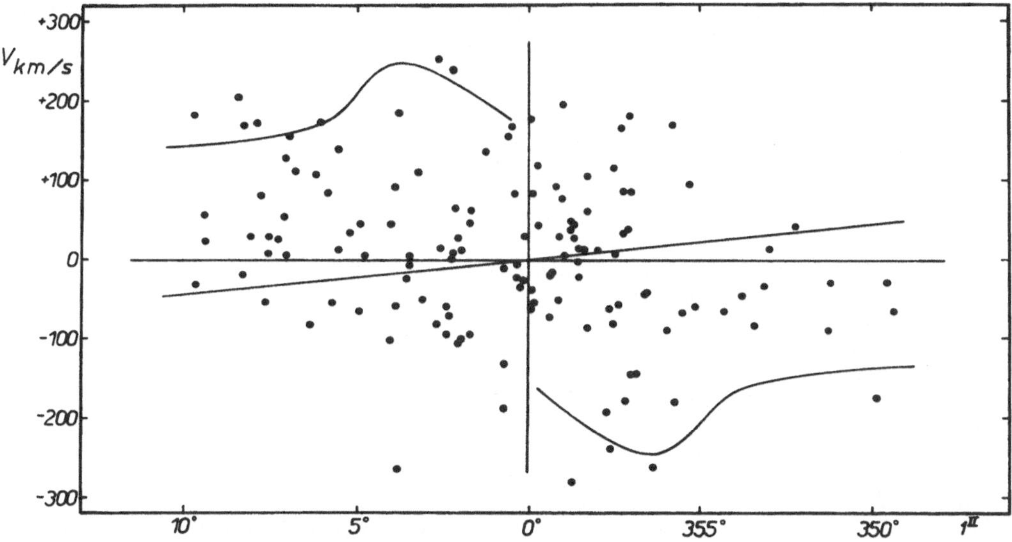

Fig. 5. *Radial velocities in the central region. The central part of Figure 3 with the scale of abscissa blown up.*

Let us discuss the planetaries exceeding the sine curve and lying thus in the 'wrong quadrants' in more detail. These planetaries are mostly small, under 10″ or even under 5″, and stellar images are frequent. The distances of stellar planetaries cannot be determined at all, and those of small objects as determined by Perek (1963) are very uncertain. They are subject to errors in measuring very small diameters affected by seeing and in measuring surface brightnesses heavily affected by interstellar extinction. The photometric distances range between 3 and 10 kpc and a small weight should be attached to them.

Since the large velocity dispersion shows only in the direction to the centre and not in other directions, we conclude that this is due to planetaries situated close to the centre. Then we have a velocity dispersion of about 30 km/sec (root-mean-square) everywhere with the exception of a small central region with a dispersion of 90 km/sec (root-mean-square). This is a lower limit because an admixture of foreground objects might reduce the true value. The region of the large dispersion coincides with the region of the hump on the rotation curve. Further, the value of the dispersion is close to that of population-II objects. This suggests an interesting hypothesis that the disk population in the central region moves with kinematics very similar to population II. It is tempting to extrapolate this hypothesis also to population I and to attribute the hump on the rotation curve to maximum velocities in rather eccentric orbits and not to circular velocities. The kinematic distinction between the populations, which is so prominent outside the centre, might have been prevented from being established in the centre. Should this hypothesis be supported by other evidence, we might change our ideas about the density of mass near the centre.

References

Abell, G.O. (1966) *Astrophys. J.*, **144**, 259.

Ambarcumjan, V.A. (1939) *Kurs teoretičeskoj astrofiziki*, p. 170. (In Russian.)

Berman, L. (1937) *Lick Obs. Bull.*, **18**, 57.

Campbell, W.W., Moore, J.H. (1918) *Publ. Lick Obs.*, **13**, 75.

Curtis, H.D. (1918) *Publ. Lick Obs.*, **13**, 9.

Delhaye, J. (1965) in *Galactic Structure*, Ed. by A. Blaauw and M. Schmidt, The Univ. of Chicago Press, Chicago, p. 61.

Gershberg, R.E. (1962) *Izv. Krym. Astrofiz. Obs.*, **28**, 159.

Henize, K.G. (1967) *Astrophys. J. Suppl.*, **14**, 125.

Kazarjan, M.A. (1966) *Astrofizika* (Jerevan), **2**, 371.

Kohoutek, L. (1960) *Bull. astr. Inst. Csl.*, **11**, 64.

Kohoutek, L. (1961) *Bull. astr. Inst. Csl.*, **12**, 213.

Minkowski, R. (1964) *Publ. astr. Soc. Pacific*, **76**, 197.

Minkowski, R. (1965) in *Galactic Structure*, Ed. by A. Blaauw and M. Schmidt, Univ. of Chicago Press, Chicago, p. 321.

Minkowski, R., Aller, L.H. (1954) *Astrophys. J.*, **120**, 261.

O'Dell, C.R. (1962) *Astrophys. J.*, **135**, 371.

Parenago, P.P. (1946) *Astr. Zu.*, **22**, 150; **23**, 69.

Perek, L. (1956) *Astr. Nachr.*, **283.**, 213.

Perek, L. (1963) *Bull. astr. Inst. Csl.*, **14**, 201, 218.

Perek, L., Kohoutek, L. (1963) *Preliminary edition of the Catalogue of Planetary Nebulae* (mimeographed).

Perek, L., Kohoutek, L. (1967) *Catalogue of Galactic Planetary Nebulae*, Academia, Praha.

Plaut, L. (1965) in *Galactic Structure*, Ed. by A. Blaauw and M. Schmidt, Univ. of Chicago Press, Chicago, p. 311.

Pskovskij, J.P. (1959) *Astr. Zu.*, **36**, 305.

Rougoor, G.W., Oort, J.H. (1960) *Proc. nat. Acad. Sci. Am.*, **46**, 1.

Schmidt, M. (1956) *Bull. astr. Inst. Netherl.*, **13**, 15.

Seaton, M.J. (1966) *Mon. Not. R. astr. Soc.*, **132**, 113.

Shklovsky, I.S. (1956) *Astr. Zu.*, **33**, 222.

Vorontsov-Velyaminov, B.A. (1934) *Astr. Zu.*, **11**, 40.

Vorontsov-Velyaminov, B.A. (1948) *Gaseous Nebulae and Novae*, Moscow. (In Russian.)

Vorontsov-Velyaminov, B.A. (1962) *Soobšč. gos. astr. Inst. P. K. Šternberga*, **118**, 3.

Vorontsov-Velyaminov, B.A., Parenago, P.P. (1931) *Astr. Zu.*, **8**, 206.

PLANETARY NEBULAE IN THE MAGELLANIC CLOUDS

BENGT E. WESTERLUND

(Steward Observatory, University of Arizona, Tucson, Ariz., U.S.A.)

ABSTRACT

The available photometric and spectroscopic data on the planetary nebulae in the Magellanic Clouds are discussed. Recent results regarding the distribution of the planetary nebulae in the Clouds and the kinematics of the Clouds are presented. Masses are determined for selected planetary nebulae in the Clouds, and the question about the most likely average mass is considered. The evolutionary tracks of the nuclei for two values of the average mass are compared with the actual positions in the HR diagram of the nuclei of the planetary nebulae in the Clouds.

1. Introduction

Planetary nebulae have been identified only in three external galaxies, M 31 and the Magellanic Clouds. Baade (1955) noted five possible planetary nebulae in M 31. Recently, Miss Swope (1963) measured their mean photographic magnitude as 22·3, which, with an uncorrected distance modulus of 24·8, gives an absolute blue magnitude of $-2{\cdot}5$. Their large distances have so far limited further investigation.

In the Magellanic Clouds the surveys for planetary nebulae have been uniform and may be considered reasonably complete to an apparent blue magnitude of about 17·5–18. The most likely distance moduli of the Clouds are: $m-M=18{\cdot}7$ for the Large Cloud (LMC), and $m-M=19$ for the Small Cloud (SMC). The limiting absolute blue magnitude of the surveys is then in the range $-0{\cdot}7$ to $-1{\cdot}5$.

All the true planetary nebulae identified so far in the Magellanic Clouds are stellar. With presently available telescopes it should be possible to detect objects in the Clouds with radii exceeding 0·3 parsec. From the currently accepted theory for the evolution of planetary nebulae it follows that the absolute magnitude of objects that have expanded to this size should be about $M_B=+2$.

The planetary nebulae in the SMC belong to the 'Central System' and are found within 3°5 of the optical centre of the galaxy. In the LMC they belong to the 'Disk System', which has a radius of about 7° (Westerlund, 1965, 1968). In the latter, the system is definitely flat and contains mostly 'Old Population-I' objects. In the SMC the structure is less clear, and it will be discussed below.

Osterbrock and O'Dell (eds.), Planetary Nebulae, 23–33. © *I.A.U.*

2. The Detection of Planetary Nebulae in the Magellanic Clouds

A. THE SMALL MAGELLANIC CLOUD

Emission objects were identified in the SMC by Miss Cannon (1933). No object appears to have been classified as a planetary nebula until 1955, when Lindsay (1955) suggested that 17 objects in the SMC belonged to that class. The following year Koelbloed (1956) identified 16 objects as planetary nebulae. Henize's catalogue of emission nebulae (1956) contains a large number of unresolved objects; none is, however, identified as a planetary nebula.

It is of interest to consider Lindsay's criteria for the identification of planetary nebulae in the Clouds. They are: (1) The object should not be associated with nebulosity; (2) it should be generally stellar in appearance, and (3) show little or no continuum spectrum; (4) it should show typical discrete nebular lines, and (5) have positive color index.

Lindsay expanded his catalogue in 1956, listing 20 planetary nebulae and 9 possible planetary nebulae (Lindsay, 1956). In 1961 he published a new catalogue of emission-line stars and planetary nebulae in the SMC, listing 30 planetary nebulae (P) and 20 probable planetary nebulae (P?) among the 593 'point-source' emission-line objects (Lindsay, 1961). The objects in his classes P and P? have photographic absolute magnitudes ranging between $-4{\cdot}7$ and $-0{\cdot}7$. The distribution of the absolute magnitudes differs significantly from that obtained for planetary nebulae in our Galaxy, as also noted by Švestka (1962). It is important both for the study of planetary nebulae and for studies of the nature of the Magellanic Clouds to establish whether the high-luminosity objects are indeed normal planetary nebulae. The resolution of Lindsay's plates is about 10″, or close to 3 parsec at the distance of the SMC. It is therefore possible that many of his objects are small diffuse nebulae. This has been confirmed by Henize and Westerlund (1963); of 14 objects brighter than $M_{pg} = -3{\cdot}0$, 11 are clearly resolved on large-scale plates and their estimated masses range from 2 to 33 solar masses. It is therefore unlikely that these objects can be considered planetary nebulae. The unresolved objects are probably true planetary nebulae, and $M_{pg} = -3{\cdot}0$ is the upper limit to the photographic absolute magnitude of this class, in reasonable agreement with the data for the galactic planetary nebulae.

B. THE LARGE MAGELLANIC CLOUD

Surveys for planetary nebulae in the LMC have been carried out by Westerlund and Rodgers (1959), Lindsay and Mullan (1963), and Westerlund and Smith (1964). Henize (1956) listed 97 unresolved Hα-emission nebulae; of these Lindsay and Mullan's catalogue includes 63.

There are 40 objects in common to the Westerlund-Smith (W-S) and the Lindsay-

Mullan (L-M) catalogues. The former contains 2 objects not included in the latter. There are 25 objects in the L-M Table 1 not listed in the W-S catalogue, and an additional 44 possible planetary nebulae in the L-M Table 2; they are all faint, uncertain objects and will not be considered here. We have searched our plates for the 25 additional objects in the L-M Table 1 and we conclude: 2 objects appear as planetary nebulae on our blue spectral plates (they were earlier considered as plate flaws); 5 fall outside the field searched by us; 3 are definitely stars, 5 are resolved emission nebulae, and 10 objects cannot be seen at all on our plates.

At present we may consider that about 30 true planetary nebulae have been identified in the SMC and about 45 in the LMC to the limiting magnitude given above. All these objects belong to the most luminous group of planetary nebulae, and it may be assumed that their present phase of evolution occupies about 1/10 of the lifetime of the nebulae. On this assumption, we find a total of approximately 300 planetary nebulae in the SMC and of 450 in the LMC. This may be compared with the estimated number of $4{\cdot}8 \times 10^4$ planetary nebulae in our Galaxy (O'Dell, 1962). The masses of the three galaxies are about $1{\cdot}5 \times 10^9$ solar masses for the SMC, 6×10^9 for the LMC, and 10^{11} for the Galaxy. The number of planetary nebulae per unit mass is in reasonable agreement between the SMC and our Galaxy, whereas there is a lack of planetaries in the LMC.

3. The Distribution and Motions of the Planetary Nebulae in the Clouds

A. THE SMALL MAGELLANIC CLOUD

Lindsay (1961) concluded that the planetary nebulae showed a preference for the Southern and Southwestern parts of the SMC, and that they were almost absent East of NGC 346 ($R.A. \sim 1^h$). Most of the objects rejected by Henize and Westerlund (1963) as non-planetary are in the core of the SMC. The remaining objects possibly show a slight concentration to the preceding edge of the Bar, most of the fainter objects (of less certain classification) are fairly well scattered in the outer parts (Figure 1).

Hindman (1967) has recently suggested that the SMC is a slightly flattened system, seen practically edge-on. This makes it important to re-analyze the distribution of the planetary nebulae. In Hindman's model the radio centre of rotation is at 01^h03^m – 72°45′ (1975) with the major axis in position angle 55°. We recall that the centre of the old population, represented by the red stars found on long-exposure infrared plates, appears to be near NGC 419 (01^h08^m – 73°10′), whereas the 'optical centre' (of bright blue stars) is near 00^h48^m – 73$^\circ\!.$3. If we accept Hindman's model we find that only 8 planetary nebulae fall within 500 parsec (0$^\circ\!.$5) of the equatorial plane of the SMC; they would then hardly belong to the disk population.

The line (plane) of symmetry of the planetary nebulae goes through the optical

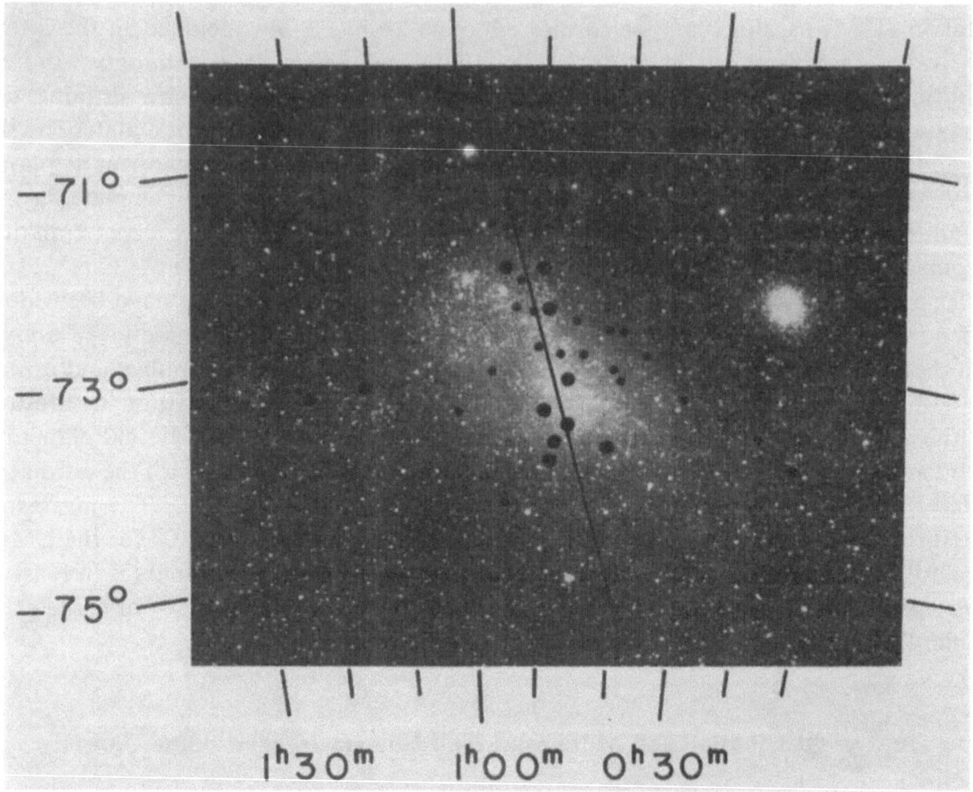

FIG. 1. *The distribution of the planetary nebulae in the Small Magellanic Cloud. Large dots represent objects confirmed to be unresolved by Henize and Westerlund; small dots represent the objects in Lindsay's catalogue which have not yet been checked. The line of symmetry for the plotted objects is drawn.*

centre of the SMC with a position angle of 16°. 25 planetary nebulae, or 68%, are within 500 parsec of this plane. Provided that the SMC is really an edge-on galaxy, we are inclined to consider this plane the true equatorial one, as it is defined with the aid of objects belonging to the old disk population in our Galaxy. This group is less likely to have been exposed to recent disturbances of the type that makes the present distribution of H I in the SMC so confusing.

Webster (1966) has determined velocities for 7 planetary nebulae in the SMC. All the velocities are smaller than the mean velocity of the SMC from Population-I objects (about 166 km/sec). They form two groups, one, consisting of 4 objects, with a mean velocity of 151 km/sec, and the second with a mean velocity of 119 km/sec. There is no connection between velocity and position in the SMC, and, consequently, no indication of any rotation. The different sub-systems in the SMC have obviously quite different characteristics in every way.

B. THE LARGE MAGELLANIC CLOUD

The surface distribution of the planetary nebulae in the LMC appears circular with the centroid at $05^h22^m - 69\overset{\circ}{.}5$ (Westerlund and Smith, 1964). 50% of the planetaries fall within 2 kpc of this centre. Table IV in the W-S paper gives the centre of the various sub-systems; the so-called radio centre of rotation has recently been redetermined to $05^h20^m - 69\overset{\circ}{.}0$ (McGee and Milton, 1966), but this does not affect the discussion.

Webster (1966) has determined radial velocities for 13 planetary nebulae in the LMC and combined her data with Feast's (1964) measurements of 6 additional objects for a discussion of the kinematics of the LMC. The corrections for solar motion and galactic rotation as well as the systematic radial velocity of the LMC, the position angle of the maximum velocity gradient, and the inclination of the LMC to the plane of the sky were chosen to conform with previous analyses (Feast *et al.*, 1961).

Rotation curves were derived about two centres: (I) the centre of symmetry of the extreme Population-I curve ($05^h20^m - 68\overset{\circ}{.}8$), and (II) the centroid of the planetary nebulae sub-system ($05^h22^m - 69\overset{\circ}{.}5$). Webster concludes that it is immediately obvious that the planetary rotational velocities deviate from those of the Extreme Population I near the centre of rotation of the latter population (Webster, 1965, 1966). They are more symmetrical around the centroid (II), which is close to the true mass centre of the LMC. Since the planetary nebulae may be expected to be less affected by gas streaming than the Extreme Population I, they are likely to give more reliable information about the gravitational field of the LMC. It is therefore suggested that the centre of rotation of the LMC coincides with the true centre of mass of the system, within the Bar. (See, however, Feast, 1964, and Feast's paper in this volume, p. 34.) No further analysis of the rotation curve is possible because of the obvious deviations from symmetry (due to the Bar) and possibly also from coplanarity (cf. McGee and Milton, 1966).

Webster suggests that three of the planetary nebulae, P 2, P 24 and P 33 (Westerlund and Smith, 1964), may move in very eccentric orbits. If these three objects are excluded, the dispersion of the remainder of the planetaries from the curve around centre (II) is 8·2 km/sec. This should be compared with the value of 9·6 km/sec for the blue supergiants when the runaway stars are excluded. We note also that McGee and Milton (1966) find that in most cases the differences between the radial velocities of the planetary nebulae and the H I in their directions are quite large.

4. Photometry of the Planetary Nebulae in the Clouds

A. NARROW-BAND PHOTOELECTRIC PHOTOMETRY

Webster (1966) has determined the fluxes in the N_1, $H\beta$, $H\gamma$, and [O II] $\lambda 3727$ lines, as well as the monochromatic magnitudes at $\lambda\lambda$ 5300, 4200 and 3500 for 10 planetary nebulae in the SMC and for 18 in the LMC. Her results are fundamental for the discussion in the following, and I am grateful for her permission to use them in advance of publication.

B. HETEROCHROMATIC MAGNITUDES

The photographic photometry of the planetary nebulae in the LMC (Westerlund and Smith, 1964) appears to be sufficiently accurate for an attempt to be made to interpret the red (R) and blue (B) magnitudes astrophysically. Comparison with Webster's data gives

$$-\log F(\mathrm{H}\beta) = 12{\cdot}33 + 0{\cdot}3(R_{\mathrm{n}} - 14). \tag{1}$$

The subscript n refers to the total nebula, an s will refer to the central star, only. The scatter is less than $\pm 0{\cdot}1$ in R_{n}.

In deriving relation (1) we have excluded 4 objects: P 7, P 9, P 17, and P 18 in the W-S catalogue. Two of these, P 7 and P 9, are known to have strong [N II] lines: this will affect the R magnitude appreciably.

B_{n} is related to m_{4200}: $m_{4200} = B_{\mathrm{n}} + 1{\cdot}03$.

The dispersion is here $\pm 0{\cdot}3$ (16 objects), which clearly shows that other factors severely influence B_{n}, notably the emission lines N_1, N_2, Hβ, and Hγ. We find

$$-\log F(\Sigma) = 11{\cdot}5 + 0{\cdot}3(B_{\mathrm{n}} - 16{\cdot}4), \tag{2}$$

where

$$\begin{aligned} F(\Sigma) &= F(\mathrm{N}_1) + F(\mathrm{N}_2) + F(\mathrm{H}\beta) + F(\mathrm{H}\gamma) \\ &\sim \tfrac{4}{3}F(\mathrm{N}_1) + \tfrac{3}{2}F(\mathrm{H}\beta). \end{aligned} \tag{3}$$

Webster's data show that for the LMC planetary nebulae we may write as a good approximation

$$\log F(\mathrm{N}_1) \sim \log F(\Sigma) - 0{\cdot}20. \tag{4}$$

From (1), (3) and (4) we conclude that

$$B_{\mathrm{n}} - R_{\mathrm{n}} \sim \log F(\mathrm{H}\beta)/F(\mathrm{N}_1). \tag{5}$$

$B_{\mathrm{n}} - R_{\mathrm{n}}$ is thus a measure of the excitation of the nebulae; low colour indicates high excitation, and high (red) colour, low excitation. R_{n} is a measure of the flux in Hβ. The excellent relation between R_{n} and $F(\mathrm{H}\beta)$ indicates that the $F(\mathrm{H}\alpha)/F(\mathrm{H}\beta)$ ratio is constant for the LMC objects.

C. THE EXCITATION, HYDROGEN FLUX DIAGRAMS

In her thesis Webster analyzes the log $F(\mathrm{N}_1)/F(\mathrm{H}\beta)$ vs. log $F(\mathrm{H}\beta)$ diagram for the planetary nebulae in the two Clouds. She concludes that all the objects have log $F(\mathrm{H}\beta)$

fainter than $-12{\cdot}3$; this defines an upper limit to the total emission of any planetary nebula. This upper limit may be expressed as a function of the excitation; there is a well-defined upper envelope on the high- as well as on the low-excitation side of $\log F(N_1)/F(H\beta)=0{\cdot}74$; the slopes are

$$\log F(H\beta) \sim \begin{cases} -0{\cdot}5 \log F(N_1)/F(H\beta) & \text{high-excitation side} \\ +0{\cdot}5 \log F(N_1)/F(H\beta) & \text{low-excitation side.} \end{cases}$$

However, it appears likely that at an $F(N_1)/F(H\beta)$ value of about 10 the Hβ flux decreases rapidly without any obvious change in the ratio. The faintest object with a known Hβ flux, P 17, falls about 0·6 in $\log F(H\beta)$ below the envelope, yet its excitation ratio is close to 10.

A certain confusion appears in the diagram as a mixture of high-excitation and low-excitation objects is found at about $F(N_1)/F(H\beta)=10$ and about 0·2 below the high-excitation envelope. A comparison with the H-R diagram for the nuclei (Figure 3) shows that 6 of the 8 objects in this group have low-temperature nuclei; the low Hβ flux is caused by the low temperature and luminosity of the star in an optically thick nebula whereas [O III] is less sensitive to this.

According to our conclusions above, an R_n, (B_n-R_n) diagram should give at least an approximation of a flux, excitation diagram. It is given for the LMC objects in Figure 2. It appears possible here also to define high- and low-excitation envelopes. We note that near the latter are the two objects, P 7 and P 9, with known strong [N II]

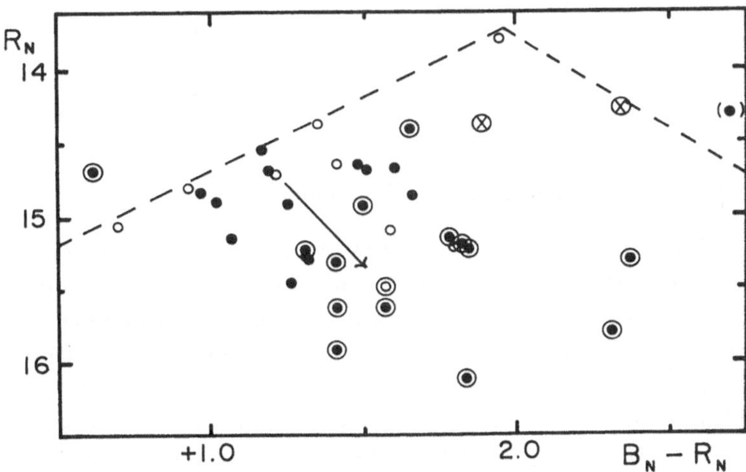

FIG. 2. *The R_n, (B_n-R_n) diagram for the planetary nebulae in the Large Magellanic Cloud. Circles represent planetary nebulae with observed He II 4686; crosses objects with known strong [N II] 6584. Large circles are drawn to identify objects situated to the West of 5^h20^m. The dashed lines are the suggested high- and low-excitation envelopes. The arrow indicates the effect of reddening.*

lines, and that 6 of the 9 objects with observed He II λ 4686 lie near the high-excitation envelope.

It is somewhat puzzling to find that most of the objects with low luminosity ($R_n > 15{\cdot}1$) lie in the preceding part of the LMC, West of 5^h20^m. This could possibly be taken as an effect of higher interstellar reddening; an additional average absorption of $A_R = 0{\cdot}6$ mag West of 5^h20^m is sufficient. However, there is no support for this in observations of other objects, and the observed Balmer decrement of the planetaries is the same; the mean value of log $F(H\beta)/F(H\gamma)$ is 0·30 in both parts of the LMC. A confirmation of the observed difference in R_n is found in the photoelectric flux measures; on the average log $F(H\beta)$ is 0·11 fainter in the preceding half of the LMC.

The difference between the planetary nebulae East and West of 5^h20^m was initially noted in the $(B_s - B_n)$ data in the W-S catalogue (Westerlund and Smith, 1964) and at first suspected as due to a systematic effect in the photographic magnitudes. There was no reason to expect that the central stars should contribute more to the total luminosity of the planetary nebulae in the preceding part than in the following. The effect is, however, obvious also in the narrow-band photoelectric colour indices:

5 of 7 objects West of 5^h20^m have $m_{4200} - m_{5300} > 0$, and only
2 of 11 objects East of 5^h20^m $m_{4200} - m_{5300} > 0$.

Furthermore, Webster (1966) has determined the apparent magnitude m_s (4861) of the central star by subtracting the nebular continuum from the 'star plus continuum' observation using its theoretical values. The average corrections are

$$m_s(4861) - m_{4200} = \begin{matrix} +0{\cdot}4 \text{ West of } 5^h20^m \\ +0{\cdot}7 \text{ East of } 5^h20^m. \end{matrix}$$

It then appears necessary to accept the results as indicating that there is a systematic difference between the 'average planetary nebula' in the two parts of the LMC.

5. The Masses of the Planetary Nebulae

It is well known that the mass of ionized hydrogen in a planetary nebula can be determined provided the distance, angular size, Hβ flux, electron temperature, and filling factor are known (Minkowski, 1965). For the planetary nebulae in the Magellanic Clouds, the distances are known, the fluxes can be observed, and reasonable assumptions may be made for the electron temperature ($T_e = 1{\cdot}5 \times 10^4$ °K) and the filling factor (ε). All the objects observed so far are stellar; this gives an upper limit to their radii of about 1″ of arc. Using this value, Webster found the maximum possible mass for the most luminous planetary in the LMC, P 25, to be 1·41 $M_\odot$ for $\varepsilon = 1$. This value is an overestimate, since the radius of P 25 is more likely to be of the order of 0·15 parsec (0″.5) from the position of its nucleus in the H-R diagram (Figure 3). Its ionized hydrogen mass is then of the order of 0.45 $M_\odot$.

We have determined the masses of two more planetary nebulae selected because of the position of their nuclei in the H-R diagram. The nucleus of the planetary N 6 in the SMC lies near the evolutionary track at log $T_s = 4{\cdot}6$. Its radius is probably smaller than 0·06 parsec. Using this value and the assumptions above, we derive a mass of 0·08 $M_\odot$. This is low and may indicate that part of the mass of N 6 is not ionized, hence that the object is still optically thick in H I. The planetary N 43 in the SMC is optically thin in H I and He I (Webster, 1966). Its central star is faint and its Hβ flux is among the strongest observed in the Cloud planetaries. We may assume that is has just reached the minimum size for optically thin nebulae, and its radius is then near 0·06 parsec. With $\varepsilon = 1$, its hydrogen mass is found to be 0·12 $M_\odot$.

Webster has applied another method for determining the most likely mass of a planetary nebula, provided there is a unique value. She assumes that the most luminous galactic planetary nebulae are identical to those in the Magellanic Clouds. They are then optically thin in hydrogen and their total emission in Hβ is log $E(\mathrm{H}\beta) = 35{\cdot}3$. The observed H$\beta$ flux permits a determination of the distance, and, consequently, of the mass of the galactic object. The true mass will be equal to the lowest mass obtained in this way for the optically thin objects. From calculations for 43 galactic planetary nebulae the best value is found to be 0·11 $M_\odot$.

We conclude that an average planetary nebula has a hydrogen mass of about 0·12 solar masses, but that masses up to about 0·5 solar masses may be expected.

6. The Evolution of the Planetary Nebulae

Webster has determined the temperatures and the luminosities of the central stars of the planetary nebulae in the Magellanic Clouds following Seaton's method. The result is given in Figure 3. The evolutionary track for a distance scale corresponding to a hydrogen mass of 0·1 $M_\odot$ is drawn, as well as Seaton's (dashed) for a hydrogen mass of 0·4 $M_\odot$. The differences on the high-temperature side are due to the scaling factors only. On the low-temperature side, the 0·1 $M_\odot$ curve is based on improved observations also of the galactic objects.

The high luminosity of the high-temperature nuclei in the Clouds is real. These objects, as well as the low-excitation nebulae in the SMC, have probably slightly higher masses (in the nuclei and in the shell?) than the average nebula. They should not be considered indicating the mean position of the evolutionary track at these temperatures. We note that in Seaton's diagram (1966, Figure 7) a similar situation exists for a group of nuclei at log $T \geq 5{\cdot}1$. It may be argued, and generally is, that the scatter in our Figure 3 and Seaton's Figure 7 simply indicates the scatter to be expected around one single average mass. However, if the theory is accepted that planetary nebulae evolve from red-giant stars, we have to bear in mind that each portion of the red-giant region in the H-R diagram is populated by stars which have evolved from two different luminosity (mass) ranges on the zero-age main sequence. It is then not unreasonable to expect to find planetary nebulae of at least two different mass groups.

7. Conclusions

The distribution of the planetary nebulae in the Small Magellanic Cloud does not contradict Hindman's model of an edge-on galaxy. The line of symmetry of the planetaries does not, however, agree with the position of the major axis in the HI model. The radial velocities of the planetary nebulae do not indicate that the SMC is a rotating system.

The distribution of the planetary nebulae in the Large Magellanic Cloud is in agreement with a system being seen nearly face-on. The system is rotating, and Webster's analysis indicates that the best centre of rotation is the centroid of the planetary nebulae system.

A number of low-excitation nebulae have been found in the SMC; their character-

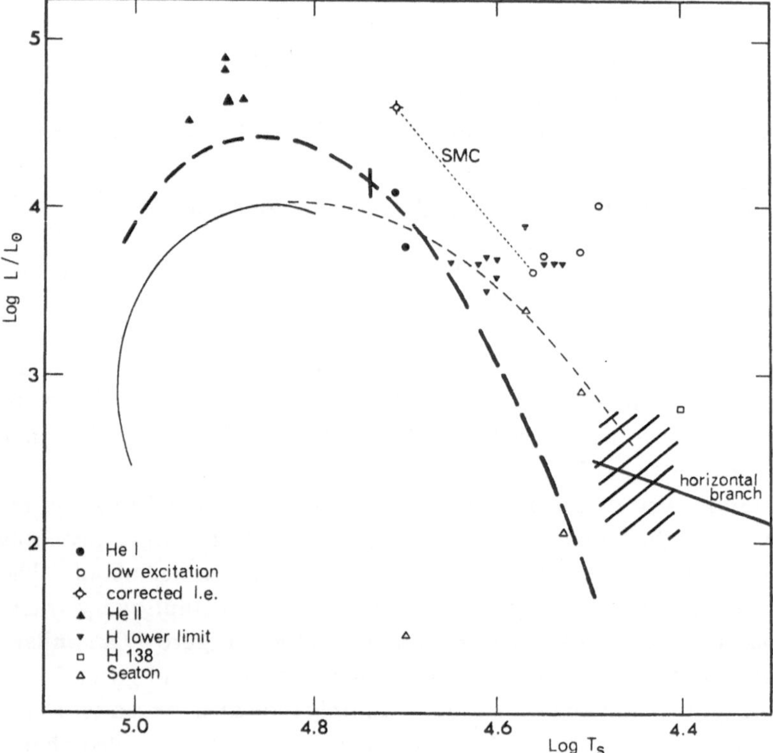

FIG. 3. *The evolutionary track of the planetary nuclei according to Webster. The heavy dashes mark Seaton's evolutionary track for a hydrogen mass of 0·4 $M_{\odot}$. The vertical line crossing this curve identifies a nebular radius of 0·06 parsec.*

istics are best understood if they have a higher than average mass. Similarly, a group of high-excitation planetary nebulae in the LMC may have masses slightly higher than the average. The best value for the average hydrogen mass of a planetary nebula is found to be near $0{\cdot}1 M_{\odot}$.

In the LMC, we have noted that the planetary nebulae to the West of 5^h20^m are on the average less luminous than those in the Eastern part. At the same time, their nuclei contribute more to their total luminosity. It is unlikely that this is an effect of interstellar reddening or an effect of different distances. A tilt of the LMC of 45° would cause a differential magnitude effect of 0·04 mag/kpc (= per degree). This could explain the lower luminosity to some extent, but not the different colors nor the different contribution of the central star to the total luminosity of the planetary nebulae.

References

Baade, W. (1955) *Astr. J.*, **60**, 151.

Cannon, A.J. (1933) *Harvard Coll. Obs. Bull.*, **891**.

Feast, M.W. (1964) *Observatory*, **84**, 266.

Feast, M.W., Thackeray, A.D., Wesselink, A.J. (1961) *Mon. Not. R. astr. Soc.*, **122**, 433.

Henize, K.G. (1956) *Astrophys. J. Suppl. Ser.*, **2**, 315.

Henize, K.G., Westerlund, B.E. (1963) *Astrophys. J.*, **137**, 747.

Hindman, J.W. (1967) *Austr. J. Phys.*, **20**, 147.

Koelbloed, D. (1956) *Observatory*, **76**, 894.

Lindsay, E.M. (1955) *Mon. Not. R. astr. Soc.*, **115**, 248.

Lindsay, E.M. (1956) *Mon. Not. R. astr. Soc.*, **116**, 649.

Lindsay, E.M. (1961) *Astr. J.*, **66**, 169.

Lindsay, E.M., Mullan, D.J. (1963) *Irish astr. J.*, **6**, 51.

McGee, R.X., Milton, J.A. (1966) *Austr. J. Phys.*, **19**, 343.

Minkowski, R. (1965) in *Galactic Structure*, Ed. by A. Blaauw and M. Schmidt, Univ. of Chicago Press, Chicago, p. 321.

O'Dell, C.R. (1962) *Astrophys. J.*, **135**, 371.

Seaton, M.J. (1966) *Mon. Not. R. astr. Soc.*, **132**, 113.

Švestka, Z. (1962) *Bull. astr. Inst. Csl.*, **13**, 35.

Swope, H.H. (1963) *Astr. J.*, **68**, 470.

Webster, B.L. (1965) in *Symposium on the Magellanic Clouds*, Ed. by J.V. Hindman and B.E. Westerlund, Canberra, p. 29.

Webster, B.L. (1966) Thesis, Austr. Nat. Univ. Canberra.

Westerlund, B.E. (1965) in *Symposium on the Magellanic Clouds*, Ed. by J.V. Hindman and B.E. Westerlund, Canberra, p. 40.

Westerlund, B.E. (1968) *Vistas in Astronomy* (in press).

Westerlund, B.E., Rodgers, A.W. (1959) *Observatory*, **79**, 132.

Westerlund, B.E., Smith, L.F. (1964) *Mon. Not. R. astr. Soc.*, **127**, 449.

SPECTROSCOPIC OBSERVATIONS OF PLANETARY NEBULAE IN THE MAGELLANIC CLOUDS

M.W. FEAST
(Radcliffe Observatory, Pretoria, South Africa)

A programme of spectroscopic observations of planetary nebulae in both Magellanic Clouds has been underway in Pretoria for some years, and some preliminary results for the LMC have been published (Feast, 1964*b*). More extensive observations are now available (altogether in both clouds 112 spectra of 39 objects), and these are being prepared for publication elsewhere. Below are summarized the main conclusions so far deduced.

The radial velocities depend very largely on 48 Å/mm plates and should be of good accuracy. Lower-dispersion spectra have less weight for velocities but are very valuable in that, in general, many of the fainter lines are observed and a better idea can be obtained of the physical nature of the object. For instance, in several objects high densities are indicated by the [O II] doublet, consistent with the classification of these objects as planetaries, and in one object a broad emission band (mainly C III) from the central star was detected.

In the LMC there appears to be no significant systematic deviation of the planetary velocities from the rotation curve defined by extreme Population-I objects (supergiant stars, diffuse nebulae and interstellar gas) despite a recent suggestion of such a deviation (Webster, 1965). It has been a problem for some while that the centre of rotation of the LMC is displaced 1 kpc from the optical centre. However, whereas the extreme Population-I objects are, in fact, distributed symmetrically about the centre of rotation, the planetaries have a centre of symmetry near the optical centre (Westerlund and Smith, 1964). Two alternative interpretations seem possible. Either, the centre of rotation is the real mass centre, in which case we must postulate high obscuration to the North of the optical bar (Wesselink, 1966), unless there are large quantities of undetected matter such as H_2 in the LMC. Or, and this is an attractive possibility, we must adopt the view (advocated by de Vaucouleurs) that we are observing non-circular motions in a barred spiral. Whichever explanation is adopted, the present work indicates that the observed systematic velocity pattern applies to objects of a wide range of ages.

For the LMC the velocity dispersion of the planetaries corrected for observational error is 22 ± 3 (s.e.) km/sec with no significant deviation from a gaussian distribution. This dispersion is significantly greater than for extreme Population-I objects ($9{\cdot}6 \pm 1{\cdot}1$

Osterbrock and O'Dell (eds.), Planetary Nebulae, 34–35. © *I.A.U.*

(s.e.) km/sec; Feast, 1964*a*) and may be taken as evidence for a collapse of the LMC to a plane (as in the case of the Galaxy). There is no large change in velocity dispersion of the planetaries from the centre to the edge of the LMC, in contrast to the results for the Galaxy. However, there is, of course, a much less marked central condensation of the planetaries in the LMC than in the Galaxy. An interesting indication has been found that the velocity dispersion of the planetaries increases with decreasing brightness. In addition there is some suggestion of a different surface distribution for bright and faint planetaries.

The radial velocities of planetary nebulae in the SMC show a clustering around two distinct values independent of position in the Cloud. These two groups are strongly reminiscent of the double peaks in the 21-cm profiles. The results suggest the same overall flow pattern for the planetaries and the HI gas though there is no detailed correlation between the positions and velocities of the planetaries and the interstellar gas. Hindman (1967) has interpreted the 21-cm results in terms of a number of large expanding shells in the SMC. However, this would require the planetaries to be young objects ($< 10^7$ years) and a detailed correlation of positions and velocities with extreme Population-I objects would then be expected. A promising approach appears to be motion in a barred spiral with the bar seen almost end on and with gas and stars flowing outwards from each end of the bar.

A distinct difference has been found between the two Clouds in the frequency of occurrence of planetaries of different excitation classes. Low-excitation planetaries predominate in the SMC. In this respect the Galaxy seems to be more similar to the LMC than to the SMC. In the LMC itself there is an indication that high-excitation planetaries are concentrated amongst the lower luminosity objects. Further study of this question should yield interesting results.

References

Feast, M.W. (1964*a*) *Mon. Not. R. astr. Soc.*, **127**, 195.
Feast, M.W. (1964*b*) *Observatory*, **84**, 266.
Hindman, J.V. (1967) *Austr. J. Phys.*, **20**, 147.
Webster, L. (1965) in *Symposium on the Magellanic Clouds*, Ed. by J.V. Hindman and B.E. Westerlund, Canberra, p. 29.
Wesselink, A.J. (1966) *Astr. J.*, **71**, 185.
Westerlund, B.E., Smith, L.F. (1964) *Mon Not. R. astr. Soc.*, **127**, 449.

A PRELIMINARY REPORT OF A SURVEY OF PLANETARY NEBULAE IN THE SOUTHERN HEMISPHERE

PIK-SIN THE
(Bosscha Observatory, Indonesia)

As you know, the new and most complete catalogue of galactic planetary nebulae, compiled by Perek and Kohoutek, has recently been published. This catalogue will certainly be very useful for astronomers working on planetary nebulae.

However, as has been pointed out by the compilers of the catalogue, the degree of completeness of the discoveries, e.g. in a 20° belt along the Milky Way, is not uniform. This inhomogeneity will affect, of course, all statistical work on planetary nebulae.

Even if we restrict ourselves to the region of the galactic centre only, say, about 30° on both sides of the galactic centre, our knowledge of the number of planetary nebulae over the field is inhomogeneous. This region has been covered by several surveys of planetary nebulae, by Henize using the 10-inch Metcalf refractor, by Abell using the Palomar Schmidt telescope, and by Haro using the Tonantzintla Schmidt telescope. Some regions have been surveyed by Blanco and myself using the Lembang Schmidt telescope. Because of the various methods and telescopes used in these surveys, it is clear that our knowledge of the distribution of planetaries in the region of the galactic centre is not homogeneous.

The situation is even worse outside the region of the galactic centre. Recently Dr Kohoutek, using the Hamburg-Schmidt telescope, had surveyed a region of the Northern Milky Way from $l^{II}=32°$ up to $l^{II}=70°$. The limiting magnitude of this survey is red magnitude 17·8. Because of this survey, the distribution of planetaries is not symmetric with respect to the galactic centre, as was pointed out by Perek and Kohoutek in their catalogue.

In view of the above-mentioned inhomogeneity and asymmetry, and in order to extend Kohoutek's survey to the South, I have started, using our Lembang Schmidt telescope, a survey of planetary nebulae. This survey will be done in two steps:

(1) The central region of the Milky Way from $l^{II}=40°$ through 0° to $l^{II}=320°$.

(2) The Southern Milky Way from $l^{II}=270°$ to $l^{II}=320°$.

The width of the survey region will be 10° on both sides of the galactic equator. The reason that I will not go far beyond Carina is because this region falls in our rainy season, during which it is very difficult to obtain plates. I am exposing my plates so that the limiting magnitude of my survey will be comparable to Kohoutek's limiting magnitude.

Osterbrock and O'Dell (eds.), Planetary Nebulae, 36–37. © *I.A.U.*

I will describe briefly the first results of my survey:

(1) The whole constellation of Scutum was covered by 103a-E survey plates, exposed behind an RG1 filter. The spectra are unwidened. The whole area surveyed is almost 120 square degrees centered at $\alpha = 18^h36^m$, $\delta = -10\overset{\circ}{.}0$ (1950), which corresponds to $l^{II} = 23\overset{\circ}{.}0$, $b^{II} = -2\overset{\circ}{.}1$. The total number of previously known planetaries is 39, of which several could not be redetected on my plates, probably caused overlapping, especially in very crowded fields. The catalogue of Perek and Kohoutek will be very useful in re-identifying these missing objects on my plates. The number of newly found planetaries is 14, more than 25 % of all the existing planetaries in this region. All the newly found planetaries in this region, except one, appear stellar on my direct plates.

(2) Furthermore, I have examined a region of 25 square degrees centered at $\alpha = 16^h11^m$, $\delta = -54\overset{\circ}{.}2$ (1950); or $l^{II} = 329\overset{\circ}{.}9$, $b^{II} = -2\overset{\circ}{.}8$. The number of known planetaries in this region is 11, while 6 new planetaries were discovered, about 35% of the total number now known. This percentage is lower than that which Kohoutek found in his Hamburg Schmidt survey, but the present survey region may perhaps be exceptional.

I think that the above preliminary result of my survey of planetaries justifies an attempt to survey the southern Milky Way for planetary nebulae. My plate material will of course be very useful for the detection of other interesting objects.

DISCUSSION

Aller: Would you expect to find more planetaries if you used an instrument with larger scale?

Pik Sin The: Yes, especially in the very crowded regions of the Milky Way.

Abell: What is the aperture of your Schmidt telescope?

Pik-Sin The: Its aperture is 70 cm.

Perek: It is pleasant to see that the number of new discoveries is smaller than the number of nebulae already known. It shows that we are approaching completeness with the present telescopes.

OBSERVED ANGULAR MOTIONS IN PLANETARY NEBULAE

MARTHA H. LILLER and WILLIAM LILLER
(Harvard College Observatory, Cambridge, Mass., U.S.A.)

ABSTRACT

Plates of planetary nebulae taken in October 1965 and June 1967 with the 100-inch and 60-inch telescopes on Mount Wilson are compared with photographs taken with the same instruments 40 or more years previously, to determine the amount of angular motion of material outward from the central stars. Densitometer tracings of the nebulae yield values of $\dot{\theta}$, the radial motion in seconds of arc per century for condensations (filaments, rings, etc.) as well as for the edges of the visible material. Preliminary data are presented for NGC 2392, 6818 and 7662. For NGC 2392 and NGC 7662 distances are derived.

1. Introduction

Several papers reporting measurements of the angular motion of material outwards from central stars of planetary nebulae have appeared in the past few years (Latypov, 1957; Čudovičeva, 1964; W. Liller, 1965; Liller *et al.*, 1966). The incomplete and, in some cases, conflicting results given in these papers suggest that additional measurements should be made on sets of plates with as large a scale as possible.

With this in mind, one of us (W.L.), in October 1965 and again in June 1967, secured a number of second-epoch plates of planetary nebulae with the 100-inch and 60-inch reflectors at Mount Wilson. These plates are being compared with first-epoch plates taken with the same instruments from 1909 to 1925.

Although the reduction of the plate material is incomplete, we can present some preliminary results of our investigation at this time.

2. Analysis of the Plates

Using a Baird microdensitometer with a precision drive, we made traces along from 1 to 5 axes of each image of the planetary nebulae. The axes were chosen to pass through the well-defined filaments of the nebula and also to give as even a coverage of position angle as possible. Projected slit sizes, the same for old and new plates taken with a given telescope and focus, ranged from 0·02 to 0·05 mm in width and 0·15 to 0·6 mm in length, depending upon the size and detail of the nebula and the plate scale.

To check for any systematic change in plate scale that may have occurred between the early- and late-epoch plates, we measured radial distances of about 6 stars from the central stars of 6 nebulae. In all cases average changes in distance amounted to less than a few microns, or less than a few hundredths of a second of arc, at the mean angular distance of the stars, and hence probably less than this amount at the distance of the outer portions of the nebula.

The densitometer tracings were measured in two ways. The first, and most satisfactory procedure when it was possible, was simply to measure the distance from the central star to the position of peak intensity of a filament or portion of a ring, or the distance from one filament peak to another diametrically opposed, if the central star was not detectable. The second procedure was to measure the distance from the central star to the 'edge' of the nebula, where edge is defined to be that point on the tracing at which a straight line, placed through the steeply decreasing trace in the outer region of the nebula, intersects the density of the sky background. The edge measures are less satisfactory than the filament measures because the edge measures are more strongly affected by seeing and because they are dependent on density or length of exposure.

The edge measures can be corrected for density and seeing effects as follows. A comparison star of suitable density is traced on all plates of a nebula in all directions in which the nebula itself was traced. Distance from edge to central star or from edge to edge for the nebula is plotted against the height of the comparison star above the sky background on the trace made for each trace direction and image. Straight-line least-squares solutions with the same slope for both epochs are fitted to the data for each epoch, and the difference in intercept gives the change in size of the nebula over the period in question. The same procedure is followed for the edge-to-center or edge-to-edge distances of the comparison star itself, and the change in size is subtracted from that for the nebula, as a rough correction for seeing and background effects.

3. The Results

Preliminary values for centennial radial angular velocity $\dot{\theta}$ (seconds of arc per 100 years) for three of the nebulae are presented below. In the case of the edges of the nebulae, $\dot{\theta}$ may refer either to true motion of the nebular material or change in radius of an ionization sphere. For the ring and filament measures, it is probable that we are measuring motion of the material. Therefore, we compare $\dot{\theta}$ from ring and filament measures with the spectroscopically determined ΔV of Wilson (1950) to derive the distances to the nebulae. We discuss each nebula in turn.

(1) *NGC 2392*: Figure 1a is a graph of $\dot{\theta}$ against the angular radius θ in seconds of arc for this planetary. Data from all position angles are presented. The discrepant point marked with an asterisk refers to a condensation at P.A. 283° that lies outside the main ring but inside the condensations marking the outer ring or 'fur', and apparently does not share in the motion shown by the rest of the nebular material. This deviating condensation is shown on Figure 2, a photograph of NGC 2392. A least-squares solution through the rest of the points in Figure 1a gives a slope of

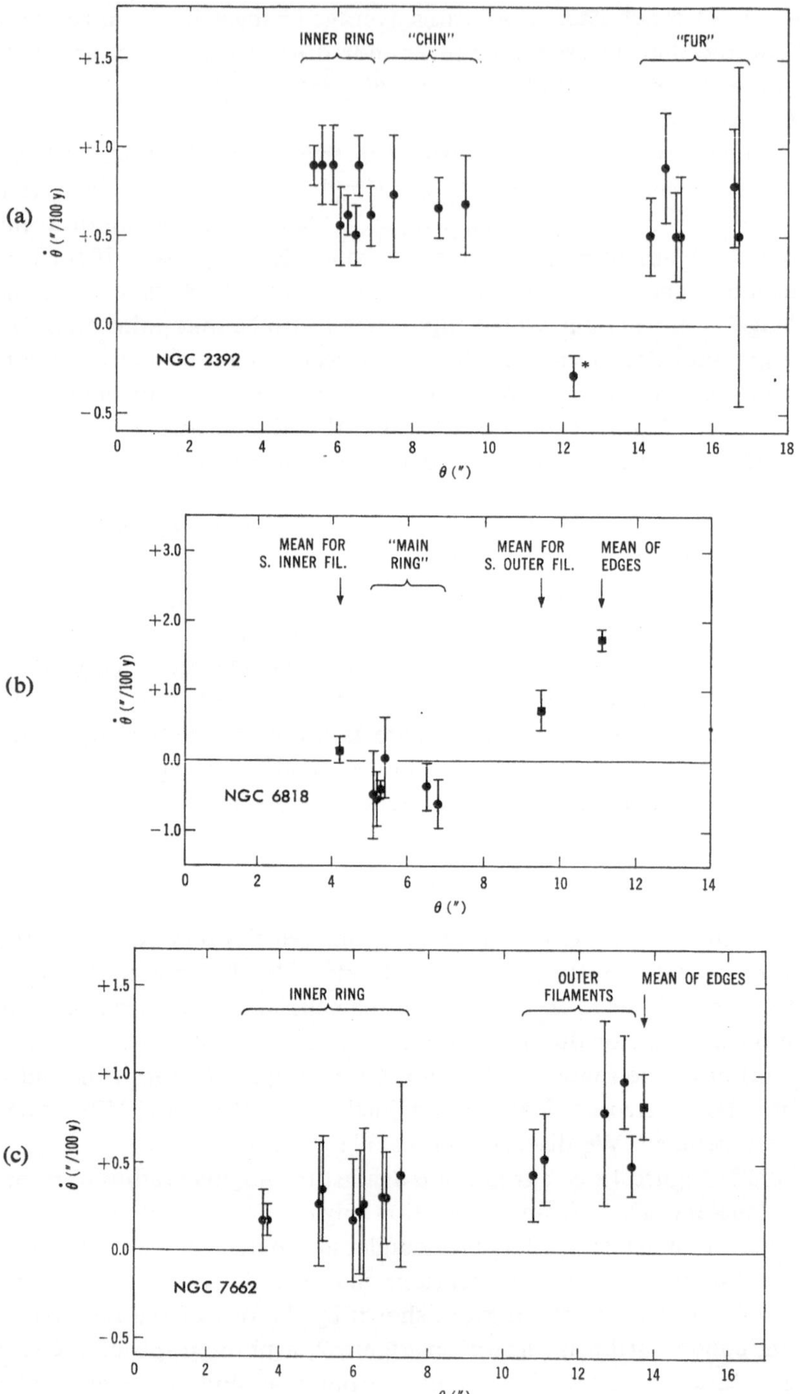
(a)
INNER RING
"CHIN"
"FUR"
NGC 2392
+1.5
+1.0
+0.5
0.0
−0.5
$\dot{\theta}$ ("/100 y)
0
2
4
6
8
10
12
14
16
18
θ (")
(b)
MEAN FOR S. INNER FIL.
"MAIN RING"
MEAN FOR S. OUTER FIL.
MEAN OF EDGES
NGC 6818
+3.0
+2.0
+1.0
0.0
−1.0
$\dot{\theta}$ ("/100 y)
0
2
4
6
8
10
12
14
θ (")
(c)
INNER RING
OUTER FILAMENTS
MEAN OF EDGES
NGC 7662
+1.5
+1.0
+0.5
0.0
−0.5
$\dot{\theta}$ ("/100 y)
0
2
4
6
8
10
12
14
16
θ (")

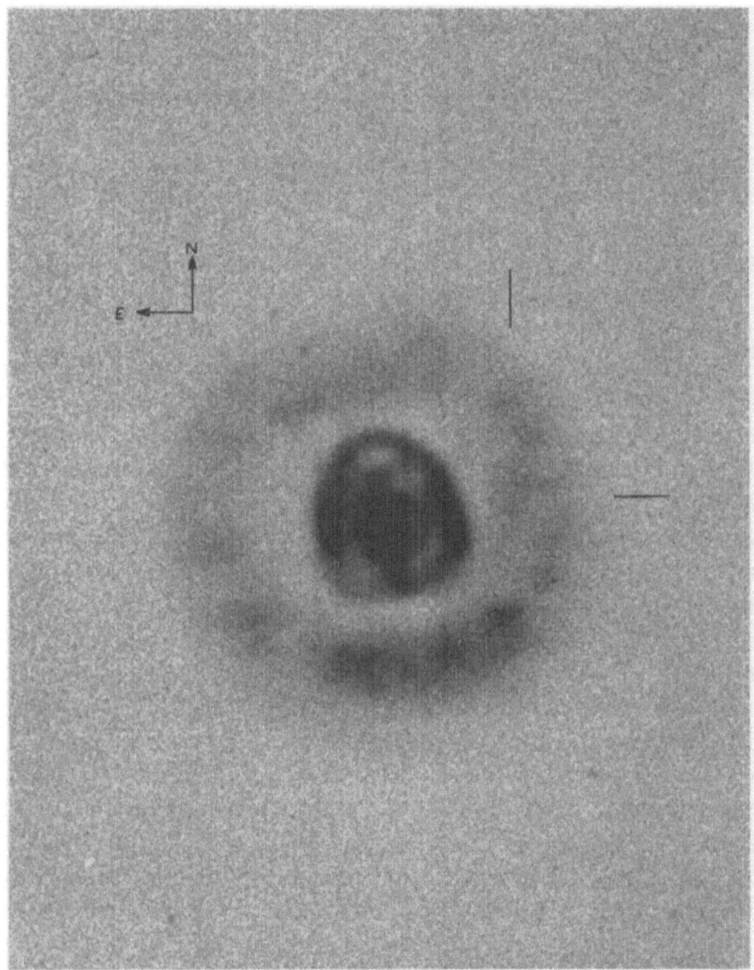

FIG. 2. *NGC 2392. The condensation producing the discrepant point in Figure 1a is marked. Mount Wilson 60-inch Cassegrain photograph.*

$-\cdot 018 \pm \cdot 012''$ of arc per 100 years per second of arc; that is, the angular outward velocity decreases slightly as distance from the central star increases. Wilson's ΔV values, with the exception of those derived from [Ne v], are reasonably constant. From the widths of emission lines in the central star spectrum, he concludes that material is being emitted from the star at nearly this same velocity.

FIG. 1a, b, c. *Radial expansion in seconds of arc per 100 years as a function of radius for three planetary nebulae. The error bars are twice the mean error in length. The point marked by an asterisk for NGC 2392 is discussed in the text. Data for NGC 2392 and NGC 6818 from 60-inch Cassegrain plates; data for NGC 7662 from 100-inch Cassegrain plates.*

If the inner ring of NGC 2392 is a projection of a spherical shell and outward velocity is constant with radius within that shell, the observed $\dot{\theta}$ will be somewhat smaller than the true one. The difference depends upon the thickness of the shell, and amounts to about 10% of $\dot{\theta}$ for a shell that has a thickness of one quarter of its radius. In the case of NGC 2392, in which the shell appears quite thin, the error is probably somewhat smaller than this value. If, however, the inner ring is a true ring, then the observed and true $\dot{\theta}$ will be identical. We combine Wilson's $\Delta V/2 = 54{\cdot}8$ km/sec with our mean $\dot{\theta}$ for the inner regions, $+0{\cdot}72 \pm {\cdot}06''$ of arc per century, to get a distance of 1600 ± 130 parsec. This distance is larger than the distance derived by most other observers, but is close to Seaton's (1966) distance of 1820 parsec.

(2) *NGC 6818*: Figure 1b shows the $\dot{\theta}$ versus θ relationship for this planetary with, again, data from five diameters presented. The mean points for the South filaments in Figure 1b refer to the section of the nebula where a doubling in the ring occurs, and are means of two measures each. Individual values for $\dot{\theta}$ are shown for the East and West portions, or the 'main' ring. Whereas the Southern filaments and the edges demonstrate a $\dot{\theta}$ that increases from zero with increasing θ, the main ring appears to exhibit a negative $\dot{\theta}$ or possible influx of material toward the central star. Wilson's ΔV measurements increase with increasing image size in somewhat the way our $\dot{\theta}$ measurements for the Southern filaments do. But with the negative $\dot{\theta}$ for the large part of the ring, it is difficult to say which of our measures correspond to his radial velocities, and thus distance is not derived here.

(3) *NGC 7662*: Figure 1c presents the $\dot{\theta}$ versus θ data for this nebula. Again, points from 5 diameters are shown. Although NGC 7662 is similar to NGC 2392 in appearance, its $\dot{\theta}(\theta)$ behaves in quite a different way. A least-squares solution through the ring and filament points yields a $d\dot{\theta}/d\theta$ of $+{\cdot}047 \pm {\cdot}007''$ of arc per 100 years per second of arc. This solution passes nearly through the origin. Wilson's ΔV measurements, in a similar fashion, increase with image size for the inner ring, again indicating that the velocity of the particles increases with increasing distance from the nucleus. Since our plates are sensitive primarily to the radiations of [O III], [Ne III], and H, we adopt Wilson's mean $\Delta V/2$ for these of 26·0 km/sec. We assign this linear velocity to our $\dot{\theta}$, $+0{\cdot}26 \pm {\cdot}09''$ of arc per century at the average radius of the ring, $5''\!.7$, to compute a distance of 2100 ± 730 parsec for NGC 7662. This value is again larger than that derived by most observers, but smaller than Seaton's distance of 2500 parsec.

The results given above are preliminary. The final results will be published when reduction of the plates is complete.

Acknowledgment

It is a pleasure to thank the National Science Foundation for their support of this research.

References

Čudovičeva, O.N. (1964) *Izv. glav. astr. Obs. Pulkove*, **23**, 154.
Latypov, A.A. (1957) *Pub. astr. Obs. Tashkent* (2), **5**, 31.
Liller, W. (1965) *Publ. astr. Soc. Pacific*, **77**, 25.
Liller, M.H., Welther, B.L., Liller, W. (1966) *Astrophys. J.*, **144**, 280.
Seaton, M.J. (1966) *Mon. Not. R. astr. Soc.*, **132**, 113.
Wilson, O.C. (1950) *Astrophys. J.*, **111**, 279.

DISCUSSION

Münch: I would like to make a remark concerning a possible systematic error in the measurement of nebular images arising from the different colour sensitivities of first- and second-epoch plates. The old Mt. Wilson plates, probably Seed 23, have a sensitivity extending into the green-yellow less than that of orthochromatic plates used at present. Consequently the relative contribution of N I and N II and Hβ may be different in two sets of plates. I have encountered this difficulty while measuring expansion rates of novae envelopes.

M. Liller: In addition, the change from silver to aluminum mirrors has increased the importance of [O II] in the newer plates. However, since in many cases individual knots and filaments were measured, the colour response should not be too critical.

O'Dell: How were the probable errors derived? Since some points fell into the negative expansion (i.e. contraction) region of your plots, can you say with certainty that such contraction occurs?

M. Liller: The probable errors were derived from internal consistency. Most measurements covered at least 3 images. However, we are not yet certain that the measured contraction is real.

Mathews: I would like to ask if the velocities of the outer blobs in NGC 7009, which are moving outward faster than the main nebula, might indicate that these blobs were ejected at a later time.

W. Liller: The outer blobs of NGC 7009 share in a quite well-defined linear relationship between angular velocity and distance from central star. Therefore, it appears that all visible mass left the star at approximately the same instant.

THE SPACE DISTRIBUTION OF PLANETARY NEBULAE

J. H. CAHN*
(University College London, England)

ABSTRACT

A punched-card catalogue of planetary nebulae has been prepared, using data extracted from all existing catalogues. A computer program calculates distances and radii using the method of Shklovsky, in which all nebulae are assumed to have the same ionized mass, and allowance for interstellar extinction is made assuming a continuous galactic-dust distribution. The assumption made in Shklovsky's method, that the nebulae are optically thin, is considered to be satisfied if the calculated radii lie within a certain well-defined interval. The reddening constants obtained are in satisfactory statistical agreement with constants determined by other methods. The local density of planetary nebulae is in agreement with estimates of local white-dwarf densities.

A study of the current data on planetary nebulae (Vorontsov-Velyaminov, 1962; O'Dell, 1962; Perek, 1963; Vorontsov-Velyaminov *et al.*, 1964*a*, *b*, 1965; Abell, 1966; Henize, 1967; Westerlund and Henize, 1967) has permitted the statistical determination of the distances of 537 planetaries where both angular size and optical flux were available. Preliminary determinations of local density and parameters describing the galactic space distribution are generally in accord with previous determinations.

The method of Shklovsky (Shklovsky, 1956) as modified by Seaton (Seaton, 1966) is used. Briefly, the luminosity Λ or the surface brightness Σ is expressed in terms of the recombination radiation for an idealized spherical geometry of radius R in the following way:

$$\Lambda = 4\pi R^2 \Sigma = \frac{4\pi R^3}{3} \varepsilon \alpha(\lambda) N_e^2 \frac{hc}{\lambda}.$$

The electron density N_e is assumed constant over the fraction ε of the volume and zero elsewhere. $\alpha(\lambda)$ is an effective recombination coefficient for line radiation of wavelength λ. In the Shklovsky method the mass M of the ionized nebula is assumed constant and replaces the electron density through the relation

$$M = \frac{4\pi R^3}{3} \varepsilon N_e m_{\text{proton}}.$$

Replacing the brightness Σ by the flux F at the telescope, corrected for atmospheric extinction, by

$$R^2 \Sigma = r^2 F \times 10^{c_\lambda (r)}$$

* Permanently at University of Illinois, Urbana, Ill., U.S.A.

Osterbrock and O'Dell (eds.), Planetary Nebulae, 44–50. © *I.A.U.*

where $c_\lambda(r)$ is the interstellar extinction and r is the nebular distance, one obtains the implicit equation for r,

$$r^5 = \frac{3}{\varepsilon}\left(\frac{M}{4\pi m_{\text{proton}}}\right)^2 \frac{hc}{\lambda\theta^3 \times 10^{c_\lambda(r)}F},$$

where θ is the angular radius. The interstellar extinction, $c_\lambda(r)$, is determined by integration of the extinction per unit distance (Perek, 1963), $\alpha_\lambda/(1+m)^2$, along the path to the planetary

$$c_\lambda(r) = \int_0^r \frac{\alpha_\lambda \, \mathrm{d}r}{(1+m^2)^2},$$

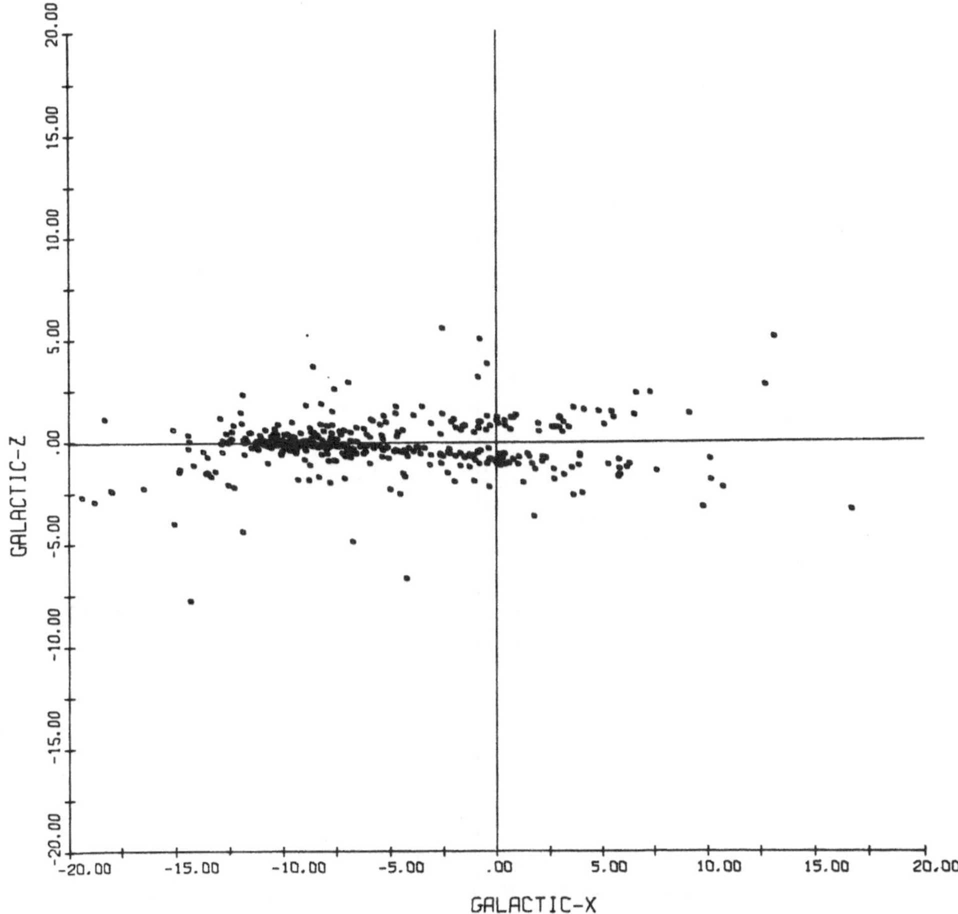

Fig. 1. *Projection on x-z axis of calculated positions of optically thin planetary nebulae. S represents Sun. The units are in kiloparsecs.*

where m^2 is the quantity $(\rho/a_d)^2+(z/c_d)^2$ determined from the galactic polar coordinates ρ, z and the scale factors a_d and c_d of the galactic dust distribution.

In order to gain an impression of the adopted distance scale which is based on Seaton's model, $M=0{\cdot}38\ m_\odot$ and $\varepsilon=0{\cdot}63$ (Seaton, 1966), Figures 1 and 2 show the calculated galactic distribution as projected on the x-z and the galactic planes. When the distance scale was varied ($k=1$ represents the scale of the present calculation) both $k=1{\cdot}5$ and $k=0{\cdot}5$ had to be rejected. The case $k=1{\cdot}5$ was untenable because far too many planetaries were placed beyond the galactic centre, while the case $k=0{\cdot}5$ gave a heliocentric distribution. The scale $k=0{\cdot}75$, which corresponds approximately to that of O'Dell (O'Dell, 1962), is hardly distinguishable from the scale $k=1$. The effect of the interstellar dust concentration in the galactic plane is shown in Figure 3 in that only planetaries relatively close to the Sun can be observed.

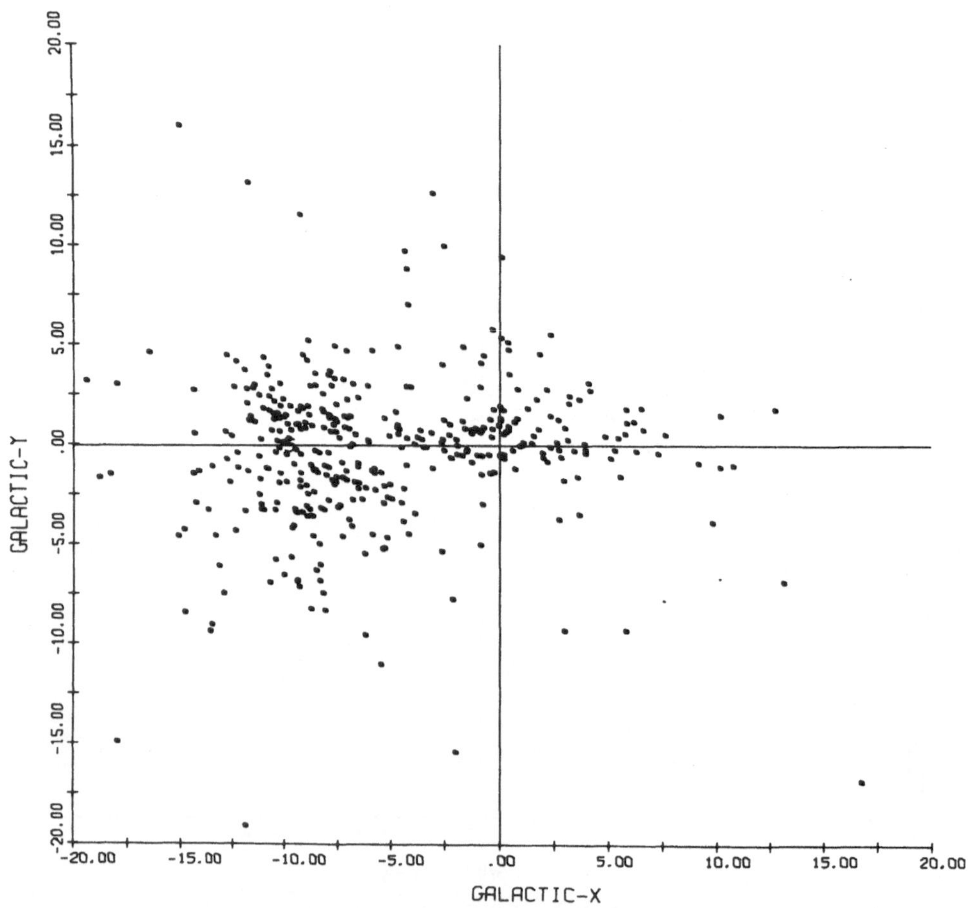

FIG. 2. *Projection of optically thin nebulae on galactic plane.*

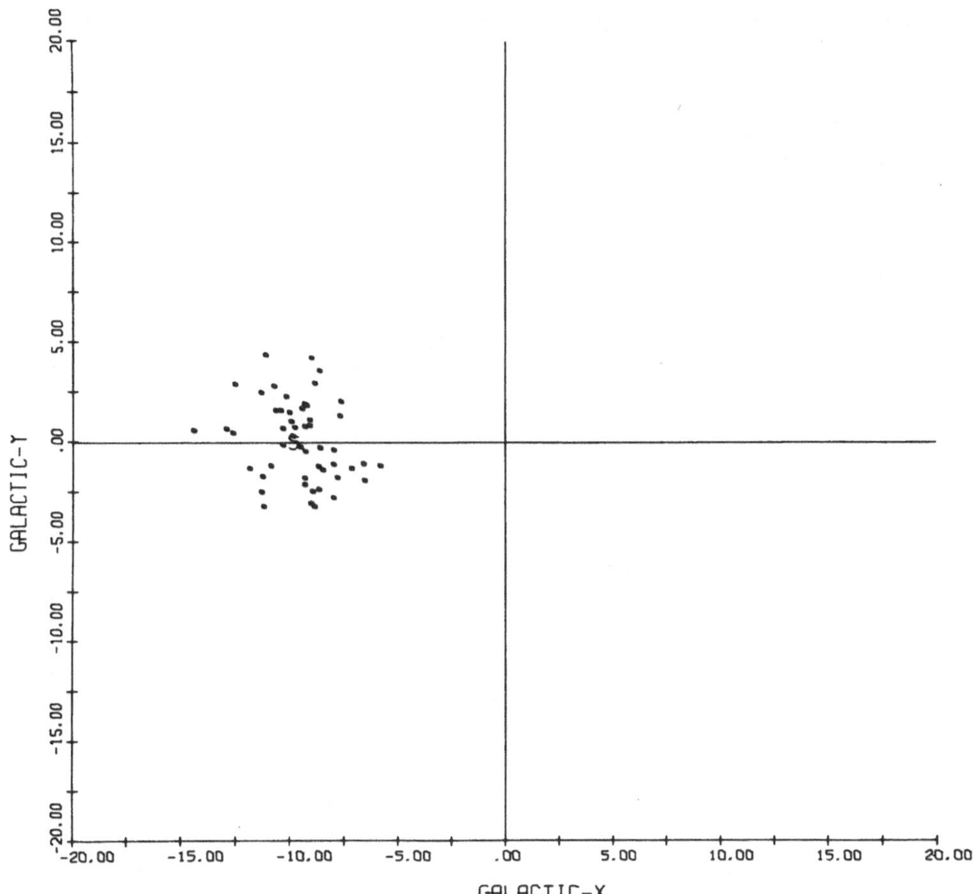

FIG. 3. *Projection on galactic plane of those optically thin nebulae within 100 parsec of the galactic plane.*

In comparing the extinctions calculated by the method of this paper with measured extinctions, we find, as shown in Figure 4, agreement to about 75% when comparing with radio and photoelectric data (O'Dell, 1962; Terzian, 1966; Thompson *et al.*, 1967). A rough check on the distance scale is provided by comparing calculated and observed extinctions for scale reductions of factors of 2. Such large reductions are clearly unacceptable when compared with observation. However, calculated extinctions are insensitive to increases of scale factor, since the elongation of the path length in general puts the planetary outside the influence of the galactic dust distribution. The extinction was found to be consistent with the galactic dust distribution adopted by Perek (Perek, 1963).

If the planetaries are undergoing a uniform expansion, there should be, in the

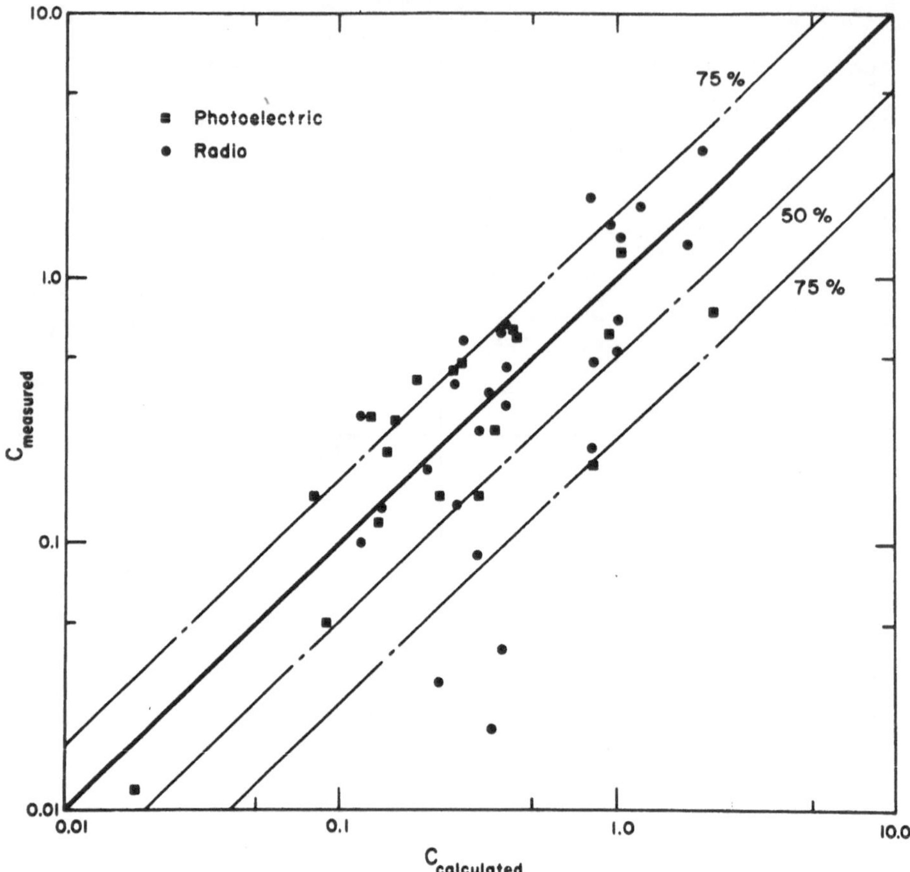

FIG. 4. *Comparison of measured and calculated extinctions.*

optically thin regime (Seaton, 1966), a constant number per unit radius interval. This is statistically borne out in Figure 5 for those radii $0{\cdot}12 < R < 0{\cdot}60$ parsec as demonstrated by Seaton (Seaton, 1966). Figure 5 shows as well the expected fall-off for larger radii, and the lack of large radii planetaries, due to low surface brightness, beyond 2 kpc. By an extrapolation procedure, it is found that the local density of optically thin planetaries in a sphere of zero radius centered at the Sun, ρ_L, is 12 kpc^{-3}. The local number per unit radius interval, $N(R)$, is obtained by dividing the local density, ρ_L, by 0·48 parsec, the radius range over which the planetaries are optically thin,

$$N(R) = \rho_L / 4{\cdot}8 \times 10^{-4}$$
$$= 2{\cdot}5 \times 10^4 \text{ kpc}^{-4}.$$

$N(R)$ is the apparent number density per unit radius interval and is equal to the physical density if the Shklovsky method is valid.

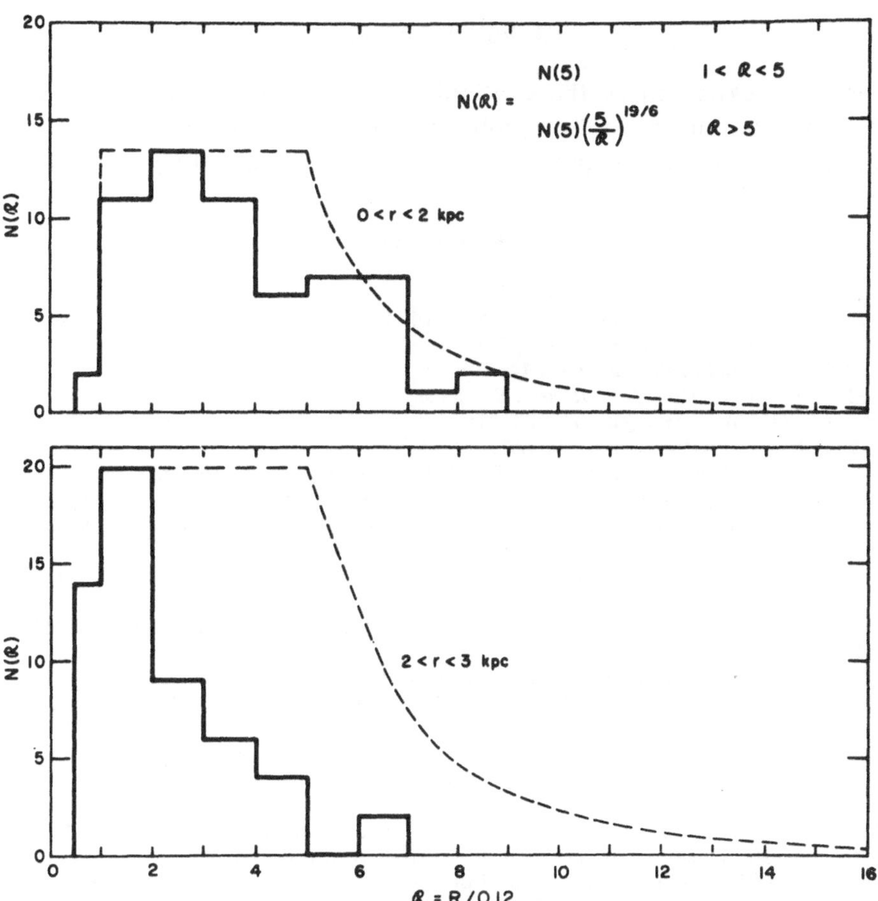

FIG. 5. *Statistical distribution of number per unit volume per unit radius, N(R), as a function of reduced radius. The expected fall-off beyond R = 5 (Seaton, 1966) is shown by the dashed line.*

From the number $N(R)$ we compute the local production rate of planetaries, χ_{PN}, from the expansion velocity $V = 20$ km sec$^{-1} = 2 \times 10^{-8}$ kpc year^{-1} to be

$$\chi_{PN} = VN(R) = 5 \times 10^{-4}\ \text{kpc}^{-3}\ \text{year}^{-1}.$$

The local white-dwarf birth rate of 2×10^{-3} kpc^{-3} year^{-1} (Weidemann, 1968) is seen to exceed the estimated production rate of planetaries by a factor 4. If scale $k = 0{\cdot}75$ is adopted, one obtains $N(R) = 7{\cdot}1 \times 10^4$ kpc^{-4} so that the corresponding production rate $\chi_{PN} = 1{\cdot}4 \times 10^{-3}$ kpc^{-3} year^{-1}. Thus, within the uncertainty of our distance scale, the white-dwarf population must have received a major contribution from the planetary nebulae.

Acknowledgments

The author acknowledges the continuing enthusiasm and advice of Professor M. J. Seaton, who suggested this problem. The hospitality of University College London and the financial support of the U.S. Navy are gratefully acknowledged.

References

Abell, G.O. (1966) *Astrophys. J.*, **144**, 259.
Henize, K.G. (1967) *Astrophys. J. Suppl. Ser.*, **14**, 125.
O'Dell, C.R. (1962) *Astrophys. J.*, **135**, 371.
Perek, L. (1963) *Bull. astr. Inst. Csl.*, **14**, 201.
Seaton, M.J. (1966) *Mon. Not. R. astr. Soc.*, **132**, 113.
Shklovsky, I.S. (1956) *Astr. Zu.*, **33**, 222, 315.
Terzian, Y. (1966) *Astrophys. J.* **144**, 657.
Thompson, A.R., Colvin, R.S., Stanley, G.J. (1967) *Astrophys. J.*, **148**, 429.
Vorontsov-Velyaminov, B.A. (1962) *Soobšč. gos. astr. Inst. P. K. Šternberga*, **118**, 3.
Vorontsov-Velyaminov, B.A., Kostjakova, E.B., Dokuchaeva, O.D., Arhipova, V.P. (1964*a*) *Astr. Cirk. SSSR*, **305**.
Vorontsov-Velyaminov, B.A., Kostjakova, E.B., Dokuchaeva, O.D., Arhipova, V.P. (1964*b*) *Astr. Zu.*, **41**, 255, 464.
Vorontsov-Velyaminov, B.A., Kostjakova, E.B., Dokuchaeva, O.D., Arhipova, V.P. (1965) *Astr. Zu.*, **42**, 730.
Weidemann, V. (1968) in the present volume, p. 423.
Westerlund, B.E., Henize, K.G. (1967) *Astrophys. J. Suppl. Ser.*, **14**, 154.

DISCUSSION

Feast: Cahn has discussed the effects of systematic changes in the distance scale. However, one also has to consider the systematic errors in the distances introduced by the random errors in individual distance estimates. This is a statistical problem which may result in considerable correction being required, especially for the mean distance of the most distant objects.

Cahn: I estimate the errors to be at most a factor of 2. The distribution function was determined by fitting only near the Sun where the discovery is reasonably complete.

GENERAL DISCUSSION – FIRST SESSION

Salpeter: We have heard an estimate for the total number of planetary nebulae in the Galaxy of about 50000. I would like to know: (a) Allowing for selection effects, etc., what is the estimate of the total number of planetary nebulae in globular clusters? (b) Is there any positive evidence for planetary nebulae in young galactic clusters (or in a binary with a young star as companion)?

O'Dell: Only a single planetary is known in a globular cluster, this being K 648 in M 15. This object was studied and found to have a common velocity and heavy-element deficiency with the cluster, and there seems to be little question of cluster membership. Many years ago Baade searched several globular clusters and found no additional objects.

Several coincidences with galactic clusters are known, but in each case studied in detail, there is a significant difference of radial velocity so that no generic relationship is indicated.

Münch: In relation to O'Dell's remark, I wish to say that two years ago I photographed 7 of the nearer globular clusters through a 50 Å passband Hα filter with the 200-inch Palomar telescope. A careful blink procedure against comparison plates taken in a passband void of emission lines failed to reveal any other planetary than the one known in M 15. It should be remarked, however, that with the estimate of 50000 in the galactic system, one would expect only 5 planetaries among the 10^7 stars involved in these globular clusters. Thus, within the uncertainty of small-number statistics, there is agreement between the expected and observed number of planetaries in globular clusters.

Perek: Planetary nebulae belong to the disk population as defined by the Vatican conference in 1957.

Osterbrock: But could not say 10% of the planetaries be Population I?

Feast: There certainly seems to be good evidence from the kinematics of the galactic planetary nebulae for a considerable range in ages. The asymmetrical drift is quite small for the planetaries near the Sun, as is the velocity dispersion. The mean age of planetaries near the Sun cannot therefore be too large. However, the velocity dispersion and hence the inferred mean age increase as one goes in towards the galactic centre.

Underhill: Is there any observational information to support the hypothesis that planetary nebulae having central stars with a Wolf-Rayet spectrum are distributed as Population I objects, whereas those with central stars having a continuous spectrum are distributed with the Disk Population?

Osterbrock and O'Dell (eds.), Planetary Nebulae, 51–53. © *I.A.U.*

Aller: In surveying central stars of as many planetary nebulae as possible, only a small number of Wolf-Rayet nuclei were found. The statistics do not suffice to decide whether they do or do not tend to favor the galactic plane.

Feast: The one planetary in the Magellanic Clouds in which it has so far been possible to detect a WR nucleus (LMC Henize N 203) is one which shows a considerable deviation from the rotation curve. My results indicate that this object would be contained with the other LMC planetaries in a single gaussian distribution, though Westerlund and Webster prefer to treat it separately as being, presumably, more like Population II.

Minkowski: If the observed proper motion of expansion of a planetary nebula is interpreted under the assumption that the nebula is spherical, an incorrect distance may result. An example is NGC 2392, which is possibly a prolate ellipsoid seen end-on, and the observed radial velocity is therefore possibly significantly larger than the tangential velocity of expansion. Furthermore, in NGC 2392 the radial velocities over the North half of the nebula are systematically positive, and in the South half, systematically negative. This cannot result from spherically symmetric expansion.

Reeves: Does rotation play a role?

Minkowski: No, the angular momentum involved would be far too large. It must result from expansion that is not spherically symmetric.

Münch: In a statistical sense, however, the method of angular motions should provide a direct estimate of distances.

Minkowski: That is true, but it is important to realize that any individual distance can be incorrect by a factor of 2 as a result of this non-spherical expansion.

Mathews: As a matter of principle, even if a planetary had perfect spherical symmetry, there are still problems in interpreting the proper motion observations of the Lillers and others. The proper motion measurements refer to the apparent contractions or expansion of the outermost layers, while the radial-velocity measurements taken at the centre of the nebula, say, represent some mean value weighted over the entire nebula. In particular a rarefaction wave moving inward into an expanding spherical nebula can give the impression in the sky that the nebula is shrinking and yet the gas velocity is everywhere outward.

Seaton: Minkowski and others have calculated distances of optically thick nebulae assuming all central stars to have equal absolute magnitudes. This is useful for galactic distribution studies but not if one wants to consider the evolution of the central stars! I think that in the latter case the only method generally available is to use forbidden-line intensity ratios to get the electron density.

For galactic-structure studies it is of interest to know the total number of planetaries per unit volume, but for evolution studies it is more important to know the number of optically thin planetaries per unit volume per unit radius interval.

Savedoff: It is dangerous for a theorist to suggest observational problems. I remember, however, that an important contribution to the resolution of the uncertainty in Cepheid and RR Lyrae distance scale (1952) was the apparent increase in the width of the galactic plane with distance. If planetaries and white dwarfs are in fact evolutionarily connected, then we should require

(a) $\overline{|z|}_{\text{w.d.}} = \overline{|z|}_{\text{planetary}}$

(b) $\overline{\dot{z}^2}_{\text{w.d.}} = \overline{\dot{z}^2}_{\text{planetary}}$

(c) that $z_{\text{planetary}}$ be free of any heliocentric peculiarities.

I recognize that the comparisons may be difficult because of selection effects, particularly in the white-dwarf data.

Cahn: An important method of calibrating the distance scale of planetary nebulae would be to measure the angular sizes of planetaries in the Magellanic Clouds. Is this observation possible?

Westerlund: So far no planetary nebula in the Magellanic Clouds has been resolved. It should be possible to resolve fainter ones ($m \leqslant 20$ mag) using existing telescopes, but the attempts have not yet been successful. Very good seeing is required.

Menon: I would like to point out that the shape of the radio-frequency spectrum, which is quite sensitive to the presence of irregularities in the nebula, can be used to obtain information about the filling factor for those cases where the radio-frequency spectra are available.

Abell: We observe fine-scale filamentary structure in many nebulae. But can we not place a lower limit to the size of such filaments? Gas at 10^4 °K has a thermal velocity of about 10 km/sec, and we should not expect it to be able to maintain very small filaments (say, less than 10^{15} or 10^{16} cm across) for periods greater than 10^2 years, should we?

Osterbrock: Filaments are there whether we understand why they are or not. They may well overlap, and therefore we cannot tell the filling factor from examination of direct photographs.

Minkowski: Though individual filaments may dissipate quickly, if some mechanism continually generates new ones, then some will always be present.

Mathews: Since planetary nebula typically expand differentially at supersonic velocities, inhomogeneities expanding at about the velocity of sound may not smooth out.

Cahn: One lifetime of a condensation will be of order: (linear dimensions)/(speed of sound), whether the flow in the large is subsonic or supersonic. In either case the continued existence of fluctuations needs to be explained.

Seaton: In principle it is possible to obtain information about fine-scale filamentary structure from studies of two or more different ratios of forbidden-line intensities. Further observational work would be of value in this connection.

Session II

OBSERVATIONS

THE ABSOLUTE SPECTROPHOTOMETRY OF 170 PLANETARY NEBULAE

B. A. VORONTSOV-VELYAMINOV, E. B. KOSTJAKOVA,
O. D. DOKUCHAEVA, V. P. ARHIPOVA
(Sternberg State Astronomical Institute, Moscow, U.S.S.R.)

An investigation of planetary nebulae has been underway at the Sternberg State Astronomical Institute in Moscow for several years. It was begun with the measurement of the emission-line intensities of planetary nebulae in an homogeneous system in absolute units. More than 300 long-exposure objective prism spectrograms were obtained with the 50-cm Maksutov telescope at the Crimean Station of the Institute and with the 70-cm meniscus telescope of the Abastumani Observatory. The dispersion of the spectrograms was 190 and 160 Å/mm at Hγ respectively.

For photometric reductions, Vorontsov-Velyaminov's wide-slit method was used. The line intensities were derived from a comparison with the spectra of A0-type stars on the same plates. An 'average A0 V star' was adopted as a standard comparison source. Its energy distribution was computed from recently published photoelectric and photographic observations. Planetary nebulae with angular dimensions less than

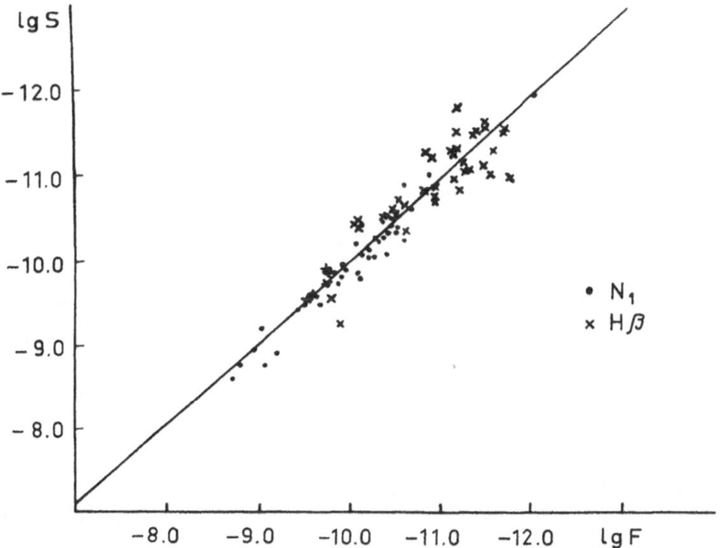

FIG. 1. *Comparison of Sternberg* (lg*S*) *measurements of planetary nebulae with O'Dell's photoelectric measurements* (lg*F*).

Osterbrock and O'Dell (eds.), Planetary Nebulae, 57–58. © I.A.U.

30″ of arc, and with total magnitudes brighter than $m = 13$ were studied. The number of lines measured for the nebulae varied from 1 to 20 (from 10 to 20 lines for most of the objects) in the spectral interval 3700–6600 Å. Comparison of the results showed good agreement with the data obtained by the photoelectric method, as can be seen in Figure 1, where our N_1 and $H\beta$ energy fluxes are plotted against those obtained by O'Dell. The continuous spectrum intensity of the planetary-nebula nucleus was also estimated if possible. For 108 planetary nebulae the absolute line intensities were determined for the first time, while for 93 objects no data at all on the line intensities had been previously published.

The most interesting group of nearly 90 planetary nebulae is situated in the region projected on the galactic centre. These nebulae have total magnitudes of about $m = 13$. The greatest majority of them have not previously been investigated by the spectroscopic method. The connection between the apparent distribution of the nebulae projected on the galactic centre region and the interstellar light absorption in this direction was also studied. It was found that these nebulae are situated farther than 2 or 3 kpc from the Sun and that they do not form a compact group in space.

The ratio of the $H\alpha$ and N_1 line intensities of the nebulae in the galactic-centre region was found to exceed systematically the ratio for nebulae in other parts of the sky. This seems to indicate the possibility of a physical difference between the nebulae in the direction of the galactic centre and the nebulae situated in the other regions of the sky.

The results of the work are published in seven papers listed below, in the References.

References

Vorontsov-Velyaminov, B. A., Kostjakova, E. B., Dokuchaeva, O.D., Arhipova, V. P. (1964) *Astr. Cirk.*, **305**.

Vorontsov-Velyaminov, B. A., Kostjakova, E. B., Dokuchaeva, O.D., Arhipova, V. P. (1964) *Astr. Zu.*, **41**, 255.

Vorontsov-Velyaminov, B. A., Kostjakova, E. B., Dokuchaeva, O.D., Arhipova, V. P. (1965) *Astr. Cirk.*, **348**.

Vorontsov-Velyaminov, B. A., Kostjakova, E. B., Dokuchaeva, O.D., Arhipova, V. P. (1965) *Astr. Zu.*, **42**, 464.

Vorontsov-Velyaminov, B. A., Kostjakova, E. B., Dokuchaeva, O.D., Arhipova, V. P. (1965) *Astr. Zu.*, **42**, 730.

Vorontsov-Velyaminov, B. A., Kostjakova, E. B., Dokuchaeva, O.D., Arhipova, V. P. (1967) *Astr. Cirk.*, **437**.

Vorontsov-Velyaminov, B. A., Kostjakova, E. B., Dokuchaeva, O.D., Arhipova, V. P. (1967) *Astr. Zu.*, **44**, 361.

DISCUSSION

Aller: To what extent do overlapping spectra of background stars complicate the measurements of nebular line intensities?

Dokuchaeva: We used only the best spectrograms in which there was no overlapping by background stars.

A PHOTOELECTRIC PHOTOMETER FOR EXTENDED SOURCES

D. MALAISE

(Institut d'Astrophysique de l'Université de Liège, Belgique)

ABSTRACT

The paper describes a photoelectric photometer especially designed for extended sources. Six bands selected by a concave grating spectrograph can be measured simultaneously. The sensitivity commutation of the electrometers is automatic and a 10′ field can be observed during measurements.

RÉSUMÉ

Nous décrivons un photomètre photoélectrique spécialement conçu pour l'observation des sources étendues. On peut effectuer les mesures simultanément dans six bandes sélectionnées au moyen d'un spectrographe à réseau concave. La commutation de sensibilité des électromètres se fait automatiquement. Un champ de 10′ peut être observé sans interrompre les mesures.

Two years ago, Professor Swings asked me to design and build a photometer for studying the monochromatic photometric profile of comets. The instrument is especially suited for the study of extended sources with emission spectra. It will be used to measure the monochromatic surface brightnesses of planetary nebulae and precise ratios of line intensities in different parts of the nebulae.

The instrument is designed as a concave grating spectrograph (Figure 1). A small circular diaphragm, defining the element of resolution in the field of view, is placed on the Rowland circle of the grating at 32° incidence angle. The grating has a radius of

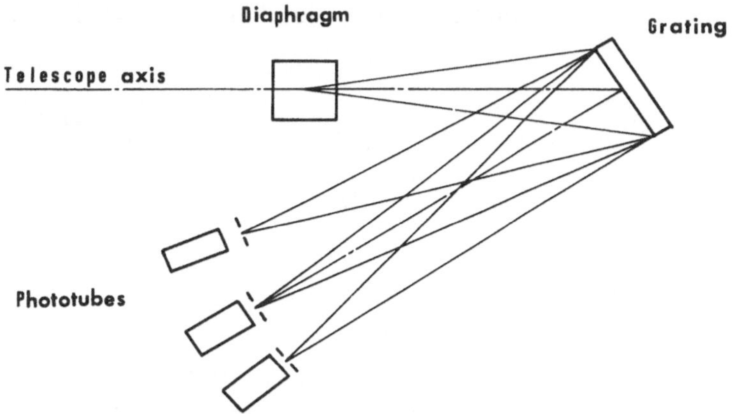

FIG. 1. *Schematic optical design of the spectrograph.*

Osterbrock and O'Dell (eds.), Planetary Nebulae, 59–61. © *I.A.U.*

750 mm and is ruled at 982 grooves/mm with a blaze angle of 18°03′ (Bausch & Lomb no. 35-52-15-65). The bands to be measured are selected by a series of adjustable slits movable on the Rowland circle. A very compact design of the mountings supporting the slits and the photomultiplier tubes enables one to select two adjacent bands only 300 Å (150 Å in the second order) apart. For reasons of compactness, we have adopted ASCOP 541A and 541E photomultiplier tubes. The dispersion is 14 Å/mm in the first order (blazed wavelength 6200 Å), and 7 Å/mm in the second order (blazed wavelength 3100 Å). The bands falling between 4000 Å and 8500 Å are picked up in the first order, while those corresponding to wavelengths between 3000 Å and 4000 Å are picked up in the second order. The diffraction angles are between +13° and −8°, which makes astigmatism nearly constant over the whole range.

The choice of the bands is very versatile and we can select any band-pass up to 70 Å at any place in the range 3000 Å to 8500 Å. The minimum width of the bands depends on the size of the entrance diaphragm and on the diffraction order. Typically, with an entrance diaphragm of 1 mm diameter and working in the first order, the minimum band-pass is 14 Å. The width of the band is set within 0·1 Å with the help of a microscope sliding in the phototube attachment in place of the phototube, and then the slit mounting is set on the Rowland circle within 0·5 Å by displacing and adjusting the mounting on the circle with respect to a calibration line delivered by the spectrograph. It is possible to work with six bands simultaneously without overcrowding the Rowland circle.

The instrument has been constructed to work at the Cassegrain focus of the 193-cm telescope of the Haute-Provence Observatory. The relative aperture is F:15 and the field scale 7″/mm. We have three easily interchangeable entrance diaphragms of 0·5, 1 and 2 mm diameter respectively. As shown in Figure 2 the entrance diaphragm is at the centre of a large diagonal mirror which folds the field rays to the field viewer. The image of a 10′ field is formed on a reticle, where it is observed through the eyepiece without interrupting the measurements.

Since we are operating with six channels simultaneously, we had to provide for

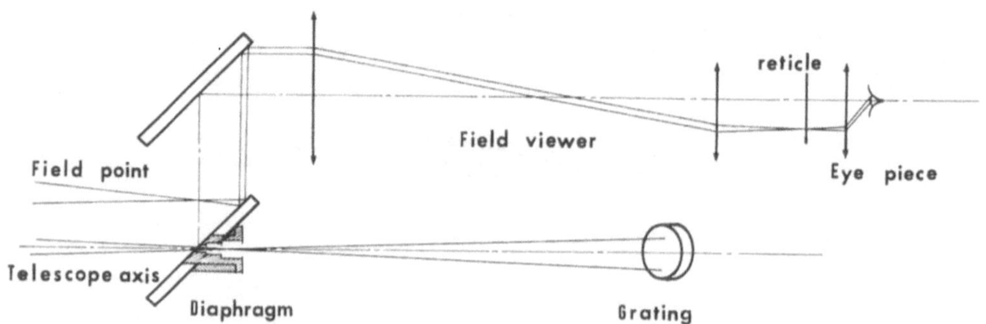

FIG. 2. *Schematic optical design of the field viewer.*

automatic commutation of the sensitivity of the electrometers. Each electrometer is equipped with a high-level and a low-level sensor, which actuate a synchronous two-way counter; the latter selects the input resistor and gain appropriate to the signal. The electrometer has been completely designed and constructed in our electronic shop. Its final form is a compact $18 \times 20 \times 7$ cm module with an 'automatic-manual' switch and an external display of the operating range. We have chosen a $10^{1/2}$ ratio between the 8 sensitivities which gives us a 4-decade total capacity. We have 6 of these electrometers mounted in an 8-case box attached directly onto the photometer. The seventh case of the box is used for interconnection of power supplies, and the last one is used to control the temperature of the whole box to within 1 °C.

The data are recorded on 6 paper chart recorders, but we plan to add an analog-to-digital converter-multiplexer and a magnetic tape recorder after the electronics have been tested in their classical form.

When the instrument is attached to the 193-cm telescope, a signal-to-noise ratio of one corresponds to a flux onto the primary mirror of 5×10^{-14} erg cm^{-2} s^{-1} at 4200 Å, and 6×10^{-13} erg cm^{-2} s^{-1} at 7300 Å.

OBSERVATIONS OF PLANETARY-NEBULA SPECTRA IN THE NEAR ULTRAVIOLET

E. B. Kostjakova

(Sternberg State Astronomical Institute, Moscow, U.S.S.R.)

ABSTRACT

Several dozen spectrograms of about 20 nebulae were obtained with the quartz spectrograph, placed at the Cassegrain focus of the 125-cm parabolic reflector of the Sternberg State Astronomical Institute Crimean Station. Nebulae with intense Balmer continuum were selected. Nearly half of them had not been studied before. Reductions were made for the nuclei spectra, for the differential atmospheric extinction and for the interstellar absorption. The A0 V stars were taken as absolute intensity standards. The work is not yet finished, but the results will be received in a short time.

OBSERVATION DES NÉBULEUSES PLANÉTAIRES DANS L'INFRAROUGE

YVETTE ANDRILLAT
(Université de Montpellier, France)

La région de l'infrarouge a été fort peu étudiée jusqu'à maintenant en astrophysique. En ce qui concerne les nébuleuses planétaires les seuls travaux publiés concernent la région de l'infrarouge photographique jusqu'à 1·2 μ.

Au delà de cette limite, on se heurte à de grosses difficultés dues à la présence de l'atmosphère terrestre et au manque de détecteurs sensibles. Cependant, depuis 5 ans environ, un important effort technique a été réalisé principalement aux U.S.A. à l'Université d'Arizona, à San Diego et à California Institute of Technology par Low et Murray et Wildey (1963) pour mettre au point des cellules photoconductrices dopées permettant d'atteindre 15 μ. Low a déjà observé quelques nébuleuses planétaires dans cette région (non publié) et un vaste programme d'observation a été mis au point par Gillet, Stein et Low. Sur le plan théorique, de nombreux et remarquables travaux ont été menés à bien par Delmer *et al.* (1967).

Ce sont les études théoriques de Burbidge *et al.* (1963) qui montrèrent que, dans les régions HII, une grande partie de l'énergie était émise sous forme de rayonnement infrarouge. Dès lors, il a paru intéressant d'étudier les nébuleuses planétaires dans cette région spectrale pour les raisons suivantes:

(1) Dans le domaine optique, l'absorption interstellaire empêche les observations à de grandes distances dans le plan galactique et au voisinage du centre galactique. Or, aux grandes longueurs d'onde, cette absorption est considérablement diminuée. En infrarouge, il est donc possible d'observer les nombreuses nébuleuses planétaires situées près du centre de la galaxie.

(2) Les raies infrarouges correspondant à des transitions interdites, le phénomène de self-absorption est négligeable. Elles fourniront donc des images monochromatiques permettant de préciser la structure de la nébuleuse.

(3) Des intensités des raies infrarouges, on déduit l'abondance des éléments. Dans les nébuleuses planétaires, la plupart d'entre eux sont présents dans plusieurs états d'ionisation qui ne sont pas tous observables dans le domaine optique. C'est p. ex. le cas du néon pour lequel NeII n'a pas de raies visibles dans le domaine optique. Par contre, une raie est observable dans l'infrarouge. La mesure de son intensité permettra d'accéder à l'abondance du néon.

(4) On trouve dans l'infrarouge quelques couples de raies d'un même élément qui

Osterbrock and O'Dell (eds.), Planetary Nebulae, 63–67. © *I.A.U.*

sont désexcitées par collisions à des densités moyennes (comme 3727–3729 Å de [O II] p. ex.). Le rapport de leurs intensités fournit la densité électronique.

Les raies infrarouges considérées dans les calculs des spectres théoriques sont les raies les plus intenses relatives à certains ions particulièrement abondants dans les nébuleuses planétaires. Elles résultent de transitions entre les niveaux de structure fine des termes de l'état fondamental. Ces transitions interdites sont de type magnétique dipolaire correspondant à $\Delta J = \pm 1$, les transitions quadrupolaires, beaucoup plus faibles, n'ont pas été envisagées. Les différences d'énergie ΔE entre les niveaux de structure fine varient de 0·01 à 0·1 eV si bien que le mécanisme principal d'excitation des niveaux supérieurs est celui des collisions inélastiques d'électrons: en effet, pour les nébuleuses, les énergies des électrons incidents sont voisines de 1 eV, quantité nettement supérieure à ΔE.

Delmer *et al.* (1967) ont calculé les intensités de ces raies infrarouges en utilisant les données déduites des observations dans le domaine optique. Ces intensités peuvent être calculées à partir de l'abondance de l'ion et de la force de collision Ω nécessaire pour exciter les niveaux supérieurs. Dans le cas des niveaux à structure fine, Ω est calculable par la 'Quantum Defect Method'. De nombreuses valeurs ont d'ailleurs été empruntées aux travaux de Seaton (1958), Osterbrock (1964, 1965) et Blaha (1964). Dans quelques cas, l'abondance peut être déduite de l'intensité des raies interdites optiques du même ion: on a alors de bonnes valeurs pour les intensités: c'est le cas notamment de S III, Ne V, A III et A V.

En général, la détermination de l'abondance est très imprécise car il y a de grosses erreurs sur les constantes atomiques utilisées, en particulier sur les 'Cross-Section' de photo-ionisation dont les calculs sont complexes. Les valeurs données par Seaton (1958), et Ditchburn et Öpik (1962) ont été adoptées. Delmer *et al.* (1967) ont étendu les calculs aux ions complexes en utilisant la 'Quantum Defect Method' indiquée par Burgess et Seaton (1960), mais les résultats obtenus sont peu précis. L'imprécision dans la détermination des intensités peut atteindre un facteur 30.

Les intensités ont été calculées pour les raies situées dans les 'fenêtres' de l'atmosphère.

Dans l'infrarouge, l'absorption atmosphérique est très importante: elle est due principalement aux nombreuses et très larges bandes de la vapeur d'eau. On trouve également les absorptions plus discrètes mais intenses de CO_2 et enfin celles plus faibles de N_2O, CH_4, O_2 et O_3 (Allen, 1964). Jusqu'à 2·4 μ environ, il existe plusieurs zônes ou 'fenêtres' dans lesquelles l'absorption atmosphérique est faible. Au delà, ces 'fenêtres' sont rares et s'étendent approximativement de 3·2 à 4·1 μ, de 4·5 à 5·2 μ, de 8 à 13·5 μ, et de 17 à 25 μ. Aux environs de 4·2 μ, 6·5–15 μ et au delà de 25 μ, l'atmosphère est à peu près opaque.

De plus, la présence d'une émission thermique de l'atmosphère perturbe fortement les mesures de photométrie photoélectrique surtout au delà de 10 μ.

Pour obtenir les intensités observées, il faut corriger les intensités calculées de l'absorption atmosphérique. Théoriquement, le problème est difficile à cause de la complexité des bandes de rotation–vibration de H_2O, CO_2. Par ailleurs, les résultats expérimentaux manquent de précision car la distribution de la vapeur d'eau dans l'atmosphère n'est pas uniforme et l'absorption varie très rapidement d'un instant à l'autre. Les valeurs adoptées pour la transmission atmosphérique T sont celles déduites des meilleurs travaux expérimentaux. Au zénith, elles sont approximativement de $T=0{\cdot}08$ pour $\lambda=4{\cdot}5\ \mu$; 0·30 pour $\lambda=7{\cdot}9\ \mu$; 0·90 de 9 à 12·8 μ; 0·33 pour $\lambda=13{\cdot}1\ \mu$; 0·38 pour $\lambda=18{\cdot}7\ \mu$; 0·02 pour 21·8 μ; 0·07 pour $\lambda=24{\cdot}2\ \mu$; et 0·02 pour $\lambda=25{\cdot}9\ \mu$ (Delmer *et al.*, 1967).

Les intensités I ont été calculées pour 11 raies observables au-dessous de l'atmosphère. On a considéré seulement celles pour lesquelles I est supérieur à 10^{-18} watts/cm^2.

[Mg IV]: transition J_1–J_2	1/2–3/2	à $\lambda=$ 4·492 μ.
[A VI]	3/2–1/2	à $\lambda=$ 4·525 μ.
[A V]	2–1	à $\lambda=$ 7·893 μ.
[A III]	1–2	à $\lambda=$ 8·990 μ.
[Si IV]	3/2–1/2	à $\lambda=$ 10·53 μ.
[Ne II]	1/2–3/2	à $\lambda=$ 12·8 μ.
[A V]	1–0	à $\lambda=$ 13·1 μ.
[S III]	2–1	à $\lambda=$ 18·68 μ.
[A III]	0–1	à $\lambda=$ 21·8 μ.
[Ne V]	1–0	à $\lambda=$ 24·2 μ.
[O IV]	3/2–1/2	à $\lambda=$ 25·87 μ.

Les spectres théoriques ont ainsi été calculés pour les 9 nébuleuses planétaires les plus brillantes: NGC 7027, IC 418, NGC 6572, NGC 6543, NGC 7662, NGC 7009, NGC 6826, NGC 6210 et NGC 6720. Pour NGC 7027, on a pû tracer le spectre de 3 à 28 μ: les raies d'émission sont nettement au-dessus du 'bremsstrahlung' continuum. Stein (1967) signale d'ailleurs que, dans cette nébuleuse, le continu serait observable.

Parmi les émissions indiquées ci-dessus, l'une d'entre elles, [Ne II] à 12·8 μ est extrêmement importante car elle permet de déterminer l'abondance totale du néon. Jusqu'ici, l'absence de raies visibles de Ne II a empêché une évaluation précise de cette quantité. Située dans une région où l'atmosphère est à peu près transparente, elle est donc observable dans de bonnes conditions à partir du sol. Particulièrement intense dans les régions H II (Gould, 1964), elle est également forte dans les nébuleuses planétaires de faible excitation. Pour IC 418, Low a mesuré entre 12·5 et 13·5 μ un flux de 1×10^{-16} watts/cm^2 en bon accord avec l'intensité théorique $0{\cdot}6\times10^{-16}$ watts/cm^2 calculée par Gould (1966) pour [Ne II] à 12·8 μ.

En dehors des 'fenêtres' de l'atmosphère, il existe d'autres raies observables (Osterbrock, 1967; Gould, 1963).

[A II]: transition J_1–J_2	3/2–1/2	à $\lambda =$ 6·983	μ.
[Ne V]	1–2	à $\lambda =$ 14·33	μ.
[Ne III]	2–1	à $\lambda =$ 15·4	μ.
[S III]	0–1	à $\lambda =$ 33·6	μ.
[Si II]	1/2–3/2	à $\lambda =$ 34·8	μ.
[Ne III]	1–0	à $\lambda =$ 36·1	μ.
[O III]	1–2	à $\lambda =$ 51·71	μ.
[N III]	1/2–3/2	à $\lambda =$ 57·31	μ.
[O III]	0–1	à $\lambda =$ 88·18	μ.
[O III]	0–1	à $\lambda =$ 113	μ.
[N II]	1–2	à $\lambda =$ 112	μ.
[C II]	1/2–3/2	à $\lambda =$ 156	μ.
[N III]	1/2–3/2	à $\lambda =$ 174	μ.
[N II]	0–1	à $\lambda =$ 204	μ.
[O III]	1–2	à $\lambda =$ 307	μ.

Leurs intensités ont été calculées par Osterbrock (1967) pour une température de 7500° et en utilisant les forces de collision données principalement par Seaton (1958). Quelques-unes de ces transitions sont particulièrement intéressantes car elles sont fonction de la densité électronique.

D'après les travaux bien connus de Seaton (1954) et Seaton et Osterbrock (1957), la mesure du rapport d'intensité de 2 raies interdites d'un même ion est un bon critère de densité électronique N_e. C'est la méthode utilisée dans le domaine optique pour déduire N_e du rapport d'intensité des 2 raies de [O II] à 3726 et 3729 Å p. ex.

Dans l'infrarouge lointain, il existe des couples de raies semblables (Osterbrock, 1967). Pour les densités électroniques échelonnées de 10^2 à 10^3 électrons par cm^3, le rapport d'intensité R des 2 raies de [N II] à 122 μ et 204 μ varie très fortement de 2·7 à 7·5 environ. Au delà de 10^3 électrons par cm^3, ce rapport est peu sensible aux variations de N_e. Au voisinage de 5×10^3, il est préférable d'utiliser celui des 2 raies de [O III] à 113 μ et 307 μ. Pour N_e de l'ordre de 10^5 electrons par cm^3 ce sont les transitions de [Ne III] à 15·4 μ et 36·1 μ qui seront employées.

Les observations dans l'infrarouge effectuées au sol permettront essentiellement de résoudre les problèmes d'abondance des éléments, en particulier celui très important de l'abondance du néon. Les observations faites au delà de notre atmosphère terrestre, complèteront les précédentes et de plus fourniront les densités électroniques des nébuleuses planétaires.

Remercîments

Je remercie vivement le Professor M. J. Seaton qui m'a communiqué très aimablement les 'Preprints' des travaux de Delmer *et al.* (1967) et de Osterbrock (1967). Que tous ces auteurs trouvent ici l'expression de ma profonde gratitude.

Bibliographie

Allen, C.W. (1964) *Astrophysical Quantities*, 2e éd., The Athlone Press, pp. 125–126.
Blaha, M. (1964) *Bull. astr. Inst. Csl.*, **15**, 33.
Burbidge, G.R., Gould, R.J., Pottasch, S.R. (1963) *Astrophys. J.*, **138**, 945.
Burgess, A., Seaton, M.J. (1960) *Mon. Not. R. astr. Soc.*, **120**, 121.
Delmer, T.N., Gould, R.J., Ramsay, W. (1967) *Astrophys. J.*, **149**, 495.
Ditchburn, R.W., Öpik, U. (1962) dans *Atomic and Molecular Processes*, D.R. Bates (Red.), Academic Press, New York, p. 79.
Gould, R.J. (1963) *Astrophys. J.*, **138**, 1308.
Gould, R.J. (1964) *Ann. Astrophys.*, **27**, 815.
Gould, R.J. (1966) *Astrophys. J.*, **143**, 603.
Murray, B.C., Wildey, R.L. (1963) *Astrophys. J.*, **137**, 692.
Osterbrock, D.E. (1964) *A. Rev. Astr. Astrophys.*, **2**, 95.
Osterbrock, D.E. (1965) *Astrophys. J.*, **142**, 1423.
Osterbrock, D.E. (1967) Expected Infrared Spectra of Gaseous Nebulae. Presented at Royal Society Discussion Meeting on Infrared Astronomy, à paraître dans *Phil. Trans. R. Soc. Lond.* A, 'Preprint' en communication privée.
Seaton, M.J. (1954) *Ann. Astrophys.*, **17**, 296.
Seaton, M.J. (1958) *Rev. Mod. Phys.*, **30**, 379.
Seaton, M.J., Osterbrock, D.E. (1957) *Astrophys. J.*, **125**, 66.
Stein, W.A. (1967) *Astrophys. J.*, **148**, 295.

SPECTRES DE NGC 1976, IC 4997, NGC 6572 ET IC 418 DANS LE PROCHE INFRAROUGE PHOTOGRAPHIQUE

YVETTE ANDRILLAT
(*Observatoire de Haute-Provence, Faculté des Sciences de Montpellier, France*)

et

LÉO HOUZIAUX
(*Institut d'Astrophysique, Université de Liège, Belgique*)

Une description qualitative des spectres d'une vingtaine de nébuleuses planétaires dans la région du proche infrarouge a été publiée en 1961 (Andrillat et Andrillat, 1961). Le spectrographe E' monté sur le télescope de 120 cm de l'Observatoire de Haute-Provence nous a permis d'entreprendre l'étude spectrophotométrique de quelques-unes de ces nébuleuses planétaires, ainsi que celle de la nébuleuse d'Orion. Ce spectrographe, pourvu d'un réseau par transmission de 300 traits au millimètre assure une dispersion pratiquement constante de 230 Å/mm dans le domaine de 5750 à 8000 Å. Les spectres obtenus sur plaques IN hypersensibilisées à l'ammoniaque, ont été calibrés photométriquement à l'aide d'un spectrographe à pénombre décrit par Barbier (1944).

Les spectres n'ont pas été élargis, la hauteur de fente correspondant à peu près au diamètre de la nébuleuse planétaire. Dans le cas de NGC 1976, la fente a été orientée suivant la direction Est–Ouest, dans une région très proche des étoiles du trapèze. Le Tableau 1 donne les renseignements habituels relatifs aux observations.

Les spectres de nébuleuses sus-mentionnés, ainsi que quelques autres, donnés aux fins de comparaison, sont reproduits sur la Figure 1.

Le continuum de NGC 1976 et de NGC 6572 a été comparé à celui de 109 Vir, obtenu sur la même plaque. La distribution d'énergie de cette étoile a été mesurée par Oke (1964). Pour IC 418 et IC 4997, ce sont respectivement 16 Ori et λ Cyg. qui ont servi d'étoiles de comparaison. Cette procédure devrait nous permettre de corriger nos mesures des facteurs instrumentaux ainsi que de l'extinction atmosphérique. Dans notre cas cependant, la précision ne peut être que médiocre par suite de:

(a) la difficulté de tracer le continuum;

(b) la présence de raies atmosphériques intenses;

(c) la longue durée de la pose, qui ne permet pas d'assigner une distance zénithale unique relative à un spectre;

(d) l'importante gamme de variation d'intensité des raies.

Pour les intensités des raies de IC 418, NGC 6572 et IC 4997, indiquées au Tableau 2, nous avons adopté l'échelle de Vorontsov-Velyaminov *et al.* (1965), en utilisant, pour effectuer le raccordement, la raie de He I à 5876 Å. Pour la nébuleuse d'Orion,

Osterbrock and O'Dell (eds.), Planetary Nebulae, 68–73. © *I.A.U.*

l'intensité est donnée dans l'échelle $H\beta = 100$, en faisant usage de la raie de [O II] à 7300 Å, dont l'intensité est donnée par Aller et Liller (1959). Les symboles utilisés ont la signification suivante: *S*: la raie est très surexposée; *s*: la raie est sous-exposée; *b*: l'intensité de la raie, qui forme un blend, n'est pas mesurable. Théoriquement, les

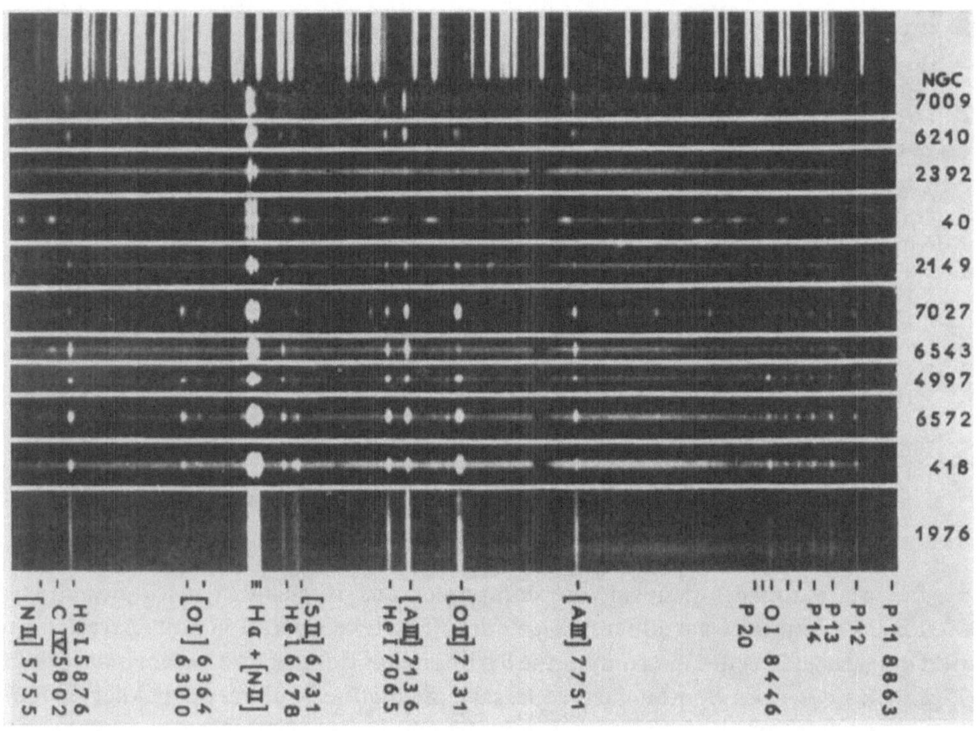

FIG. 1. *Spectres des nébuleuses planétaires dans le domaine de 5750 à 8800 Å.*

Tableau 1

Nébuleuse	Date	Pose	m_N	m	Dimensions	Fente	Étoile de comparaison
IC 418	26/11/66	4^h50^m	12·0	10·9	14″ × 11″	11 Å	16 Ori
IC 4997	6/8/60	6^h	11·4	13·7	2″	7	λ Cyg
NGC 6572	13/5/67	3^h	9·6	12	16″ × 13″	14	109 Vir
NGC 1976	19/11/66	6^h			66″ × 60″	11	16 Ori
NGC 7027	20/11/66	3^h45^m	10·4	17·1	18″ × 11″	11	16 Ori
IC 2149	17/12/63	6^h	9·9	14·0	15″ × 10″	7	ρ Leo
NGC 6543	11/7/62	6^h	8·8	11·1	22″	7	π And
NGC 7009	12/7/62	3^h	8·4	11·7	44″ × 26″	7	θ Boo
NGC 2392	20/1/62	9^h	8·3	10·5	47″ × 43″	7	ρ Leo
NGC 40	21–22/1/62	14^h45^m	10·2	11·4	60″ × 38″	7	ρ Leo
NGC 6210	24/4/61	5^h15^m	9·7	12·5	20″ × 13″	7	α Del

Tableau 2

Intensités observées

Élément	λ(Å)	NGC 1976	IC 418	NGC 6572	IC 4997
He I	5875	20	83	56	6.3
[O I]	6300	2·9	16·4	28·1	2·6
[O I]	6363	*s*	4·9	7·7	0·7
Hα	6563	*S*	*S*	*S*	*S*
He I	6678	8	27·3	27	2·5
[S II]	6717	5·2	21·7	5	0·47
[S II]	6731	8·9	42·9	13	0·75
He I	7065	16	*S*	*S*	*S*
[A III]	7136	36·7	*S*	*S*	*S*
[O II]	7330	17·3	*S*	(120)	*S*
[A III]	7751	11·9	27·2	*S*	2·15
P 20	8392	1·80	6·1	4·1	0·54
P 19	8413	1·86	8·2	4·2	0·75
P 18	8438	*b*	*b*	*b*	*b*
O I	8446	5·29	32·4	10·6	5·68
P 17	8467	2·18	12·8	6·3	1·03
P 16	8502	2·74	13·6	8·8	1·14
P 15	8545	2·86	14·2	11	1·34
P 14	8598	4·29	16·7	11·9	1·42
P 13	8665	4·37	21·3	13·9	1·64
P 12	8750	6	22·1	17	1·64

intensités ainsi données peuvent être comparées aux intensités dans le domaine de 3000 à 5000 Å publiées par différents auteurs. Une telle opération nous paraît cependant dangereuse. En effet, en comparant les intensités des raies de Paschen dans NGC 6572 à celles des raies homologues de la série de Balmer publiées par Aller et Kaler (1964), il semble que les premières soient trop intenses, si l'on adopte un rapport théorique Paschen/Balmer donné par Burgess (1958). Le rapport observé se situe aux environs de 2. En adoptant la courbe d'extinction interstellaire donnée par Allen (1963), et une extinction $A_V = 0{\cdot}9$ magnitudes kpc^{-1}, on déduit que l'extinction à Hβ devrait être de 2·4 magnitudes, alors que Aller donne pour cette extinction 1·15 magnitude (Aller, 1964).

Les intensités observées pour les raies de Paschen sont donc environ deux fois trop fortes, si l'on admet l'extinction proposée par Aller et la valeur théorique de Burgess pour le rapport Paschen/Balmer (0·356). Ce désaccord peut provenir en partie de la difficulté de situer correctement le continuum dans la région des raies de Paschen.

Appelons 1, 2, 3, les niveaux ^{4}S, ^{2}D et ^{2}P de O II. En résolvant les équations exprimant l'état stationnaire de la population N_i de ces trois niveaux, il vient:

$$\frac{N_3}{N_1} = \frac{\omega_3 b_3}{\omega_1 b_1} \exp(-\chi_{13}/kT) \tag{1}$$

où

$$b_3 = \frac{\Omega_{13}}{\omega_2}\left(\frac{\Omega_{12}\Omega_{23}}{\Omega_{13}} + \Omega_{23}\exp(-\chi_{23}/kT) + \Omega_{21} + \frac{A_{21}\omega_2}{C}\right)$$

et

$$b_1 = \frac{\Omega_{23}}{\omega_2}\left[\Omega_{12} + \Omega_{13}\exp(-\chi_{23}/kT) + \frac{\omega_3 A_{31}}{C}\exp(-\chi_{23}/kT) + \frac{A_{21}\omega_2}{C}\right]$$
$$+ \frac{\Omega_{12}}{\omega_2}\left[\frac{\omega_3 A_{32}}{C} + \frac{\omega_3 A_{31}}{C} + \Omega_{13} + \frac{\omega_3 A_{31} A_{21}\omega_2}{C^2\Omega_{12}} + \frac{A_{21}\Omega_{13}\omega_3}{C\Omega_{21}}\right].$$

Les paramètres de choc Ω peuvent être trouvés dans le travail de Seaton (1958). $C = 8{\cdot}63 \times 10^{-6} N_e T_e^{-1/2}$. Pour les probabilités de transition, nous avons adopté les valeurs suivantes: $A_{21} = 9{\cdot}7 \times 10^{-5}\ \mathrm{s}^{-1}$, $A_{31} = 0{\cdot}048\ \mathrm{s}^{-1}$, $A_{32} = 0{\cdot}17\ \mathrm{s}^{-1}$.

L'intensité des raies 7330 et 7319 peut donc s'écrire:

$$I(7330 + 7319) = N_1 \frac{\omega_3}{\omega_1}\frac{b_3}{b_1}\exp(-\chi_1/kT) \times h\nu_{7330} \times A_{32}. \tag{2}$$

D'autre part, pour calculer la population du niveau $2p^3$ 3p ^{3}P, il est nécessaire de connaître les coefficients de recombinaison radiative sur tous les niveaux de O I qui sont susceptibles d'alimenter le niveau $2p^3$ 3p ^{3}P. Il faut remarquer qu'une bonne partie des atomes qui, par recombinaison, peuplent les niveaux triplets pour lesquels $n \geqslant 3$ contribueront à émettre la raie à 8446 Å. Comme nous désirons fixer une limite supérieure à l'intensité de λ 8446, il n'est pas nécessaire de calculer en détails les coefficients de recombinaison sur les différents niveaux. D'après Allen (1963), le coefficient de recombinaison vers le niveau $2p^4$ ^{3}P vaut $8 \times 10^{-14}\ \mathrm{cm}^3\ \mathrm{s}^{-1}$ à 10000 °K. D'autre part, le coefficient de recombinaison total est estimé à $22 \times 10^{-14}\ \mathrm{cm}^3\ \mathrm{s}^{-1}$. Par conséquent, le coefficient de recombinaison aux niveaux $n \geqslant 3$ est de l'ordre de $14 \times 10^{-14}\ \mathrm{cm}^3\ \mathrm{s}^{-1}$. 3/8 seulement des recombinaisons iront aux triplets. Il est difficile d'estimer la fraction de ces recombinaisons qui donneront lieu à l'émission de λ 8446, mais comme nous désirons une limite supérieure, nous la prendrons égale à l'unité. D'autre part, presque tous les ions O II se trouvent dans le niveau $2p^3$ ^{4}S.

Par suite:

$$I(8446) < (3/8) \times 14 \times 10^{-14} \times N_e \times N_1 \times h\nu_{8446}. \tag{3}$$

Dans notre cas, $b_3/b_1 = 1{\cdot}66 \times 10^{-3}$. Ainsi donc:

$$\frac{I(8446)}{I(7330 + 7319)} < \frac{(3/8) \times 10^{-10} \times 14}{(6/4) \times 1{\cdot}66 \times 10^{-3} \times 5{\cdot}8 \times 10^{-3} \times 0{\cdot}17 \times 8446} \lesssim 2 \times 10^{-4}.$$

Or le rapport observé vaut $\simeq 10^{-1}$. Les recombinaisons radiatives ne sont donc pas suffisamment nombreuses pour expliquer l'intensité de λ 8446, et il faut faire appel à l'excitation sélective par la raie Lyman-β de l'hydrogène. Cette excitation peuple le niveau $2p^3$ 3d 3D° qui émet, avec des probabilités de transition respectivement de $2{\cdot}35 \times 10^7\ \mathrm{s}^{-1}$ et $4 \times 10^7\ \mathrm{s}^{-1}$ les raies à λ 11 287 et λ 1026. Ainsi donc, le nombre d'excitations sélectives par cm^3 et par seconde permettant de rendre compte de l'intensité de λ 8446 serait de

$$N(\mathrm{O})\,B\rho \times \frac{A(11\,287)}{A(11\,287) + A(1026)}, \tag{4}$$

où B représente la probabilité de transition par absorption du niveau $2p^4\ {}^3P$ au niveau $2p^3$ 3d ${}^3D^\circ$. ρ est la densité de rayonnement de Ly-β qui permet d'exciter le niveau ${}^3D^\circ$. $N(\mathrm{O})$ représente le nombre d'atomes dans le niveau $2p^4\ {}^3P$.

On peut se demander quelle est la valeur de ρ requise pour que le rapport $I(8446)/I(7330+7319)$ atteigne la valeur observée. Le calcul de ρ exige évidemment la connaissance de $N(\mathrm{O})/N_1$, qui peut être déduit du rapport observé $I(6300+6363)/I(7330+7319)$. En effet, l'intensité $I(6300+6363)$ peut s'écrire:

$$I(6300 + 6363) = N(\mathrm{O}) \exp(-\chi(^1\mathrm{D})/kT)\, \eta A h\nu \times \frac{\omega_2}{\omega_1} \tag{5}$$

où η s'obtient en résolvant les équations d'équilibre statistique pour les niveaux ${}^3P$, ${}^1D$ et ${}^1S$.

En négligeant les désexcitations par collisions, vu la faible densité électronique, on trouve:

$$\eta = \frac{N_e\left[\alpha_{21}(A_{32} + A_{31}) + \alpha_{31}A_{32}(\omega_3/\omega_2)\exp(-\chi_{23}/kT)\right]}{N_e\alpha_{32}(\omega_3/\omega_2)\exp(-\chi_{23}/kT)\,A_{32} + A_{21}(A_{32} + A_{31})}. \tag{6}$$

Dans cette expression, les indices 1, 2, 3 représentent les niveaux ${}^3P$, ${}^1D$, ${}^1S$. En adoptant les valeurs de α données par Seaton (1958), on trouve que $\eta = 7{\cdot}36 \times 10^{-3}$ le Tableau 2, $I(6300+6363)/I(7330+7319)$ vaut, en tenant compte de la D'après correction pour l'absorption interstellaire, 0·24.

Ainsi donc, en supposant que les volumes émetteurs sont identiques, on tire de (2) et (5) la valeur 2·36 pour le rapport $N(\mathrm{O})/N_1$. La comparaison de (4) et (2) nous permet alors de déduire une valeur de $8{\cdot}38 \times 10^{-24}$ ergs cm^{-3} pour la densité de rayonnement à la longueur d'onde de Lyman-β. L'étude des intensités relatives des raies de O I dans la région rouge et infrarouge permet ainsi de déterminer l'intensité spécifique dans la raie Lyman-β. Il est possible, à partir de cette donnée d'en déduire une valeur pour la densité en atomes d'hydrogène dans le milieu.

En conclusion, il semble bien que la raie à λ 8446 dans NGC 6572 dans les nébuleuses soit excitée par fluorescence sélective, mécanisme confirmé par l'absence de la raie λ 7772. Le rapport $I(7330)/I(8446)$ n'a pu être déterminé pour IC 418, NGC 1976, IC 4997, mais il n'est pas inférieur à sa valeur pour NGC 6572, de sorte que la présence de λ 8446 dans ces nébuleuses peut également s'interpréter de la même manière que pour NGC 6572.

Remerciements

En terminant, nous tenons à exprimer tous nos remerciements au Dr. D.E. Osterbrock et au Dr. M.J. Seaton pour de judicieuses remarques.

Bibliographie

Allen, C.W. (1963) *Astrophysical Quantities*, 2e éd., Athlone Press, London.
Aller, L.H. (1964) *Publ. astr. Soc. Pacific*, **76**, 279.
Aller, L.H., Liller, W. (1959) *Astrophys. J.*, **130**, 45.
Aller, L.H., Kaler, J.B. (1964) *Astrophys. J.*, **139**, 1074.
Andrillat, Y., Andrillat, H. (1961) *Ann. Astrophys.*, **24**, 139.
Barbier, D. (1944) *Ann. Astrophys.*, **7**, 86.
Burgess, A. (1958) *Mon. Not. R. astr. Soc.*, **118**, 477.
Oke, J.B. (1964) *Astrophys. J.*, **140**, 689.
Seaton, J.M. (1958) *Rev. mod. Phys.*, **30**, 979.
Vorontsov-Velyaminov, B.A., Kostjakova, E.B., Dokuchaeva, O.D., Arhipova, V.P. (1965) *Soviet Astr.*, **9**, 364.

THE HELIUM λ 10830 LINE IN PLANETARY NEBULAE AND THE ORION NEBULA*

ARTHUR H. VAUGHAN, JR.
(Mt. Wilson and Palomar Observatories, Carnegie Inst. of Washington, California Inst. of Technology, Pasadena, Calif., U.S.A.)

ABSTRACT

Interferometric observations of profiles of HeI λ 10830 emission lines in 11 planetary nebulae, and in selected regions of the Orion Nebula, are presented. In common with the Orion Nebula, the planetaries are shown to emit a P Cygni-like λ 10830 line, with the absorption component shifted toward the violet with respect to the laboratory wavelength in a frame of reference at rest in the centre of expansion of the gas. The emission components are shifted toward the red. In planetaries, the negative displacements of the absorption edges are, in general, approximately equal to the widths, β, of the emission components which, in turn, range from about 12 km/sec in IC 418 to about 28 km/sec in NGC 6210. The emission red-shift is about $0{\cdot}7\beta$ on the average, but individual shifts vary from $0{\cdot}5\beta$ in IC 2149 to $1{\cdot}05\beta$ in NGC 6826. The line widths and shifts tend to increase in nebulae with larger expansion velocities. In Orion, the absorption edges in the λ 10830 line coincide in velocity with those in the line HeI λ 3888 observed against the spectra of the Trapezium stars. In planetaries, the absorption edges in the λ 10830 line appear qualitatively similar to those in the line HeI λ 3888, but a coincidence in velocity could not be demonstrated.

The observed profiles indicate that the nebulae are expanding, or that they contain expanding globules or filaments. Some form of circumnebular absorbing zone may be indicated. However, it is suggested that frequency redistribution associated with resonance-like scattering in a homogeneous expanding medium might in principle (even in the absence of stratification) account for the shifted λ 10830 profiles. (See Hummer, D.G. and Rybicki, G.B. (1968), *Astrophys. J. Letters* (in press), for a further discussion of this point.)

No trace of He^3 is evident from the profiles. Quantitative conclusions are uncertain without a model which reproduces most of the phenomena, but an upper limit of $He^3/He^4 \leqslant 0{\cdot}05$ or even $0{\cdot}01$ is suggested on the basis of conservative assumptions.

DISCUSSION

Underhill: There are several good reasons for believing BD + 30° 3639, Campbell's Hydrogen Envelope Star, to be a Wolf-Rayet Star enveloped in a gaseous shell. Therefore I doubt that your observations of HeI λ 3888 absorption can be interpreted unequivocally as due to the surrounding nebula. Your He λ 10830 lines are quite broad and they appear to have P Cygni-type absorption displaced some 60 km/sec or more shortward. Can you demonstrate that this composite feature is not more closely connected to the underlying star than to the nebula? The HeI line λ 10830 is one of the first lines one would expect to appear strongly in emission in a high-temperature extended envelope.

Vaughan: Wilson's spectra of BD + 30° 3639 show an extended nebula which expands at around

* The full text of this article has been accepted for publication in the *Astrophys. J.*

16 km/sec. The violet absorption shift in 10830 is of the same order. In general, the observations do not indicate a connection between the profiles of λ 10830 and the properties of the central stars.

Münch: In a search for λ 3889 He I absorption in the spectra of central stars of planetary nebulae I made sometime ago, I convinced myself of its presence in IC 418. Unfortunately, the H8 stellar line is quite strong in IC 418, and the nebular absorption cannot be seen as clearly as in BD + 30° 3639.

INFRARED OBSERVATIONS OF THE PLANETARY NEBULA NGC 7027

F.C. GILLETT, F.J. LOW, and W.A. STEIN
(University of California, San Diego, University of Arizona, and Rice University, U.S.A.)

An infrared spectrometer has been constructed with a resolution of $\Delta\lambda/\lambda \approx 0{\cdot}02$ for observations in the wavelength range from 2·8 to 15 μ using a gallium-doped germanium bolometer (Low, 1961) as the radiation-sensitive element. Observations of the planetary nebula NGC 7027 were made with this instrument in May and June 1967 at the Catalina Observing Station of the University of Arizona. It was discovered that there was a measurable continuum flux from this object in the wavelength range from 7·5–14 μ that compared in value with the strength of stellar radiation from α Lyr (A0) at about $\lambda = 9$ μ. These observations are described in detail elsewhere (Gillett *et al.*, 1967).

The infrared observations of NGC 7027 show that the observed flux is almost two orders of magnitude larger than the expected infrared free-free continuum at $\lambda = 10$ μ. These results are perhaps similar to results of observations of the Orion nebula at $\lambda = 5$ μ (Kleinmann and Low, 1967) that showed a flux 10 times larger than the expected free-free continuum from that nebula. NGC 7027 has also recently been observed at $\lambda = 20$ μ (Low, private communication).

References

Gillett, F.C., Low, F.J., Stein, W.A. (1967) *Astrophys. J.*, **149**, L97.
Kleinman, D.D., Low, F.J. (1967) *Astrophys. J.*, **149**, L1.
Low, F.J. (1961) *J. opt. Soc. Am.*, **51**, 1300.

THE ULTRAVIOLET EMISSION SPECTRA OF PLANETARY NEBULAE

D. R. FLOWER
(Dept. of Physics, University College London, England)

ABSTRACT

Calculations show that high-excitation planetaries emit about the same total intensity of collisionally excited radiation in the unobserved ultraviolet part of the spectrum as in the visible. This ultraviolet radiation emanates from a central zone where the electron temperature is high due to absorption of He II Ly-α photons. Intensities of lines which one might expect to detect in the ultraviolet have been calculated using revised atomic data and a detailed model of the ionization and thermal structure of the central zone. Relative line intensities are presented for a wide range of excitation in the nebula. In particular, the absolute intensity of the C IV doublet at 1550 Å has been calculated for 4 selected planetaries.

Calculations have been made of the intensities of the ultraviolet lines emitted by planetary nebulae. The spectral region considered extends from 912 Å to about 3500 Å. Interstellar hydrogen and helium will absorb all radiation shortwards of 912 Å, and 3500 Å is taken as defining the short-wavelength edge of the visible part of the spectrum.

There are several reasons to justify a theoretical study of the ultraviolet spectra of planetaries. In particular, observations of these lines from space vehicles should become possible in the near future and predictions of wavelengths and absolute line intensities will be useful. Observations of B stars in the ultraviolet have already been made and the results reported at the I.A.U. Symposium No. 23 (Heddle, 1964). A comparison is made of the absolute flux at the detectors in the observations of B stars with the predicted absolute flux from high-excitation planetary nebulae.

The physical processes responsible for the production of emission lines in planetaries are well known. The lines can be due either to radiative recombination followed by cascade or due to collisional excitation followed by radiative decay. Osterbrock (1963) has already considered in some detail the ultraviolet lines produced in planetary nebulae due to both processes. However, he did not calculate the actual intensities of the lines. To do this, a model of both the ionization and thermal structure of the nebula is essential. Here we shall be concerned with the intensities of the collisionally excited lines which constitute a very large fraction of the nebular radiation both in the visible and the ultraviolet.

To illustrate the physical arguments involved in the calculations, we shall consider Ne^{3+} as an example of an ion which emits collisionally excited radiation. The expected

Osterbrock and O'Dell (eds.), Planetary Nebulae, 77–86. © *I.A.U.*

ultraviolet lines, which are produced in transitions between the terms of the ground configuration, are marked in Figure 1. The ^{2}D and ^{2}P states may be excited collisionally or by radiative recombination. In all of this work it is assumed that the rate of excitation by radiative recombination is negligible compared with the collisional excitation rate. The assumption is probably valid in almost all cases at the electron

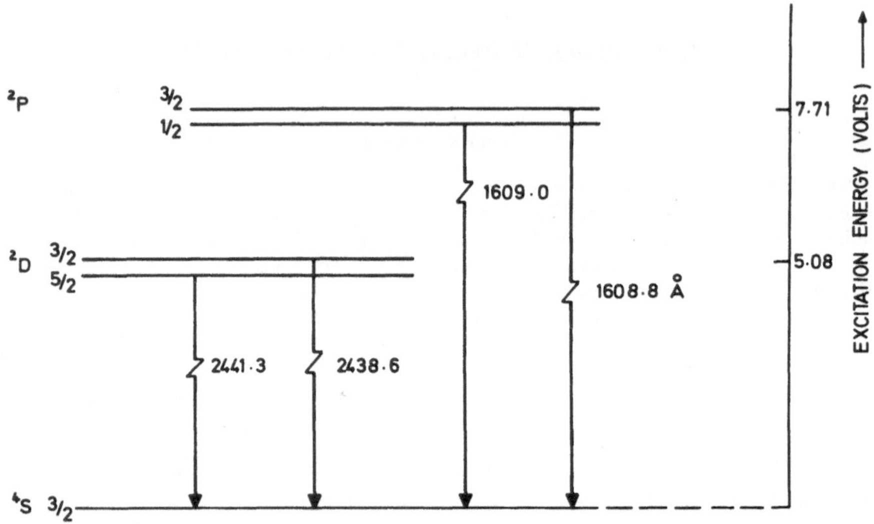

FIG. 1. *Terms of the ground configuration of* Ne^{3+}. *The wavelengths of expected ultraviolet transitions are marked in Å.*

temperatures considered. To be quite specific, we consider collisional excitation of the ^{2}P state from the ^{4}S ground state. The excitation rate is given by

$$X_c = N_e N(\mathrm{Ne}^{3+})\, q(1 \rightarrow 3) \tag{1}$$

where

$$q(1 \rightarrow 3) = 8{\cdot}63 \times \frac{10^{-6}\Omega(1,3)}{\omega_1 T_e^{1/2}} \exp\left(\frac{-\Delta E(1,3)}{kT_e}\right). \tag{2}$$

1 represents the ^{4}S state, 3 represents the ^{2}P state. N_e is the electron density, $N(\mathrm{Ne}^{3+})$ is the density of Ne^{3+} ions, $\Omega(1, 3)$ is the collision strength, and $\Delta E(1, 3)$ the energy of the transition, ω_1 is the statistical weight of the ground state, and T_e is the electron temperature.

Equation (2) clearly shows why a model of the thermal structure of the nebula is essential to any calculation of line intensities. The argument of the exponential, $-\Delta E/kT_e$, is approximately -8 for $T_e = 1 \times 10^4$ °K and -4 for $T_e = 2 \times 10^4$ °K. So a factor of 2 difference in the electron temperature introduces a factor of about 50 in the value of the line intensity.

The electron temperatures of planetary nebulae are often obtained by measuring the intensities of the nebular and auroral lines of O III. It is important to note that the value obtained yields the electron temperature only in that part of the nebula which contains O^{2+}. In high-excitation planetaries, this will be the outer part of the nebula, furthermost from the central star. It is proposed here that, in high-excitation objects, there exists a central zone where the electron temperature is high, about $1{\cdot}8$–$2{\cdot}0 \times 10^4$ °K. The value of the electron temperature in this region is not reflected in the intensities of the [O III] lines because there is very little O^{2+} in the central zone.

In using the term 'central zone', we mean quite specifically a central ionization zone. We follow the procedure of Hummer and Seaton (1964) and divide a planetary into a number of ionization zones. In any particular planetary, one or more of the possible ionization zones may not be present at all, but the general trend of the degree of excitation increasing towards the centre of the nebula is certainly supported by observations (e.g. Wilson, 1950). In the central zone, most of the helium is He^{2+} and most of the hydrogen is H^+. There is very little O^{2+} in this region of the nebula and so the electron temperature must be calculated.

The calculation of the theoretical electron temperature of the central zone involves all the physical processes of the thermal balance of that part of the nebula (Hummer and Seaton, 1964; Aller, 1956). The source of energy of all planetaries is the radiation emitted by the central star. This radiation is absorbed in the processes of ionization in the nebula when the electrons gain energy from the radiation field. For an element X in the qth state of ionization the process may be written as

$$X^{q+} + h\nu \rightarrow X^{(q+1)+} + e^- . \tag{3}$$

The electrons can lose energy in a number of ways, notably in inelastic collisions with positive ions when collisionally excited radiation can be produced:

$$X^{q+} + e^- \rightarrow X^{q+*} + e^- \tag{4}$$

$$X^{q+*} \rightarrow X^{q+} + h\nu . \tag{5}$$

In general, a balance is attained between gain of electron kinetic energy from the radiation field and loss of kinetic energy in processes such as the inelastic collisions with positive ions. However, Hummer and Seaton (1964) consider the importance of He II Ly-α radiation which is produced in the central zone. They propose that this radiation is absorbed in the central zone in maintaining the ionization of hydrogen:

$$\begin{array}{ll} \mathrm{H}^0 + h\nu(\mathrm{He\,II\ Ly}\text{-}\alpha) \rightarrow \mathrm{H}^+ + & e^- \\ \qquad\quad 40{\cdot}8\ \mathrm{eV} & 27{\cdot}2\ \mathrm{eV} \end{array} \tag{6}$$

Because of the very large energy of the He II Ly-α photon, the electron released in the process given by Equation (6) has a large energy. These electrons are thermalised by elastic collisions in the central zone and the electron temperature in this region is high as a consequence.

Computer programmes have been developed to solve the thermal balance equation in the central zone of planetary nebulae. Allowance is made for heating by radiation from the central star and by He II Ly-α radiation. The usual cooling processes of radiative recombination, free-free transitions, and inelastic collisions are taken into account. In Figure 2, electron temperature is plotted as a function of radius for two

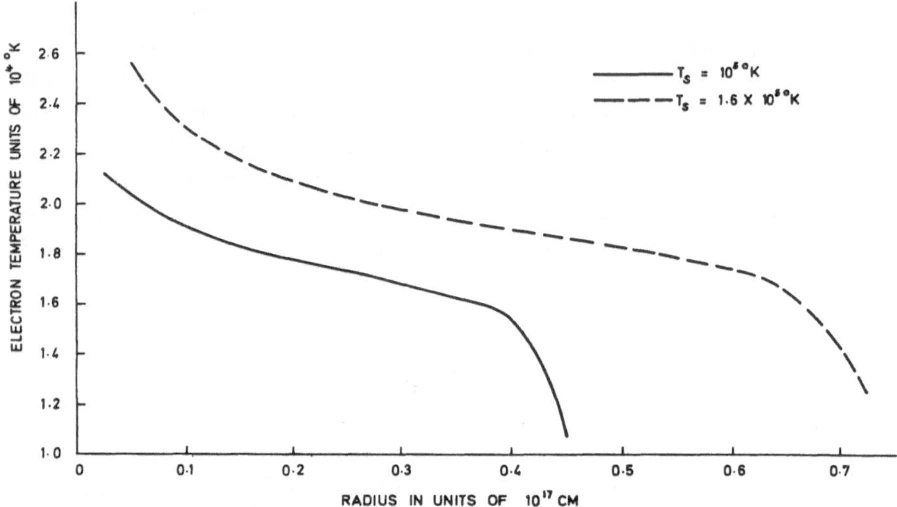

Fig. 2. *Electron temperature of the central zone as a function of radius. Plots are given for two values of central star temperature, T_s.*

values of central-star temperature, T_s, where it is assumed that the star radiates as a black-body. The electron temperature is seen to be insensitive to the value of T_s.

The computer programmes also solve for the ionization equilibrium of ions of the following elements in this zone: He, C, N, O, and Ne. Figure 3 is typical of the results obtained for oxygen. It is clear that there is very little O^{2+} in the central zone. The electron density is also obtained as a function of radius. Figure 4 shows the result for an assumed hydrogen density of $1{\cdot}0 \times 10^4$ cm^{-3} and He–H abundance ratio (Aller, 1964) of 0.18.

The results obtained are given in Tables 1–4. For each ion, the wavelengths and identifications of the ultraviolet transitions are given (Osterbrock, 1963). For Ne^{3+} and Ne^{4+}, some transitions which can be observed from the ground are also included; this makes possible a limited comparison between theory and observation. The ratio, Ω/ω, of the collision strength for the transition to the statistical weight of the ground state is also given. The best available values of collision strengths have been used (Czyzak *et al.*, 1968; Bely *et al.*, 1963). The most uncertain values are marked by an asterisk.

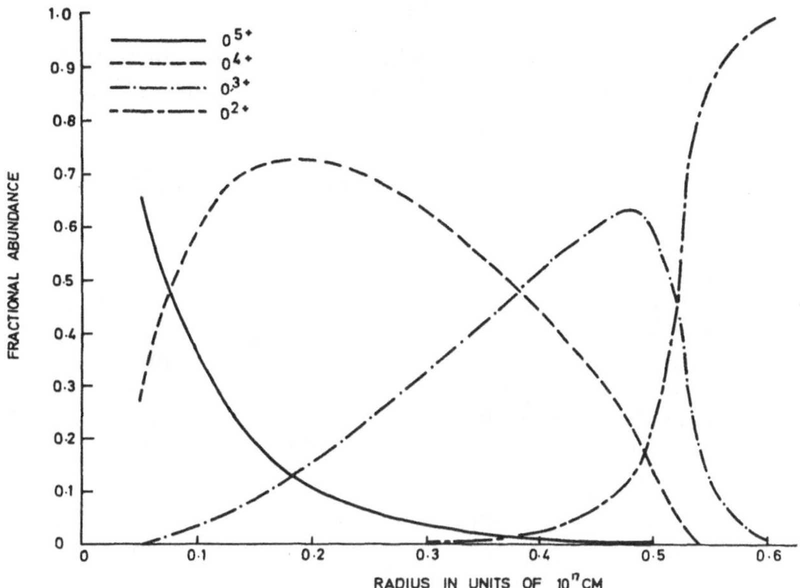

FIG. 3. *Ionization of oxygen in the central zone. The fraction of oxygen in each ionization stage is plotted as a function of radius.* $T_s = 12 \times 10^4$ °K.

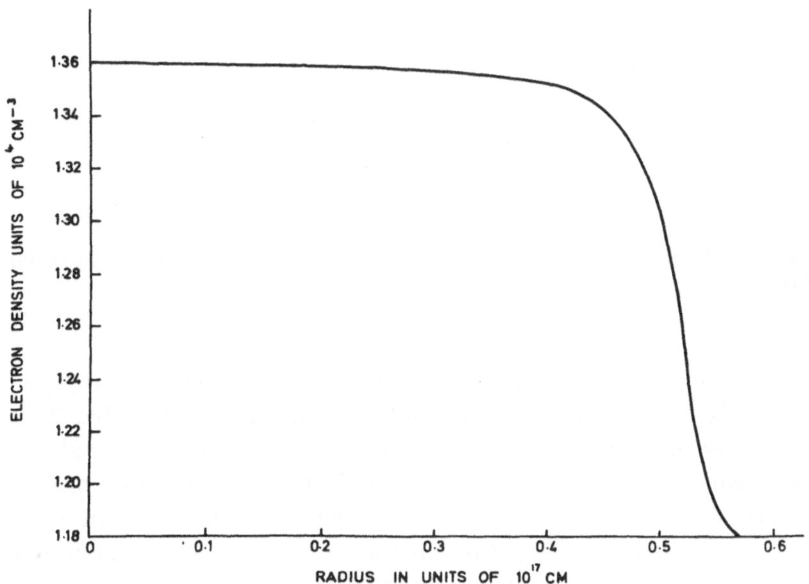

FIG. 4. *Electron density in the central zone as a function of radius.* $T_s = 12 \times 10^4$ °K.

Table 1

Ions of the C group of elements. Ω is the collision strength for the transition and ω the statistical weight of the ground state. The most uncertain values of the ratio Ω/ω are marked by an asterisk

Ion	Transition			Wavelength (Å)	Excitation Potential (volts)	Ω/ω
C^{+3}	$2s^2S_{1/2}$	–	$2p^2P_{3/2}^0$	1548·2	8·00	5·5
	$^2S_{1/2}$	–	$^2P_{1/2}^0$	1550·8		
N^{+3}	$2s^2\,^1S_0$	–	$2s2p^3P_1^0$	1488·1	8·34	0·30*
N^{+4}	$2s^2S_{1/2}$	–	$2p^2P_{3/2}^0$	1238·8	10·00	3·6
	$^2S_{1/2}$	–	$^2P_{1/2}^0$	1242·8		
O^{+3}	$2s^22p^2P_{1/2}^0$	–	$2s2p^2\,^4P_{3/2}$	1402·4:	8·81	0·20*
	$^2P_{1/2}^0$	–	$^4P_{1/2}$	1404·9:		
	$^2P_{3/2}^0$	–	$^4P_{5/2}$	1406·3:		
	$^2P_{3/2}^0$	–	$^4P_{3/2}$	1410·0:		
	$^2P_{3/2}^0$	–	$^4P_{1/2}$	1412·6:		
O^{+4}	$2s^2\,^1S_0$	–	$2s2p^3P_1^0$	1215·7	10·21	0·30*
O^{+5}	$2s^2S_{1/2}$	–	$2p^2P_{3/2}^0$	1037·6	11·99	2·6
	$^2S_{1/2}$	–	$^2P_{1/2}^0$	1031·9		
Ne^{+3}	$2s^22p^3\,^4S_{3/2}^0$	–	$2s^22p^3\,^2D_{3/2}^0$	2438·6	5·08	0·26
	$^4S_{3/2}^0$	–	$^2D_{5/2}^0$	2441·3		
	$2s^22p^3\,^4S_{3/2}^0$	–	$2s^22p^3\,^2P_{3/2}^0$	1608·8	7·71	0·107
	$^4S_{3/2}^0$	–	$^2P_{1/2}^0$	1609·0		
	$2s^22p^3\,^2D_{5/2}^0$	–	$2s^22p^3\,^2P_{3/2}^0$	4714·3	7·71	0·107
	$^2D_{5/2}^0$	–	$^2P_{1/2}^0$	4715·6		
	$^2D_{3/2}^0$	–	$^2P_{3/2}^0$	4724·2		
	$^2D_{3/2}^0$	–	$^2P_{1/2}^0$	4725·6		
Ne^{+4}	$2s^22p^2\,^3P_2$	–	$2s^22p^2\,^1S_0$	1592·7	7·83	0·024
	$^3P_1$	–	$^1S_0$	1575·2		
	$2s^22p^2\,^1D_2$	–	$2s^22p^2\,^1S_0$	2972	7·83	0·024
	$2s^22p^2\,^3P_2$	–	$2s^22p^2\,^1D_2$	3425·9	3·66	0·153
	$^3P_1$	–	$^1D_2$	3345·8		

* Where mean wavelength of a multiplet is uncertain it is marked with a colon. The spacing between the lines of the multiplet should be correct.

In all these calculations it is assumed that once a state has been collisionally excited it may decay only radiatively, i.e. collisional de-excitation is negligible. This assumption will be fairly justified for the electron-density values encountered in planetary nebulae. The function G, which is tabulated is then the product of Ω/ω with the branching ratio for the particular transition of the multiplet. Transition probabilities are taken from Allen (1964) for permitted transitions and from Garstang (1968) for forbidden transitions. All intensities are given relative to the He II λ 4686 line, which is the most intense He II line in the visible. The results for the ions of S and Ar will not

Table 2

Relative line intensity values for ions of the C group of elements. The function G is the product of Ω/ω with the branching ratio for the transition

Ion	Wavelength	G	Line intensities relative to He II λ 4686 $T_s = 6 \times 10^4$	10×10^4	14×10^4	18×10^4 °K
C^{+3}	1548·2	3·67	24	25	28	34
	1550·8	1·83	12	13	14	17
N^{+3}	1488·1	0·1	0·41	0·42	0·38	0·39
N^{+4}	1238·8	2·4	0·27	2·7	4·9	6·3
	1242·8	1·2	0·14	1·3	2·5	3·1
O^{+3}	1402·4:	0·04	0·34	0·39	0·37	0·36
	1404·9:	0·02	0·17	0·20	0·19	0·18
	1406·3:	0·07	0·51	0·60	0·57	0·55
	1410·0:	0·04	0·34	0·39	0·37	0·36
	1412·6:	0·02	0·17	0·20	0·19	0·18
O^{+4}	1215·7	0·1	0·027	0·29	0·56	0·70
O^{+5}	1037·6	1·73	$6{\cdot}2 \times 10^{-5}$	0·070	0·83	2·7
	1031·9	0·87	$3{\cdot}1 \times 10^{-5}$	0·035	0·42	1·3
Ne^{+3}	2438·6	0·22	5·3	7·6	5·6	4·1
	2441·3	0·035	0·85	1·2	0·89	0·66
	1608·8	0·053	0·27	0·44	0·34	0·25
	1609·0	0·011	0·056	0·091	0·070	0·053
	4714·3	0·016	0·081	0·13	0·10	0·076
	4715·6	0·0022	0·011	0·018	0·014	0·011
	4724·2	0·018	0·089	0·14	0·11	0·084
	4725·6	0·0078	0·040	0·064	0·050	0·037
Ne^{+4}	1592·7	$2{\cdot}4 \times 10^{-5}$	All intensities less than 10^{-4}			
	1575·2	0·015	$7{\cdot}7 \times 10^{-4}$	0·045	0·13	0·17
	2972	0·0092	$4{\cdot}7 \times 10^{-4}$	0·028	0·079	0·11
	3425·9	0·11	0·051	2·5	6·6	8·2
	3345·8	0·041	0·019	0·93	2·4	3·0

Table 3

Ions of S and Ar. The most uncertain values of Ω/ω are marked by an asterisk

Ion	Transition	Wavelength (Å)	Excitation Potential (volts)	Ω/ω
S^{+4}	$3s^2\,{}^1S_0 - 3s3p\,{}^3P_1^0$	1198·6:	10·32	0·3*
S^{+5}	$3s\,{}^2S_{1/2} - 3p\,{}^2P_{3/2}^0$	933·4	13·20	6·83
	${}^2S_{1/2} - {}^2P_{1/2}^0$	944·5		
Ar^{+4}	$3s^23p^2\,{}^3P_1 - 3s^23p^2\,{}^1S_0$	2691·4	4·70	0·0157
	${}^3P_2 - {}^1S_0$	2784·4		
Ar^{+5}	$3s^23p\,{}^2P_{1/2}^0 - 3s3p^2\,{}^4P_{3/2}$	992·0	12·49	1·2*
	${}^2P_{1/2}^0 - {}^4P_{1/2}$	1000·0		
	${}^2P_{3/2}^0 - {}^4P_{5/2}$	1001·8		
	${}^2P_{3/2}^0 - {}^4P_{3/2}$	1014·3		
	${}^2P_{3/2}^0 - {}^4P_{1/2}$	1022·6		

Table 4

Relative line intensity values for ions of S and Ar

Ion	Wavelength (Å)	G	Line intensities relative to He II λ 4686 $T_s = 10^5$ °K
S^{+4}	1198·6:	0·1	0·060
S^{+5}	933·4	4·55	0·49
	944·5	2·28	0·25
Ar^{+4}	2691·4	0·010	0·010
	2784·4	$1{\cdot}2 \times 10^{-4}$	$1{\cdot}2 \times 10^{-4}$
Ar^{+5}	992·0	0·27	0·0035
	1000·0	0·13	0·0017
	1001·8	0·40	0·0052
	1014·3	0·27	0·0035
	1022·6	0·13	0·0017

be as good as for the C group of elements because the detailed ionization structure for S and Ar has not yet been calculated.

In a high-excitation planetary, the He II λ 4686 line is about one half as intense as the Hβ line (the customary intensity standard). A simple calculation shows that the total intensity of the radiation emitted in the strongest ultraviolet lines is approximately the same as the total intensity of the radiation emitted in the visible. So high-excitation planetaries such as NGC 7027 and NGC 7662 are emitting a large fraction of their total radiation in the ultraviolet.

The absolute intensity of the C IV doublet at 1550 Å has been calculated for four planetary nebulae, and the results are presented in Table 5. The 4 objects considered

Table 5

Values of the absolute flux of the C IV doublet at 1550 Å for four selected planetaries. The flux is in photons cm^{-2} sec^{-1} rather than ergs cm^{-2} sec^{-1}

NGC	Absolute flux photons cm^{-2} sec^{-1}
1535	10
2392	26
7009	24
7662	17

were taken from a list of 47 planetary nebulae compiled by Harman and Seaton (1966). They were chosen because they satisfy the following criteria:

(a) the value of the absolute Hβ flux is fairly high and the intensity of the He II λ 4686 radiation is significant compared with the Hβ line intensity;

(b) the reddening constant for the nebula is small; and

(c) the nebula completely surrounds the central star.

The value of the reddening function, $f(\lambda)$, at 1550 Å was obtained by extrapolation of the values given by Seaton (1960). $f(\lambda)$ was plotted against $1/\lambda$ and a linear extrapolation made to 1550 Å. By this means we obtain $f(1550) = 2{\cdot}2$. The values of absolute intensity of the C IV doublet, allowing for space absorption, are seen to be of the order of 10 photons $cm^{-2}\ sec^{-1}$.

The procedure adopted in the ultraviolet observations of B stars reported by Heddle (1964) was to view the continuum of the star through a filter of typically 200 Å bandwidth. The flux at the detectors under these conditions was of the order of 10^4 photons $cm^{-2}\ sec^{-1}$. If the bandwidth of the filter was reduced to 0·1 Å, the flux would be reduced to about 10 photons $cm^{-2}\ sec^{-1}$. So the problem of observing the strongest ultraviolet lines of bright planetary nebulae seems comparable with the problem of viewing the ultraviolet continuum of B stars under quite high resolution. It is to be hoped that experiments of this sensitivity will become possible in the near future.

Acknowledgements

This work was completed whilst the author was in tenure of a Science Research Council Studentship. The calculations were performed on the I.B.M. 360/65 computer of University College London.

The author would like to express his gratitude to Professor M. J. Seaton, who made many stimulating and fruitful suggestions.

References

Allen, C. W. (1964) *Astrophysical Quantities*, The Athlone Press, London, Chapter 4.
Aller, L. H. (1956) *Gaseous Nebulae*, Chapman and Hall, London, p. 130.
Aller, L. H. (1964) *Astrophys. Norw.*, **9**, 293.
Bely, O., Tully, J., Van Regemorter, H. (1963) *Ann. Phys.*, **8**, 303.
Czyzak, S. J., Krueger, T. K., Martins, P. de A. P., Saraph, H. E., Seaton, M. J., Shemming, J. (1968) in the present volume, p. 138.
Garstang, R. H. (1968) in the present volume, p. 143.
Harman, R. J., Seaton, M. J. (1966) *Mon. Not. R. astr. Soc.*, **132**, 15.
Heddle, D. W. O. (1964) *Ann. Astrophys.*, **27**, 800.
Hummer, D. G., Seaton, M. J. (1964) *Mon. Not. R. astr. Soc.*, **127**, 217.
Osterbrock, D. E. (1963) *Planet. Space Sci.*, **11**, 621.
Seaton, M. J. (1960) *Rep. Progr. Phys.*, **23**, 313.
Wilson, O. C. (1950) *Astrophys. J.*, **111**, 279.

DISCUSSION

Aller: The calculations pertaining to [Ne IV] ultraviolet lines are amenable to direct observational check for many high-excitation planetaries. The ^2P–^2D auroral transitions fall near λ 4725 and have been measured in a number of planetaries in our program at Lick. Thus without any theoretical model at all we can predict some of the ultraviolet intensities. The electron temperatures are usually higher than 10000 °K. Values of 15000 °K – 17000 °K would be more typical.

Hummer: Collisional excitation of ultraviolet lines is of course caused by high-energy electrons.

It is important to realize that there may be some deviation from an exact Maxwell-Boltzmann distribution at high energies, because photo-ionization of H by He II Ly-α (λ 304) photons produces electrons with energy about 30 eV each, which are only very slowly thermalized.

Underhill: The He II λ 304 quanta have sufficient energy to ionize atoms in relatively low-lying levels, of the first, second, and third stages of ionization of the light elements. Such ionizations may help to create a more rapid thermalization of the energy of He II λ 304 than is possible by ionization of H only.

Seaton: However, the number of ionizations of H is larger than the number of ionizations of heavy elements, because the number of recombinations of hydrogen is greater.

OBSERVATIONS OF PLANETARY NEBULAE AT RADIO WAVELENGTHS

YERVANT TERZIAN
(Arecibo Ionospheric Observatory, Cornell University, Ithaca, N.Y., U.S.A.)*

1. Introduction

Planetary nebulae are weak radio sources. The radio emission of planetary nebulae is primarily due to free-free transitions in the ionized hydrogen cloud. Very large radio telescopes and sensitive receivers are needed to detect the weak radio emission of planetary nebulae. During the last 3 years about 80 planetary nebulae have been detected at some one radio frequency, and more than 60% of these have been detected at more than one frequency. Most of the measurements have been made to measure the flux densities of the nebulae and determine their radio spectra. A few interferometric observations have also been made to determine the radio widths of planetary nebulae. In some cases the radio observations have been used to determine the electron temperatures and emission measures of planetary nebulae, and also to derive the interstellar extinction at Hβ.

Very recently the hydrogen 109 α recombination line was detected in NGC 7027, thus opening a new radio way of studying the planetary nebulae.

2. The Radio Observations of Planetary Nebulae

One of the great difficulties of observing planetary nebulae at radio frequencies is the intrinsically weak radio energy which they emit. Planetary nebulae were first reliably detected in 1961 by Lynds (1961) who used the 85-foot Tatel radio telescope at the U.S. National Radio Astronomy Observatory (NRAO). Lynds was able to detect only the 5 strongest nebulae, IC 418, NGC 6543, NGC 6572, NGC 6853, and NGC 7293. From these observations approximate flux densities were derived for the observed nebulae at 1420 and 3000 MHz. Clearly radio telescopes with larger collecting apertures and more sensitive receivers were needed to continue the study of planetary nebulae at the radio frequencies.

In 1964, Menon and Terzian (1965) used the then newly completed 300-foot transit

* The Arecibo Ionospheric Observatory is operated by Cornell University with the support of the Advanced Research Projects Agency under a research contract with the Air Force Office of Scientific Research.

radio telescope at NRAO, and measured the flux densities of 10 planetary nebulae at 750 and 1410 MHz. These improved measurements opened up the possibility of checking the recombination theory of emission from the detected nebulae by a comparison of the observed radio-frequency fluxes with the values predicted from the Balmer-line fluxes. It was found that the fluxes at the optically thin part of the continuum spectra of the planetary nebulae agreed reasonably well with those predicted by Osterbrock (1964) from the Hβ fluxes, using the recombination theory. A number of planetary nebulae were found to be optically thick at the lower frequencies. In particular the spectrum of NGC 7027 showed that it was optically thick even at 3000 MHz. NGC 6572, NGC 6543, and IC 418 were all found to have high optical depths at 750 MHz. This fact was used to compute the electron temperatures of these nebulae since at high optical depths the planetary nebulae radiate as black bodies. The results indicated electron temperatures of the order of $10–20 \times 10^3$ °K, and were in good agreement with the optical estimates. This work was followed by more than six surveys of planetary nebulae at radio frequencies ranging from 195 to 16200 MHz.

Slee and Orchiston (1965) presented a preliminary survey of planetary nebulae at radio wavelengths south of declination +20°. The 210-foot radio telescope at Parkes (CSIRO) was used at three frequencies (620, 1420, and 2730 MHz) for these observations. About 50 planetary nebulae were detected at 2730 MHz. At the lower frequencies confusion problems were found to be serious and this made the flux measurements more difficult.

Observations at low frequencies to measure the fluxes of planetary nebulae at the optically thick part of their radio spectra were made by Terzian (1966) using the Arecibo Ionospheric Observatory's (AIO) 1000-foot spherical radio telescope. A survey of 130 planetary nebulae was made at 430 MHz between declinations −2° and +38°. A few observations were also made at 611 and 195 MHz. Figure 1 shows a sample observation at 430 MHz of NGC 6720 (the Ring Nebula) made at AIO.

Twenty-six percent of the observed planetary nebulae gave measurable radio signals at 430 MHz. The lower flux-density limit for a detectable source was about 0·1 flux unit (1 f.u. $= 10^{-26}$ Wm^{-2} Hz^{-1}). Confusion problems at these low frequencies were very serious; at least 23% of the observed nebulae were confused with nearby radio sources. This study showed that for the thermal planetary nebulae with well determined optically thin radio spectra, their Hβ emission could be predicted. These Hβ fluxes were compared with the Hβ flux measurements uncorrected for interstellar extinction, and the differences gave the extinction in the direction of individual planetary nebulae. The results were compared with the extinction estimated optically, and a good agreement was found.

Recently Thompson *et al.* (1967) and Thompson and Colvin (1967) have measured the flux densities of more than 50 planetary nebulae at 3000 MHz, and 22 nebulae at 1420 MHz. The Caltech (CIT) two 90-foot diameter antennas were used as an interferometer, and both flux densities and angular widths of the observed planetaries

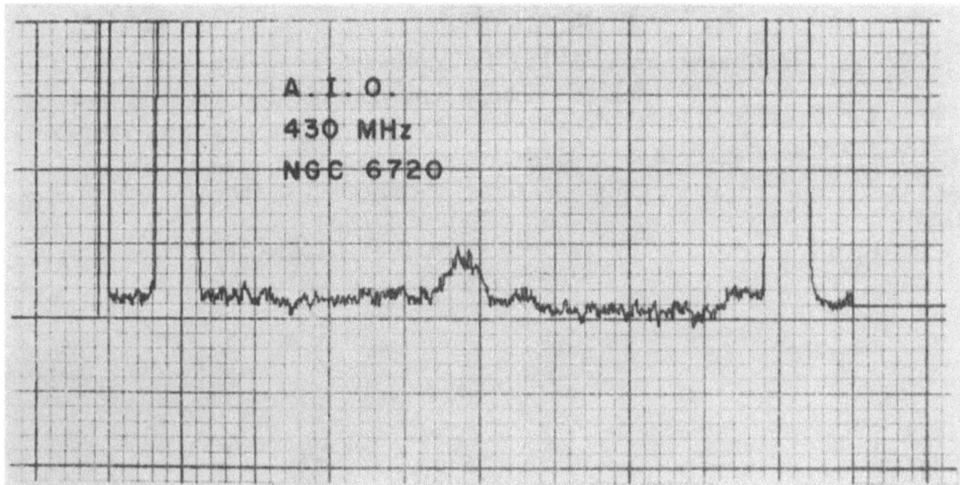

FIG. 1. *Drift curve of NGC 6720 at 430 MHz observed with the AIO 1000-foot radio telescope.*

were derived. In general the radio and optical widths were found to be in good agreement.

A radio survey of planetary nebulae was also made with the Jodrell Bank Mark-II Radio Telescope by Davies *et al.* (1967) at 1420, 2695, and 4995 MHz. Their results are in agreement with the previous surveys, indicating that most planetary nebulae are thermal sources. NGC 7008 and NGC 7635 were reported to have non-thermal spectra, characteristic of supernova remnants and extragalactic sources.

It should also be mentioned that two high-frequency radio surveys of planetary nebulae are near completion. One survey, performed by Kaftan-Kassim at NRAO, made use of the 300-foot transit radio telescope at 1400 MHz and the 140-foot radio telescope at 5000 MHz. The second survey, performed by Ehman at the University of Michigan, made use of an 85-foot radio telescope at 8000 and 16200 MHz. A few results from these surveys have been communicated to the author and are reported in the following section of this paper.

About 75% of all published flux-density measurements are in good agreement. The major uncertainties result from confusion of the planetary nebulae with non-thermal background sources. The larger the HPBW (half power beamwidth) of a radio telescope, the more serious the confusion becomes. There are about 2·5 sources per square degree with a flux density at 1400 MHz of $\geqslant$0·1 f.u. The probability of a source being confused when observed with a 300-foot radio telescope having a HPBW of 10′ at 1400 MHz is about 7%. With the exception of the 300-foot radio telescope at NRAO, all other existing radio telescopes have HPBW's $>$10′ at 1400 MHz. Table 1 shows the HPBW's of the major radio telescopes used for observations of planetary nebulae.

Table 1

Half power beamwidths (′) of radio telescopes used for planetary nebulae observations

Freq. (MHz) \ Radio Telescope	1000-ft. AIO	300-ft. NRAO	210-ft. CSIRO	140-ft. NRAO	125 × 83-ft. J. Bank
15400				2	
5000				6	7·5
3000			7·5	10	15
1400		10	14		30
750		18·8			
610	9		42		
430	10				
195	33				

With the exception of NGC 7293 (and possibly NGC 6853), all other planetary nebulae are point sources to the existing radio telescopes. NGC 7293 has a diameter of 12′ and appears as an extended source with HPBW's of 15′ or smaller.

3. Radio Spectra of Planetary Nebulae

The observations reported in the previous section clearly show that planetary nebulae have thermal radio spectra. Unlike H II regions, planetary nebulae become opti-

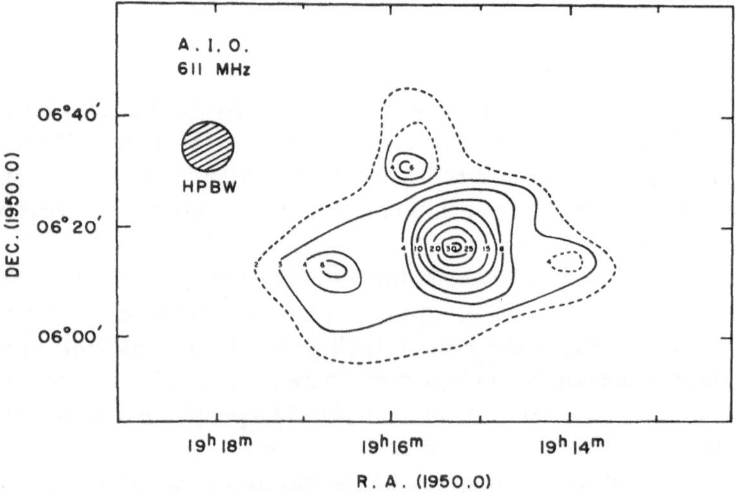

Fig. 2. *Brightness temperature distribution at 611 MHz of the region of NGC 6781. The planetary nebula coincides with the source* $19^h15^m49^s$ *and* $06°29'.5$.

Table 2

Flux densities of planetary nebulae (in units of 10^{-26} Wm^{-2} Hz^{-1})

Frequency MHz	NGC 7027	IC 418	NGC 6572	NGC 6853	NGC 7662	NGC 3242	NGC 7009	NGC 6720	References
195				0·40 ± 0·10				0·21 ± 0·05	5
430			(− 0·20 ± 0·10)	0·97 ± 0·12				0·50 ± 0·12	5
611			<0·13						5
620			<0·32	1·51 ± 0·15			0·72 ± 0·10		3
750	0·39 ± 0·06	0·43 ± 0·04		1·09 ± 0·05	0·67 ± 0·09		0·42 ± 0·05	0·46 ± 0·04	2
930		0·60 ± 0·20				1·00 ± 0·20			6
1410	1·48 ± 0·05	0·94 ± 0·07	<0·30	1·35 ± 0·07	0·84 ± 0·06		0·64 ± 0·04	0·53 ± 0·02	2
1415						1·09 ± 0·10			7
1420	1·30 ± 0·07	1·14 ± 0·06	0·42 ± 0·09	>0·51	0·51 ± 0·07	0·75 ± 0·06	0·57 ± 0·07	0·37 ± 0·12	8
		0·73 ± 0·11	0·26 ± 0·07	1·38 ± 0·09		0·89 ± 0·11	0·71 ± 0·09	0·30 ± 0·08	10
		1·02 ± 0·10	0·44 ± 0·10	1·48 ± 0·15		1·15 ± 0·12	0·65 ± 0·10		3
2695	3·44 ± 0·05	1·20 ± 0·09	0·91 ± 0·07	1·36 ± 0·06	0·61 ± 0·09	1·04 ± 0·04	0·71 ± 0·09	0·45 ± 0·03	10
2730		1·29 ± 0·12	1·07 ± 0·10	1·81 ± 0·18		0·81 ± 0·10	0·79 ± 0·08		3
3000	3·53 ± 0·30	1·41 ± 0·10	0·92 ± 0·10	1·30 ± 0·20	0·66 ± 0·05	0·72 ± 0·10	0·62 ± 0·09	0·42 ± 0·05	8
	6·20 ± 0·90	1·60 ± 0·20	1·00 ± 0·10	1·30 ± 0·10			0·80 ± 0·20		1 (ref. 2 for NGC 7027)
4995	6·50 ± 0·14	0·43 ± 0·11	1·24 ± 0·11	1·04 ± 0·14	0·78 ± 0·11	0·97 ± 0·15	0·80 ± 0·12	0·38 ± 0·09	10
5000	5·96 ± 0·35	1·45 ± 0·25	1·16 ± 0·20	1·48 ± 0·20	0·72 ± 0·30	0·76 ± 0·08	0·93 ± 0·20	0·89 ± 0·35	9 (ref. 7 for NGC 3242)
		1·66 ± 0·06	1·13 ± 0·06						11
8000	6·07 ± 0·20	1·72 ± 0·09	1·23 ± 0·05	1·15 ± 0·06	0·69 ± 0·20	0·69 ± 0·06	0·67 ± 0·04	0·39 ± 0·02	4
15400	6·00 ± 1·00								12
16200	7·42 ± 0·39	2·00 ± 0·15	1·69 ± 0·23	1·34 ± 0·24	0·78 ± 0·05	0·76 ± 0·15			4

(1) Lynds (1961). (2) Menon and Terzian (1965). (3) Slee and Orchiston (1965). (4) Ehman (1967). (5) Terzian (1966). (6) Khromov (1966). (7) Kaftan-Kassim (1966). (8) Thompson *et al.* (1967). (9) Hughes (1967). (10) Davies *et al.* (1967). (11) Kaftan-Kassim (1967). (12) Kellerman and Pauliny-Toth (1967).

cally thick at relatively high frequencies. In a few cases radio observations have indicated possible non-thermal spectra for some planetary nebulae.

Kaftan-Kassim (1966) has examined the possible non-thermal planetary nebula NGC 3242 and has found that the excess radiation at the longer wavelengths is due to a faint arc of nebulosity about 10′ SW of NGC 3242. The observations of NGC 6781 also indicated a steep non-thermal spectrum. Terzian (1967) surveyed the region of NGC 6781 and showed that the previous observations were seriously confused with a strong and complex background. Figure 2 shows the brightness temperature distribution at 611 MHz of the region of NGC 6781. Both NGC 3242 and NGC 6781 now have well-established thermal spectra. Several other nebulae, like NGC 40, NGC 7008 and NGC 7635, seem to have non-thermal spectra, and more careful examinations of such cases must be made to rule out the confusion problem, if at all possible.

Table 2 summarizes the flux-density measurements of 8 planetary nebulae observed between 195 and 16200 MHz. The spectra of these nebulae are shown in Figures 3–6. It can be seen that most of the observations are in good agreement. The low flux

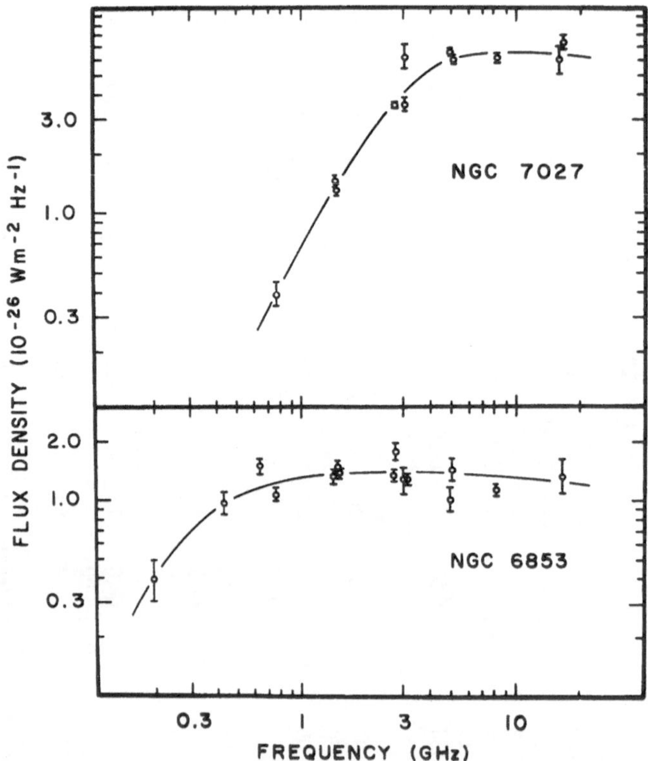

FIG. 3. *Radio spectra of NGC 7027 and NGC 6853.*

density of IC 418 at 4995 MHz reported by Davies *et al.* (1967) is due to positional errors during the observations (Davies, private communication).

The frequency at which the spectra of thermal sources turn to lower flux densities is a function of the emission measure. The higher the emission measure the higher is the turn-over frequency. The emission measure can easily be computed from the

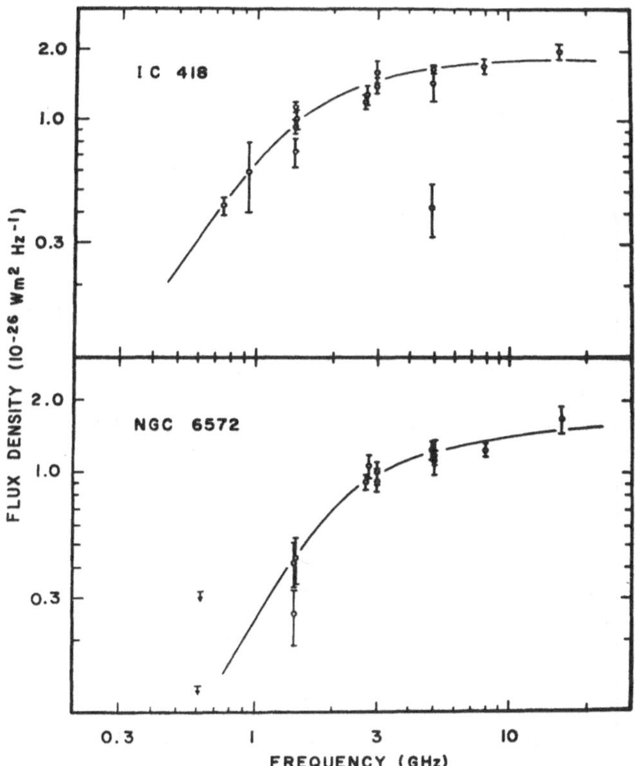

FIG. 4. *Radio spectra of IC 418 and NGC 6572.*

observed radio fluxes in the optically thin part of the spectra of planetary nebulae.

The absorption coefficient for free-free transitions as given by Oster (1961) can be written

$$\kappa_\nu = \frac{N_e N_i}{\nu^2}\left[\frac{4Z^2\,e^6}{3\,(2\pi)^{1/2}\,m^3 c}\right]\left[\frac{m}{kT_e}\right]^{3/2} 2\ln\left[\left(\frac{2kT_e}{\gamma m}\right)^{3/2}\left(\frac{m}{\pi\gamma Z e^2 \nu}\right)\right], \qquad (1)$$

where N_e and N_i are the electron and ion densities, Z is the atomic number, e and m are the electron charge and mass, c is the velocity of light, k is Boltzmann's constant, T_e is the electron temperature, ν is the frequency and γ is the base of the Napierian logarithms to the power of Euler's constant ($\gamma \simeq 1{\cdot}78$). Substituting numerical values for the constants in the above expression and assuming $N_e = N_i$, and expressing ν in MHz we have

$$\kappa_\nu = 9{\cdot}776 \times \frac{10^{-15} N_e^2}{\nu^2 T_e^{3/2}} \ln\left(49{\cdot}503 \frac{T_e^{3/2}}{\nu}\right). \tag{2}$$

We can also write

$$\kappa_\nu = \zeta \frac{N_e^2}{\nu^2 T_e^{3/2}}, \tag{3}$$

where ζ varies slowly with T_e and ν.

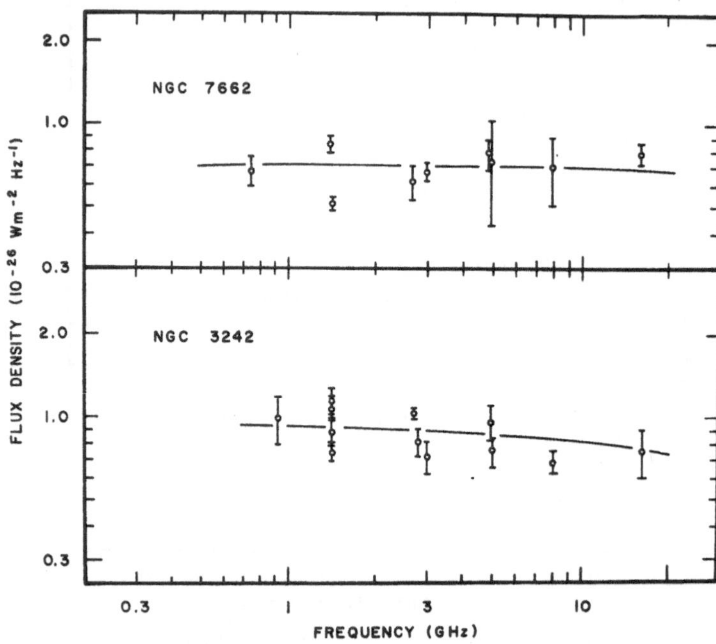

FIG. 5. *Radio spectra of NGC 7662 and NGC 3242.*

The emission measure E, defined as

$$E = \int_0^s N_e^2 \, ds, \tag{4}$$

can be derived from the radio-flux measurements of planetary nebulae and can give us some estimates of the mean densities in planetary nebulae. The flux density of a planetary nebula with an apparent size at least twice as small as the HPBW of a radio telescope is given by

$$S_\nu = \frac{2k\nu^2}{c^2} T_b \Omega , \tag{5}$$

where T_b is the brightness temperature and Ω is the apparent solid angle of the source. Assuming that the flux of a planetary nebula is being measured at a high enough frequency so that the nebula is optically thin, we can use the approximation $T_b \approx T_e \tau_\nu$, where τ_ν is the optical depth. The optical depth is defined as the integral over the absorption coefficient, and can be expressed as a function of the emission measure, i.e.

$$\tau_\nu = \frac{\zeta E}{\nu^2 T_e^{3/2}} . \tag{6}$$

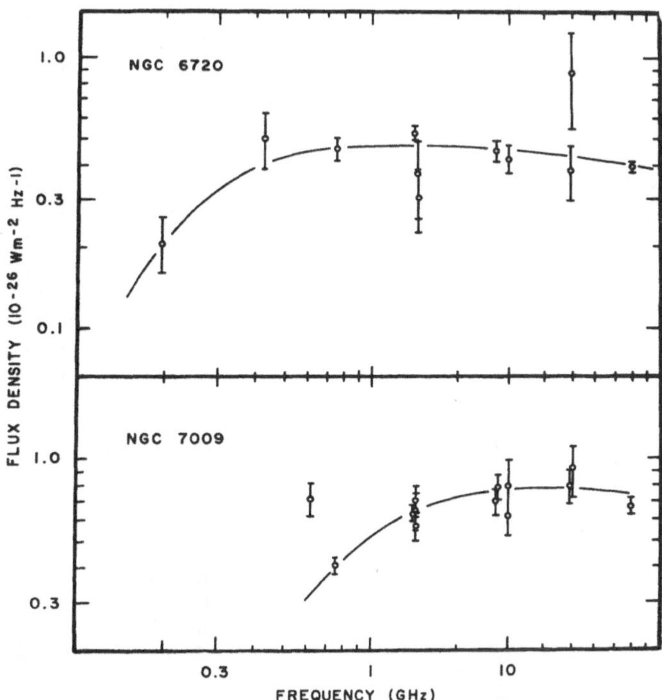

FIG. 6. *Radio spectra of NGC 6720 and NGC 7009.*

Thus the flux density expression (Equation 5) can be written as a function of the emission measure, or the emission measure can be expressed as,

$$E = \frac{S_\nu T_e^{1/2}}{\zeta \Omega} \cdot \frac{c^2}{2k} . \tag{7}$$

Table 3

Hβ and radio fluxes of planetary nebulae

Pl. Neb.		Corrected Hβ Flux (10^{-12} erg cm^{-2} sec^{-1})	Reference	Radio Flux Density (10^{-26}Wm^{-2}Hz^{-1})		
				Predicted	Observed	Reference
NGC	7027	1660	3	5·48	7·42	10
IC	418	871	3	2·88	(2·00)	10
NGC	6572	603	3	1·99	(1.69)	10
NGC	6853	513	3	1·69	1·48	6
NGC	6543	477	3	1·57	0·90	9
NGC	7293	468	3	1·54	1·60	1
HD	138403	398	5	1·31	–	–
NGC	7662	275	3	0·91	0·78	8
NGC	3242	257	3	0·85	0·97	8
NGC	246	251	5	0·83	0·24	1
NGC	7009	219	3	0.72	0·79	1
NGC	6826	209	3	0·69	0·42	4
NGC	6210	174	3	0·57	0·46	10
NGC	6804	159	5	0·53	0·23?	8
NGC	6818	126	5	0·42	0·46	8
IC	2149	126	5	0·42	0·50	2
NGC	6720	123	3	0·41	0·42	4
IC	4634	98	1	0·32	0·15	4
NGC	7026	85	1	0·28	0·27	5
NGC	6439	79	5	0·26	(–0·04)	8
NGC	6881	79	5	0·26	0·31?	8
NGC	6905	79	5	0·26	(0·32)	10
NGC	40	79	5	0·26	0·34?	5
NGC	6781	74	5	0·24	0·30	7
NGC	650-1	63	5	0·21	0·16	4
NGC	1535	63	5	0·21	0·21	1
NGC	2392	63	5	0·21	0·23	4
NGC	6803	59	1	0·20	0·11	8
IC	2165	56	1	0·19	0·22	5
NGC	3587	52	3	0·17	0·10	4
NGC	4361	50	5	0·17	0·20	1
NGC	6778	50	5	0·17	0·10	1
NGC	6741	46	1	0·15	0·24	1
IC	4593	40	5	0·13	0·12	4
NGC	6891	40	5	0·13	0·20	5
IC	4997	33	4	0·11	$<$0·09	5
IC	3568	32	5	0·11	$<$0·12	5
NGC	6894	32	5	0·11	0·12	8
J	900	30	1	0·10	0·10	9
VV	286	25	4	0·08	–	–
NGC	6751	25	5	0·08	(–0·02)	8
NGC	7354	25	5	0·08	0·51	9
NGC	2022	20	5	0·07	0·08	1
NGC	6567	20	5	0·07	0·14	1
NGC	6629	18	4	0·06	0·17	1
NGC	1501	16	5	0·05	0·19	4
NGC	6309	16	5	0·05	0·18	1

Table 3 (continued)

Pl. Neb.	Corrected Hβ Flux (10^{-12} erg cm^{-2} sec^{-1})	Reference	Radio Flux Density (10^{-26}Wm^{-2}Hz^{-1}) Predicted	Observed	Reference
NGC 7008	16	5	0·05	0·18	4
NGC 6537	14	1	0·05?	0·38?	8
NGC 2371-2	13	5	0.04	0·22	3
IC 5217	13	5	0·04	0·15	5
BD 30°3639	13	5	0·04	0·44	4
NGC 6772	10	5	0·03	0·06?	8
NGC 6563	10*	2	>0·03	0·13	5
NGC 6445	8	5	0·03	0·27	5
J 320	8	5	0·03	0·08?	8
NGC 6884	7	4	0·02	0·07?	8
IC 351	6	5	0·02	(−0·02)	8
NGC 7139	6	5	0·02	0·07?	8
NGC 6833	6	4	0·02	0·22?	8
NGC 6058	5	5	0·02	0·01	8
VV 171	5	5	0·02	–	–
NGC 6072	4*	2	>0·02	0·11	5
NGC 6886	3	4	0·01	0·13	9
NGC 6807	3	4	0·01	0·05	1
NGC 6879	2	4	0·01	0·08?	8
VV 267	2	4	0·01	–	–

( ) Optically thick.
? Flux affected by confusion.
(–) not detected.
* Hβ flux not corrected for reddening.

References for Hβ fluxes: (1) Collins *et al.* (1961). (2) O'Dell (1963). (3) Osterbrock (1964). (4) Vorontsov-Velyaminov *et al.* (1964). (5) Seaton (1966).

References for Radio Data: (1) Slee and Orchiston (1965). (2) Khromov (1966). (3) Terzian (1966). (4) Thompson *et al.* (1967). (5) Thompson and Colvin (1967). (6) Hughes (1967). (7) Terzian (1967). (8) Davies *et al.* (1967). (9) Kaftan-Kassim (1967). (10) Ehman (1967).

Substituting numerical values in the last expression, and using $T_e = 10^4$ °K, $\nu = 5000$ MHz to evaluate ζ, assuming spherical symmetry for the observed nebula, and expressing E in units of cm^{-6} parsec, S_ν in Wm^{-2} Hz^{-1}, and d (the diameter of the nebula) in degrees we have

$$E = \frac{S_\nu T_e^{1/2}}{2 \times 10^{-26}\, d^2}. \tag{8}$$

The emission measure of NGC 7027 is found to be $3{\cdot}8 \times 10^7$ cm^{-6} parsec, that of NGC 6781 is $0{\cdot}11 \times 10^5$ cm^{-6} parsec. Thus, it is possible to derive estimates of densities and masses of planetary nebulae provided good distances and filling factors (fraction of nebula filled with matter) are available.

The detection of the hydrogen 109α recombination line (5008·9 MHz) in H II regions by Höglund and Mezger (1965) has opened a new way of studying interstellar ionized hydrogen. Very recently Mezger *et al.* (1967) have succeeded in detecting the hydrogen 109α recombination line in the planetary nebula NGC 7027. From the ratio of line to continuum brightness temperatures an electron temperature of 11×10^3 °K has been derived.

4. Discussion

The ratio of the emission coefficients at a radio frequency ν and at Hβ in a thermal region as shown by Terzian (1965) is

$$\frac{j_\nu}{j_\beta} = \frac{N_i}{N_p} \frac{T_e}{\langle b_4 \rangle} \frac{1{\cdot}664 \times 10^{-19}}{0{\cdot}986 \times 10^4/T_e} \ln\left(49{\cdot}503 \frac{T_e^{3/2}}{\nu}\right). \tag{9}$$

Since the planetary nebulae are optically thin at Hβ and at high radio-frequencies, the ratio j_ν/j_β is equal to the emergent fluxes at the corresponding wavelengths. The factor $\langle b_4 \rangle$ indicates the degree of departure from thermodynamic equilibrium. The effective value of $\langle b_4 \rangle$ for the 4→2 Hβ transition is given by Burgess (1958). We shall adopt a mean value of 0·20 for $\langle b_4 \rangle$, which corresponds to Burgess' Case B (nebula optically thick in Lyman lines) at an electron temperature of 10^4 °K. Using 5000 MHz for ν, a helium abundance of 15%, and an electron temperature of 10^4 °K, the ratio of the radio to Hβ emission is

$$\frac{j_\nu}{j_\beta} = 3{\cdot}28 \times 10^{-14}. \tag{10}$$

The observed Hβ fluxes corrected for interstellar extinction taken from the literature were used to derive the radio fluxes of the bright planetary nebulae, using Equation (10). These are given in Table 3 together with the observed radio fluxes. It can easily be seen that there is good agreement between the observed and predicted flux densities. Table 3 includes 68 planetary nebulae with known Hβ fluxes. Only 9 of these have not been detected at radio frequencies. With fairly good accuracy the reported observed radio fluxes in Table 3 are the values at the optically thin part of the spectra of the planetary nebulae, except for NGC 6572, NGC 6905, and IC 418. These nebulae are still optically thick at the highest observed frequencies. In the case of NGC 6543, the radio observations at 5000, 8000 and 16200 MHz indicate that this nebula is optically thin and has a flux density of 0·9 f.u. The optical depth is unity at approximately 1500 MHz. The predicted radio flux for NGC 6543 from its Hβ flux is 1·57 f.u. This value seems high compared with the radio observations.

Since the ratio of the radio flux to that at Hβ is a function of the electron temperature we can get some estimates of the latter, by assuming the helium abundance in the nebulae. Figure 7 shows lines of constant electron temperatures in the radio and Hβ-flux plane, where a helium abundance of 15% has been assumed. The observed optically thin radio fluxes and Hβ fluxes taken from Table 3 have been plotted in Figure 7. Seventy-two percent of the nebulae show electron temperatures $> 10^4$ °K, and almost

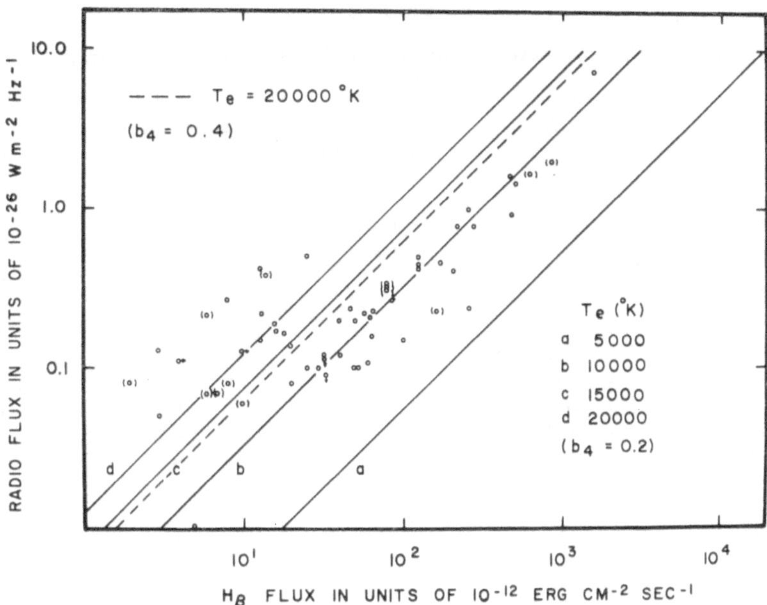

FIG. 7. *A comparison between the radio and Hβ fluxes. (The points in parenthesis have uncertain radio fluxes.)*

all nebulae have electron temperatures $>7{\cdot}5\times10^3$ °K. A few nebulae show temperatures above 2×10^4 °K; the radio-flux densities for some of these nebulae are uncertain – probably the observed radio fluxes are high due to confusion. However, NGC 7354, BD 30° 3639, NGC 2371-2, NGC 6445, NGC 6885, and NGC 6807 appear to have good radio-flux observations. One can also argue that the Hβ fluxes for these nebulae have been underestimated, particularly the estimated interstellar extinction corrections could be systematically low for these faint nebulae.

The constant-electron-temperature lines in Figure 7 were computed assuming $\langle b_4\rangle=0{\cdot}20$. This value is the best estimate for an electron temperature of 10^4 °K. However, $\langle b_4\rangle$ increases with temperature and tends to make the electron temperature dependence on j_ν/j_β smaller. The dotted line in Figure 7 was computed using $\langle b_4\rangle=0{\cdot}4$ for an electron temperature of 2×10^4 °K.

It should be pointed out that the lower limit of an accurate radio-flux measurement with present-day radio telescopes is of the order of 0·1 f.u. Figure 7 shows that most of the nebulae with very high electron temperatures were expected to have radio fluxes $\leqslant 0{\cdot}1$ f.u. predicted from their Hβ fluxes. Certainly very accurate radio and optical measurements are needed for these nebulae in order to clarify the present results.

A few nebulae which have been observed to have relatively high radio fluxes, such

Table 4

A comparison of optically estimated extinction with the extinction derived from radio data

Pl. Neb.	Measured log F(Hβ) (erg cm^{-2} sec^{-1})	Optical Extinction Δ log$F$(Hβ)	Reference	Derived Extinction Δ log$F$(Hβ)
NGC 7027	−10·12	1·34	4	1·46
IC 418	− 9·53	0·47	4	>0·31
NGC 6572	− 9·74	0·52	4	>0·45
NGC 6853	− 9·44	0·15	4	0·09
NGC 6543	− 9·60	0·28	4	>0·04
NGC 7293	− 9·35	0·02	4	0·04
NGC 7662	− 9·98	0·42	4	0·36
NGC 3242	− 9·81	0·22	4	0·28
NGC 7009	− 9·78	0·12	4	0·16
NGC 6826	− 9·92	0·24	4	0·03
NGC 6210	−10·06	0·30	4	0·21
NGC 6804	−11·28	0·93	1	1·13?
NGC 6818	−10·13	0·41	1	0·28
IC 2149	−10·50	0·68	1	0·68
NGC 6720	−10·06	0·15	4	0·17
IC 4634	−10·97	0·96	1	0·63
NGC 7026	−10·90	0·87	1	0·82
NGC 6905	−10·90	0·74	1	0·89
NGC 40	−10·64	0·46	1	0·66?
NGC 6781	−11·19	1·06	1	1·15
NGC 650-1	−10·67	0·81	1	0·36
NGC 1535	−10·36	0·28	1	0·17
NGC 2392	−10·39	0·43	1	0·24
NGC 6803	−11·15	0·92	1	0·68
IC 2165	−10·99	0·74	1	0·82
NGC 3587	−10·33	0·05	4	0·19
NGC 4361	−10·48	0·20	5	0·26
NGC 6778	−11·26	0·92	1	0·74
NGC 6741	−11·49	1·15	1	1·36
IC 4593	−10·55	0·22	1	0·11
NGC 6891	−10·60	0·63	1	0·39
IC 4997	−10·49	–	2	(−0·07)
IC 3568	−10·82	0·25	1	<0·38
NGC 6894	−11·46	1·22	1	1·02
J 900	−11·27	0·74	1	0·75
NGC 7354	−11·55	0·77	5	1·73
NGC 2022	−11·15	–	3	0·54
NGC 6567	−10·93	1·11	1	0·56
NGC 6629	−10·94	–	2	0·65
NGC 1501	−11·26	1·83	1	1·02
NGC 6309	−11·29	0·62	1	1·03
NGC 7008	−10·86	1·01	1	0·60
NGC 6537	−11·78	0·92	1	1·84?
NGC 2371-2	−10·96	0.39	5	0·79
IC 5217	−11·18	0·37	1	0·84
BD 30°3639	−11·50	0·76	1	1·63
NGC 6772	−11·65	–	3	0·91?

Table 4 (continued)

Pl. Neb.	Measured log F(Hβ) (erg cm^{-2} sec^{-1})	Optical Extinction Δ log F(Hβ)	Reference	Derived Extinction Δ log F(Hβ)
NGC 6563	−10·96	–	3	0·56
NGC 6445	−11·20	–	3	1·12
J 320	−11·37	0·46	1	0·76?
NGC 6884	−11·11	1·29	1	0·44?
NGC 7139	−11·78	0·47	5	1·11?
NGC 6833	−11·22	–	3	1·05?
NGC 6058	−11·70	0·19	1	0·18
NGC 6072	−11·37	–	3	0·90
NGC 6886	−11·50	0·90	1	1·10
NGC 6879	−11·60	0·81	1	0·99?

(1) Collins *et al.* (1961). (2) O'Dell (1962). (3) O'Dell (1963). (4) Osterbrock (1964). (5) Thompson *et al.* (1967).

as NGC 7635, NGC 6857, and IC 1470, are probably diffuse nebulae of small angular size and have not been included in Table 3. The planetary nebula NGC 6302 seems to be of particular interest since it possibly coincides with the X-ray source Sco XR-2 (Thompson and Colvin, 1967). No Hβ-flux measurement is available for this nebula; however, the radio observations indicate that NGC 6302 is a thermal source.

For most of the brighter planetary nebulae, radio fluxes can be measured with an accuracy of 10% or better. From the radio measurements the Hβ fluxes can be predicted assuming the electron temperatures and helium composition. These then can be compared with the measured Hβ fluxes uncorrected for interstellar extinction, and the latter quantity can be derived. Table 4 shows a comparison of the optically estimated Hβ extinction and the extinction derived from the radio data. Figure 8 shows the comparison of the extinctions using only the reliable points from Table 4. The correlation coefficient is 0·72. The solid straight line in Figure 8 is the expected correlation, and the dashed line is the best fit to the plotted points. The derived extinction Δ log F(Hβ) for NGC 7354, and BD 30°3639 is more than twice as high as the extinction estimated optically. As was mentioned above, this can be the result of radio fluxes measured too high, or of underestimated Hβ fluxes, and (or) extinction corrections. In some cases where the helium abundances and electron temperatures can be found from optical observations accurate Hβ extinctions can be derived from the above method.

5. Conclusion

During the last few years planetary nebulae have been observed at radio frequencies. The results of these observations confirm the recombination theory of emission and have established planetary nebulae as thermal sources.

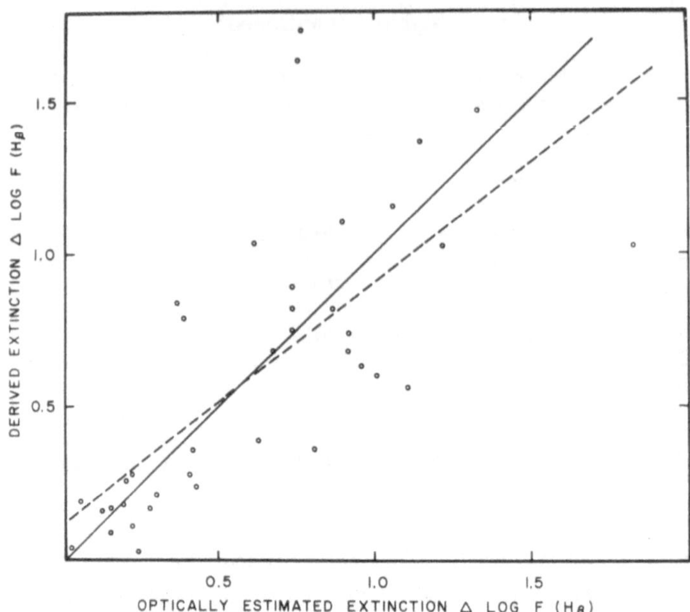

FIG. 8. *A comparison between the optically estimated Hβ interstellar extinction with the one derived from radio data.*

More accurate observations are required in some cases, specially at the low radio frequencies in order to establish the optically thick part of the spectra of these nebulae.

Observations at the hydrogen 109α recombination line should be made, as well as at other hydrogen recombination lines. Some attempts should also be made to observe the 21-cm line of neutral hydrogen, which may exist around some ionization-bounded planetary nebulae. The recently discovered strong OH emission in diffuse nebulae suggests that OH observations should be tried in planetary nebulae.

Finally it should be pointed out that observations to determine the brightness-temperature distribution in planetary nebulae do not exist. Such observations require extremely high angular resolution, which present-day radio telescopes do not have. Presently there are two ways of observing the brightness-temperature distribution of a few planetary nebulae, one using the method of aperture synthesis, and the other using the method of lunar occultations. So far no lunar occultation of a planetary has been observed, but several will be occulted in the next few years.

References

Burgess, A. (1958) *Mon. Not. R. astr. Soc.*, **118**, 477.

Collins, G.W. III, Daub, C.T., O'Dell, C.R. (1961) *Astrophys. J.*, **133**, 471.

Davies, J.G., Ferriday, R.J., Haslam, C.G.T., Moran, M., Thomasson, P. (1967) *Mon. Not. R. astr. Soc.*, **135**, 139.

Ehman, J.R. (1967) *Astrophys. J.* (in press).
Höglund, B., Mezger, P.G. (1965) *Science*, **150**, 339.
Hughes, M.P. (1967) *Astrophys. J.*, **149**, 377.
Kaftan-Kassim, M.A. (1966) *Astrophys. J.*, **145**, 658.
Kaftan-Kassim, M.A. (1967) private communication.
Kellermann, I.K., Pauliny-Toth, I.I.K. (1967) private communication.
Khromov, G.S. (1966) *Soviet Astr.*, **9**, 705.
Lynds, C.R. (1961) *Publ. nat. Radio Astr. Obs.*, **1**, 85.
Menon, T.K., Terzian, Y. (1965) *Astrophys. J.*, **141**, 745.
Mezger, P.G., Altenhoff, W., Schraml, J., Burke, B.F., Reifenstein III, E.C., Wilson, T.L. (1967) *Astrophys. J.*, **150**, L.157.
O'Dell, C.R. (1962) *Astrophys. J.*, **135**, 371.
O'Dell, C.R. (1963) *Astrophys. J.*, **138**, 293.
Oster, L. (1961) *Astrophys. J.*, **134**, 1010.
Osterbrock, D.E. (1964) *A. Rev. Astr. Astrophys.*, **2**, 95.
Seaton, M.J. (1966) *Mon. Not. R. astr. Soc.*, **132**, 113.
Slee, O.B., Orchiston, D.W. (1965) *Austr. J. Phys.*, **18**, 187.
Terzian, Y. (1965) *Astrophys. J.*, **142**, 135.
Terzian, Y. (1966) *Astrophys. J.*, **144**, 657.
Terzian, Y. (1967) *Astr. J.*, **72**, 443.
Thompson, A.R., Colvin, R.S. (1967) *Astrophys. J.*, **150**, 345.
Thompson, A.R., Colvin, R.S., Stanley, G.J. (1967) *Astrophys. J.*, **148**, 429.
Vorontsov-Velyaminov, B.A., Kostjakova, E.B., Dokuchaeva, O.D., Arhipova, V.P. (1964) *Soviet Astr.* **8**, 196.

DISCUSSION

Minkowski: You mentioned NGC 40 as possibly non-thermal. Is there not a known non-thermal source, quite close by and visible on the Sky Survey plate, which is probably a supernova remnant?

Terzian: Yes, this is so, and it may be that the radio observations are confused.

Liller: Have there been any attempts to observe by their 21-cm radiation the H I regions which must surround many planetaries?

Menon: Attempts to detect 21-cm radiation associated with planetary nebulae have not been successful so far. Observationally it is a very difficult problem.

Thompson: Terzian has mentioned NGC 7635 as a case in which the present flux-density values show some deviation from a thermal spectrum. This nebula has recently been classified as diffuse by a number of people, and thus should not be included when considering evidence for the possible existence of non-thermal planetary nebulae.

Osterbrock: It must be remembered that the 'optical extinction' values quoted by Terzian are derived in some cases from photoelectrically measured Balmer-line or Balmer/Paschen-line ratios, while in other cases they are simply estimates based on galactic latitude.

RADIO EMISSION FROM FOURTEEN PLANETARY NEBULAE AT 408 MHZ

CESARE BARBIERI and ANTONINO FICARRA
(Osservatorio Astronomico di Padova, Italy) *(Laboratorio Nazionale di Radio-astronomia, Bologna, Italy)*

The East-West arm of the 'Northern Cross' radio telescope of the Bologna University has been used to observe 14 planetary nebulae at the frequency of 408 MHz. The half-power beamwidth of the instrument is 4′ E–W by 100′ N–S. For the very weak sources studied here, confusion permits a positional accuracy of 2^s E–W by 15′ N–S. As the resolution limit is about 0·2 flux units, the measured flux densities of the weakest sources are accurate to 30%.

Constant declination scans were made through the optical positions given by Thompson *et al.* (1967), and at adjacent declinations spaced at intervals of 30′, in order to restore the N–S diagram of the sources. As calibration objects for flux density and position the sources 3C 48, 3C 71, 3C 119, 3C 216, 3C 237, 3C 245, 3C 446, and CTA 102 were used.

The results of the present observations, together with those of Slee and Orchiston (1965), Menon and Terzian (1965), Terzian (1966), Le Marne (1966), Thompson *et al.* (1967), are summarized in Table 1.

For the two common nebulae IC 418 and NGC 7009 there is excellent agreement between our measurements and those of Le Marne. On the contrary the results of Terzian are not confirmed for NGC 6853; our measurement seems very accurate for the object appears as a point source and there is no confusion in the neighbourhood of the nebula.

By comparing the present results with previous results at other frequencies a further confirmation of the thermal radiospectrum of planetary nebulae is derived.

As the results of this work have been encouraging enough, observations of about 30 other nebulae were obtained in July 1967. The reduction of data is now in progress and a fuller account will be given elsewhere (Ficarra, 1967, submitted to *Nuovo Cimento*; Barbieri and Ficarra, in preparation).

References

Le Marne, A.E. (1966) *Observatory*, **86**, 148.
Menon, T.K., Terzian, Y. (1965) *Astrophys. J.*, **141**, 745.
Slee, O.B., Orchiston, D.W. (1965) *Austr. J. Phys.*, **18**, 187.
Terzian, Y. (1966) *Astrophys. J.*, **144**, 657.
Terzian, Y. (1967) *Astr. J.*, **72**, 143.
Thompson, A.R., Colvin, R.S., Stanley, G.J. (1967) *Astrophys. J.*, **148**, 429.

Osterbrock and O'Dell (eds.), Planetary Nebulae, 104–105. © *I.A.U.*

Table 1

Nebula	Optical Position (1950) R.A.	Dec.	Average Flux Density ($10^{-26}Wm^{-2}cps^{-1}$) 3000 TCS	2730 SO	1430 TCS	750 MT	430 T	408 LM	408 BF	195 T	Notes
NGC 246	$00^h44^m30^s.8$	−12°09′	0·15	0·20					⩽0·2		
NGC 1514	04 06 08·3	+30 39	0·29		0·39		0·37		0·3		
IC 418	05 25 10·2	−12 44	1·41	1·29	1·14	0·43		⩽0·3	⩽0·3		1
NGC 2371-2	07 22 26·9	+29 36	<0·13				0·22		0·3		
NGC 2438	07 39 32·6	−14 37	0·12						<0·1		
NGC 3587	11 11 58·0	+55 17	0·10			0·22			⩽0·2		
NGC 6058	16 02 44·0	+40 49	<0·14						<0·1		
NGC 6210	16 42 23·7	+23 53	0·32	0·48	0·37	<0·85	<0·10		⩽0·2		
NGC 6720	18 51 43·3	+32 58	0·42		0·37	0·45	0·50		⩽0·4	0·21	2
NGC 6781	19 16 01·2	+06 27	0·37		0·77		0·38		0·5		3
NGC 6818	19 41 07·8	−14 16	0·34	0·34	0·45				⩽0·2		4
NGC 6853	19 57 27·0	+22 35	1·3	1·81	>0·51	1·09	0·97		1·6	0·40	
NGC 7009	23 01 27·7	−11 34	0·62	0·79	0·57	0·04		<0·3	⩽0·2		5
NGC 7662	23 23 29·5	+42 16	0·66		0·51	0·67			⩽0·4		6

[1] An unresolved source of 1·3 flux units is observed at $05^h25^m28^s$, −11°50′ (probably the same source indicated by Le Marne, 1966). A considerable broadening Southeast of the source indicates the presence of a weaker source at the optical position of IC 418.

[2] A source of 1·2 flux units is observed at $18^h51^m39^s$, +32°28′. A slight broadening North of the source indicates the probable presence of a weaker source.

[3] The flux indicated at 430 MHz is that measured by Terzian (1967). All the data now available show the spectrum is thermal. Previous evidence of a non-thermal spectrum (Terzian, 1966) comes from confusion with the source 4C 06·66.

[4] Measurement confused by the presence of a source of 1·1 flux units at $19^h41^m04^s$, −13°40′.

[5] Measurement confused by the presence of a source of 1·0 flux units at $21^h01^m12^s$, −12°00′.

[6] Confused, at high declination, by a strong source of 2·4 flux units at $23^h23^m21^s$, +43°00′.

RADIO OBSERVATIONS OF PLANETARY NEBULAE

J. G. DAVIES
(*Jodrell Bank, University of Manchester, England*)

I am to report on a rather extensive series of observations of planetary nebulae, using the Mark-II radio telescope at Jodrell Bank. The method of observation adopted involves comparing the signal received from a beam centered on the nebula with two beam areas horizontally displaced to either side of it. If the source is not too far South, the confusion effects can be reduced by making observations at different hour angles. A second source of error, particularly in the earlier results, arises from the fact that the positions adopted for some nebulae were not as accurate as had been assumed. The anomalously low 6 cm flux for *IC* 418 in particular is now believed to be due to this cause, and can be disregarded.

Observations of 21 sources at 21 cm, 65 sources at 11 cm, and 40 sources at 6 cm have already been published (Davies *et al*, *Mon. Not. R. astr. Soc.*, **135**, 1967, 139). Observations of 76 sources at 4-cm wavelength were completed in August this year, and preliminary values are now available. In addition about 12 sources have been observed at 21 cm and 73 cm using the Mark-I and Mark-II telescopes as an interferometer. The nebulae were not resolved by the instrument in most cases, but it is expected that the confusion will be greatly reduced. The results of these observations have not yet been analysed. From the results available, 35 planetary nebulae have been found to be optically thin, and the fluxes observed agree well with those obtained from the Hβ radiation, after correction for extinction.

The nebulae NGC 1501, 4634, 6307, 6543, and 6884 appear to show optically thick spectra, although in no case is this certain, since all the sources are weak and may be subject to confusion.

The nebulae NGC 6572, 6790, 7027, and IC 418, show the transition from optically thick to thin, at wavelengths of approximately 10, 6, 10 and 15 cm respectively.

Three nebulae, VV 285, NGC 6445 and 6833 appear to have non-thermal spectra, but this may be affected by confusion.

The source NGC 6857, for which Terzian reported a flux of 1·2 flux units at 70 cm, is the brightest observed planetary nebula at 4 cm, having a flux of 10·6 flux units.

Twenty-three of the planetary nebulae observed at 4 cm were undetected. Two Wolf-Rayet stars, HD 193793 and HD 16523 were detected at 11 cm. Pressure of observing time has prevented further observations of these or other Wolf-Rayet stars.

Osterbrock and O'Dell (eds.), Planetary Nebulae, 106–107. © *I.A.U.*

DISCUSSION

Thompson: The flux density of NGC 6857 as measured by Colvin and myself at 10-cm wavelength is $6{\cdot}4 \pm 0{\cdot}5$ flux units. The position of this nebula coincides with the brightness peak in the broad thermal source NRAO 621, and some diffuse nebulosity can be seen on the Palomar Schmidt plates. Some fraction of the measured radio emission may thus result from a confusing thermal source.

Terzian: NGC 6857 shows no radio signal at 195 MHz, perhaps because it is just seen in absorption.

Aller: The optical spectrum of NGC 6833 is that of a typical moderate excitation object. It has a small angular size. I would not expect it to be unusual, certainly not a non-thermal source.

HIGH-RESOLUTION RADIO OBSERVATIONS OF FIVE PLANETARY NEBULAE

B. ELSMORE
(Mullard Radio Astronomy Observatory, Cambridge, England)

ABSTRACT

Observations have been made of 5 planetary nebulae at 1407 and 408 MHz with beam widths of 23″ of arc and 80″ of arc respectively. Measurements of positions, fluxes and angular structure have been obtained; some results are presented in the form of contour maps.

The Cambridge 1-mile radio telescope has been used to map the structure of 5 planetary nebulae. The radio telescope and its mode of operation have been described elsewhere (Ryle, 1962; Elsmore *et al.*, 1966), and only a brief account is given here.

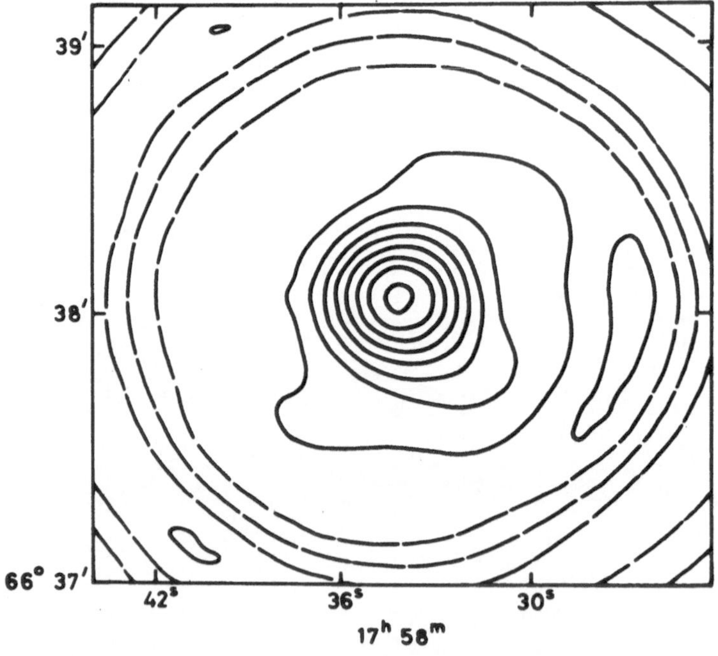

FIG. 1. *1407 MHz map of NGC 6543. (The broken contours are due to the grating side-lobe response of the instrument.)*

Osterbrock and O'Dell (eds.), Planetary Nebulae, 108–111. © *I.A.U.*

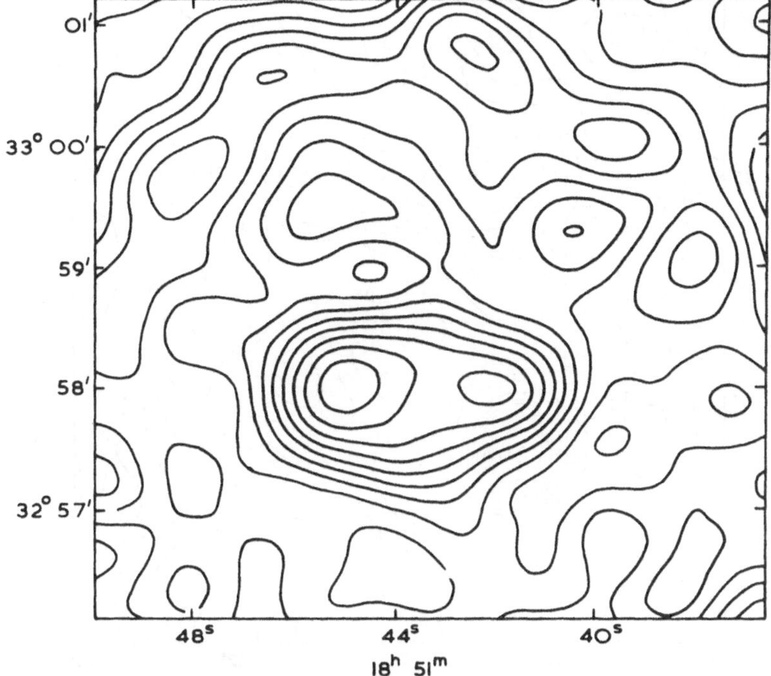

FIG. 2. *1407 MHz map of NGC 6720. The distance between the two peaks of radio emission is 36″ of arc.*

Each nebula was observed continuously for at least one 12-hour period using aerials mounted on an East-West base line. The rotation of the Earth carries one aerial around the other, enabling the amplitude and phase of the radiation from the nebula to be sampled along a circular strip. The radius of this strip is equal to the distance between the aerials.

If observations of the nebula are combined with those made at different spacings, a pencil-beam response is obtained in the vicinity of the nebula, together with a circular-grating response, the angular separation of which depends upon the number of different spacings used. Observations were made at two frequencies simultaneously, 1407 MHz and 408 MHz, giving pencil-beam widths of 23″ of arc and 80″ of arc in right ascension; the beam widths being greater in declination by a factor cosecδ.

The results appear in the form of contour maps of radio intensity of which four are shown in Figures 1–4. The declination scale of the maps has been compressed by a factor sineδ so that the beam appears circular. A summary of the observations is given in Table 1.

Generally, the radio emission originates in a region slightly smaller than that of the optical nebula except in the case of NGC 7662, where the size of the nebula at 1407

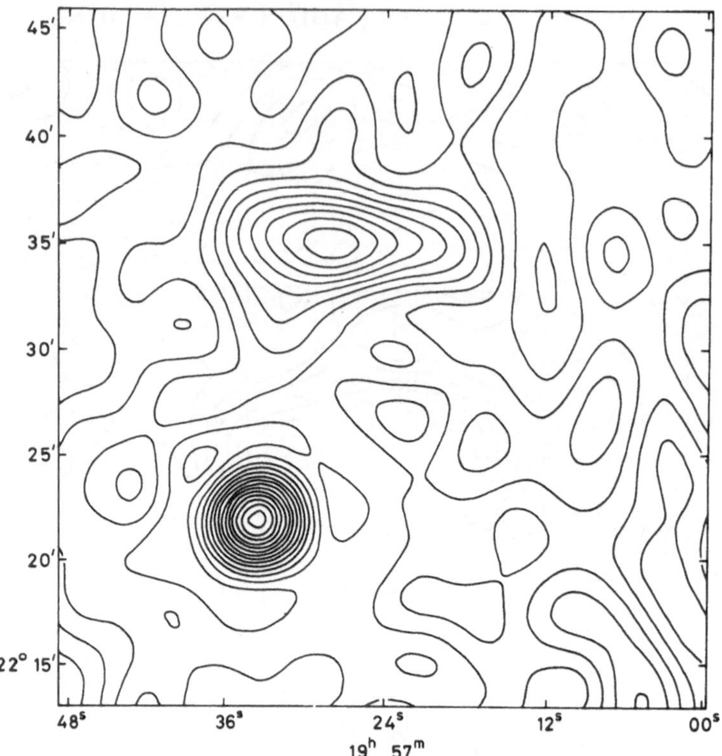

FIG. 3. *408 MHz map of NGC 6853. The object shown at declination 22°22′ is a small diameter source of 0·5 flux units, which is not related to the nebula.*

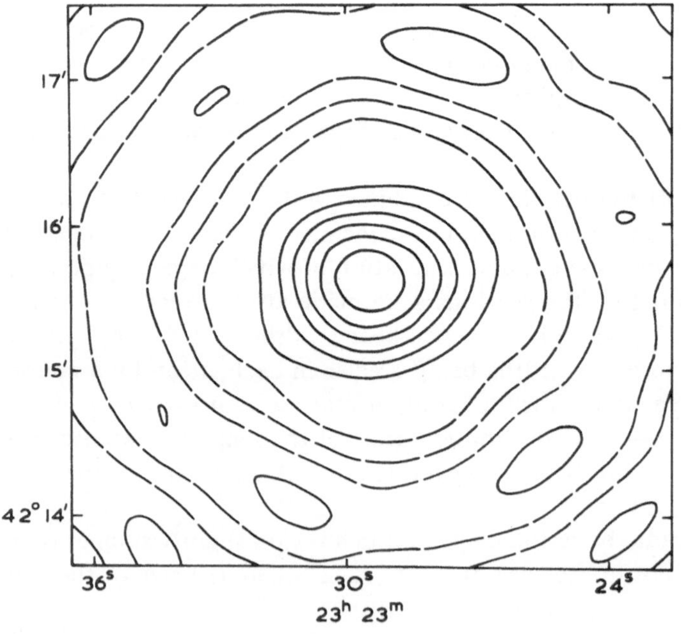

FIG. 4. *1407 MHz map of NGC 7662. (The broken contours are due to the grating side-lobe response of the instrument.)*

Table 1

NGC	Integrated Flux $\times 10^{-26}$ Wm^{-2} Hz^{-1} 1407 MHz	408 MHz	Width to Half Power at 1407MHz
6543	0·7	$\leqslant$0·2	$<10''$
6720	0·45	0·15	EW $=62''$ NS $=47''$
6853	0·9($\pm$0·3)	1·2	–
			(At 408 MHz: EW $=190''$, NS $=140''$)
7027	1·3	$\leqslant$0·2	$<5''$
7662	0·51	0·25	EW $=28''$ NS $=36''$

MHz is slightly larger than the optical size. In all cases the centre of the radio emission coincides closely with the centre of the visible nebula.

The fluxes given in Table 1 are believed to be accurate to $\pm 10\%$ and are in general agreement with previous observations, with the exception of the measurement of NGC 6720 at 408 MHz, which disagrees with the value $0{\cdot}5 \pm 0{\cdot}12$ f.u. measured at 430 MHz by Terzian (1966). There is no evidence for non-thermal emission.

References

Elsmore, B., Kenderdine, S., Ryle, M. (1966) *Mon. Not. R. astr. Soc.*, **134**, 87.
Ryle, M. (1962) *Nature*, **194**, 517.
Terzian, Y. (1966) *Astrophys. J.*, **144**, 657.

DISCUSSION

Kaftan-Kassim: I would like to ask why NGC 7293, the largest planetary nebula, was not observed. I have attempted to map it at 2 cm using the 140-ft dish at Green Bank with a 2′ beam. A double structure similar to that shown for NGC 6720 was observed.

Elsmore: NGC 7293 is too far South to be observed by the Cambridge 1-mile telescope.

Sheglov: Does the radio centre of NGC 6853 coincide with the central star of this nebula or with the brightest part of the planetary?

Elsmore: The peak of the radio emission coincides closely with the centre of the nebula rather than with the optically brightest region.

RADIO MEASUREMENTS OF PLANETARY NEBULAE

A. R. THOMPSON
(Radio Astronomy Institute, Stanford University, Stanford, Calif., U.S.A.)

ABSTRACT

Sixty-eight planetary nebulae have been investigated in a series of observations at 10-cm wavelength using the two 90-ft diameter antennas of the Owens Valley Radio Observatory. Of these, 52 were found to have flux densities greater than a minimum detectable level of approximately 10^{-27} Wm^{-2} Hz^{-1}. To indicate cases of possible confusion in the radio observations, the measured radio position of each nebula was compared with an accurate optical position. For a number of the stronger nebulae angular widths in the East–West direction and flux densities at 21 cm were also measured. The results lead to the conclusion that the radio emission is thermal, and on this basis the expected flux densities in Hβ have been calculated. A comparison with optical data shows values of the Hβ extinction $\Delta \log F_{H\beta}$, ranging from zero to approximately 2·0 for NGC 6537 and NGC 6369.

A small number of nebulae show prominently the effects of self-absorption in their radio spectra. For two of these, IC 418 and NGC 6572, an attempt has been made to derive accurate optical depths and electron temperatures using models based on Balmer-line isophotes. The temperature values have large uncertainties, but appear to be less than the values derived from forbidden-line ratios by a factor of at least 1·5. A possible explanation of this difference in terms of temperature variations within the nebulae is discussed.

Preliminary results of observations to detect absorption features at the wavelength of the 21-cm hydrogen line are described for 6 nebulae with high radio-flux densities. Two nebulae, NGC 6369 and NGC 6857 show absorption which is probably attributable to hydrogen clouds within the galaxy. No definite evidence of absorption at frequencies near the radial velocities of the nebulae was found, and an upper limit on the mass of neutral hydrogen in two nebulae is briefly discussed.

1. Flux Densities

The observations of planetary nebulae reviewed in this paper were made at the Owens Valley Radio Observatory in a program commenced in August 1965 (Thompson *et al.*, 1967; Thompson and Colvin, 1967). Flux densities of 68 planetary nebulae have been measured at 10-cm wavelength (2840 MHz and 2890 MHz) using the two 90-ft diameter antennas and interferometer receiving system. Fifty-two of these nebulae were found to have flux densities greater than the minimum detectable limit of approximately 10^{-27} Wm^{-2} Hz^{-1}. A list of the nebulae and their flux densities is given in Table 1, in which column 2 gives the designations of the nebulae in the new galactic number system of Perek and Kohoutek (1967). The list of nebulae includes almost all of those north of declination $-40°$ for which a value of $F_{H\beta}$, the integrated flux density in Hβ, greater than 10^{-11} erg cm^{-2} sec^{-1} has been measured, and should contain most of the stronger radio emitters observable from Northern latitudes.

The flux-density measurements were made with antenna spacings of 100 feet or

Osterbrock and O'Dell (eds.), Planetary Nebulae, 112–121.

200 feet East–West, and the values of right ascension deduced from the phases of the observed fringe patterns were compared with optical positions measured from the National Geographic Society – Palomar Observatory Sky Survey. Significant discrepancies between the radio and optical positions, which are taken as evidence of probable confusion in the radio measurements, were found for a small number of nebulae as indicated in Table 1. The 3 nebulae with the highest flux densities at 10-cm wavelength are NGC 6857, NGC 7027, and NGC 6302. NGC 6857 coincides with a broad thermal source, NRAO 621, and the Sky Survey plates show an overlapping diffuse nebula. A small planetary nebula, K 3-50, lies about 1′ North of NGC 6857 (Perek and Kohoutek, 1967). A detailed radio survey is required to determine the possible contributions of these other objects to the flux density of NGC 6857 given in Table 1. NGC 7027 is a well-known bright planetary nebula, and NGC 6302 is a nebula which contains unusually high expansion velocities and has been suggested as a possible X-ray source (Johnson, 1966; Minkowski and Johnson, 1967).

For 22 of the stronger nebulae, measurements of the flux density were also made at a wavelength of 21 cm (Thompson *et al.*, 1967). A comparison of the 10-cm and 21-cm results, together with measurements at other wavelengths by various observers, leads to the conclusion that the observed spectra are compatible with thermal emission for almost all nebulae. The exceptional cases usually show evidence of confusion from the position measurements, and there is no clear evidence of non-thermal emission from any of the nebulae investigated. For example NGC 6781 showed a significant deviation between the radio and optical positions at 21 cm, and the presence of a nearby confusing source was later confirmed by Terzian (1967). The broad confusing source near NGC 3242 (Kaftan-Kassim, 1966) was resolved by the interferometer, and had no significant effect on the observations.

2. Hβ Extinction

On the assumptions that the radio emission is thermal and that the nebulae are optically thin at 10-cm wavelength, it is possible to calculate the expected Hβ emission and hence to determine the extinction in Hβ between the nebula and the observer. Column 4 of Table 1 gives the logarithm of the observed value of $F_{H\beta}$ (Collins *et al.*, 1961; O'Dell, 1962, 1963; Vorontsov-Velyaminov *et al.*, 1964) and Column 5 gives the logarithm of the extinction, $\Delta \log F_{H\beta}$. The extinction was obtained using $F(\nu)/F_{H\beta}=3{\cdot}38\times10^{-14}$ sec, $F(\nu)$ being the radio-flux density as given in Column 3 of Table 1. This value of $F(\nu)/F_{H\beta}$ was calculated for the following conditions; electron temperature, T_e, $=10000\,^\circ$K, the ratio of the total ion density to the proton density $=1{\cdot}15$, the mean ionic charge $=1{\cdot}0$ and $\nu=3$ GHz. $F(\nu)/F_{H\beta}$ is proportional to $T_e^{0{\cdot}56}$, and a change in T_e by a factor of 1·5 would produce a change of only 0·10 in $\Delta \log F_{H\beta}$. Delmer *et al.* (1967) point out that for some nebulae the contribution of He^{++} ions to the radio emission may be significant. For a nebula in which the ratio

Table 1

Values of flux density at 10-cm wavelength and calculated Hβ extinction

Nebula	Perek and Kohoutek Designation	Flux Density at 10 cm Wavelength (10^{-26} $Wm^{-2}Hz^{-1}$)	Log $F_{H\beta}$ Observed (erg cm^{-2} sec^{-1})	Δ log $F_{H\beta}$	Nebula	Perek and Kohoutek Designation	Flux Density at 10 cm Wavelength (10^{-26} $Wm^{-2}Hz^{-1}$)	Log $F_{H\beta}$ Observed (erg cm^{-2} sec^{-1})	Δ log $F_{H\beta}$
NGC 40[a]	120 + 9°1	0·34 ± 0·05	−10·64		NGC 6309	9 + 14°1	0·15 ± 0·06	−11·29	0·94
NGC 246	118 − 74°1	0·15 ± 0·02			NGC 6369	2 + 5°1	1·56 ± 0·17	−11·34	2·01
NGC 650-1	130 − 10°1	0·16 ± 0·02	−10·67	0·35	NGC 6445	8 + 3°1	0·27 ± 0·06	−11·20	1·10
IC 2003	161 − 14°1	< 0·14	−11·18		H 1-40	359 − 2°3	< 0·16		
NGC 1501	144 + 6°1	0·19 ± 0·04	−11·26	1·01	NGC 6543	96 + 29°1	0·84 ± 0·11	− 9·60	0·0
NGC 1514	165 − 15°1	0·29 ± 0·04			NGC 6537[a]	10 + 0°1	0·54 ± 0·06	−11·78	1.99
NGC 1535	206 − 40°1	< 0·10	−10·36		NGC 6563	358 − 7°1	0·13 ± 0·06	−10·96	0·54
J 320	190 − 17°1	< 0·09	−11·37		NGC 6572	34 + 11°1	0·92 ± 0·1	− 9·74	0·33
IC 2120	169 − 0°1	0·17 ± 0·05			NGC 6629	9 − 5°1	0·32 ± 0·06	−10·94	0·92
IC 418	215 − 24°1	1·4 ± 0·1	− 9·53	0·21	IC 4732	10 − 6°1	< 0·10	−11·54	
NGC 2022	196 − 10°1	< 0·13	−11·15		M 1-64[a]	64 + 15°1	0·16 ± 0·06		
IC 2149	166 + 10°1	0·14 ± 0·05	−10·50	0·12	NGC 6720	63 + 13°1	0·42 ± 0·05	−10·06	0·16
SH-2-267	196 − 1°1	< 0·3	−10·50		NGC 6741	33 − 2°1	0·26 ± 0·06	−11·49	1·38
SH-2-266[a]	195 − 0°1	0·18 ± 0·05			K 1-17	51 + 6°1	< 0·14		
IC 2165	221 − 12°1	0·22 ± 0·05	−10·99	0·80	NGC 6781[a]	41 − 2°1	0·37 ± 0·05	−11·19	1·23
NGC 2346	215 + 3°1	< 0·11			A 48	53 + 3°1	0·13 ± 0·05		
NGC 2371-2	189 + 19°1	< 0·13	−10·96		NGC 6790	37 − 6°1	< 0·14	−10·75	
NGC 2392	197 + 17°1	0·23 ± 0·03	−10·39	0·23	BD 30°3639	64 + 5°1	0·44 ± 0·11	−11·50	1·62

Table 1 (continued)

NGC 2438[a]	231 + 4°2	0·12 ± 0·04	−11·02		NGC 6818	25 − 17°1	0·34 ± 0·14	−10·13	0·14
NGC 2440	234 + 2°1	0·31 ± 0·04	−10·56	0·53	NGC 6826	83 + 12°1	0·42 ± 0·05	− 9·92	0·02
M 3-4	241 + 2°1	0·13 ± 0·06			NGC 6853	60 − 3°1	1·3 ± 0·2	− 9·44	0·03
NGC 3132	272 + 12°1	0·22 ± 0·03			NGC 6857[b]	70 + 1°2	6·4 ± 0·5		
NGC 3242	261 + 32°1	0·72 ± 0·1	− 9·81	0·14	NGC 6891	54 − 12°1	0·20 ± 0·08	−10·60	0·37
NGC 3587	148 + 57°1	0·10 ± 0·03	−10·33	(−0·19)	IC 4997	58 − 10°1	< 0·09	−10·49	
NGC 4361	294 + 43°1	0·22 ± 0·03	−10·48	0·30	NGC 6905[a]	61 − 9°1	0·11 ± 0·06	−10·90	
IC 3568	123 + 34°1	< 0·12	−10·82		NGC 7008	93 + 5°2	0·18 ± 0·05	−10·86	0·59
NGC 5882[a]	327 + 10°1	0·34 ± 0·05			NGC 7009	37 − 34°1	0·62 ± 0·09	− 9·78	0·05
NGC 6058	64 + 48°1	< 0·14	−11·70		NGC 7026	89 + 0°1	0·27 ± 0·06	−10·90	0·80
IC 4593	25 + 40°1	0·12 ± 0·04	−10·55	0·10	NGC 7027	84 − 3°1	3·53 ± 0·3	−10·12	1·43
NGC 6072	342 + 10°1	0·11 ± 0·04	−11·37	0·88	NGC 7048	88 − 1°1	< 0·1	−11·39	
NGC 6153	341 + 5°1	0·51 ± 0·04			NGC 7139	104 + 7°1	< 0·16	−11·78	
NGC 6210	43 + 37°1	0·32 ± 0·04	−10·06	0·04	IC 5217	100 − 5°1	0·15 ± 0·08	−11·18	0·83
IC 4634	0 + 12°1	0·15 ± 0·05	−10·07	0·62	NGC 7354	107 + 2°1	0·66 ± 0·06	−11·55	1·85
NGC 6302	349 + 1°1	2·3 ± 0·1			NGC 7662	106 − 17°1	0·66 ± 0·05	− 9·98	0·28

[a] Radio measurements probably confused.
[b] See text.

of the He^{++} ion density to the proton density is 0·05, which is near the maximum value observed, the calculated value of $F(\nu)/F_{H\beta}$ should be increased by a factor of 1·13, and $\Delta \log F_{H\beta}$ in Table 1 decreased by 0·05. The r.m.s. error in $\log F_{H\beta}$ is 0·07 (O'Dell, 1962) and the r.m.s. errors in $F(\nu)$ are given in Table 1. The r.m.s. error in $\Delta \log F_{H\beta}$ resulting from the uncertainties mentioned above should therefore not be more than 0·15 if the accuracy of $F(\nu)$ is better than 20%. For IC 418, NGC 6572 and NGC 7027 the values of $F(\nu)$ were corrected for self-absorption of the radio emission (Thompson *et al.*, 1967), but for any other nebulae with high optical depths at 3 GHz the Hβ extinction is probably underestimated. The values of $\Delta \log F_{H\beta}$ in Table 1 range up to 2·0 for NGC 6369 and NGC 6537, corresponding to transmission of only 1% of the radiation.

3. Angular Widths

An attempt was made to compare radio and optical angular widths of 27 of the nebulae using observations of the fringe visibility at 10 cm made with spacings of 2351 and 4701 wavelengths East-West (Thompson *et al.*, 1967). High accuracy in the radio dimensions could not be achieved in most cases, since the change in the effective flux density from the shortest to the longest spacings was only a few times greater than the limiting accuracy of measurements. Further, the optical data on angular widths are not precisely defined and, except for those nebulae for which isophotes in one of the emission lines of hydrogen have been published, depend on the responses and exposure times of photographic plates. Taking into account these limitations, the agreement between the radio and optical widths was found to be generally satisfactory.

4. Optical Depths and Electron Temperatures of Two Optically Thick Nebulae

A small number of planetary nebulae show clearly the effect of self-absorption in their spectra, and in these cases an estimate of the optical depth and electron temperature can be made by fitting to the observed flux densities a calculated spectrum based on a model of the nebula which represents the variation of optical depth with solid angle. Such an analysis has been performed (Thompson, 1967) for IC 418 and NGC 6572, two nebulae for which profiles or isophotes of brightness in Hβ and other emission lines of hydrogen have been published (Berman, 1930; Wilson and Aller, 1951; Aller, 1956). The Hβ brightness at any point on the surface of a nebula is proportional to the integral along the line of sight of $N_e N_i T_e^{-0\cdot91}$ (Peimbert, 1967) and the radio frequency optical depth is proportional to $N_e N_i T_e^{-1\cdot35}$, where N_e and N_i are the electron and proton densities. Thus, if the temperature variations within a nebula are not too large, the isophotes of Hβ brightness should accurately represent the variation of relative optical depth with solid angle. On this basis a computer program was used to calculate the expected spectrum and the r.m.s. deviation from it of the measured values of flux density, for a series of values of T_e and of τ_1, the optical depth at 1GHz along a line of sight through the centre of the nebula. The measured flux densities were taken from work by Lynds (1961), Menon and Terzian (1965), Slee and Orchiston (1965), Hughes (1967), Thompson *et al.* (1967), and Kaftan-Kassim (1967). The best fit to the observed data, as indicated by the minimum value of the r.m.s. deviation, was obtained with $T_e = 7100\,°\text{K}$ and $\tau_1 = 3{\cdot}5$ for IC 418 and $T_e = 6500\,°\text{K}$ and $\tau_1 = 19{\cdot}2$ for NGC 6572. These temperatures are much lower than those derived from forbidden-line ratios, which are approximately 19000 °K for IC 418 and 13000 °K for NGC 6572 (Seaton, 1954; Liller and Aller, 1954). The temperatures derived from the radio data depend almost entirely upon the flux-density values at frequencies of 1420 MHz and lower, where the observations show appreciable self-absorption. Taking into account the errors assigned to the measured flux densities, the uncertainties in the resulting values of electron temperatures are large, and in the case of IC 418 a temperature of 10000 °K or a little higher cannot be excluded. The optical data used were corrected for the increase in angular dimensions introduced by seeing fluctuations, which would have the effect of decreasing T_e, but it is difficult to be sure that all such systematic errors were eliminated. As a general conclusion it appears, however, that this analysis indicates values of T_e which are lower than those derived from the forbidden-line ratios by a factor of at least 1·5. The emission measures obtained from the best fit values of τ_1 and T_e given above are $6{\cdot}7 \times 10^6$ cm^{-6} parsec for IC 418 and $3{\cdot}3 \times 10^7$cm^{-6} parsec for NGC 6572, and correspond to the line of sight through the centre of each nebula.

A most probable explanation of at least part of the discrepancy between the radio and forbidden-line values of T_e is that the temperature is not constant but varies significantly within each nebula. Peimbert (1967) has shown that under these circumstances the value of electron temperature determined by any particular method will be weighted towards the cooler or hotter parts of the nebula, depending upon the way in which the strengths of the measured parameters are related to T_e. A factor of 1·5 between the radio and forbidden-line values of T_e would be explained if the r.m.s. deviation of T_e, weighted in proportion to $N_e N_i$, is 0·31 of the mean value. In the case of IC 418 the isophotes of the λ5007 line of [O III], on which the forbidden-line temperature values in part depend, show a much greater concentration towards the central part of the nebula than do the Hβ isophotes (Aller, 1956). The forbidden-line temperature should thus be more strongly weighted towards the central region of the nebula than the radio value, and on this basis the present results suggest a decrease in temperature from the centre to the outer parts of IC 418.

5. The 21-cm Hydrogen Line in Absorption

At the frequency of the 21-cm hydrogen line, absorption features are to be expected in the spectra of the planetary nebulae. These would result from the presence of neutral hydrogen between the nebula and the observer, which could be either part of the galactic population of hydrogen clouds, or the neutral outer part of a planetary nebula in which the ionization boundary does not enclose the whole mass of gas. The following is a description of preliminary observations made during May 1967 by Thompson and Colvin in an investigation to detect such absorption features, again using the interferometer at the Owens Valley Radio Observatory.

For observations of hydrogen-line absorption the interferometer offers a great advantage over a single antenna, since galactic hydrogen clouds which lie within the antenna beam are highly resolved, and the response of the interferometer to the background emission can therefore be ignored. A receiver bandwidth comparable to the expected Doppler width of the absorption features is required, which limits the maximum usable bandwidth to about 100 kHz. With this bandwidth the sensitivity of the interferometer is reduced by a factor of 10 relative to that obtained with the bandwidth of 10 MHz used for continuum observations. The present investigation therefore included only the 6 nebulae listed in Table 2, which were chosen for their high

Table 2

Nebulae observed for hydrogen-line absorption

Nebula	Flux Density at 21-cm Wavelength (10^{-26} Wm^{-2} Hz^{-1})	Galactic Coordinates l^{II}	b^{II}	Total Observing Time (Hours)
IC 418	1·14 ± 0·06	215°.2	−24°.3	15·1
NGC 6302	2·02 ± 0·16	349·5	1·0	4·75
NGC 6369	1·63 ± 0·1	2·4	5·8	11·25
NGC 6543	1·13 ± 0·16	96·5	29·9	12·25
NGC 6857	5·0 ± 0·4	70·3	1·6	2·5
NGC 7027	1·30 ± 0·07	84·9	−3·5	12·0

flux densities and also for their small diameters, since nebulae in the early stages of their expansion are most likely to be incompletely ionized.

The observations were made using an interferometer system with 23 channels, each of 100-kHz bandwidth. The antenna spacing used was 400 feet East–West, and the total observation time for each nebula is given in Table 2. The results of these observations are shown in Figure 1, in which the relative amplitudes are plotted against the velocity with respect to the local standard of rest corresponding to the centre frequency of each channel. The radial velocities of the nebulae, as measured optically by Campbell and Moore (1918) or Wilson (1950), are indicated by V_R. The features of definite significance are the absorption dip near zero velocity for NGC 6369 and the broad dip near −40 km/sec for NGC 6857. The dip for NGC 6369 is well removed from V_R and is therefore attributable to galactic hydrogen clouds. The absorption for NGC 6857 is more difficult to interpret since V_R has not been measured and the effect of the

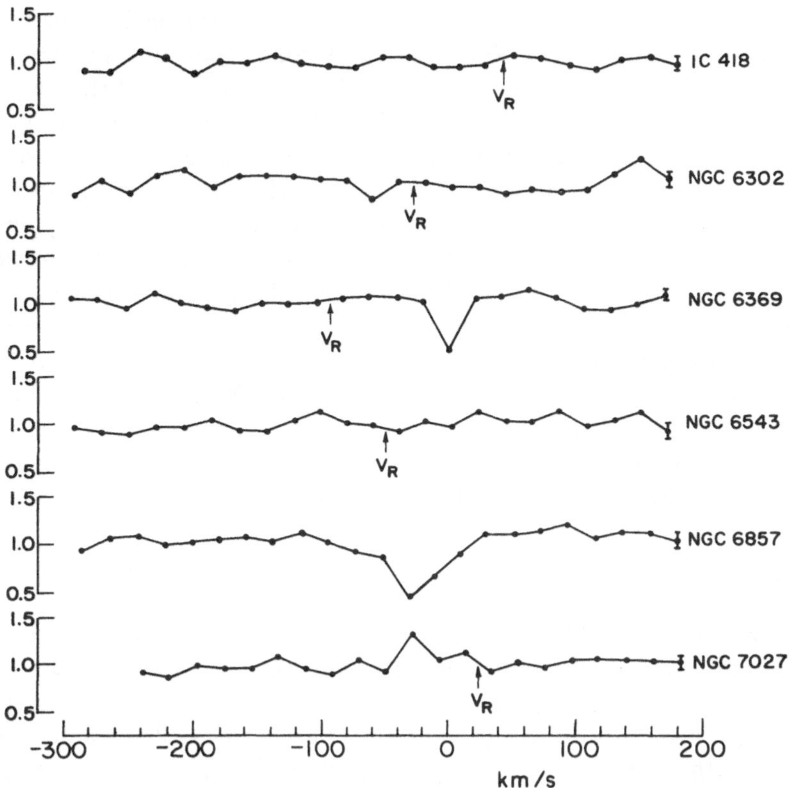

FIG. 1. *Preliminary results of a program to detect the 21-cm line in absorption. The relative signal strengths in 23 channels of width 100 kHz are plotted against velocity with respect to the local standard of rest. The radial velocities of the nebulae, determined optically, are indicated by V_R. The rms deviation of a single point is indicated by the error bar at the right-hand end of each curve.*

diffuse nebula mentioned in Section 1 is not known. It is interesting to note that NGC 6369 is also one of the nebulae for which the Hβ absorption is very large (Table 1). Absorption dips resulting from galactic hydrogen may provide information on the distances of the nebulae when examined with a narrower receiver bandwidth, so that details of the profile can be determined. Positive peaks such as those at -30 km/sec for NGC 7027 are probably caused by the residual response to galactic emission features within the antenna beam.

The absorption resulting from a neutral-hydrogen envelope around a planetary nebula would be expected to occur within a range of velocities from V_R to $V_R - V_E$, where V_E is the expansion velocity measured relative to the centre of the nebula. An expansion velocity of 21·35 km/sec has been measured for NGC 7027 and a velocity near zero for IC 418 (Wilson, 1950). The record of NGC 6302 shows a small dip near

-60 km/sec, but this cannot be regarded as significant on the basis of the present data alone. For IC 418, NGC 6369, NGC 6543 and NGC 7027 it is possible to put an upper limit of 0·1 on the fraction of power absorbed in a 100-kHz channel in the vicinity of V_R. If we assume that the width of an absorption feature is not greater than 200 kHz, which corresponds to a Doppler spread of 42 km/sec, then at least half of the absorbed power must fall within one receiving channel. Using a simple spherical model it is found that $\int \tau(v)\, dv < 2{\cdot}8$ km/sec, where $\tau(v)$ is the optical depth in the neutral-hydrogen envelope corresponding to the radial velocity v. For IC 418 and NGC 7027 the radius is approximately $1{\cdot}6 \times 10^{17}$ cm (Aller, 1956), from which one finds that the mass of neutral gas is not greater than $1{\cdot}4 \times 10^{-3} T$ solar masses, where T is the temperature of the neutral hydrogen. For $T = 100\,°K$ the upper limit on the neutral hydrogen is therefore 0·14 solar masses and for $T = 1000\,°K$, 1·4 solar masses. Since the total mass of a planetary nebula is about 0·14 solar masses (O'Dell, 1962), the upper limit is of significance only if we take a low value for the gas temperature. It is planned to make further observations to attempt to improve this limit.

Acknowledgments

The author wishes to acknowledge the cooperation of Dr. R.S. Colvin with whom the observations were carried out, and to thank Mr. G.J. Stanley, Director of the Owens Valley Radio Observatory, and Professor R.N. Bracewell for their interest and support. Research in radio astronomy at the Owens Valley Radio Observatory is supported by the U.S. Office of Naval Research under contract NONR 220(19) and at Stanford University by the U.S. Air Force Office of Scientific Research under contract AF 49(638)-1375.

References

Aller, L.H. (1956) *Gaseous Nebulae*, John Wiley & Sons, New York.
Berman, L. (1930) *Lick Obs. Bull.*, **15**, 86.
Campbell, W.W., Moore, J.H. (1918) *Publ. Lick Obs.*, **13**, 75.
Collins, G.W. II, Daub, C.T., O'Dell, C.R. (1961) *Astrophys. J.*, **133**, 471.
Delmer, T.N., Gould, R.J., Ramsay, W. (1967) *Astrophys. J.*, **149**, 495.
Hughes, M.P. (1967) *Astrophys. J.*, **149**, 377.
Johnson, H.M. (1966) *Astr. J.*, **71**, 166.
Kaftan-Kassim, M.A. (1966) *Astrophys. J.*, **145**, 658.
Kaftan-Kassim, M.A. (1967) private communication.
Liller, W., Aller, L.H. (1954) *Astrophys. J.*, **120**, 48.
Lynds, C.R. (1961) *Pub. nat. Radio Astr. Obs.*, **1**, 85.
Menon, T.K., Terzian, Y. (1965) *Astrophys. J.*, **141**, 745.
Minkowski, R., Johnson, H.M. (1967) *Astrophys. J.*, **148**, 659.
O'Dell, C.R. (1962) *Astrophys. J.*, **135**, 371.
O'Dell, C.R. (1963) *Astrophys. J.*, **138**, 293.
Peimbert, M. (1967) *Astrophys. J.*, **150**, 825.
Perek, L., Kohoutek, L. (1967) *Catalogue of Galactic Planetary Nebulae*, Academia Publ. House, Prague.

Seaton, M.J. (1954) *Mon. Not. R. astr. Soc.*, **114**, 154.
Slee, O.B., Orchiston, D.W. (1965) *Austr. J. Phys.*, **18**, 187.
Terzian, Y. (1967) *Astr. J.*, **72**, 443.
Thompson, A.R. (1967) *Astrophys. Lett.*, **1**, 25.
Thompson, A.R., Colvin, R.S. (1967) *Astrophys. J.*, **150**, 345.
Thompson, A.R., Colvin, R.S., Stanley, G.J. (1967) *Astrophys. J.*, **148**, 429.
Vorontsov-Velyaminov, B.A., Kostjakova, E.B., Dokuchaeva, O.D., Arhipova, V.P. (1964) *Soviet Astr.*, **8**, 196.
Wilson, O.C. (1950) *Astrophys. J.*, **111**, 279.
Wilson, O.C., Aller, L.H. (1951) *Astrophys. J.*, **114**, 421.

DISCUSSION

Terzian: Using your method of determining electron temperatures I find that H II regions also have lower indicated average temperatures compared with the temperatures determined by optical methods.

Aller: Walker and Kron are obtaining improved isophotes of a number of planetaries using an electronic camera. A long focal length and exquisite seeing are required. Berman's pioneer isophotometry carried out 40 years ago, was affected by the small scale of the only telescope available for his program at that time.

GENERAL DISCUSSION – SECOND SESSION

Aller: I would like to give a brief description of our spectrophotometric program on line intensities in planetaries. It is a joint effort by several investigators, including Czyzak, Walker, and Kaler. We measure the stronger lines with the photoelectric scanner on the Mt. Wilson 1·5-m reflector, comparing nebular scans with suitable standard stars. For the weaker lines, and the lines that fall close together in wavelength we rely chiefly on the Lick Observatory 3-m reflector. Prime-focus nebular spectrograph observations give relative line intensities, which can be converted to fluxes with the aid of Mt. Wilson data. Coudé spectra obtained with the image converter cover the region $\lambda\lambda$ 4700–5900. In each instance we calibrate with wide-slit spectra of suitable comparison stars. For photographic observations, a density-log$I$ relation is obtained by laboratory calibration. For the image converter, we use a combination of plate and developer that gives a linear relation between density and intensity. A few observations of bright planetaries have been obtained with the Mt. Wilson 2·5-m telescope. Kaler has secured a number of observations with the Kitt Peak 2-m telescope and coudé spectrograph.

Flower: What are the percentage errors in the line intensities obtained from high-resolution spectra? For example, what would be the probable error of a line intensity given as 10 on the scale $H\beta = 100$?

Minkowski: About 5% if the photometry is good.

Pottasch: What are the probable errors of the published $H\alpha/H\beta$ line-intensity ratios?

O'Dell: About 10%.

Terzian: I suggest that authors publishing Balmer-line fluxes should give the probable errors of these measurements. In the past few errors have been reported.

Gurzadian: It is necessary to make spectrophotometric measurements of the outer envelopes of planetary nebulae whenever possible.

Osterbrock: The similarity of the planetary nebula NGC 7027 and the Seyfert galaxy NGC 1068 in infrared emission is interesting, because these two objects also have very similar emission-line spectra.

Westerlund: Webster's survey of Southern planetary nebulae led to the detection of many peculiar objects. One of them, H 141, dominated by 2 bright knots, has a very flat Balmer decrement, $\log F(H\beta)/F(H\gamma) = 0{\cdot}28$, so the lines may be self-absorbed, presumably in the knots. It has very strong He II λ 4686 and [O II] λ 3727. No central star is seen on the direct plates. The object has a strong continuum with the color of a G9 dwarf and a strong ultraviolet excess. Interstellar reddening cannot be the explanation as the decrement is too flat. The nebula falls in a color-color diagram in the region of the white dwarfs, the nuclei of Seyfert galaxies, etc. It may be similar to NGC 7027.

Menon: The 109α line from NGC 7027 shows that it is an ionized-hydrogen region with density and temperature both higher than in a normal H II region. There is no similarity between NGC 7027 and a Seyfert galaxy as far as the radio spectra are concerned.

O'Dell: Seaton has shown that the observations by Bowen, Aller, and Minkowski of the ultraviolet continuum longward of the Balmer discontinuity in NGC 7027 do not fit thermal theory. This argues that an additional anomaly in the continuum exists.

Reeves: We should take into consideration the possibility that high-energy protons are irradiated from the central star into the nebula, and calculate if the infrared radiation and other phenomena could not be explained in this way.

Seaton: We have to consider whether the observed ionization in the nebulae can be understood in terms of purely radiative processes. If it turns out that this cannot be done, then we may have to suppose that fast particles also contribute to the ionization.

Menon: I would like to report on some measurements of the structure of 5 planetary nebulae made by Colomb at Green Bank using the interferometer at 10 cm and at spacings of 1·2 km and 2·7 km. The data consist of visibilities at various hour angles and at both spacings. On the basis of this data Colomb has constructed a model for each nebula, giving the best agreement with the observations. The parameters of the models are as follows:

(1) IC 418. A ring of inner radius 2″.1, outer radius 6″.0.

(2) NGC 6543. A uniform disk of radius 8″.5 × 10″.4 extended in position angle 15°.

(3) NGC 6572. A uniform disk of radius 3″.3 × 4″.7 extended in P.A. 10°.

(4) NGC 7027. A gaussian of 5″.4 × 8″.5 extended in P.A. 40°.

(5) NGC 7662. A uniform disk of radius 10″.4 × 12″.2 in P.A. 35°.

Gurzadian: If the existence of planetary nebulae with non-thermal radio emission is real, then the energy spectra of the relativistic electrons in some of these objects must be surprisingly steep. For example, in the cases of the nebulae NGC 6153 and NGC 6445, the observed frequency spectra ($F_\nu \propto \nu^{-\gamma}$) have observed exponents γ_{obs} = 4·4 and 3·0 respectively. These exponents must however be corrected to remove the effects of the underlying thermal emission, and when this is done the exponents of the pure non-thermal spectra are $\gamma = 8$ and $\gamma = 5$ respectively. From these values we must conclude that the energy spectra of the relativistic electrons in non-thermal planetaries differ from the spectra of the electrons in other well-known non-thermal sources. In fact the law $N(E) \propto E^{-\beta}$ cannot be used for the planetary nebulae. I think that a gaussian spectrum of electron energies is more probable.

There is some basis for thinking that the non-thermal and thermal emissions are generated in different parts of the same nebula: non-thermal in the centre, thermal, mostly in the other ring-like part of the nebula. Hence, it is very desirable to make special measurements with large angular resolution in order to obtain the spectra of radio emission in the central part of the nebulae as well as in the outer part.

It can be shown theoretically that the electron temperature in the central regions of a nebula must be larger at least by a factor of 1·5–2 relative to the electron temperature in the outer ring. However, it would be interesting to have observational evidence on this question from detailed measurements of the radio brightness across the nebula.

In many cases the outer envelopes of the two-envelope planetary nebulae are too faint to observe in the optical region. Perhaps these outer envelopes may be detected in some cases more easily by radio observations (some trace of the outer envelope of the nebula NGC 6543 may be seen in the results of Elsmore).

Thompson: I would like to emphasize that several planetaries, originally reported to have non-thermal spectra, have been shown to have been subject to confusion by nearby non-thermal sources. There is no evidence at the present time that any planetary has a non-thermal spectrum.

Flower: There may be considerable electron temperature gradients in planetary nebulae. When quoting a value for the electron temperature in a nebula it is important to state the part of the nebula which is the source of the radiation used to determine the electron temperature. I think this effect may be at least partly responsible for the discrepancies in electron temperatures that we have heard this morning.

Thompson: One can reasonably well understand the difference between the electron temperatures which I have deduced from the radio data and the forbidden-line values, in terms of temperature variations. Temperatures deduced from radio observations of the hydrogen recombination lines should, according to Peimbert, be weighted towards the cooler regions of nebulae. It therefore seems a little surprising that the temperature obtained by Mezger *et al.* from the 109α line in NGC 7027 is as high as the forbidden-line value.

Menon: I wish to emphasize that the determination of temperatures from radio data is very much handicapped by the lack of good-quality high-resolution radio and optical isophotes. The presence of a large number of small-scale irregularities could introduce serious errors in the temperatures determined by Thompson's procedure.

Thompson: I agree that the presence of filaments would produce fine structure in the distribution of optical depth over the nebulae, and that omission of such detail from the isophotes could decrease the values of electron temperature deduced from the radio data. In IC 418 such filaments must be small since they are not detected optically, and the effect of many small filaments would tend to average out over the volume of the nebula. Improved data may well increase by one or two thousand degrees the values that I have given, but I think that the radio temperatures will still be significantly lower than the forbidden-line temperatures.

Osterbrock: Are there any negative results on attempted observations of radio-frequency H I lines in other planetaries? Though NGC 7027 is the brightest in the continuum, the theories show there is a complicated dependence of the strength of the lines on density, temperature, and radiation field, so NGC 7027 may not necessarily be the brightest planetary in the 109α line.

Kaftan-Kassim: NGC 7027 has a flux of over 5 f.u. at 6 cm, which makes it the only suitable planetary in the Northern Hemisphere, since all the others have fluxes of 1 f.u. or lower.

Session III

PHYSICAL PROCESSES

REVIEW OF ATOMIC COLLISION PROCESSES

M.J. SEATON
(Dept. of Physics, University College London, England)

1. Recombination Spectra

1.1. GENERAL THEORY

The following processes have to be considered.

(a) Radiative capture,

$$X^{+} + e \rightarrow X_i + h\nu .$$

The number of captures on the level i of the atom X, per unit volume per unit time, is $N_e N(X^+)\alpha_i(T_e)$, where the recombination coefficient $\alpha_i\,(\mathrm{cm}^3\ \mathrm{sec}^{-1})$ may be expressed in terms of the cross-section for photo-ionization from level i.

(b) Cascade,

$$X_{i'} \rightarrow X_i + h\nu_{i'i} .$$

The coefficients $\alpha_i^{(c)}$ and $\alpha_{i'i}$ are defined to be such that $N_e N(X^+)\ \alpha_i^{(c)}$ is the number of atoms entering i due to capture and cascade, and that $N_e N(X^+)\ \alpha_{i'i}$ is the number of photons emitted in the $i' \rightarrow i$ line.

(c) Collisional transitions between excited states, including: collisional redistribution of angular momentum; collisional redistribution of energy between highly excited states; and collision processes which depopulate metastable states.

(d) Effects of radiative transfer. The following limiting cases have been considered:

Case A: Nebulae optically thin in all spectrum lines.

Case B: Nebulae optically thick for transitions to the ground state.

Case C: As Case B, but with absorption of stellar quanta in transitions from the ground state, assuming the central star to radiate as a black body.

1.2. HYDROGENIC SYSTEMS

Transition probabilities (Baker and Menzel, 1938; Green *et al.*, 1957) and recombination coefficients (Burgess, 1964) are known exactly for hydrogenic systems.

1.2.1. *Collisional Redistribution of Angular Momentum*

In all earlier calculations (Baker and Menzel, 1938) it was assumed implicitly that angular momentum is completely redistributed before radiation is emitted. With this

Osterbrock and O'Dell (eds.), Planetary Nebulae, 129–137. © *I.A.U.*

assumption, the most accurate calculations are those of Seaton (1959). Calculations assuming no redistribution of angular momentum have been made by Searle (1958) for $n \leqslant 10$, by Burgess (1958) for $n \leqslant 12$, and by Pengelly (1964*a*) for an infinite number of levels. These calculations give stronger intensities for transitions of the type $n \to n-1$. Pengelly and Seaton (1964) have shown that collisional redistribution of angular momentum, due to processes such as

$$X_{nl} + \mathrm{H}^+ \to X_{nl\pm 1} + \mathrm{H}^+ ,$$

is important for the higher states. With $N_e = 10^4\ \mathrm{cm}^{-3}$ collisional redistribution occurs before emission of radiation for $n \gtrsim 15$ in H and for $n \gtrsim 22$ in He^+. Table 1 gives effective recombination coefficients for Hβ and for He II λ 4686 ($n=4 \to n=3$), assuming

Table 1

Effective recombination coefficients for hydrogenic ions

	$10^{14}\alpha(\mathrm{H}\beta)$		$10^{14}\alpha(\mathrm{He\,II}\ \lambda\ 4686)$	
$10^{-4}\ T_e =$	1	2	1	2
(a) High density	3·00	1·60	20·8	11·8
(b) Low density	3·07	1·64	37·8	18·0
(c) $N_e = 10^4 \mathrm{cm}^{-3}$	3·02	1·60	29·4	14·8

(a) complete redistribution of angular momentum (limit of high density),

(b) no redistribution of angular momentum (limit of low density),

(c) taking explicit account of redistribution of angular momentum for $N_e = 10^4\ \mathrm{cm}^{-3}$ (Pengelly, 1964*b*).

It is seen that the Hβ line is not sensitive to the assumptions made, but that the He II line is much more sensitive.

The calculations of Burgess (1958) allow for complete redistribution of angular momentum for $n \leqslant 12$, and no redistribution for $n > 12$. They are the best results to use for H I at densities of order $10^4\ \mathrm{cm}^{-3}$.

1.2.2. *Collisional Redistribution of Energy*

Allowance for collisional redistribution of energy must be made in calculations of the populations of highly excited states. Approximate calculations (Seaton, 1964) for $N_e = 10^4\ \mathrm{cm}^{-3}$ indicate that such processes are important only for $n \gtrsim 50$ and may therefore be neglected in considering observed intensities of resolved hydrogen lines in the visible spectrum. A check on the theory should be possible from observations of the intensity due to blended lines within a few Ångstroms of the Balmer limit. Populations of highly excited states are also of interest in connection with the lines observed at radio wavelengths (Goldberg, 1966). Further calculations have been made by Hayler (1967) and by McCarroll and Hoang Binh-Dy (1968).

1.3. NON-HYDROGENIC SYSTEMS

In calculating recombination spectra one is concerned with atomic systems having a single valence electron in excited states. Transition probabilities can be calculated using the method of Bates and Damgaard (1949) and photo-ionization cross-sections can be obtained using the method of Burgess and Seaton (1960*a*). Improved tabulations of the quantities required for the photo-ionization calculations are given by Peach (1967).

1.3.1. *Helium*

Complications arise from the metastability of 2 ^{1}S and 2 ^{3}S.

(a) *Determination of helium abundances:* Effects arising from the metastability of 2^1S and 2^3S have little influence on the lines due to transitions $nd \rightarrow 2p$, and these lines should therefore be used for determinations of helium abundances. Table 2 gives

Table 2

Effective recombination coefficients for He I lines

10^{-4} T$_e$ =	1	2
$10^{14}\alpha$(He I λ5876)	5·21	2·30
$10^{14}\alpha$(He I λ4471)	1·35	0·644

recombination coefficients for 3 ^{3}D→2 ^{3}P λ 5876 and 4 ^{3}D→2 ^{3}P λ 4471, calculated by Pengelly (1964*b*) using accurate transition probabilities and photo-ionization cross-sections, and neglecting collisional redistribution of angular momentum. The calculated intensity ratio $I(5876)/I(4471)$ is 2·9 at 1×10^4 °K and 2·7 at 2×10^4 °K, compared with a mean value of 2·5 for 7 nebulae observed by O'Dell (1963), corrected for reddening. Neglect of collisional redistribution will lead to α(3 ^{3}D→2 ^{3}P) being slightly overestimated but will have little effect on α(4 ^{3}D→2 ^{3}P).

(b) *Self-absorption:* Effects due to absorption of radiation in helium lines ending on 2 ^{3}S have been considered by Pottasch (1961), Osterbrock (1964), and Capriotti (1967). When the optical depth is suitably chosen there is good agreement with observations for the intensities of 3 ^{3}P→2 ^{3}S λ 3889 and 3 ^{3}S→2 ^{3}P λ 7065, relative to λ 5876, but self-absorption effects do not explain the very large intensities observed for 2 ^{3}P→2 ^{3}S λ 10830.

(c) *Collisional effects:* In order to calculate the population of 2 ^{3}S, and to explain the intensity of λ 10830, the following processes must be considered: capture and cascade; collisional transitions between all $n=2$ states; depopulation of 2 ^{3}S by two-photon emission and by photo-ionization. Accurate cross-sections for transitions between the $n=2$ states, produced by electron impact, have been calculated by Burke *et al.* (1967*a*) and these have been used to obtain the rate coefficients q (cm^3 sec^{-1}) given in Table 3. In this table we include values of the recombination coefficients

Table 3

Reaction rates required for the calculation of He I λ 10830 intensities

	$10^{-4} T_e =$ 1·0	1·5	2·0
$10^8 q(2^3S \rightarrow 2^1S)$	3·1	3·5	3·4
$10^8 q(2^3S \rightarrow 2^1P)$	1·4	2·4	3·1
$10^8 q(2^3S \rightarrow 2^3P)$	20	38	53
$10^8 q(2^1S \rightarrow 2^1P)$	79	122	153
$10^8 q(2^1S \rightarrow 2^3P)$	9·9	10·1	10·1
$10^{14} \alpha^{(c)}(2^3S)$	21·0	15·0	12·0
$10^{14} \alpha^{(c)}(2^3P)$	14·4	10·0	7·8
$\frac{I(10830)}{I(5876)}$ $N_e \rightarrow 0$	1·50	1·66	1·84
$\frac{I(10830)}{I(5876)}$ $N_e \rightarrow \infty$	11·1	17·6	24·7

$\alpha^{(c)}(2\ ^3S)$ and $\alpha^{(c)}(2\ ^3P)$, and values of the ratio $I(10830)/I(5876)$ calculated in the limit of low density (collisional excitation neglected) and the limit of high density (two-photon emission and photo-ionization neglected).

1.3.2. *Ions of Oxygen and Carbon*

Calculations of recombination spectra have been made by Burgess and Seaton (1960*b*) for O III, O IV and O V and by Pengelly (1964*b*) for C II, C III and C IV.

(a) *Oxygen ions:* In Table 4 we give relative abundances of ions of hydrogen, helium

Table 4

Ion abundances in NGC 7027

Ion	Relative abundance	
H^+	100000	
He^+	10900	
He^{+2}	5100	
O^0	0·4	from forbidden lines
O^+	1·0	from forbidden lines
O^{+2}	15	from forbidden lines
O^{+3}	64	from permitted lines
O^{+4}	28	from permitted lines
O^{+5}	8	from permitted lines

and oxygen in NGC 7027. The abundances of O^0, O^+ and O^{+2} are obtained from forbidden lines, and the abundances of O^{+3}, O^{+4} and O^{+5} are obtained on interpreting the observed permitted lines of O III, O IV and O V using recombination theory. Since the ionization potential of O^{+2} (54·89 eV) is close to the ionization potential of He^+ (54·40 eV) we would expect the ion abundances to be such that

$$\frac{N(He^+)}{N(He^{+2})} = \frac{N(O^+) + N(O^{+2})}{N(O^{+3}) + N(O^{+4}) + N(O^{+5})}.$$

From Table 4 we obtain 2·1 for the ratio on the left-hand side and 0·16 for the ratio on the right-hand side. The helium-ion abundances, and the abundances from forbidden lines should be reliable, and it therefore appears that the abundances of O^{+3}, O^{+4} and O^{+5} obtained from recombination theory are too large by an order of magnitude.

(b) *Carbon ions:* A similar result is obtained from the calculations for C ions. Interpreting observed C II, C III and C IV intensities in NGC 2392, 7027 and 7662 using recombination theory, Pengelly (1964*b*) obtains carbon/hydrogen abundance ratios of about 5×10^{-3} compared with ratios of about 2×10^{-4} for typical stellar atmospheres (Aller, 1961). Unless it is assumed that all of these nebulae are carbon-rich, it must be concluded that the abundances obtained using recombination theory are much too large.

1.4. CASE C

Since the original work of Baker *et al.* (1938), little attention has been paid to Case C. It was pointed out by Aller *et al.* (1939) that the surface brightness must be low if Case C has to be considered for the hydrogen lines. We consider Case C for heavier elements.

Let us define a ratio

$$R(X^{+m}) = \frac{(\text{number of excitations of } X^{+m} \text{ by line absorption of stellar quanta})}{(\text{number of excitations of } X^{+m} \text{ by recombination})}.$$

For an ion such as O^{+2} it is shown by Burgess and Seaton (1960*b*) that optical depths will be large for transitions ending on the ground state. The number of excitations by line absorption of stellar quanta is then largely determined by the Doppler widths of the lines and is insensitive to the ion abundance. The number of recombinations is proportional to the abundance of X^{+m+1} and to the electron density. It follows that the ratio $R(X^{+m})$ for a heavy element will be much larger than the ratio $R(H^0)$

Table 5

NGC 7027, Ratio, $R(O^{2+})$, of number of O^{2+} excitations by line absorption of stellar quanta to number by recombination

$\tau(H)$	$T_s = 1 \times 10^5$	$T_s = 2 \times 10^5$
1	12	48
5	4	16
10	3	11
∞	2	5

for hydrogen. Table 5 gives the results of approximate calculations of $R(O^{+2})$ for NGC 7027, as a function of optical depth at the Lyman limit, $\tau(H)$, and of the star temperature, T_s (Seaton, 1968). It appears that this ratio may be of order 10, and hence that excitation by absorption of stellar radiation in spectrum lines is much more important than excitation by recombination. A similar result may be expected for other ions, and this would seem to give a satisfactory explanation of the results for oxygen and carbon ion abundances discussed in Section 1.3.

Effects of Case C will be much less important for hydrogen lines but should be considered if calculations are being compared with observations of high accuracy.

2. Collisional Excitation

Collisional excitation is responsible for the production of the forbidden lines in the visible spectrum, and may also be expected to be responsible for excitation of permitted lines in the ultraviolet (Osterbrock, 1963; Hummer and Seaton, 1964; Flower, 1968).

Cross-sections $Q(i \to j)$ are conveniently expressed in terms of collision strengths $\Omega(i,j)$, first introduced by Hebb and Menzel (1940):

$$Q(i \to j) = \frac{\pi}{\omega_i}\left(\frac{\hbar}{mv_i}\right)^2 \Omega(i,j),$$

where v_i is the velocity of the incident electron and ω_i is the statistical weight of ion level i. The collision strengths are dimensionless and symmetrical in initial and final states. Put $E_i = \frac{1}{2}mv_i^2$ and $E_j = \frac{1}{2}mv_j^2$, where v_j is the velocity of the scattered electron. The rate coefficient for de-excitation is

$$q(j \to i) = \frac{8{\cdot}63 \times 10^{-6}\Upsilon(i,j)}{\omega_j T_e^{1/2}} \text{ cm}^3 \text{ sec}^{-1} \quad (E_i < E_j),$$

where T_e is in °K and

$$\Upsilon(i,j) = \int_0^\infty \Omega(i,j) \exp(-E_i/kT_e)\, \mathrm{d}(E_i/kT_e).$$

If Ω is independent of energy, $\Upsilon = \Omega$. The rate coefficient for excitation is

$$q(i \to j) = \frac{\omega_j}{\omega_i} q(j \to i) \exp[-(E_j - E_i)/kT_e] \quad (E_i < E_j)$$

2.1. THE FORBIDDEN LINES

A formulation of the collision problem was given by Seaton in 1953 and calculations were made for O^0, O^+, O^{+2}, N^+, Ne^{+2} and S^+. The results obtained, together with approximate estimates for other ions, were summarized by Seaton in 1958 and have been used extensively for the interpretation of forbidden-line intensities. All of this earlier cross-section work is now superseded by the results of more recent calculations.

2.1.1. *Results for Neutral Atoms*

Smith *et al.* (1967) have given a formulation of the collision problem which is similar to that of Seaton (1953), and have made calculations for C^0, N^0 and O^0 which are much more complete and accurate. Their results, which are given in Table 6, are not very different from the earlier results.

Table 6

Collision strengths for neutral C, N and O

(Levels 1, 2 and 3 are the ground configuration terms: ^3P, ^1D and ^1S for C and O; ^4S, ^2D and ^2P for N. Values of E_1 in Rydbergs.)

E_1	C			N			O		
	$\Omega(1,2)$	$\Omega(1,3)$	$\Omega(2,3)$	$\Omega(1,2)$	$\Omega(1,3)$	$\Omega(2,3)$	$\Omega(1,2)$	$\Omega(1,3)$	$\Omega(2,3)$
·2	2·94	·01	·05	–	–	–	·29	–	–
·3	4·69	·46	·40	1·01	–	–	·92	–	–
·4	5·72	·73	·64	1·78	·21	·79	1·40	·08	·14
·5	6·42	·90	·84	2·18	·56	1·69	1·73	·15	·21
·6	6·89	1·00	·99	2·46	·79	2·40	1·97	·21	·27
·8	7·56	1·14	1·19	2·83	1·07	3·45	2·28	·27	·38
1·0	8·03	1·23	1·29	3·08	1·24	4·13	2·52	·32	·45

2.1.2. *Results for Positive Ions*

Results obtained in an extensive programme of new calculations are reported in a separate paper (Czyzak *et al.*, 1968).

Calculations for transitions between fine structure levels, using methods of quantum defect theory, have been made by Osterbrock (1966).

2.2. THE PERMITTED LINES

It is to be expected that lines due to resonance transitions in Li-like ions and Na-like ions will be strong in the ultraviolet spectra of high excitation planetaries. Accurate collision strengths are available and are quoted by Flower (1968). Approximate estimates of collision strengths for other ultraviolet lines are given by Osterbrock (1963); some of these are very uncertain.

As a consequence of the *l*-degeneracy of the H atom, cross-sections for electron-impact excitation remain finite at threshold. In the near-threshold region the best estimates of the cross-sections are given by Burke *et al.* (1968).

A more detailed discussion of the subject matter of this review is being published elsewhere (Seaton, 1968).

References

Aller, L.H. (1961) *The Abundance of the Elements*, Interscience, Publ., New York and London, p. 115.

Aller, L.H., Baker, J.G., Menzel, D.H. (1939) *Astrophys. J.*, **89**, 587.

Baker, J.G., Menzel, D.H. (1938) *Astrophys. J.*, **88**, 52.

Baker, J.G., Menzel, D.H., Aller, L.H. (1938) *Astrophys. J.*, **88**, 422.

Bates, D.R., Damgaard, A. (1949) *Phil. Trans. R. Soc. Lond.*, **A**, **242**, 101.

Burgess, A. (1958) *Mon. Not. R. astr. Soc.*, **118**, 477.

Burgess, A. (1964) *Mem. R. astr. Soc.*, **69**, 1.

Burgess, A., Seaton, M.J. (1960*a*) *Mon. Not. R. astr. Soc.*, **120**, 121.

Burgess, A., Seaton, M.J. (1960*b*) *Mon. Not. R. astr. Soc.*, **121**, 471.

Burke, P.G., Cooper, J.W., Ormonde, S., Taylor, A.J. (1967*a*) *5th Int. Conf. on Physics of Electronic and Atomic Collisions*, Abstracts of Papers, Publishing House Nauka, Leningrad, p. 376.

Burke, P.G., Taylor, A.J., Ormonde, S., Whitaker, W. (1967*b*) *ibid.*, p. 368.

Capriotti, E.R. (1967) *Astrophys. J.* (in press).

Czyzak, S.J., Krueger, T.K., Martins, P. de A.P., Saraph, H.E., Seaton, M.J., and Shemming, J. (1968) in the present volume, p. 138.

Flower, D.R. (1968) in the present volume, p. 77.

Goldberg, L. (1966) *Astrophys. J.*, **144**, 1225.

Green, L.C., Rush, P.P., Chandler, C.D. (1957) *Astrophys. J. Suppl. Ser.*, **3**, 37.

Hayler, D. (1967) *Astrophys. J.*, **150**, 95.

Hebb, M.H., Menzel, D.H. (1940) *Astrophys. J.*, **92**, 408.

Hummer, D.G., Seaton, M.J. (1964) *Mon. Not. R. astr. Soc.*, **127**, 217.

McCarroll, R., Hoang Binh Dy (1968) in the present volume, p. 153.

O'Dell, C.R. (1963) *Astrophys. J.*, **138**, 1018.

Osterbrock, D.E. (1963) *Planet. Space Sci.*, **11**, 621.

Osterbrock, D.E. (1964) *Rev. Astr. Astrophys.* **2**, 95.

Osterbrock, D.E. (1966) *Astrophys. J.*, **142**, 1423.

Peach, G. (1967) *Mem. R. astr. Soc.*, **71**, 1.

Pengelly, R.M. (1964*a*) *Mon. Not. R. astr. Soc.*, **127**, 145.

Pengelly, R.M. (1964*b*) Thesis, London University.

Pengelly, R.M., Seaton, M.J. (1964) *Mon. Not. R. astr. Soc.*, **127**, 165.

Pottasch, S.R. (1961) *Astrophys. J.*, **135**, 385.

Searle, L. (1958) *Astrophys. J.*, **128**, 489.

Seaton, M.J. (1953) *Phil. Trans. R. Soc. Lond.* **A 245**, 469.

Seaton, M.J. (1958) *Rev. mod. Phys.*, **30**, 979.

Seaton, M.J. (1959) *Mon. Not. R. astr. Soc.*, **119**, 90.

Seaton, M.J. (1964) *Mon. Not. R. astr. Soc.*, **127**, 177.

Seaton, M.J. (1968) in *Advances in Atomic and Molecular Physics*, Ed. by D.R. Bates and I. Estermann (in press).

Smith, K., Henry, R.J.W., Burke, P.G. (1967) *Phys. Rev.*, **157**, 51.

DISCUSSION

Aller: Hydrogen recombination rates and level populations have been calculated by Clarke. Three cases of different values of n_{coll} (the value of the principal quantum number, n, at which collisional redistribution is important) viz. $n_{\text{coll}} = 0$, 20, and 60 are considered, for both He II and H I. Some results are reported in Vol. 7 of *Stars and Stellar Systems* (in press).

The C II λ 4267 line served to suggest a reasonable carbon abundance in the Orion Nebula (W. Liller and L.H. Aller, *Astrophys. J.*, **130**, 1959, 45) but carbon recombination calculations for lines of C II, C III, C IV in high-excitation planetaries suggested high abundances of C. I thought these results were due to our rather primitive calculations. Clarke has re-examined the problem but his results are not yet available. It should be pointed out that in C IV transitions like 3^2S–3^2P and 5^2F–6^2G, which arise from very different levels of excitation, are observed. Accurate intensities interpreted by theory might discriminate between the two rival hypotheses of excitation by fluorescence and by recombination.

Intensities of the O II lines interpreted as arising from recombination give the concentration of O^{++} ions which can also be calculated from the intensities of the [O III] lines. Cox has been looking into this problem, and should be able to obtain a direct check on the mechanism by computing the O^{++} ionic concentration in the two ways.

COLLISION STRENGTHS FOR EXCITATION OF FORBIDDEN LINES

S. J. Czyzak and T. K. Krueger
(Ohio State University, and
Aerospace Research Laboratory,
Wright-Patterson Air Force Base, U.S.A.)

and

P. de A. P. Martins, H. E. Saraph,
M. J. Seaton and J. Shemming*
(Dept. of Physics,
University College London, England)

New calculations of collision strengths are being made for ions in configurations np^q, $n=2$ and 3 and $q=1$–5. Wave functions for the entire system are represented by sums over anti-symmetrized products of ion functions and orbitals for the colliding

Table 1
Calculated collision strengths for $k_3^2 = 0$

Ion	$\Omega(1,2)$	$\Omega(1,3)$	$\Omega(2,3)$
N^+	3·14	0·342	0·376
O^{+2}	2·39	0·335	0·310
F^{+3}	1·93	0·279	0·237
Ne^{+4}	1·38	0·218	0·185
O^+	1·43	0·428	1·70
F^{+2}	1·25	0·461	1·67
Ne^{+3}	1·04	0·427	1·42
Na^{+4}	0·836	0·359	1·22
F^+	1·34	0·147	0·193
Ne^{+2}	1·27	0·164	0·188
Na^{+3}	1·14	0·163	0·157
Mg^{+4}	0·973	0·146	0·129
P^+	6·31	1·12	1·11
S^{+2}	4·97	1·07	0·961
Cl^{+3}	1·99	0·328	1·03
Ar^{+4}	1·19	0·141	0·945
S^+	3·07	1·28	6·22
Cl^{+2}	3·19	1·97	6·64
Ar^{+3}	1·43	0·645	4·92
K^{+4}	0·751	0·256	4·24
Cl^+	3·94	0·412	0·749
Ar^{+2}	4·75	0·724	0·665
K^{+3}	1·92	0·296	0·681
Ca^{+4}	0·908	0·115	0·777

* Now at Norwich City College.

Osterbrock and O'Dell (eds.), Planetary Nebulae, 138–142. © *I.A.U.*

Table 2

Energy variation of collision strengths for N^+, O^{+2}, O^+ and Ne^{+2}

Ion	k_3^2	k_2^2	$\Omega(1,2)$	$\Omega(1,3)$	$\Omega(2,3)$
N^+	−0·240	0·000	2·81	–	–
	−0·120	0·120	2·99	–	–
	−0·000	0·240 −	3·15	–	–
	+0·000	0·240 +	3·05	0·342	0·376
	0·200	0·440	3·20	0·391	0·424
	0·400	0·640	3·29	0·428	0·457
O^{+2}	−0·307	0·000	2·45	–	–
	−0·1535	0·1535	2·48	–	–
	−0·000	0·307 −	2·50	–	–
	+0·000	0·307 +	2·39	0·335	0·310
	0·200	0·507	2·40	0·345	0·319
	0·400	0·707	2·39	0·351	0·326
O^+	−0·272	0·000	1·36	–	–
	−0·136	0·136	1·47	–	–
	−0·000	0·272 −	1·55	–	–
	+0·000	0·272 +	1·43	0·428	1·70
	+0·050	0·322	1·46	0·445	1·90
	+0·100	0·372	1·48	0·462	1·99
Ne^{+2}	−0·247	0·000	1·24	–	–
	−0·124	0·124	1·29	–	–
	−0·000	0·247 −	1·33	–	–
	+0·000	0·247 +	1·27	0·164	0·188
	+0·200	0·447	1·31	0·173	0·194
	+0·400	0·647	1·34	0·180	0·201

Table 3

Transitions between O^+ fine-structure states

i	j	$\Omega(i,j)$	
		$k_3^2=0{\cdot}0$	$k_3^2=0{\cdot}1$
$^4S_{3/2}$	$^2D_{3/2}$	0·568	0·589
$^4S_{3/2}$	$^2D_{5/2}$	0·852	0·884
$^4S_{3/2}$	$^2P_{1/2}$	0·144	0·156
$^4S_{3/2}$	$^2P_{3/2}$	0·288	0·311
$^2D_{3/2}$	$^2D_{5/2}$	0·890	0·924
$^2D_{3/2}$	$^2P_{1/2}$	0·353	0·378
$^2D_{3/2}$	$^2P_{3/2}$	0·481	0·543
$^2D_{5/2}$	$^2P_{1/2}$	0·342	0·398
$^2D_{5/2}$	$^2P_{3/2}$	0·909	0·999
$^2P_{1/2}$	$^2P_{3/2}$	0·257	0·283

Table 4

Transitions within ^{3}P terms

Ion	$k_3{}^2$	$\Omega(^3P_0, {}^3P_1)$	$\Omega(^3P_0, {}^3P_2)$	$\Omega(^3P_1, {}^3P_2)$
N^+	0·0	·401	·279	1·128
	0·1	·424	·282	1·164
O^{+2}	0·0	·376	·213	·948
	0·4	·391	·207	·954
Ne^{+2}	0·0	·185	·131	·527
	0·4	·200	·134	·551

electron, and sets of coupled integro-differential equations are obtained for the colliding electron radial functions. The main approximation is neglect of coupling to configurations other than np^q. Calculations are made using improved forms of the exact resonance and distorted wave approximations. For the $2p^q$ ions it is found that the dominant contributions come from the p-waves; the exact resonance approximation is used for the p-waves and the distorted wave approximation is used for all other partial waves. It is considered that the results obtained should approximate closely to those which would be obtained from full solutions of the coupled equations. For the $3p^q$ ions the dominant contributions come from the d-waves; for these ions the distorted wave approximation is used for all partial waves.

We give some results which have been obtained to date for $q=2$, 3 and 4. These configurations have three terms (^{3}P, ^{1}D and ^{1}S for $q=2$ and 4; ^{4}S, ^{2}D and ^{2}P for $q=3$) denoted, in order of increasing excitation energy, by $i=1$, 2 and 3. Energies of the colliding electron, in Rydbergs, are denoted by k_i^2. Table 1 gives the calculated collision strengths for $k_3^2=0$. The collision strength $\Omega(1, 2)$ has a complicated resonance structure for $k_3^2<0$; we average over the resonances using a formula of Gailitis (1963). Table 2 gives collision strengths for N^+, O^{+2}, O^+ and Ne^{+2} as functions of energy. Table 3 gives results for transitions between the fine structure states of O^+, which are required for the interpretation of [O II] λ 3726/ λ 3729 intensity ratios; further calculations of $\Omega(^2D_{3/2}, {}^2D_{5/2})$ are being made for energies $k_3^2<0$. Table 4 gives results for transitions between the fine structure states in the ^{3}P terms of N^+, O^{+2} and Ne^{+2}.

The reactance matrices calculated for collisions with an ion X^{+m} can be extrapolated to negative energies and used to calculate the positions of bound states of $X^{+(m-1)}$. A comparison with observed positions gives a guide to the accuracy of the calculations. As a final step, parameters may be introduced into the expressions for the reactance matrices and adjusted so as to obtain an improved agreement with observed bound states, and the adjusted matrices may be used to obtain improved collision strengths. This method has, so far, been applied successfully only to the p-wave contributions for the $2p^q$ ions. The calculations of bound-state energy levels are made in intermediate coupling and it is found that the differences between observed and calculated energies are practically the same for all levels within any one SL term. Table 5

Table 5

Effective quantum numbers ν for $2p^2$ np SL states of O^+ referred to the O^{+2} $2p^2$ ^3P series limit

n	SL	$\nu_{calc} - \nu_{obs.}$	$\nu_{adj.} - \nu_{obs.}$
3	^2S	·0432	·0078
3	^2P	·0323	− ·0077
3′	^2P	·0632	− ·0090
4	^2P	·0361	− ·0027
3	^2D	·0328	·0017
3′	^2D	·0728	·0043
4	^2D	·0365	·0035
5	^2D	·0327	·0005
3′	^2F	·0410	− ·0170
3	^4S	·0652	·0270
3	^4P	·0280	− ·0070
5	^4P	·0253	− ·0047
3	^4D	·0299	− ·0054
4	^4D	·0278	− ·0037
5	^4D	·0269	− ·0033

gives the differences between calculated and observed effective quantum numbers for 15 $2p^2$ np SL terms in O^+, and differences obtained when 4 parameters are introduced into the expressions for the reactance matrices and adjusted so as to give a least squares fit in the energy levels. Table 6 gives some results for the calculated and adjusted *p*-wave contributions to the collision strengths of various ions. It should be noted that, for the energy differences between the np^q terms, we use calculated values for Tables 1 and 2 and experimental values for Tables 3, 4, 5, and 6.

Table 6

***p*-Wave contributions to collision strengths, $k_3{}^2 = 0$**

Ion	$\Omega^{pp}(1, 2)$ Calc.	$\Omega^{pp}(1, 2)$ Adj.	$\Omega^{pp}(1, 3)$ Calc.	$\Omega^{pp}(1, 3)$ Adj.
N^+	2·657	2·678	0·288	0·480
O^{+2}	1·658	1·582	0·204	0·195
F^{+3}	1·045	0·802	0·131	0·098
O^+	1·302	1·449	0·381	0·423
Ne^{+2}	0·979	1·001	0·116	0·157
Na^{+3}	0·730	0·543	0·088	0·069

Table 7 gives a summary of results obtained in different calculations for O^{+2} (Seaton, 1953, 1955, 1958; Saraph *et al.*, 1966). It is seen that there is little change in the ratio $\Omega(1, 2)/\Omega(1, 3)$, which is of importance for the determination of electron tem-

Table 7

Comparison of different calculations for O^{+2}, $k_3^2=0$

	$\Omega(1,2)$	$\Omega(1,3)$	$\Omega(2,3)$	$\Omega(1,2)/\Omega(1,3)$
Calculated, 1953	1·73	0·195	–	8·87
Calculated, 1955	–	–	0·61	–
Recommended, 1958	1·59	0·22	0·64	7·23
Calculated, 1966	2·39	0·335	0·310	7·14
Adjusted, 1967	2·32	0·326	0·309	7·12

peratures. The present results for $\Omega(2, 3)$ differ from the earlier results due to the inclusion of exchange for the higher order partial waves.

Some further details on the formulation employed in this work are given by Saraph *et al.* (1966), together with results for $2p^2$ ions. Results for $3p^3$ ions have been given by Czyzak and Krueger (1967) and results for $2p^3$, $2p^4$, $3p^2$ and $3p^4$ have been reported by Czyzak *et al.* (1967). Our programme of calculations is being continued and a full account of the work is being prepared for publication.

References

Czyzak, S.J., Krueger, T.K. (1967) *Proc. Phys. Soc., London*, **90**, 623.

Czyzak, S.J., Krueger, T.K., Saraph, H.E., Shemming, J. (1967) *Proc. Phys. Soc., London*, **92**, 1146.

Gailitis, M. (1963) *Soviet Phys. JETP*, **17**, 1328.

Saraph, H.E., Seaton, M.J., Shemming, J. (1966) *Proc. Phys. Soc.*, **89**, 27.

Seaton, M.J. (1953) *Proc. R. Soc. London*, **A218**, 400.

Seaton, M.J. (1955) *ibid.*, **231**, 37.

Seaton, M.J. (1958) *Rev. mod. Phys.*, **30**, 979.

DISCUSSION

Garstang: The variation of the collision strength as a function of Z along an isoelectronic sequence is of interest. Usually this variation is smooth. Have you found any cases where this variation is not smooth, e.g. by means of irregularities of quantum defects, or any cases where the square roots of the collision strengths pass through a zero?

Seaton: One considers $Z^2\Omega$, where Z is the charge; this remains finite as $Z\to\infty$. More or less irregular variations in $Z^2\Omega$ can occur along an isoelectronic sequence.

TRANSITION PROBABILITIES FOR FORBIDDEN LINES

R. H. GARSTANG
(Joint Institute for Laboratory Astrophysics and University of Colorado, U.S.A.)*

ABSTRACT

A compilation is given of transition probabilities of forbidden lines which occur in the spectra of gaseous nebulae.

The spectra of gaseous nebulae contain both permitted and forbidden lines of many elements in various stages of ionization. These are of great importance because of the information they can provide on the physical conditions and chemical composition of the nebulae. The forbidden lines belong to both non-metals and metals, and indeed for some elements the only observable spectral lines are forbidden lines. Transition probabilities are now known for essentially every forbidden line which has been observed with reasonable certainty in gaseous nebulae, including the Orion Nebula and the very rich-spectrum planetary nebula NGC 7027.

Table 1

Transition probabilities for the $2p^2$ configuration**

Transition	C I	N II	O III	F IV	Ne V
$^1D_2$–$^1S_0$	0·50	1·08	1·60	2·10	2·60
	8727·4	5754·6	4363·2	3532·2	2972
$^3P_2$–$^1S_0$	$1·9 \times 10^{-5}$	$1·6 \times 10^{-4}$	$7·1 \times 10^{-4}$	$2·3 \times 10^{-3}$	$6·8 \times 10^{-3}$
	4627·3	3070·8	2331·6	1889·3	1592·7
$^3P_1$–$^1S_0$	$2·6 \times 10^{-3}$	0·034	0·23	1·1	4·2
	4621·5	3063·0	2321·1	1875·5	1575·2
$^3P_2$–$^1D_2$	$2·3 \times 10^{-4}$	$3·0 \times 10^{-3}$	0·021	0·098	0·38
	9849·5	6583·4	5006·8	4060·2	3425·9
$^3P_1$–$^1D_2$	$7·8 \times 10^{-5}$	$1·03 \times 10^{-3}$	0·0071	0·034	0·138
	9823·4	6548·1	4958·9	3997·4	3345·8
$^3P_0$–$^1D_2$	$5·5 \times 10^{-8}$	$4·2 \times 10^{-7}$	$1·9 \times 10^{-6}$	$6·4 \times 10^{-6}$	$1·9 \times 10^{-5}$
	9808·9	6527·4	4931·0	3960·7	3300·0
$^3P_1$–$^3P_2$	$2·7 \times 10^{-7}$	$7·5 \times 10^{-6}$	$9·8 \times 10^{-5}$	$7·9 \times 10^{-4}$	$4·6 \times 10^{-3}$
$^3P_0$–$^3P_2$	$2·0 \times 10^{-14}$	$1·3 \times 10^{-12}$	$3·5 \times 10^{-11}$	$5·0 \times 10^{-10}$	$5·2 \times 10^{-9}$
$^3P_0$–$^3P_1$	$7·9 \times 10^{-8}$	$2·1 \times 10^{-6}$	$2·6 \times 10^{-5}$	$2·1 \times 10^{-4}$	$1·3 \times 10^{-3}$

** Compiled from Wiese *et al.* (1966), which is based on work by Garstang; Naqvi; and Yamanouchi and Horie.

* Of the National Bureau of Standards and the University of Colorado.

Osterbrock and O'Dell (eds.), Planetary Nebulae, 143–152. © *I.A.U.*

Table 2

Transition probabilities* for the $2p^3$ configuration

Transition	N I	O II	F III	Ne IV	Mg VI
$^2P_{\frac{1}{2}}$–$^2P_{1\frac{1}{2}}$	(very small)	$6{\cdot}0 \times 10^{-11}$	(very small)	$2{\cdot}3 \times 10^{-9}$	$1{\cdot}6 \times 10^{-5}$
$^2D_{2\frac{1}{2}}$–$^2P_{1\frac{1}{2}}$	0·054 10395·4	0·115 7319·4	0·18 5721·2	0·40 4714·3	2·4 3485·5
$^2D_{1\frac{1}{2}}$–$^2P_{1\frac{1}{2}}$	0·025 10404·1	0·061 7330·7	0·114 5733·0	0·44 4724·2	3·8 3488·1
$^2D_{2\frac{1}{2}}$–$^2P_{\frac{1}{2}}$	0·031 10395·4	0·061 7318·6	0·088 5721·2	0·11 4715·6	0·15 3500·4
$^2D_{1\frac{1}{2}}$–$^2P_{\frac{1}{2}}$	0·047 10404·1	0·100 7329·9	0·16 5733·0	0·39 4725·6	2·5 3503·0
$^4S_{1\frac{1}{2}}$–$^2P_{1\frac{1}{2}}$	$6{\cdot}2 \times 10^{-3}$ 3466·4	0·060 2470·4	0·26 1939·6	1·33 1608·8	13·
$^4S_{1\frac{1}{2}}$–$^2P_{\frac{1}{2}}$	$2{\cdot}5 \times 10^{-3}$ 3466·4	0·0238 2470·3	0·10 1939·6	0·53 1609·0	5·3
$^2D_{2\frac{1}{2}}$–$^2D_{1\frac{1}{2}}$	$1{\cdot}3 \times 10^{-8}$	$1{\cdot}3 \times 10^{-7}$	$7{\cdot}6 \times 10^{-7}$	$1{\cdot}4 \times 10^{-6}$	$1{\cdot}5 \times 10^{-7}$
$^4S_{1\frac{1}{2}}$–$^2D_{2\frac{1}{2}}$	$6{\cdot}9 \times 10^{-6}$ 5200·4	$4{\cdot}8 \times 10^{-5}$ 3728·8	$1{\cdot}3 \times 10^{-4}$ 2933·1	$5{\cdot}9 \times 10^{-4}$ 2441·3	0·0054
$^4S_{1\frac{1}{2}}$–$^2D_{1\frac{1}{2}}$	$1{\cdot}6 \times 10^{-5}$ 5197·9	$1{\cdot}70 \times 10^{-4}$ 3726·0	$1{\cdot}3 \times 10^{-3}$ 2930·0	$5{\cdot}6 \times 10^{-3}$ 2438·6	0·12

* Compiled for N I, O II, F III and Ne IV from Wiese *et al.* (1966), which is based on work by Ufford and Gilmour; Naqvi; Garstang; and Seaton and Osterbrock. Mg VI recalculated by Garstang for inclusion here.

Table 3

Transition probabilities* for the $2p^4$ configuration

Transition	O I	F II	Ne III	Na IV
$^1D_2$–$^1S_0$	1·34 5577·4	2·1 4157·5	2·8 3342·5	3·5 2803·3
$^3P_2$–$^1S_0$	$3{\cdot}7 \times 10^{-4}$ 2958·4	$1{\cdot}6 \times 10^{-3}$ 2225·5	$5{\cdot}1 \times 10^{-3}$ 1793·8	0·012 1503·7
$^3P_1$–$^1S_0$	0·067 2972·3	0·49 2246·6	2·2 1814·8	7·6 1529·1
$^3P_2$–$^1D_2$	$5{\cdot}1 \times 10^{-3}$ 6300·3	0·038 4789·5	0·17 3868·8	0·66 3241·7
$^3P_1$–$^1D_2$	$1{\cdot}64 \times 10^{-3}$ 6363·8	0·012 4869·3	0·052 3967·5	0·20 3362·2
$^3P_0$–$^1D_2$	$1{\cdot}1 \times 10^{-6}$ 6391·6	$4{\cdot}1 \times 10^{-6}$ 4904·8	$1{\cdot}2 \times 10^{-5}$ 4012·7	$3{\cdot}0 \times 10^{-5}$ 3416·2
$^3P_1$–$^3P_0$	$1{\cdot}7 \times 10^{-5}$	$1{\cdot}8 \times 10^{-4}$	$1{\cdot}2 \times 10^{-3}$	$5{\cdot}5 \times 10^{-3}$
$^3P_2$–$^3P_0$	$1{\cdot}0 \times 10^{-10}$	$1{\cdot}8 \times 10^{-9}$	$2{\cdot}0 \times 10^{-8}$	$1{\cdot}5 \times 10^{-7}$
$^3P_2$–$^3P_1$	$9{\cdot}0 \times 10^{-5}$	$9{\cdot}0 \times 10^{-4}$	$6{\cdot}0 \times 10^{-3}$	0·030

* Compiled for O I, F II and Ne III from Wiese *et al.* (1966), based on work by Garstang; Naqvi; Yamanouchi and Horie and for (O I) Omholt; Stoffregen and Derblom. Na IV from Malville and Berger (1965).

Table 4

Transition probabilities* for the $3p^2$ configuration

Transition	S III	Cl IV	Ar V	K VI	Ca VII
$^1D_2-{}^1S_0$	2·54	3·15	3·78	4·1	4·3
	6312·1	5323·3	4625·5	4097:	3688:
$^3P_2-{}^1S_0$	0·016	0·038	0·081	0·14	0·25
	3796·7	3203·2	2784·4	2471·7	2226:
$^3P_1-{}^1S_0$	0·85	2·6	6·8	16·	34·
	3721·7	3118·3	2691·4	2366·8	2112:
$^3P_2-{}^1D_2$	0·064	0·20	0·51	1·1	2·5
	9532·1	8045·6	7005·7	6228·4	5614·7
$^3P_1-{}^1D_2$	0·025	0·080	0·22	0·53	1·2
	9069·4	7530·5	6435·1	5603·2	4939:
$^3P_0-{}^1D_2$	$9·1 \times 10^{-6}$	$2·2 \times 10^{-5}$	$4·9 \times 10^{-5}$	$1·1 \times 10^{-4}$	$2·1 \times 10^{-4}$
	8831·5	7262·3	6131·0	5269·2	4571:
$^3P_1-{}^3P_2$	$2·4 \times 10^{-3}$	$8·2 \times 10^{-3}$	0·027	0·076	0·20
$^3P_0-{}^3P_2$	$4·7 \times 10^{-8}$	$2·8 \times 10^{-7}$	$1·3 \times 10^{-6}$	$5·4 \times 10^{-6}$	$1·9 \times 10^{-5}$
$^2P_0-{}^3P_1$	$4·7 \times 10^{-4}$	$2·1 \times 10^{-3}$	$8·0 \times 10^{-3}$	0·026	0·076

* S III, Cl IV and Ar V from Czyzak and Krueger (1963), K VI and Ca VII from Malville and Berger (1965).

Table 5

Transition probabilities* for the $3p^3$ configuration

Transition	S II	Cl III	Ar IV	K V
$^2P_{\frac{1}{2}}-{}^2P_{1\frac{1}{2}}$	$1·0 \times 10^{-6}$	$7·6 \times 10^{-6}$	$5·2 \times 10^{-5}$	$2·8 \times 10^{-4}$
$^2D_{2\frac{1}{2}}-{}^2P_{1\frac{1}{2}}$	0·21	0·36	0·67	1·5
	10320·6	8481·6	7237·3	6317:
$^2D_{1\frac{1}{2}}-{}^2P_{1\frac{1}{2}}$	0·17	0·39	0·91	2·3
	10287·1	8433·7	7170·6	6223:
$^2D_{2\frac{1}{2}}-{}^2P_{\frac{1}{2}}$	0·087	0·108	0·122	0·19
	10372·6	8550·5	7332·0	6447:
$^2D_{1\frac{1}{2}}-{}^2P_{\frac{1}{2}}$	0·20	0·35	0·68	1·5
	10338·8	8501·8	7262·8	6349:
$^4S_{1\frac{1}{2}}-{}^2P_{1\frac{1}{2}}$	0·34	0·96	2·55	6·5
	4068·6	3342·9	2854·8	2494·5
$^4S_{1\frac{1}{2}}-{}^2P_{\frac{1}{2}}$	0·134	0·37	0·97	2·4
	4076·4	3353·3	2869·1	2514·5
$^2D_{2\frac{1}{2}}-{}^2D_{1\frac{1}{2}}$	$3·3 \times 10^{-7}$	$3·2 \times 10^{-6}$	$2·3 \times 10^{-5}$	$1·4 \times 10^{-4}$
$^4S_{1\frac{1}{2}}-{}^2D_{2\frac{1}{2}}$	$4·7 \times 10^{-5}$	$1·01 \times 10^{-3}$	$2·2 \times 10^{-3}$	$6·9 \times 10^{-3}$
	6716·4	5517·2	4711·3	4122·6
$^4S_{1\frac{1}{2}}-{}^2D_{1\frac{1}{2}}$	$3·0 \times 10^{-4}$	$7·0 \times 10^{-3}$	0·028	0·11
	6730·8	5537·7	4740·2	4163·3

* From Czyzak and Krueger (1963) for S II, Cl III and Ar IV. K V calculated by Garstang for inclusion here.

Table 6

Transition probabilities* for the $3p^4$ configuration

Transition	S I	Cl II	Ar III	K IV	Ca V
$^1D_2$–$^1S_0$	1·78 7724·7	2·3 6152·9	3·1 5191·8	3·9 4510·9	4·6 3996·3
$^3P_2$–$^1S_0$	$7·3 \times 10^{-3}$ 4506·9	0·018 3583·0	0·043 3005·1	0·086 2593·5	0·16 2280·0
$^3P_1$–$^1S_0$	0·35 4589·0	1·3 3675·0	4·0 3109·0	10·4 2711·2	24· 2412·4
$^3P_2$–$^1D_2$	0·028 10819·8	0·10 8579·5	0·32 7135·8	0·83 6101·8	1·9 5309·2
$^3P_1$–$^1D_2$	$8·0 \times 10^{-3}$ 11305·8	0·029 9125·8	0·083 7751·0	0·20 6795·8	0·43 6086·9
$^3P_0$–$^1D_2$	$5·0 \times 10^{-6}$ 11540·1	$1·2 \times 10^{-5}$ 9381·8	$2·9 \times 10^{-5}$ 8036·4	$6·0 \times 10^{-5}$ 7110·4	$1·1 \times 10^{-4}$ 6428·2
$^3P_1$–$^3P_0$	$3·0 \times 10^{-4}$	$1·4 \times 10^{-3}$	$5·1 \times 10^{-3}$	0·015	0·035
$^3P_2$–$^3P_0$	$7·1 \times 10^{-8}$	$4·8 \times 10^{-7}$	$2·7 \times 10^{-6}$	$1·2 \times 10^{-5}$	$4·5 \times 10^{-5}$
$^3P_2$–$^3P_1$	$1·4 \times 10^{-3}$	$7·5 \times 10^{-3}$	0·031	0·10	0·31

* S I, Cl II and Ar III from Czyzak and Krueger (1963), K IV and Ca V from Malville and Berger (1965).

Table 7

Transition probabilities for selected* lines of Mn VI and Fe VII

Transition	J–J′	Mn VI λ	Mn VI A	Fe VII λ	Fe VII A
a^3F–a^1D	2–2	6518·3	0·14	5721·1	0·30
	3–2	6852:	0·23	6086·9	0·49
	4–2	7315:	$9·0 \times 10^{-4}$	6598·8	$1·6 \times 10^{-3}$
a^3F–a^3P	2–0	5622:	0·087	4989:	0·11
	2–1	5536:	0·031	4893·4	0·043
	3–1	5776·4	0·050	5159·0	0·063
	2–2	5367:	$6·3 \times 10^{-3}$	4699·8	0·012
	3–2	5591:	0·030	4944·0	0·065
	4–2	5894·0	0·050	5277·7	0·060
a^3F–a^1G	3–4	4036·8	0·12	3587·8	0·26
	4–4	4193·1	0·17	3760·3	0·37

* From Garstang (1964) for Mn VI and Pasternack (1940) for Fe VII, with the electric quadrupole contributions in Fe VII reduced by the appropriate factor given by Garstang (1964). Results for many additional lines can be found in these references.

Tables 1–11 give the transition probabilities of spontaneous emission in sec^{-1}. The transitions (with one exception specially noted) take place by magnetic dipole radiation, by electric quadrupole radiation, or by both. When both are possible one type usually predominates, but there are some cases where the two types of radiation have comparable probabilities. In our tables we have given the total transition probability; anyone interested in the type of radiation involved in a particular line and in the individual magnetic dipole or electric quadrupole-transition probabilities is referred to the original papers or compilations mentioned in the notes to the tables. The wavelengths are given (in Ångstroms unless microns are indicated) for lines in the observable spectral region and for some infrared and ultraviolet transitions. Many of the

Table 8

Transition probabilities for selected* lines of Mn V and Fe VI

Transition	J–J′	Mn V		Fe VI	
		λ	A	λ	A
$a^4F–a^4P$	$4\frac{1}{2}–2\frac{1}{2}$	6393·6	0·041	5677·0	0·048
	$3\frac{1}{2}–2\frac{1}{2}$	6166·2	0·016	5426·6	0·021
	$2\frac{1}{2}–2\frac{1}{2}$	5991:	$4{\cdot}1 \times 10^{-3}$	5233·9	$5{\cdot}9 \times 10^{-3}$
	$1\frac{1}{2}–2\frac{1}{2}$	5868:	$5{\cdot}0 \times 10^{-4}$	5097·5	$7{\cdot}9 \times 10^{-4}$
	$3\frac{1}{2}–1\frac{1}{2}$	6346:	0·031	5630·8	0·036
	$2\frac{1}{2}–1\frac{1}{2}$	6159:	0·026	5423·9	0·032
	$1\frac{1}{2}–1\frac{1}{2}$	6030:	$9{\cdot}3 \times 10^{-3}$	5277·5	0·014
	$2\frac{1}{2}–\frac{1}{2}$	6218·6	0·026	5484·8	0·031
	$1\frac{1}{2}–\frac{1}{2}$	6088:	0·044	5335·2	0·055
$a^4F–a^2G$	$4\frac{1}{2}–4\frac{1}{2}$	5891·1	0·24	5176·4	0·56
	$3\frac{1}{2}–4\frac{1}{2}$	5695:	0·096	4967·3	0·22
	$2\frac{1}{2}–4\frac{1}{2}$	5544:	$1{\cdot}7 \times 10^{-6}$	4805·4	$3{\cdot}1 \times 10^{-6}$
	$4\frac{1}{2}–3\frac{1}{2}$	6069:	$5{\cdot}9 \times 10^{-3}$	5370·5	0·012
	$3\frac{1}{2}–3\frac{1}{2}$	5862·3	0·096	5145·8	0·22
	$2\frac{1}{2}–3\frac{1}{2}$	5703:	0·088	4972·1	0·20
	$1\frac{1}{2}–3\frac{1}{2}$	5592:	$5{\cdot}6 \times 10^{-6}$	4849·0	$1{\cdot}1 \times 10^{-5}$

* From Pasternack (1940) with the electric quadrupole contributions reduced by appropriate factors given by Garstang (1964). There are many other lines of these ions for which data can be found in these references.

wavelengths are quoted from Bowen (1955, 1960). Some wavelengths are uncertain by several tenths of an Ångstrom in cases where they have not been directly observed and reliance is upon predictions based on ultraviolet permitted-line spectroscopy.

A compilation of data for all atoms up to neon has been given by Wiese *et al.* (1966). Where the data we quote are the same as theirs, we reference only their book. References may be found in their book to the original papers by Garstang; Naqvi; Yamanouchi and Horie; Ufford and Gilmour; Seaton and Osterbrock; Omholt; and Stoffregen and Derblom upon which their compilation is based. For atoms heavier than

Table 9

Transition probabilities for selected* lines of Fe III and Fe V

Transition	J–J′	Fe III λ	Fe III A	Fe V λ	Fe V A
$a^5D–a^5D$	0–1		$1{\cdot}4 \times 10^{-4}$		$1{\cdot}6 \times 10^{-4}$
	1–2		$6{\cdot}7 \times 10^{-4}$		$1{\cdot}2 \times 10^{-3}$
	2–3		$1{\cdot}8 \times 10^{-3}$		$2{\cdot}6 \times 10^{-3}$
	3–4		$2{\cdot}8 \times 10^{-3}$		$3{\cdot}0 \times 10^{-3}$
$a^5D–a^3P$	1–0	4930·5	0·67	4180·9	1·3
	2–0	4884·5	$2{\cdot}4 \times 10^{-4}$	4229·3	$2{\cdot}8 \times 10^{-4}$
	0–1	5084·8	0·091	4003·0	0·13
	1–1	5060·5	$1{\cdot}5 \times 10^{-4}$	4026·4	$2{\cdot}1 \times 10^{-4}$
	2–1	5011·3	0·53	4071·3	1·1
	3–1	4936·4	$3{\cdot}2 \times 10^{-5}$	4136·2	$4{\cdot}1 \times 10^{-5}$
	0–2	5439·9	$1{\cdot}5 \times 10^{-5}$	3777·2	$4{\cdot}1 \times 10^{-5}$
	1–2	5412·2	0·038	3798·0	0·036
	2–2	5355·9	$1{\cdot}1 \times 10^{-4}$	3838·0	$2{\cdot}0 \times 10^{-4}$
	3–2	5270·3	0·40	3895·5	0·71
	4–2	5151·9	$7{\cdot}1 \times 10^{-6}$	3970·0	$1{\cdot}5 \times 10^{-5}$
$a^3P–a^3P$	0–1		$7{\cdot}5 \times 10^{-3}$		0·014
	1–2		0·047		0·045
$a^5D–a^3H$	4–4	4881·1	$4{\cdot}8 \times 10^{-3}$	4227·5	$1{\cdot}1 \times 10^{-3}$
$a^3H–a^3H$	4–5		$1{\cdot}9 \times 10^{-4}$		$6{\cdot}5 \times 10^{-4}$
	5–6		$4{\cdot}1 \times 10^{-4}$		$5{\cdot}8 \times 10^{-4}$
$a^5D–a^3F$	0–2	4799·4	$9{\cdot}3 \times 10^{-6}$	3735·7	$2{\cdot}2 \times 10^{-5}$
	1–2	4777·9	0·049	3756·1	0·10
	2–2	4733·9	0·10	3795·2	0·20
	3–2	4667·0	0·026	3851·4	0·047
	4–2	4573·8	$2{\cdot}1 \times 10^{-6}$	3924·2	$1{\cdot}5 \times 10^{-6}$
	1–3	4814·1	$1{\cdot}1 \times 10^{-5}$	3744·8	$8{\cdot}6 \times 10^{-6}$
	2–3	4769·6	0·087	3783·6	0·16
	3–3	4701·6	0·27	3839·5	0·40
	4–3	4607·1	0·038	3911·9	0·066
	2–4	4824·2	$7{\cdot}6 \times 10^{-6}$	3764·4	$8{\cdot}6 \times 10^{-7}$
	3–4	4754·8	0·081	3819·8	0·16
	4–4	4658·1	0·44	3891·3	0·74
$a^5D–a^3G$	3–3	4046·4	$8{\cdot}0 \times 10^{-3}$	3445·4	0·017
	4–4	4008·4	0·019	3463·4	0·032
$a^5D–a^3D$	0–1	3366·2	0·13	?	0·22
	1–1	3355·6	0·15	?	0·19
	1–2	3356·6	0·095	?	0·20
	2–2	3334·9	0·11	?	0·18
	3–2	3301·6	0·027	?	0·11
	2–3	3319·3	0·044	?	0·097
	3–3	3286·2	0·047	?	0·089
	4–3	3239·7	0·23	?	0·37

* From Garstang (1957), where results are given for many additional lines.

Table 10

Transition probabilities for selected* lines of Fe II

Transition	J–J′	λ	A
$a^6D–a^6S$	4½–2½	4287·4	1·12
	3½–2½	4359·3	0·82
	2½–2½	4413·8	0·58
	1½–2½	4452·1	0·37
	½–2½	4474·9	0·18
$a^4F–a^4G$	4½–5½	4244·0	0·90
	3½–5½	4346·9	0·21
	4½–4½	4177·2	0·14
	3½–4½	4276·8	0·65
	2½–4½	4352·8	0·31
	4½–3½	4146·6	$8{\cdot}7 \times 10^{-3}$
	3½–3½	4244·8	0·25
	2½–3½	4319·6	0·53
	1½–3½	4372·4	0·28
	4½–2½	4134·0	$2{\cdot}0 \times 10^{-4}$
	3½–2½	4231·6	0·024
	2½–2½	4305·9	0·31
	1½–2½	4358·4	0·73
$a^4F–b^4F$	4½–4½	4814·6	0·40
	3½–4½	4947·4	0·050
	2½–4½	5049·3	$7{\cdot}2 \times 10^{-4}$
	4½–3½	4774·7	0·13
	3½–3½	4905·3	0·22
	2½–3½	5005·5	0·071
	1½–3½	5076·6	$1{\cdot}6 \times 10^{-5}$
	4½–2½	4745·5	0·013
	3½–2½	4874·5	0·17
	2½–2½	4973·4	0·14
	1½–2½	5043·5	0·065
	3½–1½	4852·7	0·022
	2½–1½	4950·7	0·17
	1½–1½	5020·2	0·18
$a^6D–b^4F$	4½–4½	4416·3	0·46
	3½–4½	4492·6	0·060
	2½–4½	4550·5	$2{\cdot}6 \times 10^{-7}$
	4½–3½	4382·8	0·055
	3½–3½	4458·0	0·29
	2½–3½	4514·9	0·066
	1½–3½	4555·0	$4{\cdot}8 \times 10^{-8}$
	4½–2½	4358·1	$1{\cdot}6 \times 10^{-5}$
	3½–2½	4432·5	0·054
	2½–2½	4488·8	0·15
	1½–2½	4528·4	0·046
	½–2½	4552·0	$2{\cdot}0 \times 10^{-6}$
	3½–1½	4414·5	$5{\cdot}9 \times 10^{-6}$
	2½–1½	4470·3	0·029
	1½–1½	4509·6	0·058
	½–1½	4533·0	0·016

* From Garstang (1962), where data for many other lines may be found.

neon references are given to the original papers containing the data we have quoted. In each case we have quoted what we believe to be the best available results.

In a few cases we have given original data calculated for inclusion here. Mg VI and K V were originally calculated by Pasternack (1940). We have recalculated these, using the best available technique. The principal change results from the use of improved quadrupole radial integrals, which we have estimated by extrapolation. The changes

Table 11

Transition probabilities for some miscellaneous lines

Ion	Transition	J–J′	λ	A	Notes
Mg I	$3s^2\ ^1S$–$3s3p\ ^3P$	0–2	4562·5	$2{\cdot}0 \times 10^{-4}$	1
Ni III	^3F–^3P	3–1	6401·5	0·038	2
	^3F–^3P	2–0	6682·2	0·046	
	^3F–^3P	3–2	6533·7	0.12	
	^3F–^1G	3–4	4596·8	0·18	
C II	$2p\ ^2P$	½–1½	156 μ	$2{\cdot}4 \times 10^{-6}$	3
N III	$2p\ ^2P$	½–1½	57·3 μ	$4{\cdot}8 \times 10^{-5}$	3
O IV	$2p\ ^2P$	½–1½	25·9 μ	$5{\cdot}2 \times 10^{-4}$	3
Ne II	$2p^5\ ^2P$	1½–½	12·8 μ	0·0086	3
Mg IV	$2p^5\ ^2P$	1½–½	4·49 μ	0·20	4
Si II	$3p\ ^2P$	½–1½	34·8 μ	$2{\cdot}1 \times 10^{-4}$	4
S IV	$3p\ ^2P$	½–1½	10·6 μ	0·0077	4

[1] This line arises partly from magnetic quadrupole radiation and partly from nuclear-spin-induced electric dipole radiation. See Garstang (1967).
[2] Possible identification of 6401·5 by Flather and Osterbrock (1960). Other Ni III lines have not been seen in gaseous nebulae. Transition probabilities from Garstang (1958).
[3] From Wiese *et al.* (1966).
[4] Calculated by Garstang for inclusion here.

in the magnetic dipole results and in the relative electric quadrupole results are fairly small. We have also revised Pasternack's (1940) results for Mn VI, Fe VII, Mn V and Fe VI by introducing revised quadrupole radial integrals as described by Garstang (1964). We included Mg VI because of its possible, as yet unconfirmed, identification by Gauzit (1966). In Table 11 we have included some infrared transitions; other such transitions appear in many of the other tables, or can be found in the references (e.g. for Fe II and other heavy ions).

Acknowledgment

The preparation of this paper was supported by the National Science Foundation under Grant GP-6595.

References

Bowen, I.S. (1955) *Astrophys. J.*, **121**, 306.
Bowen, I.S. (1960) *Astrophys. J.*, **132**, 1.
Czyzak, S.J., Krueger, T.K. (1963) *Mon. Not. R. astr. Soc.*, **126**, 177, with an important correction in 1965, *ibid.*, **129**, 103.
Flather, E., Osterbrock, D.E. (1960) *Astrophys. J.*, **132**, 18.
Garstang, R.H. (1957) *Mon. Not. R. astr. Soc.*, **117**, 393.
Garstang, R.H. (1958) *Mon. Not. R. astr. Soc.*, **118**, 234.
Garstang, R.H. (1962) *Mon. Not. R. astr. Soc.*, **124**, 321.
Garstang, R.H. (1964) *J. Res. nat. Bur. Stand. Sec. A.*, **68**, 61.
Garstang, R.H. (1967) *Astrophys. J.*, **148**, 579.
Gauzit, J. (1966) *C. r. hebd. Séanc. Acad. Sci. Paris*, **262**, 523.
Malville, J.M., Berger, R.A. (1965) *Planet. Space Sci.*, **13**, 1131.
Pasternack, S. (1940) *Astrophys. J.*, **92**, 129.
Wiese, W.L., Smith, M.W., Glennon, B.M. (1966) *Atomic Transition Probabilities, Vol. I: Hydrogen through Neon*, Nat. Stand. Ref. Data Series – Nat. Bur. Stand. **4**, U.S. Government Printing Office, Washington, D.C.

DISCUSSION

Garstang: I would like to open the discussion myself by asking the question I am frequently asked – Do I believe the results? The answer is – Yes, I do. While one cannot entirely exclude the possibility that some unsuspected configuration interaction may produce significant perturbations, this seems unlikely for the transitions of interest for forbidden lines. There is now some substantial evidence for the basic correctness of the results. The comparison of [Fe II] lines in η Carinae with theoretical values (Thackeray, *Mon. Not. R. astr. Soc.*, **135**, 1967, 23) shows astonishingly good agreement, confirming the broad overall accuracy of the relative line strengths. Recent experimental work on interference effects in Zeeman components of two lines of mixed magnetic-dipole and electric-quadrupole radiation of [Pb I] and [Pb II] by Hults (*J. opt. Soc. Am.*, **56**, 1966, 1298) shows excellent agreement with the relative contributions of the two kinds of radiation predicted by Garstang (*J. Res. nat. Bur. Stand., Sec. A*, **68**, 1964, 61). Finally, experiments (Husain and Wiesenfeld, *Nature*, **213**, 1967, 1227) on flash photolysis of trifluoro-iodomethane have led to an estimate of the lifetime of the upper state of the lowest doublet in I I within a factor 3 of the theoretical lifetime given by Garstang (*J. Res. nat. Bur. Stand., Sec. A*, **68**, 1964, 61). In view of the experimental difficulties in handling a state whose lifetime is of the order of 0·1 sec this agreement must be considered satisfactory. Taking all these results together I think we must regard the transition probabilities of forbidden lines as reasonably well established.

Menzel: Some mention has been made of high-level transitions in hydrogen. I have derived an asymptotic formula for the f-values of such transitions between levels of quantum numbers n and n' with $n - n' = c$.

The f-value is

$$f_{nn'} = \frac{4n'}{3c^2} J_c(c)\, J_c'(c) = n' M(c),$$

where $J_c(c)$ and $J_c'(c)$ are respectively the Bessel functions of equal argument and order and its derivative. Examples of $M(c)$ follow:

$$M(1) = 1{\cdot}9077 \times 10^{-1}, \quad M(3) = 8{\cdot}1056 \times 10^{-3}$$
$$M(2) = 2{\cdot}6332 \times 10^{-2}, \quad M(4) = 3{\cdot}4917 \times 10^{-3}.$$

Note that for $n' = 100$, $f_{101,100} = 19$. The question might be asked, How can one reconcile such large f-values with the well-known f-sum rule $f = 1$? The chief point is that one must sum over all transitions, both upward and downward, from level n'. The former are counted positive, the latter, negative. Downward f-values can be calculated from upward ones by the formula $f_{n''n'} = -n''^2 f_{n'n''}/n'^2$. With these formulas, the negative contributions nearly cancel the large positive contributions, leaving the exact remainder

$$4 \sum_{c=1}^{\infty} \frac{J_c(c)\, J_c'(c)}{c} = 1$$

a new theorem in the theory of Bessel functions.

POPULATION DE NIVEAUX FORTEMENT EXCITÉS

HOANG BINH DY et R. MCCARROLL
(Observatoire de Paris, France)

Depuis peu, on témoigne un grand intérêt aux régions nébulaires, suscité par la découverte des raies d'émission de radiofréquence dues aux transitions entre niveaux atomiques fortement excités. Etant donné que la fréquence ν de ces raies est telle que $h\nu/kT \ll 1$, il est évident qu'il faut tenir compte de l'émission stimulée. Mais, comme a remarqué Goldberg (1966), ceci s'avère difficile en raison de la grande sensibilité de l'émission stimulée aux écarts à l'équilibre thermodynamique. Par exemple, considérons le facteur de correction pour l'émission stimulée

$$1 - \frac{b_n}{b_{n'}} \exp(-h\nu/kT), \tag{1}$$

où b_n est l'écart à l'équilibre thermodynamique du niveau n, ν est la fréquence de la raie émise par la transition $n \rightarrow n'$. Dans les conditions typiques d'une région nébulaire, l'hypothèse que l'on puisse remplacer $b_n/b_{n'}$ par l'unité est loin d'être justifiée. Il nous a donc semblé intéressant d'étudier la variation de la quantité $\mathrm{d}b_n/\mathrm{d}n$ avec la densité et avec n.

Suivant Seaton (1964), nous considérons les processus, qui peuplent et dépeuplent le niveau n. Faisant l'hypothèse que b_n varie uniformément avec n, on aboutit facilement à l'équation pour b_n

$$\frac{\mathrm{d}^2 b_n}{\mathrm{d}t^2} + f(n) \frac{\mathrm{d}b_n}{\mathrm{d}t} + g(n)\, b_n + k(n) = 0, \tag{2}$$

où $t = R_\infty/n^2$ et R_∞ est la constante de Rydberg. Les fonctions f, g, k dépendent essentiellement des taux de réactions des processus de peuplement. Nous avons pris les mêmes taux radiatifs que Seaton; quant aux taux collisionnels, nous avons utilisé d'une part ceux proposés par Seaton et d'autre part ceux obtenus par la méthode de Gryzinski (1959).

L'Équation (2) est sujet aux conditions suivantes

(i) $b_n = 1$ pour $t = 0$;

(ii) b_n tend vers la solution radiative lorsque n devient petit – de l'ordre de 40.

Nous avons résolu l'Équation (2) en prenant une température électronique de 10^4 °K, pour une gamme de densités allant de 10 jusqu'à 10^4 cm^{-3}. Les résultats montrent en particulier que $\mathrm{d}b_n/\mathrm{d}n$ est très sensible à la densité électronique, ce qui suggère que l'on puisse utiliser les valeurs de $\mathrm{d}b_n/\mathrm{d}n$ à déterminer la densité électronique des ré-

Osterbrock and O'Dell (eds.), Planetary Nebulae, 153–154. © *I.A.U.*

gions nébulaires. Les valeurs calculées d'une part avec les sections proposées par Seaton et d'autre part avec les sections de Gryzinski, sont assez voisines.

Les détails de ces travaux seront publiés ailleurs.

Bibliographie

Goldberg, L. (1966) *Astrophys. J.*, **144**, 1225.
Gryzinski, M. (1959) *Phys. Rev.*, **115**, 374.
Seaton, M.J. (1964) *Mon. Not. R. astr. Soc.*, **127**, 177.

DISCUSSION

Osterbrock: I should like to describe briefly the theoretical work of Dyson on the populations of the highly excited levels of hydrogen. In solving for the level populations, he took account of collisional transitions of the type $\Delta n = \pm 1$ only, and in addition took into account absorptions and induced emissions of the type $\Delta n = \pm 1$. The bulk of the radiation is due to the free-free continuum, so the results depend on electron density, temperature, and emission measure. For small radiation fields, as in H II regions, collisional effects dominate, and Dyson's results are in close agreement with the recently published work of Hoang Binh Dy and McCarroll. At high radiation fields, as in bright planetaries, the effect of the radiative transitions is to decrease the differences in population between adjacent levels, and therefore to weaken the radio-frequency emission lines, or put another way, to decrease the temperature computed from the observed strength of a line below the temperature that would be computed neglecting radiative transitions. As a specific example, at $N_e = 10^4$ cm^{-3}, and for the 109α line, T_L is decreased by about 15% for emission measure 10^7 parsec cm^{-6}, and by about a factor 5 for emission measure 10^8 parsec cm^{-6}. The results cannot be considered numerically accurate, because all collisional and radiative transitions have not been included, but they indicate schematically the effects expected in planetaries.

Seaton: I think that it may be necessary to take additional collision processes into account in calculating populations of highly excited states.

THE DETERMINATION OF ELECTRON DENSITIES, ELECTRON TEMPERATURES AND CENTRAL-STAR TEMPERATURES FOR A SERIES OF PLANETARY NEBULAE

V. P. Arhipova and E. B. Kostjakova
(Sternberg State Astronomical Institute, Moscow, U.S.S.R.)

From theoretical treatments of the physical processes in planetary nebulae, the *average* physical conditions in these objects are known with some certainty. However, the physical parameters of the *individual* nebulae have not been determined exactly. This may be due to the different methods used for determining these quantities, and to the inhomogeneous data available. Therefore, it seems to be of great importance to obtain individual physical parameters for a large number of planetaries in a homogeneous system.

In our preceding paper (Vorontsov-Velyaminov *et al.*, 1968) we measured the intensities of emission lines, the intensities of the continuous spectra of the nuclei, and the diameters of the monochromatic images of several dozen planetaries by the same method. In the present paper we used the data obtained for the determination of the following parameters: the electron temperature, T_e, the electron density, n_e, and the temperature of the central star, T_* for 65 planetary nebulae.

The temperatures of the central stars, T_* were obtained by the well-known Zanstra methods. We applied the 'hydrogen' method to each of the first four lines of hydrogen, Hα, Hβ, Hγ and Hδ, and derived an average hydrogen temperature, T_H. The 'nebulium' method was applied to all the strong forbidden lines seen in our spectra. The same was done for the He I lines λλ 5876, 4471, and for the He II line λ 4686. The necessary values of $\mathbf{q} = \sum_2^\infty C_j N_e N_+ / A_{i2} N_i$ $(i=3, 4, \ldots)$ for hydrogen and helium were taken from the works of Zanstra, and Zanstra and Aller. It should be noticed that for about 20 of these planetaries either the continuous spectrum of the nucleus was not drawn exactly or the presence of the nuclear spectrum on the spectrograms was rather doubtful. In such cases, using the Zanstra method, we can obtain only a lower limit of the star temperatures.

The electron densities, n_e, were obtained from the measured nebular surface brightness in the hydrogen lines. As this method requires the knowledge of the distance to the nebula, we used three distance scales for the calculations, the scales of Vorontsov-Velyaminov, Shklovsky, and O'Dell. The derived density of a particular nebula, varies according to which distance is adopted, but by no more than a factor 3 even in the worst cases.

Osterbrock and O'Dell (eds.), Planetary Nebulae, 155–158. © *I.A.U.*

The electron temperatures were obtained by the following four methods:

The first is the well-known method, which uses the ratio of intensities of the nebular lines of [O III] to the auroral line;

The second is the method suggested by Sobolev, based on the energy balance of the free electrons in the nebula;

The third is the method based on the energy distribution in the Balmer continuum of the nebular spectra;

Finally, the fourth is the method of the Balmer discontinuity. The electron densities, necessary for this method, were taken from our results described above.

The results obtained are shown in Figure 1, where the planetary nebulae are arranged in order of decreasing Zanstra hydrogen temperatures. In the upper part of the figure the dots represent the hydrogen temperatures of the central stars, the open

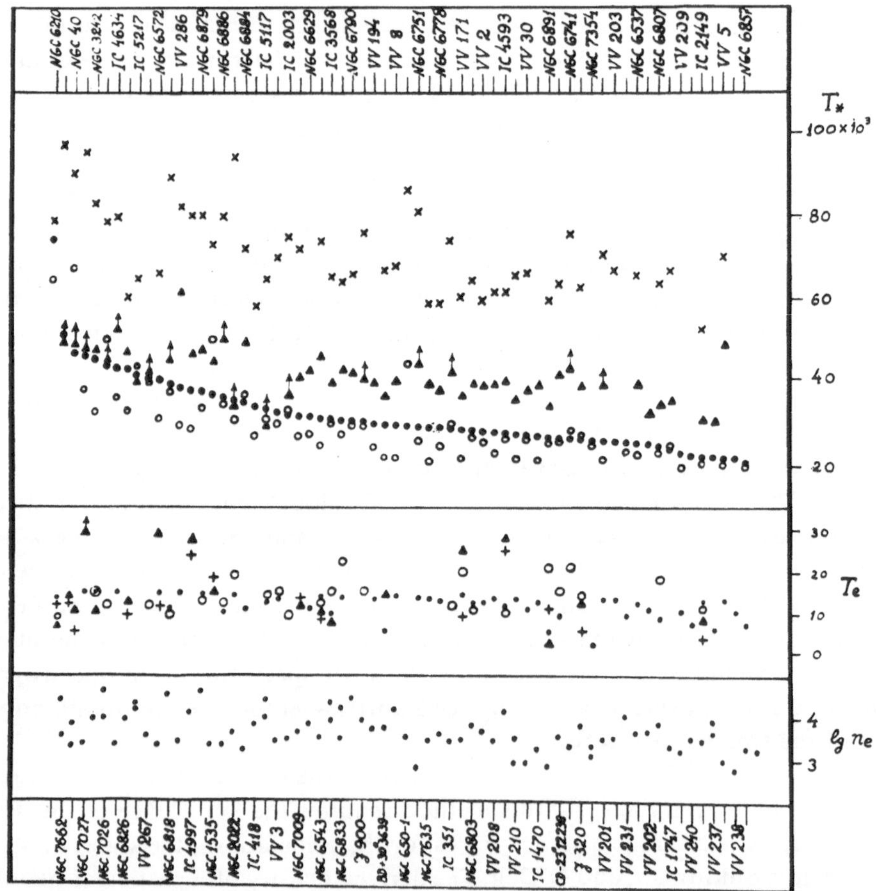

FIG. 1. *Temperature of central star, electron temperature, and electron density for 65 planetary nebulae, arranged in order of decreasing Zanstra hydrogen temperature of central star.*

circles show their 'nebulium' temperatures, T_{neb}, the triangles show the nuclear temperatures obtained by means of He I lines, and those obtained by means of He II lines are represented by crosses. In the middle of the figure the electron temperatures obtained for the same planetaries by the four methods described above are plotted; the electron densities are plotted at the bottom of the figure.

The same notations are used in Figure 2, but the planetaries are arranged in order of decreasing He II star temperatures.

Figures 1 and 2 show that for the majority of planetaries the 'nebulium' star temperatures, T_{neb}, are lower than the Zanstra temperatures, T_H. This may be due to the fact that not all of the forbidden lines were considered, particularly in the infrared and perhaps also in the ultraviolet. It may also testify to the complete absorption of

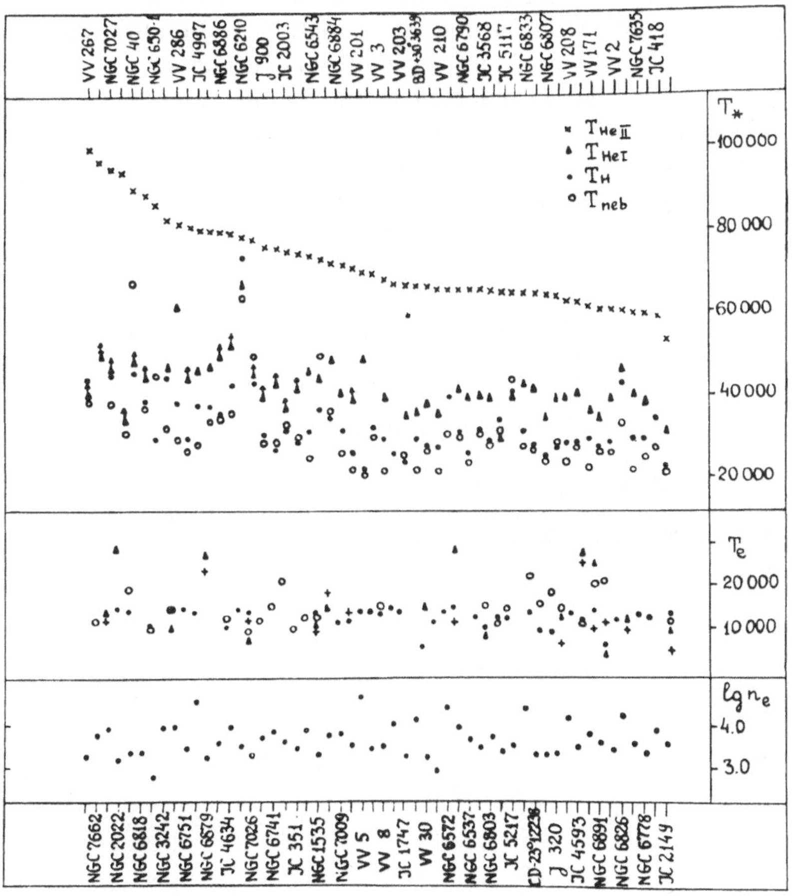

FIG. 2. *Temperature of central star, electron temperature, and electron density for 65 planetary nebulae, arranged in order of decreasing Zanstra He II temperature of central star.*

the ionizing ultraviolet quanta by the hydrogen atoms in the nebula. Only a few planetaries show 'nebulium' temperatures exceeding their 'hydrogen' temperatures, $T_{neb} > T_H$, most of them being in agreement with Zanstra's results. The nuclear temperatures obtained from the He I lines are on the average about 10000° higher than T_H, and the He II temperatures are still higher.

The ratios of line intensities $I(\lambda\, 4686)/I(\mathrm{H}\beta)$ and $I(\lambda\, 4686)/I(\lambda\, 5876)$ were obtained, and when interpreted according to Seaton's criteria they showed that in the large majority of these planetaries there is complete absorption of the ionizing quanta by hydrogen atoms. Also, according to the second criterion of Seaton, many planetaries, especially those with the hottest nuclei, have incomplete absorption of quanta $\lambda < 504$Å, mainly by He I atoms. This means that we have obtained only a lower limit to the He I temperatures for them. In Figures 1 and 2 such cases are marked by arrows.

We prefer to explain the differences between the derived star temperatures of a particular nebula by a deviation of the star radiation in the ultraviolet from black-body radiation, rather than by the influence of optical thickness.

Finally, we can see in Figure 1 that the electron temperatures (and even the electron densities) of the planetaries studied show on the average a slight increase with increasing nuclear temperature, which probably means that conditions in planetary nebulae are not entirely thermostatic.

Reference

Vorontsov-Velyaminov, B.A., Kostjakova, E.B., Dokuchaeva, O.D., Arhipova, V.P. (1968) in the present volume, p. 57.

DISCUSSION

Seaton: I agree with the authors of this paper that the 'nebulium' method of determining star temperatures may be in error due to emission in the infrared and ultraviolet. The only sure procedure is to make a detailed study of the thermal balance.

ON THE METHODS OF CORRECTION FOR THE INTERSTELLAR REDDENING OF PLANETARY NEBULAE

V. P. ARHIPOVA
(Sternberg State Astronomical Institute, Moscow, U.S.S.R.)

Knowledge of the interstellar extinction is of great importance for the study of the properties of planetary nebulae. An estimate of the interstellar reddening for an individual nebula may be obtained by means of three independent methods. In the first method the observed Balmer decrement is compared with the theoretical decrement (case B of Baker and Menzel). The second method consists of a comparison of the observed ratio of Balmer and Paschen intensities of lines, arising from a common upper level, with the theoretically predicted ratio. Finally, for thermal planetary nebulae with well-determined radio spectra, the optically thin part of their spectra can be used for a determination of interstellar reddening. Comparison of the radio flux from an optically thin nebula with the observed Hβ flux (in absolute units) allows us to evaluate the amount of extinction. Radio observations may thus give the interstellar extinction with high accuracy. It may especially be noted that the Balmer-decrement method, which is widely applied, gives satisfactory accuracy only if highly accurate H-line intensities are used. In all other cases it gives the most uncertain results, since its wavelength base-line is small. One must also take into account the fact that the Hα-line intensity is often not measured.

In the present work a comparison has been made among the values of the interstellar extinction at the Hβ-line derived for 28 planetary nebulae by the above-mentioned methods. For each nebula the observed Balmer decrement was deduced using intensities of the hydrogen lines taken from the literature. The relative intensities of the Hα, Hβ, Hγ, and Hδ lines were used for the determination of the interstellar extinction. The theoretical Balmer decrement was taken from Burgess, and the reddening law of Whitford was used. Published radio flux densities of planetary nebulae at frequencies 430–3000 MHz taken from the works of Terzian and Menon, Lynds, Slee and Orchiston, Davies *et al.*, and absolute Hβ fluxes from O'Dell and Vorontsov-Velyaminov, Kostjakova, Dokuchaeva and Arhipova, were used for the determination of the extinction by the third method.

The results are given in Table 1. The table includes all nebulae for which the interstellar extinction may be computed by at least two of the three methods mentioned. The first column of the table gives the nebula, and the following three columns show its interstellar extinction at Hβ, in stellar magnitudes, obtained by the three methods.

Osterbrock and O'Dell (eds.), Planetary Nebulae, 159–161.

Table 1

Interstellar extinction $A\beta$

Nebula	From Balmer decrement	From Paschen/Balmer	From radio method	Average
NGC 1535	0·25		0·48	0·36
2022	1·3		1·3	1·3
2392	0·6	0·5	0·91	0·66
3242	0·65		0·70	0·67
6210	0·4		0·44	0·42
6543	0·57	0·75	0·35	0·56
6572	0·6	0·85	0·87	0·77
6741	1·4		1·01	1·1
6818	0·0		0·41	0·2
6826	0·70	0·72	–	0·71
7009	0·06		0·30	0·18
7027	3·0	3·10	3·6	3·25
7662	0·8	0·9	0·83	0·84
IC 418	1·2	0·96	1·12	1·1
2149	1·6	1·6	1·5	1·57
4593	0·35		0·46	0·40
4634	1·7		1·7	1·7
4997	1·3	1·53	–	1·41
BD + 30°3639	1·7	1·6	–	1·65
NGC 246	2·2		0·0	0·0
1514	2·8		1·2	1·2
3587	1·4		0·30	0·30
4361	1·9		0·73	0·73
6720	1·0		0·58	0·6:
6853	0·9		0·23	0·23
7293	1·55		0·12	0·1

The values of the interstellar extinction derived from the Paschen/Balmer ratio method were taken from the work of O'Dell. The last column gives the average interstellar extinction for each nebula. As the table shows, the agreement of results is generally good within the random errors in line intensities and those in the radio fluxes. One can also see that the extended nebulae with lower surface brightness show a noticeable discrepancy between the interstellar reddening values obtained by the optical method and by the radio method. These seven nebulae are listed at the foot of the table. The discordance is too large, being beyond the error limits.

The relative intensities of the $H\beta$, $H\gamma$ and $H\delta$ lines used for the determination of extinction of these nebulae were measured chiefly by Minkowski. The extinction derived from the radio-observation method is considerably smaller than the extinction derived from the Balmer-decrement method. There are several independent reasons for adopting the smaller values of interstellar extinction for these particular planetaries: their high galactic latitudes, their relatively small distances from the Sun, etc.

One can show that the Balmer decrements of the nebulae with low surface brightness observed by Minkowski have a systematic error; his relative intensities of the Hγ and Hδ lines are too low. Therefore the interstellar extinction for these nebulae should be determined from the radio data. Furthermore, it is very desirable that the line-intensity measurements for these nebulae be repeated.

The good numerical agreement of interstellar extinction values derived by the Balmer-decrement method and by the radio method means that the theoretical Balmer decrement seems to be true (at least for the first four lines).

DISCUSSION

Minkowski: My observations, reported in 1942, were based on a few slit spectrograms per object. They therefore cover a negligibly small part of the area of any of these large, low-surface brightness nebulae, and the decrements derived from them may not necessarily be representative of the nebulae as a whole.

Osterbrock: Was the extinction determined from the radio/Hβ fluxes calculated under the assumption $T_e = 10^4$?

Arhipova: Yes.

ON THE DETERMINATION OF PHYSICAL CONDITIONS IN THE NEBULAE

A. A. Boyarchuk, R.E. Gershberg, N. V. Godovnikov, and V. I. Pronik
(Crimean Astrophysical Observatory, U.S.S.R.)

Following the well-known physical theory of recombination and forbidden-line emission, we have carried out calculations which may be useful for a quantitative analysis of the observations of planetary nebulae and other emission objects (diffuse nebulae, emission details in extragalactic nebulae, symbiotic and flare stars). As the result, we have a set of four types of graphs.

In the first-type graphs the intensity and spectral distribution of continuous hydrogen plasma radiation (free-free, free-bound and two quanta transitions) are given for different T_e and n_e. The Balmer and Paschen discontinuities as functions of T_e and n_e are also given.

In the second-type graphs, the ratios of auroral and nebular lines of [O III], [N II], [Ne III], [Ne V], [O II] and [S II] ions are shown. These ratios are given in the plane ($\log T_e$, $\log n_e$) as curves of constant ratios and can be used for the determination of n_e and T_e by Seaton's curve-intersection method.

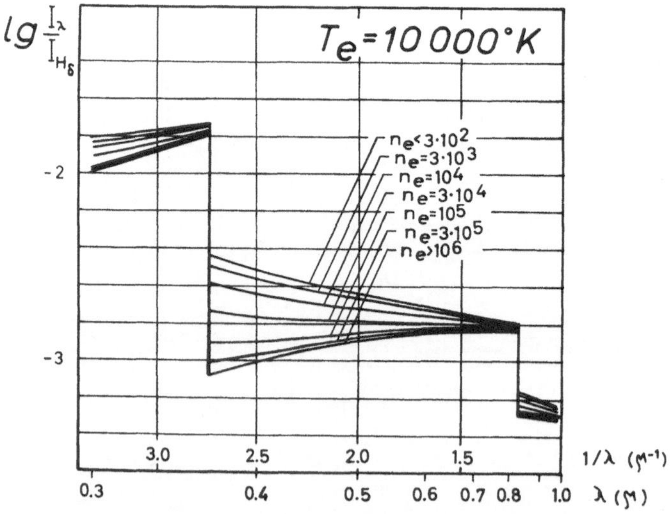

FIG. 1. *Intensity distribution in continuum at $T_e = 10000°$ as a function of n_e.*

Osterbrock and O'Dell (eds.), Planetary Nebulae, 162–165. © *I.A.U.*

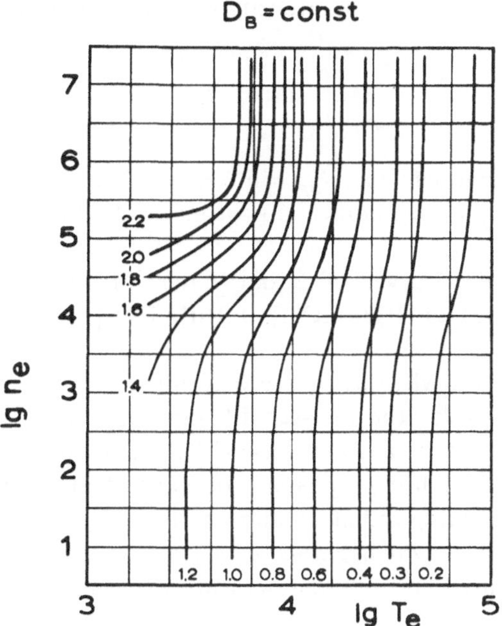

FIG. 2. *Balmer discontinuity as a function of* n_e, T_e.

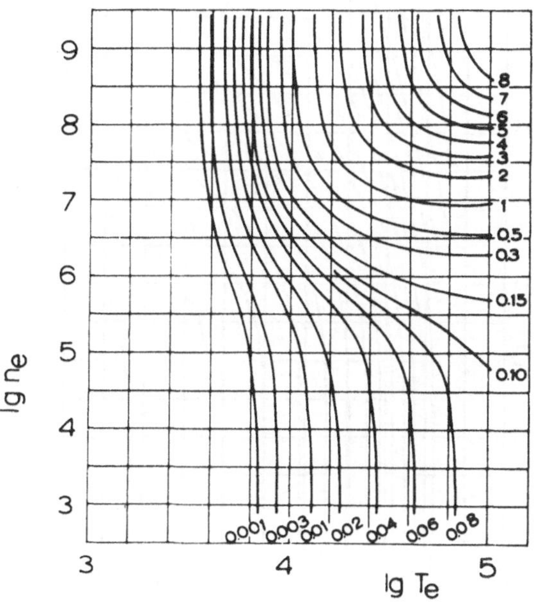

FIG. 3. *[O III]* $I_{\lambda 4363}/(I_{\lambda 4959}+I_{\lambda 5007})$ *as a function of* n_e, T_e.

For objects optically thin in Hβ the intensity of a forbidden line is given by

$$\frac{I_i}{I_{H\beta}} = \frac{n_1}{n_p} \theta_i(n_e, T_e),$$

where I_i=intensity of forbidden line, $I_{H\beta}$=intensity of Hβ line, n_1=the number of ions in the ground state, n_p= the number of protons, and $\theta_i(n_e, T_e)$=known functions of T_e and n_e.

In the third-type graphs, curves of equal values of θ_i are given in the plane ($\log T_e$, $\log n_e$) for 14 lines of [OI], [OII], [OIII], [NII], [SII], [SIII], [NeIII] and [NeV].

For stationary conditions, when n_1/n_p are determined by the abundances and the Saha-Boltzmann equations for the non-equilibrium case, the ratios $I_i/I_{H\beta}$ and the intensities I_i can be determined as functions of T_e, n_e, temperature of radiation T_* and dilution factor W. These last graphs can be used to analyze some objects with secular variations of the exciting radiation and/or other physical conditions. All these graphs will be published in *Publications of Crimean Astrophysical Observatory*, **38** and **39**. All formulae used and references of atomic constants are given there.

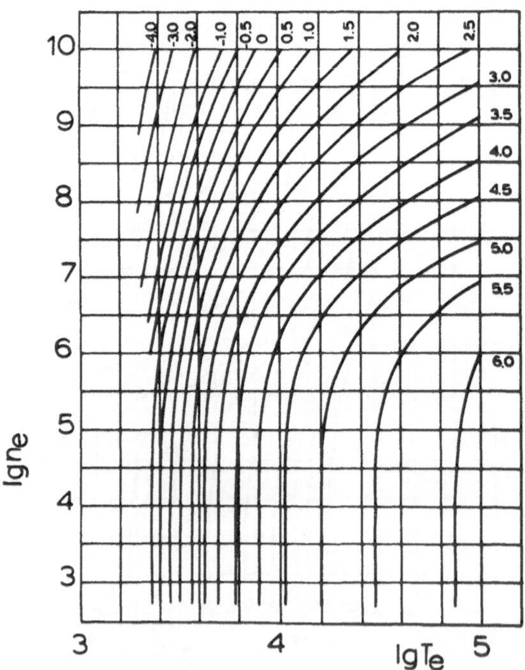

FIG. 4. *[OIII] lg* $\theta_{\lambda 4959+\lambda 5007}$ *as a function of* n_e, T_e.

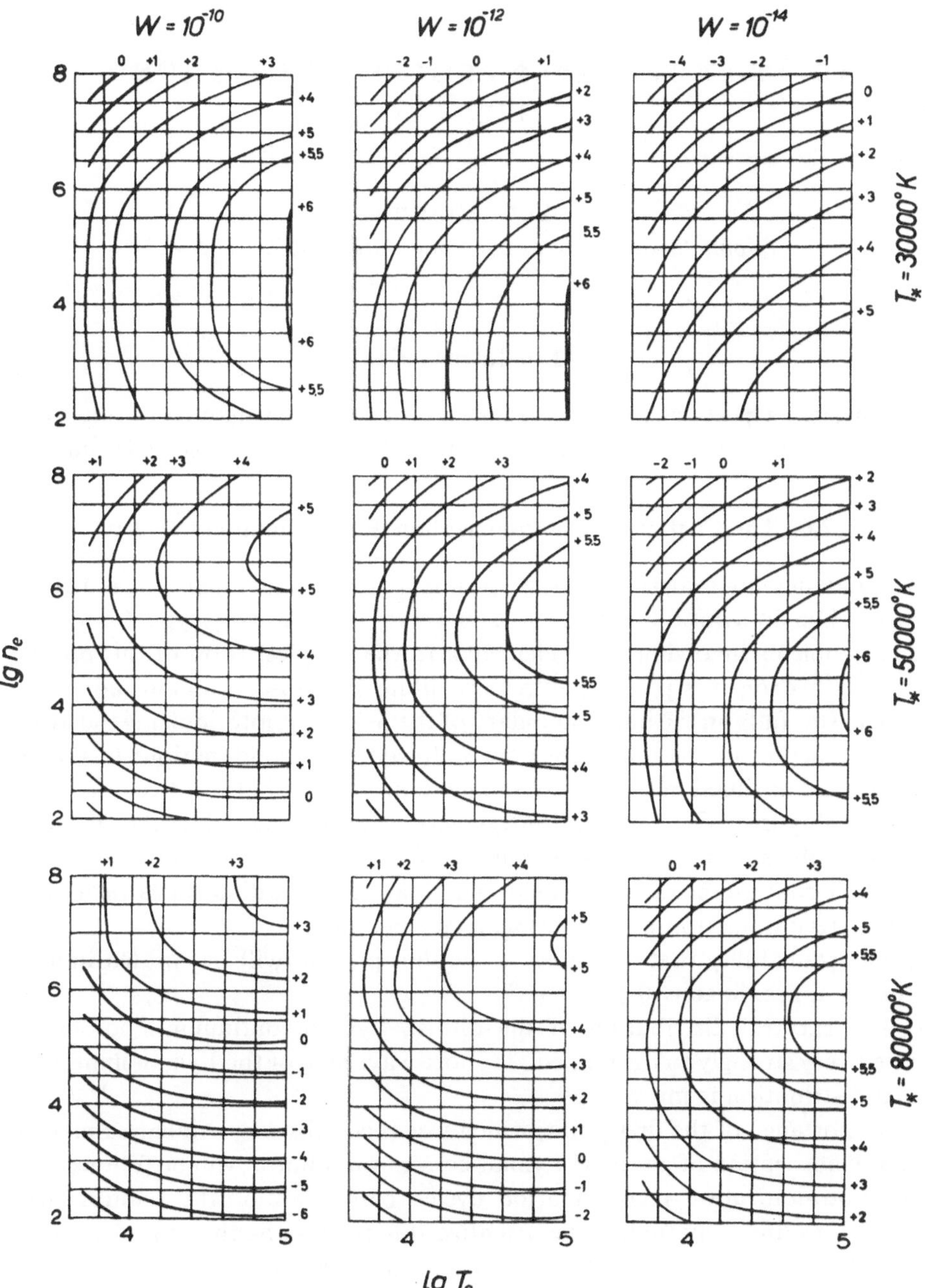

FIG. 5. $lg[\{(I_{\lambda 4959}+I_{\lambda 5007})/I_{H\beta}\}\times\{N(H)/N(O)\}]$ *as a function of* n_e, T_e *for various dilution factors and various stellar temperatures* T_*.

RADIATIVE TRANSFER PROCESSES IN PLANETARY NEBULAE

D. G. Hummer
(Joint Institute for Laboratory Astrophysics, U.S.A.)

1.0. Introduction

The physical aspects of radiative transfer processes that may be of importance in planetary nebulae are discussed, and recent work on these problems is summarized.

2.0. The Continuum Photo-ionization–Recombination Problems

The fundamental radiative transfer problem in planetary nebulae is that of the hydrogen Lyman continuum, which also serves as the prototype for continuum transfer problems involving other elements. In the simplest form of this problem, in which only hydrogen is present, extremely dilute Lyman-continuum radiation from the central star falls on the inner boundary of the nebula, photo-ionizing the hydrogen. A certain fraction of the ionizations are followed by recombinations to the ground state, with the production of 'diffuse' Lyman-continuum photons. Because the diffuse field is roughly isotropic and has sources within the nebula, while the stellar field is radial and satisfies a simple transfer equation, it is convenient to maintain this division of the total continuum radiation field. There are two cardinal facts relating to the ionization-recombination process in planetary nebulae:

(1) The probability that a Lyman-continuum photon will be emitted in a recombination is approximately 0·4.

(2) The spectral distribution of the diffuse Lyman-continuum photons is concentrated very strongly to the immediate short-wave side of the Lyman limit, while the original distribution is much wider.

The importance of the first point, which was recognized by Ambartsumian (1932), is that the absorption of a typical Lyman-continuum photon will be followed by only a few photo-ionization–recombination cycles before all of the photon energy is degraded into line radiation. If we visualize this process as the repeated absorption and emission of a photon, we can say that a diffuse photon can move only a few free paths from its point of creation. This diffuse photon will, of course, change its frequency at each scattering because of the 'reshuffling' of the continuum states, a process first treated in this context by Baker *et al.* (1939). The second point indicates

Osterbrock and O'Dell (eds.), Planetary Nebulae, 166–184. © *I.A.U.*

that diffuse photons see the maximum photo-ionization cross-section, i.e. their free paths are as short as possible.

Zanstra (1951*a*), recognizing that these points together imply that a diffuse photon does not travel very far before being converted into line photons, introduced a very useful approximation: that the diffuse photons are degraded at the point where they are created. This so-called *on-the-spot* (OS) approximation reduces the transfer problem to the evaluation of simple integrals, which have been tabulated by De Jong (1951) and Hummer and Seaton (1963) for hydrogen, and by Hummer and Seaton (1964) for ionized helium. In particular, the OS approximation is much easier to apply than is the method used by Aller *et al.* (1939), in which an integro-differential equation for the diffuse radiation field is solved approximately by iteration. The OS approximation can also be applied to the ionization–recombination problem for other elements. Necessary conditions for its validity are

(1) The probability per recombination that a continuum photon, as opposed to line photons, is produced must be small, and

(2) The electron temperature must be low enough for kT_e to be small in comparison to the ionization energy.

A further condition for the validity of the OS approximation may be obtained by estimating the Lyman-limit optical distance between the point where a typical diffuse photon is created and the point where it finally is degraded into line photons. We refer to this distance as the 'break-up length' and represent it by Λ. Obviously photons created within an optical distance Λ of a boundary will most likely escape before being degraded, so the OS approximation fails there. If the optical thickness of the nebula is of the same order as Λ, the OS approximation fails everywhere. Thus a third condition for the validity of the OS approximation is that the optical thickness be larger than Λ.

Van Blerkom and Hummer (1967), in analogy with an unpublished expression derived by Rybicki for line scattering, have obtained the result

$$\Lambda = l/\sqrt{\varepsilon},$$

where

$$l^2 = \frac{2}{3}\int_{\nu_1}^{\infty} [f(\nu/\nu_1)]^{-1} B_\nu(T_e)\, d\nu \Big/ \int_{\nu_1}^{\infty} f(\nu/\nu_1)\, B_\nu(T_e)\, d\nu$$

and

$$\varepsilon = 1 - (\alpha_1/\alpha_{tot}).$$

Here $f(\nu)$ is the photo-ionization cross-section normalized to unity at the Lyman limit ν_1; and α_1 and α_{tot} are, respectively, the recombination coefficients to the ground state and to all states. Using Kramers' approximation, $f = (\nu/\nu_1)^{-3}$, we have

$$l^2 \simeq \frac{2}{3}\left\{1 + 6\left(\frac{kT_e}{h\nu_1}\right) + 24\left(\frac{kT_e}{h\nu_1}\right)^2 + \cdots\right\}, \quad h\nu_1 \gg kT_e.$$

If the nebula is sufficiently thin, the diffuse radiation is much weaker than the direct stellar radiation and the ionization balance is reasonably accurate, even if the OS approximation is not. Because of the evidence given by Harman and Seaton (1966), that planetary nebulae are not completely opaque in the Lyman continuum for a substantial fraction of their lives, Van Blerkom and Hummer (1967) have derived a modified OS approximation which makes some allowance for the boundaries and can be used for all optical thicknesses. Comparison with accurate numerical solutions indicates that, for the conditions of planetary nebulae, the error in the radiation field in the improved approximation is less than 10%. This approximation turns out to coincide with that obtained by bringing the source function through the integral sign in the integral form of the transfer equation, an approximation used earlier by Biberman (1948) and by Sobolev (1957) for line transfer problems (a rather unreliable procedure) and by Biberman *et al.* (1965) for continuum problems. This work has also confirmed that diffuse radiation entering the nebula from the opposite hemisphere is generally unimportant.

For nebulae sufficiently thick so that all Lyman-continuum radiation is absorbed, the OS approximation is considerably more accurate than the well-known formula developed by Strömgren (1939). Although the radius of the ionized region agrees with Strömgren's estimate, the very sharp transition region is smoothed out because the stellar photons with high energies see a comparatively small photo-ionization cross-section and also because of the diffuse radiation. A further point, which has recently emerged, is that the size of the ionized region and the degree of ionization depend primarily on the *total number* of ionizing photons entering the nebula and only weakly on the spectral distribution.

The effects of electron collisions on the ionization and thermal equilibrium and on the distribution of excited states have been considered by a number of authors, the most recent being Chamberlain (1953), Hummer (1963*a*), and Parker (1964). When the nebula is excited by a central star, then collisional excitations and ionization are unimportant compared to photo-ionization and recombination, except perhaps in the thermal balance of the outermost regions. In situations where the primary excitation mechanism is collisional, collisions from the ground state determine the state of excitation and ionization in optically thin nebulae. However, Van Blerkom (1968) has shown that diffuse radiation becomes important in populating excited states if the nebula is thick.

2.1. Models including Helium and Heavy Elements

The picture becomes considerably more complicated when helium and heavier elements are added to the hydrogen. In the first place, the ionization continua of the different elements overlap, so that in many spectral regions two or more elements are competing for the same radiation. The second problem is that radiation emitted in the recombination and cascade of highly ionized systems can ionize species in lower stages of ionization, the most important example of which is the ionization of hydrogen and helium by He II Ly-α. From a technical point of view, both of these effects lead to severely non-linear transfer problems. Additional complications arise from fluorescent mechanisms, such as that discovered by Bowen, caused by the coincidence of lines of different elements.

The first solution for a combined hydrogen-helium model was obtained by Eberlein (1955, 1957), who used a differential formulation of the Strömgren theory with a mean absorption coefficient to determine the ionization balance between He^{+} and He^{++}, along with that for hydrogen. He obtained the radii of the ionization zones and the λ 4686 Å/Hβ ratio as a function of the relative abundance of helium. Osaki (1962) also discussed a multi-component model, but as he was interested primarily in the thermal balance rather than the ionization balance, he used rather crude estimates for the ion densities.

Hummer and Seaton (1964) gave an extensive, though approximate, discussion of the hydrogen-helium problem. The He^{+}–He^{++} balance was solved using the OS approximation. Because of the fortunate circumstance that the Ly-α line, the two-quantum continuum, and the Balmer continuum of He II were sufficiently strong to keep the hydrogen in the He^{++}-zone fully ionized, the stellar radiation with $\nu < 4\nu_1$ could, to a first approximation, be regarded as passing through this zone unimpeded. Although the spectral distribution would be altered somewhat, the number of photons available for the ionization of H and He beyond the He^{++} zone would be correct. Beyond the He^{++} zone, the ionization equilibria involving neutral hydrogen and neutral helium in this radiation field were solved using a differential formulation of the OS approximation (capable of handling any number of elements and stages of ionization). Probably the most important result of this investigation was that for star temperatures below about 50000 °K the helium spectrum reflects the stellar temperature, while at higher temperatures the helium spectrum depends only on the helium abundance.

The most complete treatment published to date is that of Goodson (1967), who solved the combined ionization and thermal balance problems, including H, He, C, N, O and Ne, using the stellar fluxes computed by Böhm and Deinzer (1965, 1966). Goodson used the OS approximation as an initial estimate for an iterative procedure, although the results of the iterative calculation did not differ greatly from the OS results. Probably the least satisfactory aspect of Goodson's work is his treatment of He II Ly-α, which he assumes to be absorbed on the spot by hydrogen alone. This assumption considerably overestimates the rate of ionization of hydrogen in the He^{++} region and underestimates it elsewhere. It is possible that He II Ly-α diffuses out of the region where it is created and is absorbed mainly in ionizing helium and hydrogen at the places where they are becoming neutral, thereby altering the conditions in the transition zone. Goodson's neglect of the ionization of helium by He II Ly-α is probably also unjustified, since for He II Ly-α the helium photo-ionization cross-section is about five times that of hydrogen, compensating for the lower abundance of helium. For both of these reasons, Goodson overestimates the energy of the free electrons in the He^{++} region at the expense of the electrons elsewhere. Although Goodson's temperatures are probably somewhat in error, the ionization equilibria are probably quite good, in the sense that they represent the stellar fluxes assumed.

More recent calculations of ionization equilibria are reported in these Proceedings by Flower and by Williams. Ionization equilibria including heavy atoms have been computed for *dynamical* models of H II regions by Hjellming (1966), who employed the OS approximation.

3.0. Lyman-α Lines

The most important of the resonance-line transfer problems are those of H I, He I and He II Ly-α. It is convenient to regard the hydrogen line as the prototype, if only because it has received the most attention. These lines are distinguished by their large optical thickness and by the very strong sources feeding energy into them. This combination of factors leads to the possibility of very large intensities. For example, the line centre optical thickness of a planetary nebula in H I Ly-α is about 10^4 times that at the Lyman limit (assuming $T_e = 10^4$ °K) and, in the simplest picture, the line receives about $\frac{2}{3}$ of the photons in the stellar Lyman continuum.

Generally, the intensity of radiation in a line is determined by the relative rates at which photons enter and leave the line. While the rate at which photons are fed into these lines is given simply by recombination theory, the rate at which they leave depends on many factors. The most obvious method is to escape through the boundary of the nebula. However, if the optical thickness of the nebula in the line is very large, the number of scatterings necessary to escape can be extremely large, so that processes with a small probability per scattering of destroying photons can become important.

3.1.0. Escape mechanisms

The rate at which photons escape from the nebula is determined mainly by four factors: noncoherent scattering, velocity gradients, thermal gradients, and blanketing by interstellar hydrogen.

3.1.1. Noncoherent scattering

Henyey (1941) first pointed out that when a photon is scattered by a moving atom, its initial and final frequencies as seen by a stationary observer will in general differ because the change in the photon's direction will cause the Doppler shift to vary. This so-called *noncoherence* was recognized by Sobolev (1944, 1947) (in a paper written in 1941) and independently by Zanstra (1949, 1951*a*) as being of primary importance for the Ly-α problem, since it allows photons created in the very opaque line core to be scattered into the relatively transparent wings where they can more readily escape. A general discussion of noncoherent scattering is given by Hummer (1962).

Two forms of noncoherent scattering are encountered in planetary nebulae. At a kinetic temperature of $10^4\,°K$ (assumed equal to the electron temperature), the ratio of the natural width to the Doppler width is $3{\cdot}7 \times 10^{-4}$ for H I Ly-α and $3{\cdot}8 \times 10^{-3}$ for He II Ly-α. Within the core of the line, out to some three Doppler half-widths, the absorption coefficient is dominated by Doppler broadening, and the emission coefficient is approximately proportional to the absorption coefficient, i.e. there is no correlation between the absorption and emission frequencies. In the line wings, which are dominated by natural broadening, the scattering has a strong coherent component, because photons lying far from the line centre are scattered mainly by atoms in the low-velocity part of the Maxwellian distribution. These two cases are referred to as *complete* and *partial redistribution*, respectively. If the line-centre optical thickness is not too large, less than e.g. about 10^4 for H I Ly-α, then the line becomes optically thin at frequencies which are still within the Doppler core and complete redistribution can safely be assumed for the entire line.

Complete solutions of the radiative transfer problems can be obtained only by numerical methods; a very comprehensive review of these techniques is given by Hummer and Rybicki (1967). The first accurate numerical solution for complete redistribution was obtained by Koelbloed (1956) for isothermal plane-parallel nebulae with line-centre optical thickness as large as 2×10^4. When partial redistribution must be taken into account, the computational problem is much more difficult. The earliest attempts were too crude to yield anything but a qualitative picture of the radiation field. Unno (1951, 1955) obtained the first solutions for H I and He II Ly-α which were essentially correct, although not very accurate, because of the number of approximations he was forced to make. His results demonstrated clearly that while the mean intensity in the line core was quite large the flux was large only in the far wings. Hearn (1964) and Hummer (1968) have obtained accurate numerical solutions using a Čebyšev expansion technique and a generalized discrete-ordinate method, respectively, although these solutions were not obtained in the context of the nebular line problem. Auer (1968) has employed a Monte-Carlo technique specifically to study the Ly-α line in planetary nebulae.

For many problems it is sufficient to know the average number of scatterings experienced by a photon before leaving the nebula, or alternatively, the mean probability of escape. When complete redistribution obtains, a simple argument given by Zanstra (1949) may be used. The basic assumption is that a photon will escape if its monochromatic optical distance from the nearest surface is less than unity and that it will be re-scattered otherwise. If $\phi(\nu)$ is the normalized emission coefficient, equal by assumption to the normalized absorption coefficient, the probability of escape from a point with line-centre optical depth τ_0 is

$$p(\tau_0) = 2 \int_{\nu_c}^{\infty} \phi(\nu)\, d\nu\,,$$

where ν_c is defined by

$$\frac{\phi(\nu_c)}{\phi(\nu_0)}\,\tau_0 = 1\,.$$

For Doppler scattering, $\phi(\nu) = \exp(-(\nu - \nu_0/\Delta)^2)/\Delta\sqrt{\pi}$ and

$$p(\tau_0) = \frac{2}{\sqrt{\pi}} \int_{\sqrt{\ln \tau_0}}^{\infty} e^{-x^2}\, dx\,, \quad \tau_0 \gg 1\,.$$

Strictly speaking, this argument applies only to a one-dimensional nebula, since the relevant quantity is the monochromatic optical depth measured along the photon's direction of flight and not, as assumed above, the optical depth normal to the surface. Obviously, to obtain the *mean* escape probability, $p(\tau_0)$ should be multiplied by the depth distribution of emission and the quantity averaged over depth. The objections are partially overcome by Hummer (1964), and very recently Sobolev (1967) has obtained rigorous bounds and asymptotic formulae for the mean number of scatterings. Capriotti (1965) has applied Zanstra's argument to a uniform spherical nebula expanding with a constant velocity gradient and has given a number of approximate expressions for mean escape probabilities.

When partial redistribution is important, calculation of escape probabilities or the mean number of scatterings is much harder. Osterbrock (1962), in a very important paper, has taken advantage of the fact that a typical Ly-α photon does not move very far from its point of creation to the point where it experiences its *last* scattering before escaping, i.e. the photon remains at roughly the same depth while undergoing enough frequency changes to get it into the wing from which it escapes. Osterbrock also assumes that no photons escape in the line core, which he takes as extending 3·25 Doppler half-widths on either side of the line centre. Photons arriving in the wing escape with a certain probability, otherwise scattering back into the core. By following the history of each photon on the basis of an approximate frequency-diffusion theory, Osterbrock is able to calculate the mean number of scatterings experienced by a photon created at a depth τ_0. His results differ at most by a factor of 3 from Zanstra's conclusion that $\langle N \rangle \sim \tau_0 \sqrt{\pi \ln \tau_0}$.

It would appear that Osterbrock's estimates are probably larger than the exact values. In the first place, Auer (1968) finds that approximately $\frac{1}{4}$ of the escaping photons have frequencies in the core. If the number of scatterings is large, a core photon can diffuse a considerable distance, especially at frequencies near the somewhat artificial division between core and wing. The outward decrease of the kinetic temperature in real nebulae also provides a bias towards preferential outward scattering. Some error also arises from Osterbrock's assumption that all photons are created at depth τ_0, which he takes to be the optical thickness of the nebular shell. If the nebula is optically thick in the Lyman continuum, then most of the Ly-α photons are produced near the inner boundary and identifying τ_0 with the optical thickness is correct. However, if the Lyman continuum is optically thin, then Ly-α photons are created more or less uniformly throughout the shell, so that Osterbrock's (and Zanstra's) estimates should be averaged over the shell.

The time development of the Ly-α line in a very thick shell has been studied in a very general manner by Field (1959) and by Ivanov (1967), on the basis of partial and complete redistribution, respectively. Ivanov has shown that the leading term in the asymptotic solutions for long times is identical for both kinds of redistribution.

3.1.2. Velocity Gradients

If a nebula is expanding with a velocity gradient, the line opacity will be shifted in frequency by increasing amounts as one proceeds outwards through the nebula. After Ambartsumian's (1932, 1933) and Chandrasekhar's (1935) pioneering investigations of the Ly-α radiation field in static and *uniformly* expanding nebulae led to catastrophic radiation pressure arising from the unrealistic assumption of coherent scattering, Zanstra (1934, 1936) introduced the effects of velocity gradients. He also assumed coherent scattering and found a drastic reduction in the radiation pressure. Sobolev (1947) treated this problem in a more general and practical way and found that Zanstra's treatment was in error.

Sobolev's original investigation included the effects of both velocity gradients and noncoherent scattering, but was based on the rather unsatisfactory assumption of a rectangular absorption profile. However, in a later paper, which does not seem to have attracted the attention it deserves, Sobolev (1957) generalized his treatment to include an arbitrary absorption coefficient, with complete redistribution, and allowing for an arbitrary distribution of expansion velocities with depth. Most of Sobolev's results are presented for a *constant* velocity gradient.

Unfortunately, no numerical results are given in Sobolev's paper. It is easy to see, however, that for typical expansion velocities of two or three times the mean thermal velocity, the magnitude of the average internal radiation field will not be seriously affected, although the detailed frequency dependence will be modified. For example, if the expansion velocity increases linearly with optical depth from the inner boundary and becomes three times the mean thermal velocity at the outer boundary, then roughly speaking, the optical thickness at any frequency in the line cannot be less than $\frac{1}{3}$ of its value in a static nebula, assuming Doppler broadening. Since the scattering is, in fact, noncoherent, the mean number of scatterings and therefore the average radiation field, is reduced by about the same factor. On the other hand, if the scattering were coherent, then this reduction would be by roughly a factor of 9, which shows why velocity gradients appeared to be so important in the early studies. Capriotti (1965) has estimated that the mean probability of escape from a nebula with a line-centre optical thickness 10^4 increases by only 50% as the expansion velocity at the outside increases from zero to three times the mean thermal velocity.

If one adopts an isothermal model of a planetary nebula and ignores the natural line width, the conclusion is that velocity gradients of the magnitude observed do not seriously alter the physical conditions within the nebula, although the details of the radiation field are affected.

Note added in proof. The accurate numerical solutions of the line transfer problem in cluding arbitrary velocity gradients recently obtained by Hummer and Rybicki (*Astrophys. J.*, in press) verify the predictions of this section. Moreover, nebular shells expanding differentially towards the observer were found to cause red-shifted line profiles. This effect is probably that seen in Vaughan's observations of the λ 10830 line (these Proceedings and *Astrophys. J.*, in press).

3.1.3. Thermal Gradients

According to presently accepted ideas, the electron and kinetic temperatures decrease toward the outer edge of the nebula, slowly in the H II region and very rapidly at the boundary between the H II and H I regions. The general effect of such a temperature gradient is to reduce the opacity in the wings, thereby increasing the rate at which photons escape from the nebula. The effects of thermal gradients on resonance line transfer have been studied in detail by Hummer and Rybicki (1966) and Rybicki and Hummer (1967), who present extensive numerical results.

From an approximate solution of the nebular Ly-α problem with a thermal gradient, Pleškova (1962) found that such gradients as may plausibly exist in H II regions have little influence on the Ly-α radiation field in planetary nebulae. Even when an H I region is present, it can exert very little influence on the Ly-α radiation field, because the line opacity is concentrated strongly in the region of the line centre, while the radiation leaving the H II region is appreciable only in the line wings. Although Pleškova ignores the natural broadening wings, her conclusion is still essentially correct, for the following reason:

If we take the temperatures of the H II and H I regions to be 10^4 °K and 10^2 °K, respectively, then the Doppler width in the H I region is 1/10 that in the H II region, and the ratios of natural to Doppler width are $a_{\mathrm{II}} = 3{\cdot}7 \times 10^{-4}$ and $a_{\mathrm{I}} = 3{\cdot}7 \times 10^{-3}$. Since the optical thickness in Ly-α of the H II region is about 10^4, most of the flux lies beyond three Doppler widths, measured in the H II region, from the line centre and therefore beyond 30 Doppler widths, as measured in the H I region. Using the asymptotic form of the Voigt function, it is easy to show that the absorption coefficient for these frequencies is less than 2×10^{-6} the line-centre value. Since there are no sources of Ly-α in the H I region, we can conclude that the H I region has very little influence on the Ly-α field in planetary nebulae. This fact is also important in assessing the effects of velocity gradients, for it is sometimes argued that because scattering in the H I region, occurring as it does predominantly in the far wings, is coherent, velocity gradients in the H I region have an important effect on the radiation field. However, since the line opacity in the relevant region is so small, this argument is irrelevant unless the extent of the H I region is enormous, i.e. $\tau_0 \gg 10^6$. Even the shift of the line-centre opacity into the region of large flux by velocity gradients has little effect because the core in the H I region is so narrow.

3.1.4. Interstellar hydrogen

If the interstellar neutral-hydrogen density is much lower than that in the outermost region of the nebula, the free path of an escaping photon will be much larger than it would be in the nebula, and simple geometrical arguments show that the photon has little chance of re-entering the nebula. On the other hand, if the nebula is completely ionized, then the *neutral*-hydrogen density may be of the same order of magnitude, so that a substantial effect could arise from the 'blanketing' by interstellar hydrogen.

3.2.0. Absorption mechanisms for H I Ly-α

The radiation field in a line varies inversely as the sum of the mean probabilities per scattering of absorption and of escape. Thus, if the optical thickness of the nebula in the line is increased indefinitely, it is the most probable absorption process that finally limits the intensity of radiation in the line. As was first pointed out by Thomas (1949) and later by Burgess (1958), the assumptions of Baker and Menzel's (1938) Case B are equivalent to setting both escape and absorption probabilities to zero, with the result that the populations of the $n=2$ levels are infinite. Provided that one is *a priori* sure that no upwards transitions occur from $n=2$, then no harm results from these assumptions in calculating the populations of the higher levels. On the other hand, if one wishes to check this point theoretically it becomes necessary to include the dominant escape and absorption processes in calculating the population of the $n=2$ levels.

It is also clear that the condition for 'detailed radiative balance' to occur, i.e. for upwards and downwards *radiative* transitions to balance sufficiently well for them to be regarded as exactly cancelling, is that the mean escape probability be smaller than the mean absorption probability. For the purposes of this discussion, we shall here take as standard conditions $T_e = 10^{4}$°K and $N_e = 10^4$ °K.

3.2.1. Mechanisms destroying Ly-α in scattering

Let us first examine possible mechanisms for depopulating the 2p level before the emission of a further Ly-α photon occurs, i.e. for destroying a photon during the act of scattering. Spitzer and Greenstein (1951) first pointed out that collisional transitions from the 2p to 2s states, followed by two-quantum decay, could be important in this respect (cf. Seaton, 1955). The probability per scattering is

$$p_1 = \frac{N_e\, q\,(2\text{p} \rightarrow 2\text{s})}{A\,(2\text{p} \rightarrow 1\text{s})} \cdot \frac{A_{2q}(2\text{s} \rightarrow 1\text{s}) + N_e\, q\,(2\text{s} \rightarrow 1\text{s})}{A_{2q}(2\text{s} \rightarrow 1\text{s}) + N_e\big(q\,(2\text{s} \rightarrow 1\text{s}) + q\,(2\text{s} \rightarrow 2\text{p})\big)},$$

where we have assumed $N_e = \mathrm{N}^+$. Here the q's are collisional rate constants and the A's are Einstein coefficients, with the subscript $2q$ denoting two-quantum processes. Seaton (1955) gives $q(2\text{s}\rightarrow 2\text{p}) = 3q(2\text{p}\rightarrow 2\text{s}) \approx 5\times 10^{-4}\ \text{cm}^3\ \text{sec}^{-1}$ for $1 \leqslant 10^{-4}\ T_e \leqslant 2$. In the same temperature range, the expression given by Hummer (1963*a*) reduces to $q(2\text{p}\rightarrow 1\text{s}) \sim 8\times 10^{-9}\ \text{cm}^3\ \text{sec}^{-1}$ and $q(1\text{s}\rightarrow 2\text{p}) \sim 2{\cdot}4\times 10^{-8}\ \exp(-11.8\times 10^4/T_e)\ \text{cm}^3\ \text{sec}^{-1}$. Using the values $A\,(2\text{p}\rightarrow 1\text{s}) = 6{\cdot}26\times 10^8\ \text{sec}^{-1}$ and $A_{2q}(2\text{s}\rightarrow 1\text{s}) = 8{\cdot}227\ \text{sec}^{-1}$, we find

$$p_1 \simeq \frac{2{\cdot}7\times 10^{-13}\, N_e}{1 + 0{\cdot}6\times 10^{-4}\, N_e}, \quad 1 \leqslant 10^{-4}\ T_e \leqslant 2\,.$$

At standard conditions, $p_1 \simeq 2\times 10^{-9}$. The failure of Yada (1955*a*, *b*) to obtain any reduction in the Ly-α flux by including this process in an approximate solution of the transfer problem for Lyman continuum optical thickness up to 10 is understandable, since the number of scatterings would have been at most 10^6.

Collisional de-excitation of the 2p level is also possible. The probability per scattering for this process is

$$p_2 = \frac{N_e\, q\,(2\text{p} \rightarrow 1\text{s})}{A\,(2\text{s} \rightarrow 2\text{p})} \simeq 1{\cdot}3\times 10^{-17}\, N_e\,, \quad 1 \leqslant 10^{-4}\ T_e \leqslant 2\,.$$

At standard conditions, $p_2 \simeq 1{\cdot}3\times 10^{-13}$, which is negligible compared to p_1.

Kipper and Tiit (1958) have shown that in addition to the allowed 2p→1s transition there is also a small probability of two-quantum decay. They find, for the probability per scattering of a photon being destroyed in this way, $p_3 \simeq 3\times 10^{-14}$. Gurzadian (1961) has shown that Ly-α can be scattered in magnetic dipole 1s→2s transitions, since the energy difference between the 2s and 2p states is less than the energy corresponding to a Doppler width. For the photon to be destroyed, two-photon decay must then occur. The probability per scattering for this mechanism is $p_4 \simeq 5\times 10^{-15}$. Gurzadian's original proposal involved a much larger decay rate for 2s, arising from a hypothetical electric dipole moment of the electron which was being discussed at the time. Subsequent experimental work has eliminated this possibility.

Collisional and radiative transitions from the 2p to higher states are possible, although an atom thus excited will emit a photon that will, with high probability be degraded into another Ly-α photon, unless the nebula is very thin. The probability per scattering of collisional transitions from 2p to 3s and 3d is about 10^{-12}, and that for radiative excitation by black-body radiation at $2{\cdot}5 \times 10^5$ °K and a dilution factor of 10^{-14} is about a factor of 10 smaller. Another possibility is the absorption of H I Ly-α in 2→4 transitions in He^+. However, as the frequencies of the two lines differ by about 10 hydrogen Doppler widths, the probability of this process is also extremely small.

Kahn (1962) has suggested that since an H I Ly-α photon loses 10^{-8} of its energy, on the average, per scattering to atomic recoil, a photon could eventually diffuse out the red wing of the line. Under typical conditions, this would occur in about 10^4 scatterings. However, because of the Doppler redistribution, as soon as the red side of the line becomes more intense, the number of red-to-blue scatterings increases to balance out the net energy loss to zero.

3.2.2. Mechanisms Destroying Ly-α between Scatterings

Ly-α photons will be destroyed between scatterings by photo-ionizing atoms in states having an ionization potential less than 10·15 eV. The highly populated metastable states of hydrogen and helium are the most likely to be important in this respect.

If N_A and N_S are the densities of absorbing and scattering atoms respectively, and $R_A(\nu)$ and $R_S(\nu)$ are the rates per unit frequency of absorbing and scattering radiation of frequency ν, we can write

$$R_A(\nu)\,d\nu = 4\pi N_A\, a_\nu \frac{J_\nu}{h\nu}\, d\nu\,,$$

and

$$R_S(\nu)\,d\nu = N_S B_{12} \phi(\nu)\, J_\nu\, d\nu$$

where a_ν is the photo-ionization cross-section; $\phi(\nu)$ is the line-scattering coefficient normalized to unity; B_{12} is the Einstein coefficient for the scattering process; and J_ν is the mean intensity. For simplicity, assume that J_ν is constant over a frequency interval 2δ and zero elsewhere. Then the total rates are

$$R_A = \int_{\nu_0-\delta}^{\nu_0+\delta} R_A(\nu)\,d\nu \simeq 4\pi N_A \frac{a_{\nu_0}}{h\nu_0}\, 2\delta\, J_{\nu_0}\,,$$

and

$$R_S = \int_{\nu_0-\delta}^{\nu_0-\delta} R_S(\nu)\,d\nu \simeq N_S B_{12} J_{\nu_0}\,,$$

and the probability of absorption per scattering is

$$p_A = \frac{R_A}{R_S} = \frac{8\pi\delta\, a_{\nu_0}}{h\nu_0 B_{12}} \cdot \frac{N_A}{N_S}\,.$$

The half-width δ is typically 3 Doppler half-widths. Using the numerical values for H I Ly-α, $\Delta_0 = 1{\cdot}06 \times 10^{11}$ sec^{-1}; $h\nu_0 B_{12} = 0{\cdot}14$ cm^2 sec, we have

$$p_A = 5 \times 10^{13} \frac{N_A}{N_S} a_{\nu 0}\,.$$

The photo-ionization cross-sections at Ly-α of H(2s) and He(2^3S) are approximately 6×10^{-19} cm^2 (Seaton, 1960) and 1.5×10^{-18} cm^2 (Huang 1948), respectively. We can estimate the H(2s) population by balancing recombination against two-quantum decay and 2s→2p transitions, assuming $N_e = N_+$,

$$\tfrac{1}{3}\alpha_B N_e^2 = N(2s)\left[A_{2q}(2s \to 1s) + N_e q(2s \to 2p)\right],$$

or using the numerical values given above,

$$N(2s) = \frac{1 \times 10^{-14} N_e^2}{1 + 0{\cdot}6 \times 10^{-4} N_e}\,.$$

As the population of He(2^3S) is affected by a number of processes, we shall infer its value from O'Dell's (1965) measures of the ratio $I(10\,830)/I(5876)$. Since the 2^3P level is populated by collisional excitations from 2^3S and by recombination, and depopulated mainly by radiative transitions to 2^3S, we have (cf. O'Dell, 1965)

$$I(10\,830)/I(5876) = \frac{5876}{10\,830} \frac{N(2^3S)\, q(2^3S \to 2^3P) + \tfrac{1}{2}\alpha_B N(He^+)}{N(He^+)\, \alpha_{eff}(5876)}\,.$$

Using the values given by Seaton (1968) for $T_e = 10^4$ °K, $q(2^3S \to 2^3P) \sim 2 \times 10^{-7}$ cm^3 sec^{-1}, and $\alpha_{eff}(5876) = 5{\cdot}2 \times 10^{-14}$ and taking $\tfrac{1}{2}\alpha_B = 1{\cdot}3 \times 10^{-14}$, we have

$$\frac{N(2^3S)}{N(He^+)} = 2{\cdot}6 \times 10^{-7}\left[1{\cdot}8 \frac{I(10\,830)}{I(5876)} - 2.5\right].$$

O'Dell's values of $I(10\,830)/I(5876)$ lie between 1·9 and 9·6, so that $N(2^3S)/N(He^+)$ lie between 4×10^{-6} and 2×10^{-7}. In the region where the helium lines are formed $N(He^+) \sim N_{He} \sim \tfrac{1}{5} N_H \sim 2 \times 10^3$ cm^{-3}. Thus for the probabilities per scattering of absorption of H I Ly-α by H(2s) and He(2^3S), we have

$$p_A(H2s) = 2 \times 10^{-11}/N_s\,,$$

and

$$3 \times 10^{-8} \leqslant N_s p_A(He\ 2^3S) \leqslant 6 \times 10^{-7}\,.$$

Since $N_S = N$(H 1s) is hardly smaller than 1 cm^{-3}, absorption by H(2s) is unimportant, while absorption by He(2^3S) appears to be the dominant loss mechanism, at least in regions where helium is mostly singly ionized and $N(H\ 1s) \lesssim 10^2$ cm^{-3}.

Finally, there is the possibility of extinction by dust grains. Although there is at present no direct observational evidence for dust in planetary nebulae, quantities of dust otherwise unobservable could still play a role in destroying Ly-α because of the very long paths of these photons in the nebula. If planetary nebulae do, in fact, evolve from red giants, then it is not surprising that dust should exist in at least the H I regions of planetaries. Any energy absorbed by dust would be re-emitted in the infrared; whether or not the strong infrared continuum observed in NGC 7027 by Gillett *et al.* (1967) has its origin in this mechanism remains a matter of speculation.*

3.2.3. Absorption mechanisms for He I and He II Ly-α

The Ly-α lines of both He I and He II are capable of ionizing hydrogen, and that of He II can ionize helium. Using the formula for p_A derived above, and the numerical values for He II Ly-α; $\Delta = 2{\cdot}1 \times 10^{11}$ sec^{-1}, $h\nu_0 B_{12} = 0{\cdot}14$ cm^2 sec^{-1}; and $a_{\nu_0} = 2{\cdot}9 \times 10^{-19}$ cm^2, we obtain for the probability per scattering that a He II Ly-α photon is absorbed by hydrogen,

$$p_A(\mathrm{H}) \sim 3{\cdot}3 \times 10^{-5} \frac{N(\mathrm{H\ 1s})}{N(\mathrm{He^+\ 1s})}.$$

From Goodson's (1967) ionization equilibria we find values of $N(\mathrm{H\ 1s})/N(\mathrm{He^+\ 1s})$ on the order of 10^{-4}–10^{-2} throughout the region of interest, so that $p_A(\mathrm{H}) \sim 10^{-9}$–$10^{-7}$. The probability per scattering of absorption of He I Ly-α by hydrogen is similar. The probability that a He II Ly-α photon is absorbed in a $2p^2\ {}^3P_2 \rightarrow 2p3d\ {}^3P_2^0$ (the initial transition in the Bowen mechanism) is difficult to estimate because it will depend on the details of the radiation field which can be obtained only by a solution of the transfer equation, but it could easily be important because of the resonant nature of the absorption. On the other hand, the probability per scattering of loss by 2p→2s transitions in He^+ is about a factor of 70 smaller than for H (Hummer, 1963*b*).

3.3.0. Possible consequences of large Ly-α intensities

Historically, the first important consequence of Ly-α intensity to be considered was the catastrophic radiation pressure. The calculations of Koelbloed (1956) and of Zanstra (1956) show that for a total thickness of 2×10^4, the inclusion of noncoherent scattering reduces the radiation pressure by a factor between 40 and 500, depending on position, from the value with coherent scattering, with the result that Ly-α radiation pressure is not of dynamical importance. Moreover, as we have argued above, any increase in the optical thickness in Ly-α beyond 10^4 must come from an H I region, in which the hydrogen line is so narrow as to miss most of the radiation escaping from the H II region. It would appear that H I Ly-α radiation pressure is unimportant until the Ly-α optical thickness exceeds 10^7.

If the Ly-α radiation becomes very intense, the population of the 2p level may become sufficiently large so that absorption of the Balmer lines and collisional transi-

* Krishna Swamy and O'Dell (*Astrophys. J.*, **151**, L61, 1968) have presented impressive evidence in support of this mechanism.

tions from this level become important. H I Ly-α could also play a role in depopulating the 2^3S state in He I.

3.3.1. Absorption in the Balmer lines

In his original investigations of the Ly-α radiation field, Ambartsumian (1932, 1933) found that the 2p population was large enough for the Balmer α and β lines to be optically thick. Pottasch (1960*a*, *b*), who included in his calculations the effects of noncoherent scattering, obtained smaller excited state populations than Ambartsumian, but concluded that self-absorption could occur in the early Balmer lines. However, because his frequency-quadrature scheme had artificially impaired the escape of Ly-α radiation, Pottasch obtained values of the 2p population considerably too large. Osterbrock's (1962) calculations showed that for typical nebulae self-absorption was probably negligible. Gershberg (1961) and Mathis (1962) reached the same conclusion from observational evidence.

Capriotti (1964*a*, *b*) has calculated the Balmer decrement as a function of the optical thickness in the 2p→3s, 3d transitions and the 2s→3p transition. An examination of the observed Balmer α, β, γ ratios in the light of Capriotti's results indicates that self-absorption in the Balmer lines is completely unimportant for the overwhelming majority of nebulae for which we have observations.

The effect of self-absorption on the triplet system of helium has been discussed most recently by Pottasch (1962) and by Osterbrock (1964), who give references to earlier work. It appears that self-absorption is important in some objects, although other mechanisms may also be operative.

3.3.2. Collisional processes from $n=2$

Jefferies and Pottasch (1959) and Pottasch (1960*c*) found that for optically thick nebulae the populations of the excited states of hydrogen are so large that collisional ionization from these states are comparable in number with ground state photo-ionizations. Consequently the generation of diffuse photons is increased substantially by the inclusion of these processes, and the ionized region is considerably larger than the Strömgren estimate. However, if this effect exists, it must arise primarily from the $n=2$ level, since the populations of the higher states are much smaller, while their collisional ionization rate constants are not much larger than those for $n=2$. An upper limit to the 2p population is obtained by setting the escape probability to zero and assuming that 2p→2s transitions followed by two-quantum decay are the dominant absorption mechanism, with a probability per scattering of 10^{-9} (see Section 3.2.0). Then the maximum density of atoms in the 2p state is given approximately by (Osterbrock, 1964)

$$10^{-9} A(2\text{p} \to 1\text{s})\, N_{\max}(2\text{p}) = N_e N_+ \tfrac{2}{3}\alpha_B ,$$

or assuming $N_e = N_+ = 10^4$,

$$N_{max}(2p) \sim 2 \times 10^{-5}\ \text{cm}^{-3}.$$

The 2s population is about 10^{-6} (Section 3.2.2) in these conditions. Rudge and Schwartz (1966) give the ionization rate constant for 2s as $6{\cdot}8 \times 10^{-9}$ cm^3 sec^{-1} at $T_e = 10^4$, and since Prasad (1966) has shown that the 2s and 2p ionization cross-sections are not very different at low energies, we see that the maximum collisional ionization rate from $n=2$ is about 10^{-9} sec^{-1}, while the photo-ionization rate from 1s is about 10^{-5} sec^{-1}. It would appear that collisional ionization from excited states is completely unimportant for planetary nebulae.

Pottasch (1960*a*) has also investigated the possible role of collisional excitation from the $n=2$ states and finds that it is negligible by several orders of magnitude, even using his overestimated $n=2$ populations.

3.3.3. Photo-ionization of He 2^3S by H I Ly-α

The photo-ionization of He(2^3S) by H I Ly-α has already been discussed (Section 3.2.2) as a destruction mechanism for Ly-α. Münch has suggested that these processes could be important in controlling the population of He(2^3S). O'Dell (1965) has investigated this problem on the assumption that 2^3S is populated by recombination (including cascade) and is depopulated by two-photon transitions to 1^1S, photo-ionization by stellar radiation, collisional transitions to 2^1S, and photo-ionization by Ly-α. From the observed intensity ratios of the lines He I $\lambda\lambda$ 10830 Å and 3889 Å to λ 5876 Å, O'Dell derives a measure of the distance traveled by a typical Ly-α photon in the nebula, much larger than would be estimated from Osterbrock's mean number of scatterings for the H II zone. O'Dell suggests that this is evidence for an extensive neutral-hydrogen region blanketing the H II region. However, O'Dell's $2^3\text{S} \rightarrow 2^1\text{S}$ collision rate constant is too small by a factor of 5, and his value for the $2^3\text{S} \rightarrow 2^3\text{P}$ is too large by a factor of 1·7 (Seaton, 1968). He also neglected collisional transitions to 2^1P and what is perhaps the most important mechanism,* photo-ionization by the entire Balmer spectrum of He II. All of these factors tend to make O'Dell's estimate of the ionizing power of H I Ly-α too large.

3.3.4. Effects of He II Ly-α

Probably the most important role played by this radiation is in ionizing hydrogen and helium. It is clear from Goodson's (1967) work that He II Ly-α makes a major contribution to the energy balance, so that a careful solution of this transfer problem

* Robbins (*Astrophys. J.*, **151**, L35, 1968) has suggested that excitation of doublyexcited auto-ionizing states in helium by ultraviolet lines of C III and N IV may be important in depopulating the 2^3S level of He I.

along with the ionization balance of H and He is important in determining the correct run of electron temperature with depth. In particular, it is important to know whether He II Ly-α is absorbed in the He^{+2} region, or whether it escapes and is absorbed in regions where the densities of H and He are larger. Since the electrons produced by the photo-ionization of H and He have energies of 27·2 and 16·2 eV respectively, which are much larger than the mean electron energy (~ 1 eV), their thermalization will occur much more slowly than would be expected from the estimates of Böhm and Aller (1947). If these electrons suffered inelastic collisions with atoms before being thermalized, a substantial part of their energy could escape as radiation and should not be included in the thermal balance.

4.0. Transfer Effects and the Hydrogen Recombination Spectrum

The Lyman lines beyond Ly-α have an entirely different character from Ly-α because their photons have a large probability per scattering ($\gtrsim 0{\cdot}5$) of being degraded into photons in a lower Lyman line and a subordinate line. For this reason the population of the excited levels other than $n=2$ varies little with optical thickness.

Since the optical thicknesses of the Lyman lines have fixed ratios, the populations of the excited states should increase monotonically from their Case A to Case B values as the optical thickness is increased, assuming self-absorption in the Balmer line is negligible. Since, however, the level populations will increase at different rates, the *ratios* of Balmer line intensities need not necessarily lie between their Case A and Case B values. Capriotti (1966) has investigated the effect on the relative intensities of Balmer α, β, γ and δ of allowing for the partial escape of Lyman photons. The escape probabilities was estimated by an extension of Zanstra's argument (Section 3.1.1) to a nebula expanding with a constant velocity gradient. The line-intensity ratios, displaced by an arbitrary amount along the reddening curve, show better agreement with the observed ratios than do either the Case A or Case B values, although by no means all observed ratios lie within Capriotti's bounds. It is also not clear that arguments based on averages of quantities that vary by orders of magnitude are an entirely reliable basis on which to discuss variations in line ratios on the order of 0·03 in the logarithm. Since upward transitions from excited states are negligible, each Lyman line transfer problem* can be solved successively, starting with a high, optically thin line, and working towards Ly-α. This is quite feasible with modern computing technique and a definite answer to this problem should be available.

* Van Blerkom and Hummer (*Astrophys. J.*, in press) have shown the importance of overlapping on the transfer problem of the higher Lyman lines and have developed a simple band model for the inclusion of these effects.

Acknowledgements

I am grateful to Professor M.J. Seaton, Dr John Stewart, Dr G. Khromov, and Mr D. Van Blerkom for many discussions which have helped to clarify my understanding of various aspects of the material discussed here, and to Dr L.H. Auer for giving me his Monte-Carlo results prior to publication.

References

Aller, L.H., Baker, J.G., Menzel, D.H. (1939) *Astrophys. J.*, **90**, 601.
Ambartsumian, V.A. (1932) *Mon. Not. R. astr. Soc.*, **93**, 50.
Ambartsumian, V.A. (1933) *Bull. Obs. centr. Poulkovo*, **13**, 3.
Auer, L.H. (1968) *Astrophys. J.*, **153** (in press).
Baker, J.G., Menzel, D.H. (1938) *Astrophys. J.*, **88**, 52.
Baker, J.G., Aller, L.H., Menzel, D.H. (1939) *Astrophys. J.*, **90**, 271.
Biberman, L.M. (1948) *Dokl. Akad. Nauk. SSSR*, **49**, 659.
Biberman, L.M., Vorobyev, V.S., Lagarkov, A.N. (1965) *Optika Spektrosk.*, **19**, 326; trans. *Optics Spectrosc.*, **19**, 186.
Böhm, D., Aller, L.H. (1947) *Astrophys. J.*, **105**, 1.
Böhm, K.-H., Deinzer, W. (1965) *Z. Astrophys.*, **61**, 19.
Böhm, K.-H., Deinzer, W. (1966) *Z. Astrophys.*, **63**, 177.
Burgess, A. (1958) *Mon. Not. R. astr. Soc.*, **118**, 477.
Capriotti, E.R. (1964*a*) *Astrophys. J.*, **139**, 225.
Capriotti, E.R. (1964*b*) *Astrophys. J.*, **140**, 632.
Capriotti, E.R. (1965) *Astrophys. J.*, **142**, 1101.
Capriotti, E.R. (1966) *Astrophys. J.*, **146**, 709; *cf. erratum, ibid.*, **148**, 1967, 318.
Chamberlain, J.W. (1953) *Astrophys. J.*, **117**, 387.
Chandrasekhar, S. (1935) *Z. Astrophys.*, **9**, 266.
Eberlein, K. (1955) *Z. Astrophys.*, **38**, 14.
Eberlein, K. (1957) *Z. Astrophys.*, **41**, 271.
Field, G.B. (1959) *Astrophys. J.*, **129**, 551.
Gershberg, R. (1961) *Astr. Zu.*, **38**, 250; trans. *Soviet Astr.*, **5**, 188.
Gillett, F.C., Low, F.J., Stein, W.A. (1967) *Astrophys. J.*, **149**, L97.
Goodson, W.L. (1967) *Z. Astrophys.*, **66**, 118.
Gurzadian, G.A. (1961) *Dokl. Akad. Nauk. SSSR.*, **141**, 1061; trans. *Soviet Phys. Dokl.*, **6**, 1031, 1962.
Harman, R.J., Seaton, M.J. (1966) *Mon. Not. R. astr. Soc.*, **132**, 15.
Hearn, A.G. (1964) *Proc. Phys. Soc., London*, **84**, 11.
Henyey, L.G. (1941) *Proc. nat. Acad. Sci. Am.*, **26**, 50.
Hjellming, R. (1966) *Astrophys. J.*, **143**, 420.
Huang, S.S. (1948) *Astrophys. J.*, **108**, 354.
Hummer, D.G. (1962) *Mon. Not. R. astr. Soc.*, **125**, 21.
Hummer, D.G. (1963*a*) *Mon. Not. R. astr. Soc.*, **125**, 461.
Hummer, D.G. (1963*b*) *Thesis*, Univ. of London.
Hummer, D.G. (1964) *Astrophys. J.*, **140**, 276.
Hummer, D.G. (1968) *Mon. Not. R. astr. Soc.*, **138** ,73.
Hummer, D.G., Rybicki, G.B. (1966) *J. quantit. Spectrosc. radiat. Transfer*, **6**, 661.
Hummer, D.G., Rybicki, G.B. (1967) Computational Methods for Non-LTE Line-Transfer Problems, in *Methods in Computational Physics*, **7**, Ed. by B. Alder, S. Fernbach, M. Rotenberg, Academic Press, New York, p. 53.
Hummer, D.G., Seaton, M.J. (1963) *Mon. Not. R. astr. Soc.*, **125**, 437.
Hummer, D.G., Seaton, M.J. (1964) *Mon. Not. R. astr. Soc.*, **127**, 217.

Ivanov, V.V. (1967) *Bull. astr. Inst. Netherl.*, **19**, 192.
Jefferies, J.T., Pottasch, S.R. (1959) *Ann. Astrophys.*, **22**, 318.
De Jong, J.H. (1951) *Bull. astr. Inst. Netherl.*, **11**, 345.
Kahn, F. (1962) Dynamics of Interstellar Gas, in *Interstellar Matter in Galaxies*, Ed. by L. Woltjer, W.A. Benjamin, Inc., New York, p. 164.
Kipper, A.Ya., Tiit, V.M. (1958) *Vop. Kosmog.*, **6**, 99.
Koelbloed, D. (1956) *Bull. astr. Inst. Netherl.*, **12**, 341.
Mathis, J. (1962) *Astrophys. J.*, **136**, 374.
O'Dell, C.R. (1965) *Astrophys. J.*, 142, 1093.
Osaki, T. (1962) *Publ. astr. Soc. Japan*, **14**, 111.
Osterbrock, D.E. (1962) *Astrophys. J.*, **135**, 195.
Osterbrock, D.E. (1964) *A. Rev. Astr. Astrophys.*, **2**, 95.
Parker, R.A.R. (1964) *Astrophys. J.*, **139**, 208.
Pleškova, T.F. (1962) *Astr. Zu.*, **39**, 235; trans. *Soviet Astr.*, **6**, 182.
Pottasch, S.R. (1960*a*) *Astrophys. J.*, **131**, 202.
Pottasch, S.R. (1960*b*) *Ann. Astrophys.*, **23**, 749.
Pottasch, S.R. (1960*c*) *Astrophys. J.*, **132**, 269.
Pottasch, S.R. (1962) *Astrophys. J.*, **135**, 385.
Prasad, S.S. (1966) *Proc. phys. Soc., London*, **87**, 393.
Rudge, M.R.H., Schwartz, S.B. (1966) *Proc. phys. Soc.*, **88**, 563.
Rybicki, G.B., Hummer, D.G. (1967) *Astrophys. J.*, **150** , 607.
Seaton, M.J. (1955) *Mon. Not. R. astr. Soc.*, **115**, 279.
Seaton, M.J. (1960) *Rep. Prog. Phys.*, **23**, 313.
Seaton, M.J. (1968) *Review* of *Atomic Collision Processes*, in the present volume, p. 129.
Sobolev, V.V. (1944) *Astr. Zu.*, **21**, 143.
Sobolev, V.V. (1947) *Dvižuščiesja Oboločki Zvezd*, Leningrad; trans. *Moving Envelopes of Stars*, translated from the Russian by S. Gaposchkin, Harvard University Press, Cambridge, Mass., 1960.
Sobolev, V.V. (1957) *Astr. Zu.*, **34**, 694; trans. *Soviet Astr.*, **1**, 678.
Sobolev, V.V. (1967) *Astrofizika*, **3**, 137.
Spitzer, L., Greenstein, J.L. (1951) *Astrophys. J.*, **114**, 407.
Strömgren, B. (1939) *Astrophys. J.*, **89**, 526.
Thomas, R.N. (1949) *Astrophys. J.*, **109**, 480.
Unno, W. (1951) *Publ. astr. Soc. Japan*, **3**, 158.
Unno, W. (1955) *Publ. astr. Soc. Japan*, **7**, 81.
Van Blerkom, D. (1968) *Astrophys. J.*, **152** (in press).
Van Blerkom, D., Hummer, D.G. (1967) *Mon. Not. R. astr. Soc.*, **137**, 353.
Yada, B. (1955*a*) *Publ. astr. Soc. Japan*, **5**, 128.
Yada, B. (1955*b*) *Publ. astr. Soc. Japan*, **6**, 76.
Zanstra, H. (1934) *Mon. Not. R. astr. Soc.*, **95**, 84.
Zanstra, H. (1936) *Mon. Not. R. astr. Soc.*, **97**, 37.
Zanstra, H. (1949) *Bull. astr. Inst. Netherl.*, **11**, 1.
Zanstra, H. (1951*a*) *Bull. astr. Inst. Netherl.*, **11**, 341.
Zanstra, H. (1951*b*) *Bull. astr. Inst. Netherl.*, **11**, 359.
Zanstra, H. (1956) *Bull. astr. Inst. Netherl.*, **12**, 349.

DISCUSSION

Böhm: I agree with Hummer's statement that in principal the treatment of the He II resonance radiation in Goodson's paper is not satisfactory. On the other hand the numerical values calculated by Goodson for the electron temperature in the core of the nebula lie between 2×10^4 and $2{\cdot}2 \times 10^4$ °K, and essentially agree with the numbers given by Flower.

Seaton: The agreement between Goodson and Flower is a consequence of their having made the same assumptions.

Ly-α RADIATION DENSITIES IN PLANETARY NEBULAE

EUGENE R. CAPRIOTTI

(Perkins Observatory, Ohio State and Ohio Wesleyan Universities, U.S.A.)

1. Introduction

The average Ly-α radiation densities are calculated for 9 planetary nebulae from the observed $\lambda 10830/\lambda 5876$ line-intensity ratios under the assumption that the 2^3S state of He I is depopulated via photo-ionization by Ly-α radiation. Three of the nebulae are probably optically thick in their respective hydrogen Lyman continua. For these objects, the Ly-α radiation densities as determined from the $\lambda 10830/\lambda 5876$ line-intensity ratios may represent the actual Ly-α radiation densities. If they do, then the Ly-α radiation pressures exceed the respective gas pressures in these objects. For the remaining nebulae, it seems that the observationally determined Ly-α radiation densities are representative only if either one or both of the following conditions is fulfilled in each case:

(1) The object is essentially optically thick in the hydrogen Lyman continuum when other properties show it to be optically thin in the hydrogen Lyman continuum.

(2) We are viewing the object during a specific 100-year time span during the course of its evolution.

For these objects, the gas pressures probably exceed the respective Ly-α radiation pressures.

2. Lyman-α Radiation Densities

In recent years, much attention has been turned to the dynamical structure and the origin of the condensations of the planetary nebulae. Mathews (1966) and Sofia and Hunter (1967) have constructed theoretical dynamical models under the assumption that the gas pressure is the driving force. The studies of the origin of the condensations as a thermal instability phenomenon by Zanstra (1955), Daub (1963), Sofia (1966), and Harrington (1967) are all based on the assumption that the total pressure is essentially equal to the gas pressure. However, Ly-a radiation is imprisoned in the planetary nebulae to the extent that it may, in some cases, be greater than the gas pressure. Here, we report on estimates of the ratios of the Ly-a radiation pressure and the gas pressure for 9 planetary nebulae.

Münch (1963, private communication) suggested that photo-ionization by Ly-α radiation is the most important mechanism working in the depopulation of the 2^3S state of helium in the planetary nebulae. O'Dell (1965), following this suggestion, was

Osterbrock and O'Dell (eds.), Planetary Nebulae, 185–189. © *I.A.U.*

able to estimate for each of 9 planetary nebulae the mean distance (in units of the nebular radius) traveled by a Ly-α photon before it is lost by the radiation field in the nebula if the 2^3S state of helium is depopulated through photo-ionization by Ly-α radiation. These estimates were obtained indirectly from the observed $\lambda 10830/\lambda 5876$ line intensity ratios. The $\lambda 10830$ line is excited by collisions to the 2^3P state from the 2^3S state as well as by capture-cascade processes, while the $\lambda 5876$ line is excited almost entirely by capture-cascade processes. The observed $\lambda 10830/\lambda 5876$ line-intensity ratios are then measures of the relative population densities of the 2^3S state of neutral helium. O'Dell also showed that two-photon transitions to the ground state and collisional transitions to the ground state and the 2^1S state are too slow to account for the observed values of the relative population densities of the 2^3S state. In this manner, support was lent to the idea that photo-ionization by Ly-α radiation is the chief mechanism for the depopulation of the 2^3S state.

We denote the mean distance (in units of the nebular radius) traveled by a Ly-α photon before it is lost by the nebular radiation field by f. f is also the mean Ly-α radiation density in units of the Ly-α radiation density for the case where the mean transit time of a photon is R/c. R is the nebular radius and c is the speed of light. Column 2 of Table 1 lists values of f calculated by O'Dell (1965).

It may appear that one could apply any theoretical model to the objects listed in Table 1 and obtain matching values of f by simply adjusting the optical thickness in

Table 1

Interpretations of the data under the assumption that the 2^3S state of helium is depopulated through photo-ionization by Ly-α radiation

Object	f	$1/\lambda$	$P(\text{Ly-}\alpha)/P(\text{gas})$	t
IC 418*	$1{\cdot}3 \times 10^2$	$4{\cdot}0 \times 10^8$	15	32
IC 2149	$7{\cdot}3 \times 10^2$	$3{\cdot}9 \times 10^8$	31	41
NGC 2392	$4{\cdot}0 \times 10^2$	$6{\cdot}6 \times 10^8$	16	120
IC 4997*	$5{\cdot}3 \times 10$	$2{\cdot}5 \times 10^8$	10	0·25
BD + 30°3639	$5{\cdot}2 \times 10^3$	$6{\cdot}6 \times 10^8$	80	310
NGC 6572	$8{\cdot}3 \times 10^2$	$6{\cdot}6 \times 10^8$	7·5	480
NGC 7009	$1{\cdot}5 \times 10^3$	$5{\cdot}8 \times 10^8$	46	110
NGC 7027*	$3{\cdot}8 \times 10^2$	$5{\cdot}8 \times 10^8$	5·2	230
NGC 7662	$6{\cdot}7 \times 10^2$	$8{\cdot}1 \times 10^8$	9·0	430

* Has measurable [OI] lines in spectrum.

the Ly-α line. In attempting to represent real planetary nebulae by theoretical models, however, one must keep in mind the fact that Ly-α photons can be lost by the radiation field through conversion to the two-photon continuum. The third column in Table 1 lists values of the inverse of the probability that a Ly-α photon will be converted to the two-photon continuum as a result of a single scattering. These values are denoted by $1/\lambda$. The number of scatterings suffered on the average by a Ly-α photon before it is lost by the radiation field is bounded by the value of $1/\lambda$. Therefore, for a given model, there is a largest possible value of f, and this value of f is set by the value of $1/\lambda$. Of the objects listed in Table 1, only IC 418 and IC 4997 can be represented by isothermal models, although the value of f for NGC 7027 is close to the range of values allowed by an isothermal model. In all other cases, the values of f listed in Table 1 are far too large to be consistent with an isothermal model. It is interesting to note that IC 418, IC 4997 and NGC 7027 have measurable [O I] lines in their spectra. The existence of these lines indicate that neutral oxygen atoms co-exist with electrons having thermal velocities corresponding to a temperature of the order of 10^4 °K. Hydrogen has the same ionization potential as oxygen and the presence of [O I] lines indirectly indicates that substantial amounts of neutral hydrogen co-exist with electrons near 10^4 °K. For example, Osterbrock (1964) estimated that roughly 26% of the hydrogen in NGC 7027 is neutral in regions where the electron temperature is of the order of 10^4 °K. It could well be that isothermal models represent fairly well the conditions in IC 418, IC 4997 and NGC 7027, and that the values of f listed in Table 1 are representative of the Ly-α radiation densities in these objects.

O'Dell (1965) suggested that a Yada-Osaki (1957) type model might represent the objects under consideration. This model consists of an H II region at 10^4 °K and a surrounding H I region at 10^2 °K. Indeed, the maximum values of f allowable for Yada-Osaki models are greater than the respective values of f listed in Table 1.

Let us suppose for a moment that the values of f computed from the He I lines are really representative of the Ly-α radiation densities in the 9 objects under consideration. Column 4 of Table 1 lists the values of the ratios of the mean Ly-α pressure and the mean gas pressure estimated under the latter supposition. In each case, the Ly-α pressure is greater than the gas pressure. One must conclude, therefore, that if the values of f estimated under the assumption that the 2^3S state of helium is depopulated through photo-ionization by Ly-α radiation are representative of the true Ly-α radiation densities, then the Ly-α pressure dominates over the gas pressure in each case.

There is reason to believe, however, that the values of f listed in Table 1 are not representative of the Ly-α radiation densities for the objects under consideration with the possible exceptions of IC 418, IC 4997, and NGC 7027. Suppose that Yada-Osaki models apply to the objects in Table 1. One can then estimate for each object the length of time t, in years, in which it would take the surrounding H I region to be accelerated to a given speed, say 200 km/sec, under the driving force of the Ly-α pressure. The fifth column of Table 1 lists values of t. For a given object, the velocity gradient would become so large in a time t that the effective optical thickness in Ly-α would be greatly reduced. The Ly-α radiation density would then be much smaller than the initial value of f. The largest value of t listed in Table 1 is 480 years. We conclude that if the blanketing of Ly-α radiation by a surrounding H I region is the cause of the large observed values of f for a nebula, then this blanketing could occur for a short time when compared to the lifetime of the nebula. We feel that the probability of viewing a planetary nebula during this stage of development would be too small to account for the number seen to have the observed values of f.

For IC 418, IC 4997 and NGC 7027, isothermal models may be applicable. The Ly-α pressure gradient for isothermal models is small except in the outermost layers. These outermost layers of small mass can be accelerated to great speeds in a short time. However, contrary to the case of the Yada-Osaki model, the outer layers in the case of an isothermal model play an insignificant role in the trapping of Ly-α radiation. Therefore, the length of time involved before the Ly-α pressure is greatly reduced is much larger in the case of an isothermal model than in the case of a Yada-Osaki model having the same optical thickness in Ly-α. Furthermore, as pointed out above, evidence that large amounts of neutral hydrogen co-exist with electrons at temperatures of the order of 10^4 °K in IC 4997, IC 418 and NGC 7027 appears in the form of [O I] lines in the spectra. One knows that each of these objects is very probably optically thick in the Lyman continuum and that the optical thickness in Ly-α could well be large enough so that the values of f and P(Ly-α)/P(gas) listed in Table 1 are representative for IC 4997, IC 418, and NGC 7027.

The work of Harman and Seaton (1966) provides us with relevant results. They have estimated the black-body temperatures of the central stars of 35 planetary nebulae under the assumption that in each case there is a complete absorption of the radiation emitted by the central star in the Lyman continuum of He^+. NGC 2392, BD +30°3639, NGC 6572, NGC 7009 and NGC 7662 are among these objects. For each of these 5 objects, estimates of the black-body temperatures of the central stars under the assumption that there is a complete absorption in the hydrogen Lyman continuum yield values that are lower than the temperatures determined by Harman and Seaton. One interpretation is that there is incomplete absorption by neutral hydrogen (Würm and Singer, 1952; Harmon and Seaton, 1966). Capriotti (1967*a*) estimated upper and lower limits to the optical thicknesses in the hydrogen continua for NGC 2392, BD + 30°3639, NGC 6572, NGC 7009, and NGC 7662 as well as for other objects under the latter assumption. The upper limits to the optical thicknesses in the hydrogen Lyman continua are used in order to estimate the values of f and P(Ly-α)/P(gas) for each of the 5 objects. These values are listed in Table 2. In each case, the gas pressure dominates.

Table 2

Characteristic values of f and P(Ly-α)/P(gas)

Object	f	P(Ly-α)/P(gas)
NGC 2392	13	0·49
BD + 30°3639	12	0·19
NGC 6572	40	0·37
NGC 7009	27	0·85
NGC 7662	24	0·34

We conclude that high surface brightness objects, like NGC 7027, IC 418 and IC 4997 with [O I] lines in their spectra, are probably optically thick in their Lyman continua. The trapping of Ly-α radiation in such an object could be strong enough so that the Ly-α pressure exceeds the gas pressure. For objects that show evidence that they are optically thin in the hydrogen Lyman continua, we conclude that the gas pressure exceeds the Ly-α radiation pressure.

The values of the parameters (electron density, collision cross-sections, etc.) needed in order to make the calculations reported in this work were taken from O'Dell (1965) and Osterbrock (1964). The theoretical models were constructed by using available theories of Ly-α radiation transfer. In particular, we adopted the treatments of Osterbrock (1962) and Yada and Osaki (1957) (Capriotti, 1967*b*).

References

Capriotti, E.R. (1967*a*) *Astrophys. J.*, **147**, 979.
Capriotti, E.R. (1967*b*) *ibid.*, **150**, 79.
Daub, C.T. (1963) *Astrophys. J.*, **137**, 184.
Harman, R.J., Seaton, M.J. (1966) *Mon. Not. R. astr. Soc.*, **132**, 15.
Harrington, J.P. (1967) unpublished Thesis, Ohio State University.
Mathews, W.G. (1966) *Astrophys. J.*, **143**, 173.
O'Dell, C.R. (1965) *Astrophys. J.*, **142**, 1093.
Osterbrock, D.E. (1962) *Astrophys. J.*, **135**, 195.
Osterbrock, D.E. (1964) *A. Rev. Astr. Astrophys.*, **2**, 95.
Sofia, S. (1966) *Astrophys. J.*, **145**, 84.
Sofia, S., Hunter, J.H. (1967) *Astr. J.*, **72**, 830.
Würm, K., Singer, O. (1952) *Z. Astrophys.*, **30**, 387.
Yada, B., Osaki, T. (1957) *Publ. astr. Soc. Japan*, **9**, 82.
Zanstra, H. (1955) in *Vistas in Astronomy*, Vol. 1, Ed. by A. Beer, Pergamon Press, New York, p. 256.

THE IONIZATION OF PLANETARY NEBULAE

ROBERT E. WILLIAMS
(Steward Observatory, University of Arizona, U.S.A.)

ABSTRACT

The ionization of the most abundant elements in planetary nebulae has been determined for a number of models of nebulae at different epochs in their expansion. The values used for the temperatures and radii of the central stars and the sizes and densities of the shells have come from Seaton's evolutionary sequence. The ionizing radiation field has been taken from model atmosphere calculations of the central stars by Gebbie and Seaton, and Böhm and Deinzer. Emission-line fluxes have been calculated for the models and compared with observations of planetary nebulae by O'Dell, Osterbrock's group, and Aller and his collaborators. Results indicate that the central stars have strong He^+ Lyman continuum excesses, similar to those predicted by Gebbie and Seaton. The mean abundance determinations for the nebulae made by Aller are confirmed, with the exception of nitrogen, which appears to be 3 or 4 times more abundant than his value. It is also seen that the electron temperatures of the nebulae are higher than previous theoretical determinations, providing better agreement with empirically derived values.

1. Introduction

The basic processes which govern the ionization structure of planetary nebulae are believed to be reasonably well understood. Ionization of the elements is caused by the far-ultraviolet radiation from the central star, hence the fundamental problem to be solved in determining the statistical equilibrium is that of the radiative transfer of this radiation within the gaseous shell. A knowledge of the state of ionization of all the elements is useful because this information can be used to predict the intensities of the emission lines of the nebulae. Comparison of these theoretical intensities with those actually observed then provides a check on the validity of the theory, and gives information on the physical parameters of the gas and the radiation field of the central star.

Hummer and Seaton (1963, 1964) have recently considered in detail the ionization of planetary nebulae for nebulae composed of hydrogen and helium, and excited by a star radiating like a black body. They did not calculate any emission-line intensities because the heavy elements were not included in their work, and also because little was known at that time about the far-ultraviolet distribution of radiation from the central stars and the physical conditions within the nebulae at different stages in their expansion. However, our present knowledge of these quantities is greatly improved. O'Dell (1963*a*), Harman and Seaton (1964), and Seaton (1966) have delineated an evolutionary sequence for planetary nebulae which allows us to attach some statistical-

Osterbrock and O'Dell (eds.), Planetary Nebulae, 190–204. © *I.A.U.*

ly meaningful size and density to these objects, in addition to the temperatures and radii of the central stars, at various stages of evolution. In addition, Gebbie and Seaton (1963) and Böhm and Deinzer (1965, 1966) have computed model atmospheres for the central stars, giving the emergent intensity of radiation at frequencies above the hydrogen Lyman limit. In view of these developments, and the improved techniques used by O'Dell (1963*c*), Osterbrock's group at Wisconsin, and Aller and his collaborators to obtain emission-line fluxes of planetary nebulae, it seems worthwhile to consider the ionization of the elements in these objects.

2. Theory

The theory of the determination of the ionization and electron temperature of a gas excited by a known radiation field has been discussed by Hummer and Seaton (1963, 1964) for the elements hydrogen and helium, and extended to include the heavy elements by Williams (1967; referred to hereafter as Paper I). It will not be repeated here except to point out the modifications necessary to determine the mean intensity of ionizing radiation in the gas for a curved geometry, rather than the semi-infinite plane-parallel configuration used in Paper I.

It is well known that the equation of transfer is sufficiently simple for a plane-parallel gas so that the diffuse radiation of such a nebula may be calculated straightforwardly. The introduction of curvature terms into the transfer equation, as would be required in the case of planetary nebulae, makes these calculations very difficult. The traditional approach to this problem for planetaries, therefore, has been to make the on-the-spot approximation, which assumes that the diffuse ionizing radiation produced by the gas is immediately re-absorbed. This makes it possible to equate the mean intensity of diffuse radiation at any point with the source function. This is tantamount to requiring the ionization of H and He to remain approximately constant over a distance where the gas is optically thick to the radiation causing the ionization – a situation that is fairly well satisfied in planetary nebulae (Hummer and Seaton, 1963, 1964). In this approximation, the mean intensity of radiation from both the star and the gas is

$$J_\nu = \frac{R_s^2}{4r^2} I_\nu^0 e^{-\tau_\nu} + \sum_i N_e \frac{x(i+1)}{x(i)} \frac{2h^4}{(2\pi mkT_e)^{3/2}} \frac{\nu^3}{c^2} e^{\frac{h(\nu_i - \nu)}{kT_e}}, \tag{1}$$

in the notation of Paper I, and I_ν^0 is the emergent intensity of radiation from the central star, R_s is the stellar radius, $x(i)$ is the relative abundance of the ion i, and the summation is performed over the ions H^0, He^0, and He^+.

Since the on-the-spot approximation requires detailed balancing to hold in the nebula between the ground state and continuum of the H^0, He^0, and He^+ ions, it can be shown upon substitution of Equation (1) into the ionization equation [Eqn. (1) in Paper I] that the ionization of the elements H and He may be computed by considering photo-ionizations from the dilute stellar radiation to be balanced by recombinations to their excited states (cf. Hummer and Seaton, 1963). This information can then be used in Equation (1) to compute the radiation field in the nebula, from which the ionization of the remaining elements in the gas, which are not abundant enough to contribute to the opacity or scattered radiation field of the gas, may be determined. Except for the fact that the ionization caused by the helium resonance lines is ignored and the diffuse radiation is accounted for by the on-the-spot approximation in the present investigation, the procedure used to calculate the ionization of the elements and the electron temperature at each point in the gas is the same as in Paper I. Ion abundances are initially calculated by assuming the gas to have some fixed value of the temperature. These results are then used in the equation of thermal equilibrium [Eqn. (7) in Paper I] to compute the temperature at all points in the gas. Once the temperature is known, the statistical equilibrium is again determined for each of the elements. This iteration procedure is continued until the values of the $x(i)$ and T_e do not change by more than several percent after one iteration.

3. The Models

The evolutionary sequence for planetary nebulae is one in which the shell continually expands from a small, dense, optically thick object to one which becomes optically thin, while the star increases in temperature and luminosity. At a point when the nebula becomes optically thin in the hydrogen Lyman continuum, the star undergoes a contraction and decrease in luminosity until the gas eventually becomes optically thick again, in spite of its much lower density. Böhm and Deinzer (1965, 1966) have taken the effective temperatures and surface gravities of the central stars as determined by O'Dell (1963*a*) and Harman and Seaton (1964) at different epochs in their evolution and have computed model atmospheres for the stars. They give the emergent flux of radiation for several non-gray models at different effective temperatures, including in their calculations the contribution made to the opacity by the heavier elements. Prior to the time when the evolutionary tracks of the central stars in the H-R diagram had been outlined, Gebbie and Seaton (1963) had performed similar model-atmosphere calculations. Since little was known about the surface gravity of the stars at the time of their work, they assumed g to have the minimum value capable of keeping the atmosphere stable to radiation pressure. Furthermore, unlike Böhm and Deinzer, they did not include the heavier elements as sources of opacity. As a result, the principal difference in the frequency distribution of radiation between the models of Gebbie and Seaton and those of Böhm and Deinzer is that the smaller gravity used by the former enhances the effect of electron scattering and results in a considerable He^+ Lyman excess for the hot models. On the other hand, the latter find that the consideration of a higher gravity and the appreciable absorption of the ions of nitrogen, oxygen, and neon at high frequencies leads to a large He^+ Lyman deficiency.

We have taken both types of models and have fitted interpolation formulae to the curves giving the flux as a function of frequency, and have used these data to calculate the ionization in the shell. For lack of a better geometrical representation that can be applied generally to planetaries, we have assumed the gas in the shell to be distributed spherically symmetrically at constant density. If we assume the 'filling factor' ε has the average of the values found by O'Dell (1962) and Seaton (1966), $\varepsilon=0{\cdot}67$, then for a uniform shell of gas $\varepsilon=1-(R_1/R_2)^3$, where R_1 and R_2 are the inner and outer radii of the nebula, respectively, and we have the condition that $R_1=0{\cdot}69\ R_2$. With the exception of the density, N, of the gas, all the parameters necessary for the calculations may be obtained from Seaton's (1966, Table IV) article. He gives the Zanstra temperature of the star T_S, the stellar radius R_S, and the outer radius of the nebular shell. At any epoch in the expansion, the density of the gas can be determined by assuming that planetary nebulae are all objects of the same mass, at different stages of expansion. Under this assumption, it is obvious that the quantity $(N_e\varepsilon R_2^3)=$constant for optically thin nebulae, where the constant can be obtained from observations of the surface brightness of the nebulae. We use Seaton's value of $\log(N_e\varepsilon R_2^3)=0{\cdot}47$, derived for $T_e=10^4\,°$K, where R_2 is in parsecs.

Two different effective temperatures of the central stars have been used in calculating the models of planetary nebulae: 63100°K, which is representative of cooler, low-excitation objects, and 100000°K, which is typical of the older, higher-excitation nebulae. The exact temperatures used were, of course, dictated by the available model atmospheres. Böhm and Deinzer have central star models at both of the temperatures given; however, Gebbie and Seaton have one at only the higher temperature. The latter have two other models published, but neither could be used in the present calculations. Their hottest model, with $T_S=200000\,°$K, is considerably hotter than any of the temperatures normally encountered on the evolutionary track, whereas their coolest model, with $T_S=41700\,°$K, does not give the flux above the ionization limit of He^+. Consequently, a low-temperature model has been built assuming the ionizing radiation field to be Planckian. In all models the abundances of the elements have been taken to be those found by Aller (1961) for planetary nebulae, where, by number, the logarithms of the relative abundances of H, He, N, O, and Ne are 12·00, 11·18, 8·37, 8·77, and 8·05, respectively.

The results of the ionization and electron temperature calculations for several of the models are shown in Figures 1–3. The ionization curves give the relative abundance of the ions of each of the elements at every point in the gas. Also plotted are the optical depths of the gas at the ionization edges of the ions H^0, He^0, and He^+. With the exception of the second model, the incident ionizing stellar radiation has been taken from the model atmospheres previously mentioned. All other pertinent information concerning the nebulae have been obtained from the data given by Seaton (1966). The parameters used for each of the models are as follows: (1) Stellar radiation from Böhm and Deinzer (1966, Figure 3b) model atmosphere, $T_S=63100\,°$K, $R_S=1{\cdot}20\ R_\odot$,

$N = 9{\cdot}0 \times 10^3$ cm^{-3}, $R_2 = 0{\cdot}080$ parsec; (2) Stellar radiation black body, $T_S = 63\,100\,°$K, remaining quantities the same as Model 1; (3) Stellar radiation from Böhm and Deinzer (1966, Figure 4b) model atmosphere, $T_S = 100000\,°$K, $R_S = 0{\cdot}10\,R_\odot$, $N = 1{\cdot}50 \times 10^2$ cm^{-3}, $R_2 = 0{\cdot}31$ parsec; (4) Stellar radiation from Gebbie and Seaton (1963, Figure 1) model atmosphere, $T_S = 100000\,°$K, remaining quantities the same as Model 3.

The scale of the graphs does not permit the abundances of the neutral atoms of the heavy elements to be shown. In all of the models, the abundance ratio of singly ionized

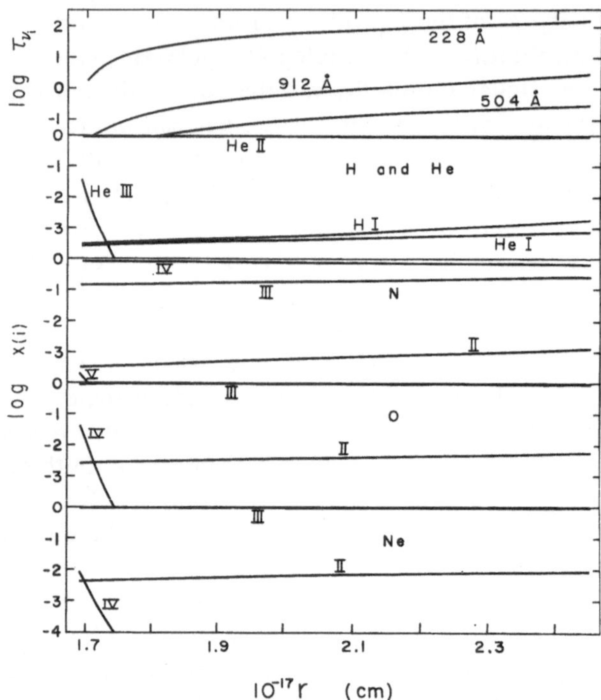

FIG. 1. *The ionization curves and optical depth of the gas for Model 1 (Böhm and Deinzer model atmosphere, $T_S = 63\,100\,°K$).*

ions to neutral atoms for these elements is fairly uniform throughout the gas, and approximately equal to the value 3×10^3. Therefore, the fractional abundance of N^0, O^0, and Ne^0 in the models varies from 10^{-7} to 10^{-5}.

4. Discussion of the Results

The ionization curves show that for the lower-temperature model the elements H and He are primarily singly ionized, while the heavy elements are predominantly doubly and triply ionized. At the higher temperature, the ionization is greater because

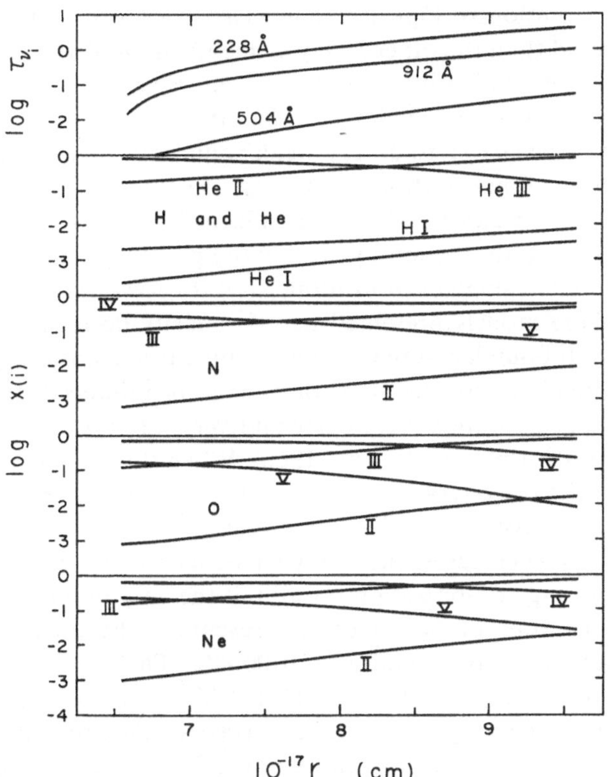

FIG. 2. *The ionization curves and optical depth of the gas for Model 4 (Gebbie and Seaton model atmosphere,* $T_S = 100\,000\,°K$*).*

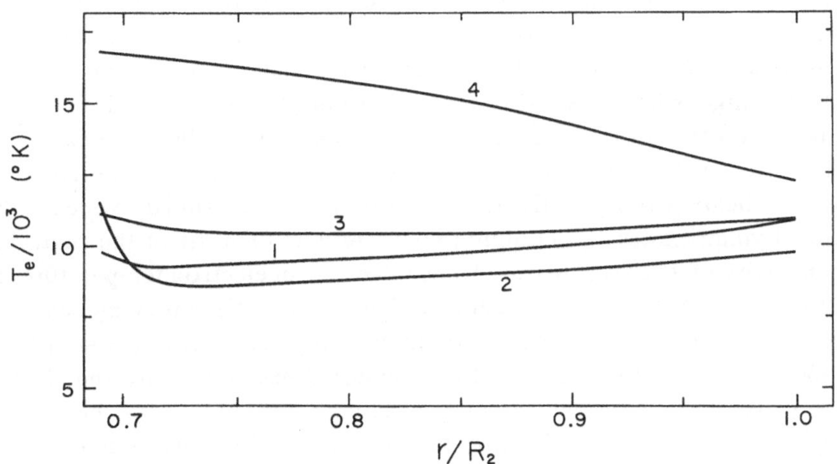

FIG. 3. *The electron temperature at each point in the gas for the four models. The distances are given relative to the outer radius of the nebular shell.*

of the shift in the radiation to higher frequencies. In general, one finds nitrogen to be more highly ionized than the other heavy elements in all of the models. This is due to the fact that nitrogen, unlike the other heavy elements, has an ionization potential for the third stage of ionization which is less than that of He^+. The strong absorption of He^+ for $\lambda < 228$ Å gives rise to a very weak radiation field for this region of the spectrum for all but the inner portions of the gas. Because the N^{+2} can be ionized by radiation with 261 Å$>\lambda>$228 Å, which is unaffected by any He^+ Lyman discontinuity in the central star or absorption by He^+ in the gas, the nitrogen has a large fraction of ions in the fourth stage of ionization, even for low-excitation nebulae.

Because each of the models presented here is density-bounded – the nebulae do not become ionization-bounded at these stellar temperatures until $\tau_{912\,\text{Å}} \sim 100$ – except for some of the more highly ionized ions, there is no pronounced stratification of the ions. This feature appears only at very early and very late evolutionary stages, when the nebulae are optically thick (cf. Seaton, 1966). When this does occur, there appears a small region near the periphery of the shell, where the heavy elements are virtually all singly ionized. However, even then, this region occupies less than 15% of the volume of the shell. In such cases, the heavy elements are generally in the third (and fourth, for nitrogen) stage of ionization for all but the very outer portions of the gas.

The electron temperatures of the models are presented in Figure 3. Except for Model 4, all have temperatures in the vicinity of 10000°K. These values are about 2000° higher than those obtained by Osterbrock (1965) from similar calculations. Part of this difference is due to the rough estimates Osterbrock made for the ionization of the heavy elements; however, the principle cause of the discrepancy is the use of different element abundances in the two investigations. The primary coolants in the nebulae are neon and oxygen, and our abundances of these elements are 5 and 1·5 times less, respectively, than those used by Osterbrock. In order to determine the effect this has on the resultant temperatures, some of our thermal equilibrium calculations were repeated using Osterbrock's abundances. The values of T_e obtained were in the 7000°–9000°K range, which is what Osterbrock found from his calculations.

The higher electron temperature of Model 4 is caused by the large excess in the emergent intensity of the stellar radiation for this model at high frequencies, which is very efficient in heating the gas. Evidence will be presented shortly which indicates that the He^+ Lyman excess is a common characteristic of most of the central stars. If so, the sensitivity of T_e to this feature leads us to expect electron temperatures in the range of 10000°–17000°K for most nebulae. This is in satisfactory agreement with the temperatures of planetary nebulae found by Liller and Aller (1954, 1963) and O'Dell (1966) from measurements of the emission lines. Certainly the theoretical calculations of electron temperatures in the nebulae need no longer be considered significantly lower than one should expect on the basis of empirical determinations.

It should be pointed out that since the ionization of H^0 by He II Lyman-α has been ignored in the heating of the gas, the electron temperature is unrealistically low in those regions of our models where the He^{+2} abundance is high. Hummer and Seaton (1964) and Goodson (1967) have shown that $T_e \sim 20000$ °K in such regions. This should affect our results only slightly, however, since there is relatively little He^{+2} in our models. Because we have taken the inner radius of the nebular shell to be determined by the condition that the filling factor $\varepsilon = \frac{2}{3}$, rather than $\varepsilon = 1$, as used by both Hummer and Seaton and Goodson, the radiation field is diluted by a factor of $\sim 10^{13}$ before any ionization can occur. As a result, the level of ionization in our calculations is considerably less than that found by Goodson.

Little can be said about the validity of the individual models from a knowledge of only the ionization of the gas. Most of the available information concerning conditions in nebulae comes from observations of the emission-line fluxes. In recent years, photoelectric techniques have been used to good advantage to secure intensities of the lines in a number of planetary nebulae. Consequently, there are sufficient data with which to compare theoretical results such that some conclusions can be drawn about the over-all structure of the nebulae. We have calculated the emergent flux of radiation, πF_ν, in a number of the emission lines for the models that have been computed, using the equation which is derived for an optically thin medium possessing spherical symmetry,

$$\pi F_\nu = \frac{4\pi}{R_2^2} \int_{R_1}^{R_2} j_\nu(r)\, r^2\, \mathrm{d}r\,, \tag{2}$$

where $j_\nu(r)$ is the volume emission coefficient in the line at a distance r from the centre of the nebula. The emission coefficients used for the lines of the H I Balmer series and He II Paschen series were those determined most recently by Pengelly (1964), while the emissivity of the He I recombination lines and the collisionally excited lines of the heavy elements has been taken from Seaton (1960). The collision strengths used in the calculations were the same as those given in Paper I. The resultant fluxes of the various emission lines are listed in Table 1, and are given relative to Hβ, whose absolute flux is then given for each of the models as the last entry in the table.

In order for the comparison between the theoretical and observed fluxes to be meaningful, we must have a number of observations of planetary nebulae at the same expansion epoch as those of our models. Because of the limited data available for nebulae at any given epoch, and the uncertainty in the physical parameters of the individual nebulae, the selection of specific representative objects with which to compare the models is impractical. Instead, we have chosen to plot the intensities of the lines of as many nebulae as possible for which observations exist as a function of some parameter which characterizes the stage in the evolution. The most widely determined parameter that fulfills this purpose is the Zanstra temperature of the central star. The recent evaluations of central star temperatures for a number of the nebulae by Harman and Seaton (1966) are particularly useful in this connection be-

Table 1

Emission-line fluxes computed from the models

Model	[Ne V] λ3426	[O II] λ3727	[Ne III] λ3869	He I λ4471	He II λ4686	[O III] λ5007	[N II] λ6584	[Ne II] λ12·79μ	Absolute Emergent Hβ Flux (ergs cm^{-2} sec^{-1})
	Emergent Line Fluxes Relative to Hβ								
1	9·95 (−11)	6.73 (−2)	6·12 (−1)	6·71 (−2)	8·73 (−4)	1·14 (1)	8·16 (−3)	5·92 (−3)	5·97 (−3)
2	3·24 (−6)	1·02 (−1)	4·17 (−1)	6·67 (−2)	2·43 (−2)	8·30 (0)	2·58 (−2)	1·57 (−2)	6·39 (−3)
3	6·35 (−7)	2·00 (−1)	7·21 (−1)	6·59 (−2)	2·91 (−2)	1·37 (1)	3·45 (−2)	9·75 (−3)	6·89 (−6)
4	1·78 (−1)	2·85 (−1)	7·79 (−1)	3·71 (−2)	7·89 (−1)	1·20 (1)	8·02 (−2)	7·56 (−3)	4·58 (−6)

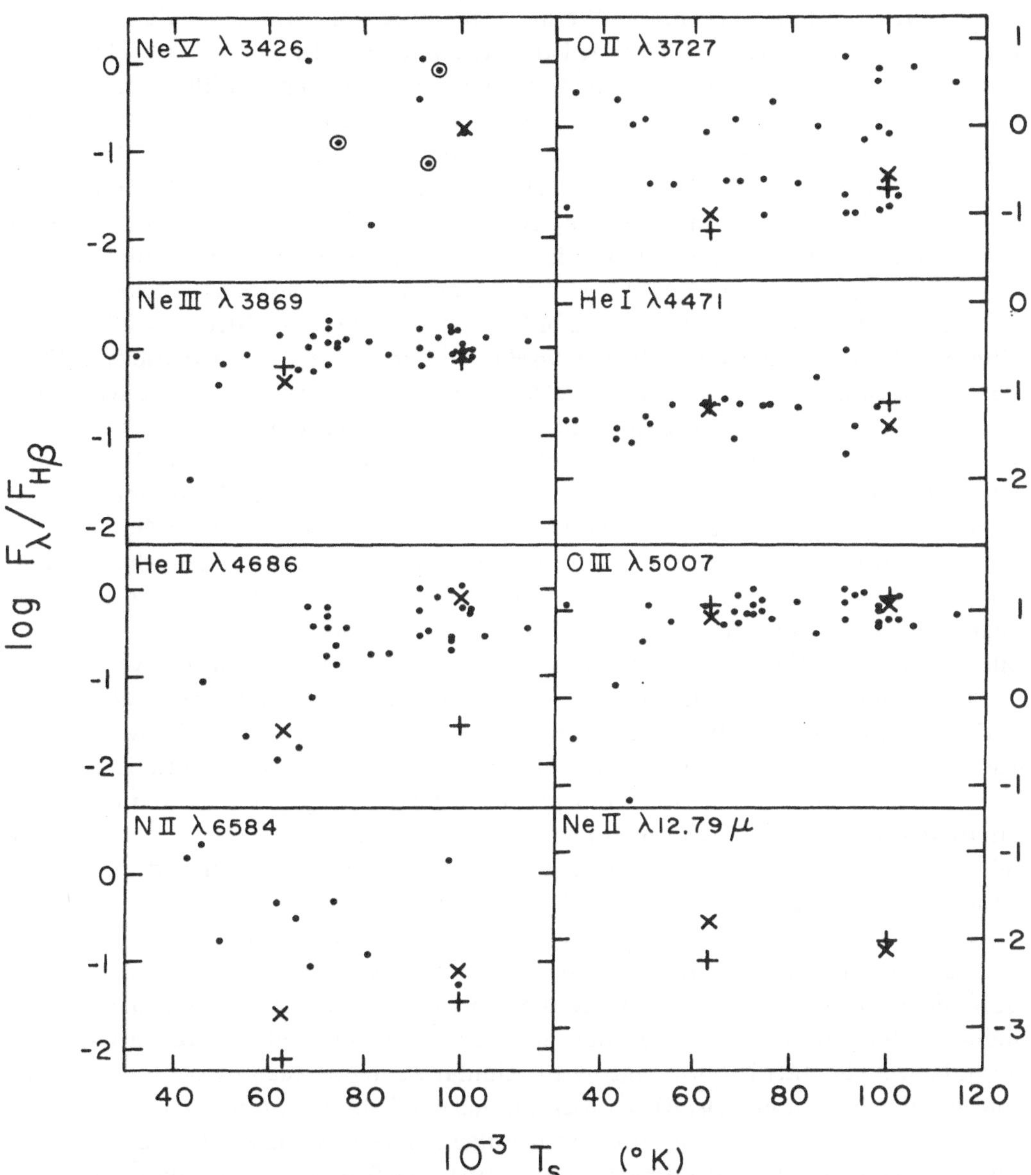

FIG. 4. *Emission-line fluxes of planetary nebulae plotted as a function of the temperature of the central star. Only those nebulae are considered for which Harman and Seaton have determined Zanstra temperatures. The filled circles denote observational data. The three measurements of λ3426 which are encircled are uncertain because of blends with* O III *λ3444. The computed fluxes for the models are also plotted, using the following symbols: + for Models 1 and 3, and × for Models 2 and 4. Several of the calculated fluxes for the lines λ3426 and λ4686 fall outside the scale of the graph, and are not shown. However, for these cases the numerical values can be readily obtained from Table 1. The* [Ne II] *λ12·79 μ line is shown, although no published observations of this line are yet available.*

cause of their improved accuracy. We have therefore taken those nebulae for which Harman and Seaton have determined the stellar temperatures and have plotted all observed fluxes of the lines for which calculations were made with respect to the temperature of the star, as shown in Figure 4. Also depicted, with different symbols, are the flux calculations from the models, as given in Table 1. The flux of each line is given relative to Hβ. All of the uncorrected data have been corrected for interstellar reddening using values of the reddening constant, c, given by Harman and Seaton for each nebula. With few exceptions, the observed data have been taken from the following sources: all He I λ4471 and He II λ4686 measurements are those given by Harman and Seaton (1966). Intensities of [O III] λ5007 are those listed by Collins *et al.* (1961) and, for several cases, O'Dell (1963*b*). All but one of the observations of [N II] λ6584, which are few because of the relative scarcity of work done in the red region of the spectrum, come from Osterbrock *et al.* (1963). The remaining line fluxes were obtained from a number of sources, including Minkowski (1942), Wyse (1942), Aller (1941, 1951), Minkowski and Aller (1956), O'Dell (1963*c*), and the recent series of papers by Aller and co-workers entitled 'Spectrophotometric Studies of Gaseous Nebulae' (cf. Aller *et al.*, 1966, for the most recent paper). In all instances where several observations are available for the same object, an average of the measurements has been used, except in cases where photoelectric observations have been made. In the latter event, only the photoelectric measurements are considered in the average.

The two most important factors that govern the intensity of a line are the abundance and ionization concentration of the element. Since we have no *a priori* knowledge that any of our models give a correct representation of the ionization conditions in planetary nebulae, differences between our predicted fluxes and those observed cannot be attributed simply to the use of incorrect element abundances. The relative importance of the two factors must be established. Some statement concerning the effect of abundances can be made by considering the line radiation from ions whose abundances are insensitive to different physical conditions in the gas. It is seen from the ionization curves of the models that the ions He^{+}, N^{+2}, N^{+3}, O^{+2}, and Ne^{+2} satisfy this requirement reasonably well – they are the most abundant ions of the respective elements in both high- and low-excitation objects, showing only minor variations in abundance. This fact is borne out by the similar strengths each of the lines of these ions has for the four models. Also, observational confirmation of the relative constancy in abundance of these ions in different nebulae is given by the smaller scatter of the fluxes of He I λ4471, [O III] λ5007, and [Ne III] λ3869 in Figure 4 compared to the other lines. Consequently, we have some assurance that a direct comparison of the theoretical and observational fluxes of these lines will yield information on the abundances of these particular elements. It is seen that there is excellent agreement between the observed and predicted fluxes of the He I and [O III] lines. The agreement is likewise good for the [Ne III] line, however, it would be improved if the logarithmic abundance of neon in the models were to be increased by an amount +0·20. Therefore, we find general agreement with the abundances of these elements as derived by Aller (1961) for the nebulae. Because there are no strong N III or N IV lines in the visible, the abundance of nitrogen must be deduced from [N II] λ6584, which is sensitive to ionization conditions. On the other hand, the confirmation of the abundances of helium, oxygen, and neon enables the lines from other ions of these elements to be used to evaluate the ionization equilibrium of the models.

Let us consider the strength of [O II] λ3727. In spite of a considerable amount of scatter in the observations, it is evident that all of our models predict too small a flux for λ3727 by a factor of about 3, due to an underestimate of the abundance of O^+ by a similar amount in the computations. In each instance, the models using the atmosphere calculations of Böhm and Deinzer show poorer agreement than the other models, although the differences are slight. The disparity in the O^+ abundance between our models and typical nebulae must be due to our choice of the parameters affecting the statistical equilibrium – the sizes and densities of the nebulae, or perhaps the spectral distribution of the central stars. In view of the uncertainties in the determination of these quantities, the discrepancy should not be considered too serious. It is of some interest to note in this connection, that as a result of the study of planetary nebulae in the Magellanic Clouds, Webster (1966) has criticized the value of the mean mass of the nebulae (0·6 $M_\odot$) found by Seaton (1966). She believes this is caused by an underestimate of the mean density of several nebulae which were used by Seaton to calibrate his distance scale, resulting in an overestimate of their sizes. If this is correct, the underabundance of O^+ in the models could be explained, since model calculations made with higher densities and smaller dimensions would result in an increased abundance of the singly ionized ions, as long as there is no appreciable stratification.

It is seen from Figure 4 that the calculated strength of [N II] λ6584 is also less than it should be, by one order of magnitude for models of both temperatures. As was true in the case of the theoretical [O II] λ3727 fluxes, part of this discrepancy may be attributed to a deficiency in the computed abundance of N^+. Because of the similar ionization potentials of the first two stages of ionization of nitrogen and oxygen, it is probable that the abundance deficiencies of N^+ and O^+ are approximately the same. Correcting for this effect, however, accounts for only half of the difference between the fluxes. The [N II] fluxes of the models are still three times smaller than the observations require. The only plausible explanation for the remaining disagreement that is consistent with the conclusions reached from the earlier analysis of other spectral lines is that too small a nitrogen abundance has been used in the calculations. On this basis, we suggest that the normal nitrogen abundance in planetary nebulae should be increased to the value $\log[N(\mathrm{N})/N(\mathrm{H})] = -3{\cdot}15$.

Finally, we consider the fluxes of the lines He II λ4686 and [Ne V] λ3426. The relative abundances of these two elements have already been ascertained, consequently any discrepancies between theory and observation may be attributed to the choice of incorrect parameters which govern the ionization. A great deal of variation in the line strengths of the different models is evident from Figure 4. It is caused by the widely differing intensities of the model atmospheres at frequencies above the He^{+} Lyman limit, since it is the radiation from this region which produces He^{+2} and Ne^{+4}. In each of the two models which use the stellar atmospheres of Böhm and Deinzer, both lines have unacceptably small fluxes. On the other hand, at the higher temperature the strengths of the lines for the model which is based upon the atmosphere calculations of Gebbie and Seaton show very good agreement with the observations. The combination of a lack of good data and a considerable amount of scatter in the intensities of the [Ne v] line at lower stellar temperatures makes it difficult to pick a representative flux of this line for the low-excitation models. The value $\log [F(\lambda 3426)/F(\mathrm{H}\beta)] = -3{\cdot}0$ is certainly a lower limit. The black-body model at $T_S = 63\,100\,°\mathrm{K}$ fails to produce this much $\lambda 3426$ radiation, indicating that the central stars of the nebulae must have a considerable He^{+} Lyman excess at high frequencies even at this lower stellar temperature. The fact that the strength of $\lambda 4686$ for this model agrees well with the observational data indicates that the stellar continuum does not differ markedly from that of a Planckian distribution at wavelengths near 228 Å. Actually, this behavior of the spectral distribution is quite similar to that found by Gebbie and Seaton for the central stars. In view of this fact, the evidence suggests that the spectral distribution of the central stars of both high- and low-excitation nebulae are similar in nature for $\lambda < 228$ Å, and may best be depicted qualitatively in this spectral region by the Gebbie and Seaton model for which $T_S = 1{\cdot}0 \times 10^5\,°\mathrm{K}$.

There is an alternative explanation that might account for the observed He II and [Ne v] line strengths without requiring a strong far-ultraviolet radiation field. The reason Böhm and Deinzer's model atmospheres have so little emergent radiation below 228 Å is because this radiation is absorbed in the atmosphere by He^{+2} and Ne^{+4}, among other ions, whose abundances are high. Recently, evidence has been presented by Mathews (1966) in support of a continuous loss of mass occurring from the central stars in the form of a stellar wind. Could it be that the ions He^{+2} and Ne^{+4}, which are produced in the stars, are injected by such a wind into the nebulae, where the emission lines from these ions are then produced? In order to test this hypothesis, calculations were made to determine the maximum amount of radiation that could be produced in each of the lines in this manner, using several different sets of values for the density and velocity of the stellar wind. It was found that once the ejected gas reaches the nebular shell, recombination of the ionized material occurs in a time-scale much too short to enable the required amount of radiation to be emitted. In no case was the flux of either the He II or [Ne v] line found to be within 4 orders of magnitude of the observed fluxes. Furthermore, the distance over which the recombination takes place is $\lesssim 10^{14}$ cm, which means the radiation would come from a very thin rim around the inside boundary of the shell. Yet, monochromatic images of planetary nebulae obtained by Wilson (1950) and Aller (1956) show that the radiation from $\lambda 3426$ and

λ4686 usually extends an appreciable distance into the shell. These considerations make it difficult to associate any He II and [Ne V] emission with a stellar wind.

From the standpoint of attempting to correctly account for the ionization structure of planetary nebulae, there is evidence for an ultraviolet excess at frequencies above the He^+ ionization limit. It is an entirely different matter to account for this behavior by constructing model atmospheres of the central stars. The most thorough work of this nature has been done by Böhm and Deinzer, and they predict a strong ultraviolet deficiency. The fact that the emission-line flux predictions for the model using the Gebbie and Seaton atmosphere agree fairly well with observations must be considered fortuitous since they did not consider the heavy elements as sources of opacity. Presumably, the inclusion of these elements into their calculations will also tend to depress the stellar continuum at high frequencies. The possible resolution of this dilemma is discussed elsewhere in this volume in the article by Professor Böhm. It may be that strict radiative equilibrium does not apply in the very outer layers of the star because of stellar mass loss, thereby decreasing the steep temperature gradient and minimizing the importance of absorption with respect to scattering in the atmosphere. At the same time, however, further investigation must be made into the possible effects of a density gradient, nebular condensations, and the use of a temperature scale for the central stars that is based upon the models and not upon black bodies, on the ionization results that have been presented here.

References

Aller, L.H. (1941) *Astrophys. J.*, **93**, 236.
Aller, L.H. (1951) *Astrophys. J.*, **113**, 125.
Aller, L.H. (1956) *Gaseous Nebulae*, Chapman & Hall, Ltd., London.
Aller, L.H. (1961) *The Abundance of the Elements*, Interscience Publ., New York.
Aller, L.H., Kaler, J.B., Bowen, I.S. (1966) *Astrophys. J.*, **144**, 291.
Böhm, K.-H., Deinzer, W. (1965) *Z. Astrophys.*, **61**, 1.
Böhm, K.-H., Deinzer, W. (1966) *Z. Astrophys.*, **63**, 177.
Collins, G.W., Daub, C.T., O'Dell, C.R. (1961) *Astrophys. J.*, **133**, 471.
Gebbie, K.B., Seaton, M.J. (1963) *Nature*, **199**, 580.
Goodson, W.L. (1967) *Z. Astrophys.*, **66**, 118.
Harman, R.J., Seaton, M.J. (1964) *Astrophys. J.*, **140**, 824.
Harman, R.J., Seaton, M.J. (1966) *Astrophys. J.*, **132**, 15.
Hummer, D.G., Seaton, M.J. (1963) *Mon. Not. R. astr. Soc.*, **125**, 437.
Hummer, D.G., Seaton, M.J. (1964) *Mon. Not. R. astr. Soc.*, **127**, 217.
Liller, W., Aller, L.H. (1954) *Astrophys. J.*, **120**, 48.
Liller, W., Aller, L.H. (1963) *Proc. Nat. Acad. Sci.*, **49**, 675.
Mathews, W.G. (1966) *Astrophys. J.*, **143**, 173.
Minkowski, R. (1942) *Astrophys. J.*, **95**, 243.
Minkowski, R., Aller, L.H. (1956) *Astrophys. J.*, **124**, 93.
O'Dell, C.R. (1962) *Astrophys. J.*, **135**, 371.
O'Dell, C.R. (1963*a*) *Astrophys. J.*, **138**, 67.
O'Dell, C.R. (1963*b*) *Astrophys. J.*, **138**, 293.
O'Dell, C.R. (1963*c*) *Astrophys. J.*, **138**, 1018.
O'Dell, C.R. (1966) *Astrophys. J.*, **143**, 168.

Osterbrock, D.E. (1965) *Astrophys. J.*, **142**, 1423.
Osterbrock, D.E., Capriotti, E.R., Bautz, L.P. (1963) *Astrophys. J.*, **138**, 62.
Pengelly, R.M. (1964) *Mon. Not. R. astr. Soc.*, **127**, 145.
Seaton, M.J. (1960) *Repts. progr. Phys.*, **23**, 313.
Seaton, M.J. (1966) *Mon. Not. R. astr. Soc.*, **132**, 113.
Webster, B.L. (1966) unpublished Ph.D. dissertation, Australian National University.
Williams, R.E. (1967) *Astrophys. J.*, **147**, 556.
Wilson, O.C. (1950) *Astrophys. J.*, **111**, 279.
Wyse, A.B. (1942) *Astrophys. J.*, **95**, 356.

DISCUSSION

Hummer: I would like to question the assumption by Williams of one value of an inner radius, for we have found that the ionization balance is extremely sensitive to the assumed dilution factor, given the flux distribution of the star.

Williams: The ratio of the inner to the outer radius of a nebula cannot be varied without changing the filling factor. The latter quantity is very sensitive to this ratio. Since the ionization results have meaning only in a statistical sense, I do not feel one is at liberty to abandon the concept of an average filling factor for the nebulae.

Seaton: In making models of ionization structure one should try ultimately to fit to observations for the nebular line intensities and for the observed intensity of the radiations of the central star. When model-atmosphere central-star fluxes are used, it is not a consistent procedure simply to adopt Zanstra temperatures obtained assuming black-body fluxes.

THE IONISATION EQUILIBRIUM FOR HEAVY ELEMENTS

D. R. Flower
(Dept. of Physics, University College London, England)

ABSTRACT

Calculations are being made of the distribution of the ions of heavy elements in planetary nebulae. Initial work has been concentrated on the central or He^{2+} zone of planetaries. The optical depths of ions of C, N, O, and Ne have been computed using ground state ionisation cross-sections and using approximations which should be substantially better than hydrogenic. A comparison has been made between the combined optical depth of the heavy elements and the optical depth of He^+ in the far ultraviolet. The optical depths of the heavy elements in this spectral region may become significant, but a reasonable first approximation to the radiation field may be obtained by neglecting the absorption of all ions except He^+. The distribution of the ions of the heavy elements has been calculated on this assumption.

Calculations have been made of the distribution of ions of heavy elements in the central zone of planetary nebulae. The basic problem is to solve the ionisation-equilibrium equation at each point in the central zone. If we consider an element X, the equilibrium between the qth and $(q+1)$th stages of ionisation may be written

$$4\pi N(X^{q+}) \int J_\nu a_\nu \, \mathrm{d}\nu = N_e N(X^{(q+1)+}) \, \alpha . \tag{1}$$

$4\pi J_\nu$ is the mean flux of photons, N represents a number density, N_e is the electron density, a_ν the ground state photo-ionisation cross-section, and α is the total radiative recombination coefficient. If the optical depth associated with the ion X^{q+} is assumed to be small, then the total recombination coefficient is a sum of coefficients for radiative recombination to all excited states and to the ground state:

$$\alpha = \sum_{n=n'}^{\infty} \alpha_n \tag{2}$$

where n' is the principal quantum number of the ground state of the ion X^{q+}. The recombination coefficients are slowly varying functions of the electron temperature, T_e. If equations of the form of (1) can be solved, then the ratio, $N(X^{(q+1)+})/N(X^{q+})$, of the concentrations of successive stages of ionisation of the element X can be obtained at each point in the central zone of the nebula. Given the abundance of X relative to hydrogen, it is then possible to calculate the abundance of each ion of X relative to hydrogen at each point.

Initial calculations have been restricted to the central ionisation zone of planetaries

Osterbrock and O'Dell (eds.), Planetary Nebulae, 205–208. © *I.A.U.*

where most of the helium is fully ionised, i.e., most of the helium is He^{2+}. This zone will be a significant fraction of the total nebular volume only in high-excitation nebulae. All the ions present in this region of the nebula will have ionisation potentials above the He^+ threshold value at 228 Å. The radiation field shortwards of 228 Å is calculated at each point of the central zone on the assumption that He^+ dominates the absorption of this radiation in the central region. This would appear to be a reasonable first approximation because the photo-ionisation cross-section of hydrogen is very small at such short wavelengths and the abundance of helium is much greater than the abundances of heavier elements. Essentially, we assume initially that only He^+ contributes significantly to the optical depth in the central zone. The photon flux at some point in the central zone distant R from the nuclear star is then given by

$$4\pi J_\nu = \frac{L_\nu}{4\pi R^2} e^{-\tau_\nu(\mathrm{He}^+)} \tag{3}$$

on a spherically symmetric model. τ_ν (He^+) is the optical depth of He^+ at frequency ν, and L_ν the photon luminosity of the star at that frequency.

The ground-state photo-ionisation cross-sections, a_ν, required for a solution of Equation (1) are taken from a paper by Seaton (1958) for ions with electron configurations $1s^2\ 2s^2\ 2p^q$, where q is any integer from 1 to 6 inclusive. Seaton fits his calculations to a function of the form

$$a_\nu = BC \times 10^{-18} \left\{ \alpha \left(\frac{\nu}{\nu_0}\right)^{-s} + (1-\alpha)\left(\frac{\nu}{\nu_0}\right)^{-s-1} \right\} \tag{4}$$

where B, C, s, and α are tabulated quantities, ν_0 is the ionisation threshold frequency, and a_ν is in units of cm^2. Similar calculations have been completed for the configurations $1s^2\ 2s$ and $1s^2\ 2s^2$, and the results are presented in Tables 1 and 2. The calculations were based on the method of quantum defects described by Burgess and Seaton (1960) using revised tabulations of necessary functions given by Peach (1967). The calculations for the $1s^2\ 2s$ configuration presented no difficulty and the results should be substantially better than hydrogenic approximations. Calculations for the

Table 1

Photo-ionisation cross-section parameters for the lithium sequence, $1s^2\ 2s$, for which $C=1$. The cross-section is then given by Equation (4)

Ion	B	s	α
Li	1.3	1·4	3·5
Be^+	1·4	2·0	2·2
B^{2+}	0·98	1·8	1·0
C^{3+}	0·68	2·0	1·0
N^{4+}	0·48	2·0	1·0
O^{5+}	0·36	2·1	1·0

Table 2

Photo-ionisation cross-section parameters for the beryllium sequence, $1s^2\ 2s^2$, for which $C=2$. The cross-section is then given by Equation (4). Be^0 is omitted because the results obtained were not accurate for that particular case

Ion	B	s	α
B^+	1·5	3·0	2·6
C^{2+}	0·80	3·0	2·6
N^{3+}	0·54	3·0	2·6
O^{4+}	0·39	3·0	2·6
F^{5+}	0·29	3·0	2·6

$1s^2\ 2s^2$ sequence are complicated by the presence of equivalent electrons in the outer shell. The results presented are accurate only at threshold and the frequency variation is taken to be hydrogenic. The results for the neutral in the $1s^2\ 2s^2$ sequence, namely Be^0, were poor and have been omitted from Table 2. Nevertheless, figures presented for higher members of this sequence should be significantly better than hydrogenic approximations.

Having solved for the distribution of ions of the heavy elements in the central zone it is then possible to calculate the optical depths of these ions and compare their combined optical depth with that of He^+. At wavelengths just shortwards of 228 Å the optical depth of He^+ will indeed be much greater than the combined optical depth of other ions. However, as Khromov (1967) has pointed out, this may not be the case at much shorter wavelengths because the ground-state photo-ionisation cross-section of He^+ decreases inversely as the third power of the frequency. In the far ultraviolet the combined optical depth of ions of the heavy elements may become comparable with or even greater than that of He^+.

Table 3 shows the results of calculations of optical depths at 90 Å. The inner radius of the nebular shell is taken as 5×10^{15} cm and the central star is assumed to radiate

Table 3

Optical depths of He^+ and ions of heavier elements at a wavelength of 90 Å. The absolute hydrogen density is taken as $1{\cdot}0 \times 10^4\ cm^{-3}$ and the relative element abundances are taken from Aller (1964)

Radius ($\times 10^{-16}$ cm)	Optical depth of He^+ at 90 Å	Optical depths of ions of the heavy elements at 90 Å					Combined optical depth of ions of the heavy elements
		C^{3+}	N^{4+}	O^{4+}	Ne^{3+}	Ne^{4+}	
4·5	0·120	0·005	0·017	0·134	0·073	0·111	0·340
5·0	0·229	0·008	0·018	0·142	0·093	0·118	0·379
5·5	0·632	0·011	0·019	0·145	0·112	0·121	0·408
5·81	1·201	0·013	0·019	0·146	0·118	0·122	0·418

as a black body at 12×10^4 °K. The electron temperature of the central zone is assumed constant at 2×10^4 °K (Flower, 1968). It appears that absorption of far-ultraviolet radiation by ions of the heavy elements is at least as important as absorption by He^+. More refined models of the ionisation structure of planetary nebulae should take this absorption into account.

Acknowledgements

This work was completed whilst the author was in tenure of a Science Research Council studentship. The author would like to thank Professor M.J. Seaton for his interest in this work.

References

Aller, L.H. (1964) *Astrophys. Norw.*, **9**, 293.
Burgess, A., Seaton, M.J. (1960) *Mon. Not. R. astr. Soc.*, **120**, 121.
Flower, D.R. (1968) in the present volume, p. 77.
Khromov, G.S. (1967) *Mon. Not. R. astr. Soc.*, **137**, 175.
Peach, G. (1967) *Mem. R. astr. Soc.*, **71**, 13.
Seaton, M.J. (1958) *Rev. Mod. Phys.*, **30**, 979.

THE CHEMICAL COMPOSITION OF PLANETARY NEBULAE

LAWRENCE H. ALLER and STANLEY J. CZYZAK
(University of California, Los Angeles, Calif., U.S.A.) *(The Ohio State University, U.S.A.)*

The problem of the determination of the chemical compositions of planetary and other gaseous nebulae constitutes one of the most exasperating problems in astrophysics. On the one hand, the problem appears to be conceptually simple – the mechanisms of excitation of the various lines appear to be well understood and the necessary physical parameters can be obtained by quantum mechanical theory. Yet the task is a difficult one and we want to explore some of the significant features.

Three distinct problems are involved:

(1) *Basic physical parameters and excitation theory.* These include the transition probabilities for forbidden lines, and collision strengths, not only for the forbidden lines but also for transitions from low levels that are necessarily metastable – as e.g. in Mg I. Particularly urgently needed – not just for planetaries, but also for objects such as η Carinae, novae, and combination variables – are collision strengths for iron atoms in various stages of ionization. Ubiquitous iron shows up, in one form or another, in almost every celestial source.

Also needed are appropriate theories for numerous weak permitted lines that appear in a number of nebulae. For example, carbon is represented only by permitted lines of C II, C III, and C IV. Some work on this element is being done by Clarke. Most nitrogen ions in planetaries are revealed only by permitted lines. Both these elements – so important in hydrogen-burning processes in stars – need our urgent attention. Seaton has suggested that the permitted lines of oxygen formerly attributed to recombination actually are excited by the central star. Hence, they cannot be used to obtain ionic abundances. Therefore it would be worthwhile to study the permitted lines of O II and Ne II, since the ionic concentrations of O^{++} and Ne^{++} can be obtained from their forbidden lines; thus a check on the theory is possible, and for these ions at least it would be possible to assess the relative importance of direct line excitation and recombination.

(2) *Adequate observational data on a good representative set of nebulae.* In practice, we are pretty well limited to nebulae of high or moderately high surface brightness. We need to know the integrated brightness of the nebula in its strongest lines in absolute units. These data are best obtained by photoelectric photometry. We also need the relative intensities of the weaker lines and of the nebular continuum over as long a wavelength range as possible. Radio-frequency data help us to assess space

Osterbrock and O'Dell (eds.), Planetary Nebulae, 209–223. © *I.A.U.*

absorption, and we hope that ultimately observations from above the Earth's atmosphere will enable us to fill in the blank places in the spectrum.

(3) We must make allowance for the *structural features and stratification effects* in nebulae and for the *distribution of atoms among various stages of ionization*. The first of these effects can sometimes be handled if we can obtain nebular spectra on a sufficient scale, with the slit of the spectrograph placed in a known position in the object. Examples will be described shortly.

The classical investigations of Bowen and Wyse (1939) and of Wyse (1942) constituted the first definitive study of the chemical compositions of the planetaries. Looking back, after nearly a generation, one is impressed by their observational skill and the enduring character of their essential conclusions. Theoretical studies by Menzel and his associates enable certain refinements to be introduced, but the problem of collision strengths remained as a roadblock until it was successfully resolved by Seaton (1953) and his associates and others who followed the trail he blazed. See Czyzak (1967), Czyzak and Krueger (1967), and Seraph *et al.* (1966).

Using the notation of Seaton (1960) and the equations of statistical equilibrium (Menzel *et al.*, 1941; Aller, 1956, p. 192) and modern cross-section data reviewed here recently by Seaton, we may write down some typical equations. In each instance, we express the concentration of the ion in terms of that of the O^{++} ion, and the intensities of the relevant forbidden lines in terms of the sum of the intensities of the green nebular lines.

We define $x = 0{\cdot}01\, N/\sqrt{T}$, so that $x = 1$ for a 'standard' bright nebula with $N = 10000$ cm^{-3} and $T = 10000$ °K. Let I_n denote the sum of the intensities of the nebular transitions of the ion in question, I_a the sum of the intensities of the auroral transitions and I_{TA} the sum of the intensities of trans-auroral transitions. Let I_0 denote the sum of the intensities of the green nebular lines. Let $D(y) = 10^y$. We assume that the ionic and [O III] lines are produced in the same layers, so that the same values of x and T_ε are applicable.

Nitrogen:

$$\frac{N(\mathrm{N}^+)}{N(\mathrm{O}^{++})} = 0{\cdot}149\, \frac{3770\left(\frac{1}{x} + 0{\cdot}14\right)}{1 + 0{\cdot}29x}\, \frac{D\left(\frac{7800}{T}\right)}{1 + 67/x}\, \frac{I_a(\mathrm{N\,II})}{I_0}$$

$$\frac{I_{\mathrm{neb}}}{I_a} = \frac{8{\cdot}5 D(10\,800/T)}{1 + 0{\cdot}29\,x} \qquad [\mathrm{N\,II}]$$

Oxygen:

$$\frac{I_0}{I(\lambda 4363)} = \frac{7{\cdot}15}{1 + 0{\cdot}028x}\, D\left(\frac{14\,300}{T}\right) \qquad [\mathrm{O\,III}]$$

New fluorine cross-section data yield:

$$\frac{N(\mathrm{F}^{+3})}{N(\mathrm{O}^{++})} \sim 1{\cdot}0\, \frac{I_{\mathrm{neb}}(\mathrm{F\,IV})}{I_0}\, D(2980/T) \qquad [\mathrm{F\,IV}]$$

For the ions of neon we have the following relation:

$$\frac{N(\mathrm{Ne}^{++})}{N(\mathrm{O}^{++})} = 0{\cdot}0983\,\frac{(1+1020/x)}{(1+67/x)}\,D\left(\frac{3480}{T}\right)\frac{I_n(\mathrm{Ne\,III})}{I_0} \qquad [\mathrm{Ne\,III}]$$

$$\frac{N(\mathrm{Ne}^{+4})}{N(\mathrm{O}^{++})} = 0.0365\,\frac{(1+2190/x)}{(1+67/x)}\,D\left(\frac{6100}{T}\right)\frac{I_n(\mathrm{Ne\,V})}{I_0}, \qquad [\mathrm{Ne\,V}]$$

while at low densities ($x \leqslant 1{\cdot}0$)

$$\frac{N(\mathrm{Ne}^{+3})}{N(\mathrm{O}^{++})} \simeq 9{\cdot}8\,\frac{I_a(\lambda 4724{,}5)}{I_0}\,D\left\{\frac{26\,400}{T}\right\} \qquad [\mathrm{Ne\,IV}]$$

The only ion of sodium observed is Na^{+++}; we have

$$\frac{N(\mathrm{Na}^{+3})}{N(\mathrm{O}^{++})} = 0{\cdot}020\,\frac{(1+4510/x)}{(1+67/x)}\,D\left\{\frac{5800}{T}\right\}\frac{I_n(\mathrm{Na\,IV})}{I_0} \qquad [\mathrm{Na\,IV}]$$

Singly ionized sulphur presents a difficult problem because of collisional interlocking between levels of the same term. The most elementary approach is to ignore the fine structure and assume that collisions distribute the ions between levels in proportion to their statistical weights.

For 10000 °K $< T <$ 20000 °K and $0{\cdot}2 < x < 10$, the factor expressing deviation from thermodynamic equilibrium can be put in the form $(b_3/b_1) = 2{\cdot}6 \times 10^{-4}x$.

Then, from the trans-auroral lines

$$\frac{N(\mathrm{S}^{+})}{N(\mathrm{O}^{++})} = \frac{48}{x+67}\,D\left(\frac{2820}{T}\right)\frac{I_{\mathrm{TA}}(\mathrm{S\,II})}{I_0} \qquad [\mathrm{S\,II}]$$

The red nebular and violet trans-auroral lines are sensitive to both temperature and density:

$$\frac{I_{\mathrm{TA}}(\mathrm{S}^{+})}{I_{\mathrm{neb}}(\mathrm{S}^{+})} = 0{\cdot}164\left\{3{\cdot}8 + x\left[1 + 1{\cdot}32\,D\left(-\frac{6000}{T}\right)\right]\right\}D\left(-\frac{6000}{T}\right) \qquad [\mathrm{S\,II}]$$

For the [S III] lines we have the following relationships:

$$\frac{N(\mathrm{S}^{++})}{N(\mathrm{O}^{++})} = 0{\cdot}60\left[\frac{1+101/x}{1+67/x}\right]D\left(-\frac{5800}{T}\right)\frac{I_n(\mathrm{S\,III})}{I_0}$$

$$\frac{N(\mathrm{S}^{++})}{N(\mathrm{O}^{++})} = 0{\cdot}070\left(\frac{1+\dfrac{3700}{x}}{1+67/x}\right)D\left(\frac{4320}{T}\right)\frac{I_a(\mathrm{S\,III})}{I_0}. \qquad [\mathrm{S\,III}]$$

An expression similar to the last holds for the trans-auroral line near the Balmer limit, λ3722, except that the constant 0·070 is replaced by 0·124.

The [Cl III] λ5517/5537 line ratio should be a sensitive indicator of the density, but it will be necessary to compute the collision strengths for $^2D_{3/2}$–$^2D_{5/2}$ and $^2P_{1/2}$–$^2P_{3/2}$ exchanges. Similar remarks apply to [Ar IV]. The [Cl III] λ5517 line intensity is related to the concentration of Cl^{++} ions by:

$$\frac{N(\mathrm{Cl}^{++})}{N(\mathrm{O}^{++})} = \frac{11{\cdot}5D\left(\frac{-1220}{T}\right)}{1+\frac{67}{x}}\left\{\frac{1+\frac{3{\cdot}63}{x}+1{\cdot}7D\left(\frac{-7300}{T}\right)}{1+0{\cdot}193D(-7300/T)}\right\}\frac{I(5517)}{I_0} \qquad [\mathrm{Cl\,III}]$$

with an analogous expression for $I(5537)$.

Ions of Cl^{+++} are represented by lines in the green (auroral transition) and near infrared

$$\frac{N(\mathrm{Cl}^{+++})}{N(\mathrm{O}^{++})} \simeq 19D\left(\frac{-4250}{T}\right) I_{\mathrm{n}}\frac{(\mathrm{Cl\,IV})}{I_0} \qquad [\mathrm{Cl\,IV}]$$

$$\frac{N(\mathrm{Cl}^{+++})}{N(\mathrm{O}^{++})} \simeq 14{\cdot}2D\left(\frac{7300}{T}\right)\frac{I_{\mathrm{a}}(\mathrm{Cl\,IV})}{I_0}.$$

Argon is represented by forbidden lines of [Ar III], [Ar IV] and [Ar V]. We have

$$\frac{N(\mathrm{Ar}^{++})}{N(\mathrm{O}^{++})} \sim 79D\left(\frac{8050}{T}\right)\frac{I(\lambda 5191)}{I_0} \qquad \text{auroral line [Ar III]}$$

$$\frac{I_{\mathrm{a}}(\mathrm{Ar}^{++})}{I_n(\mathrm{Ar}^{++})} \simeq 0{\cdot}091D\left(\frac{-12\,000}{T}\right). \qquad \text{ratio of auroral and nebular lines}$$

The most frequently observed line of [Ar IV] is λ4740; $^2D_{3/2}$–$^4S_{3/2}$ its companion λ4711 is often blended with He I λ4713:

$$\frac{N(\mathrm{Ar}^{+++})}{N(\mathrm{O}^{++})} = \frac{0{\cdot}520}{1+67/x}D\left(\frac{705}{T}\right)\left\{\frac{1+\frac{229}{x}+2{\cdot}8D\left(\frac{-8660}{T}\right)}{1+0{\cdot}214D(-8660/T)}\right\}\frac{I(4740)}{I_0}.$$

The red auroral transitions of [Ar IV] are often observed, although less conveniently since they fall in the λ6700 region.

For the [Ar V] lines we may write

$$\frac{N(\mathrm{Ar}^{+4})}{N(\mathrm{O}^{++})} \simeq 2{\cdot}71D\left(\frac{-3180}{T}\right)\frac{I_n(\mathrm{Ar\,V})}{I_0}, \quad \frac{N(\mathrm{Ar}^{+4})}{N(\mathrm{O}^{++})} \sim 41{\cdot}0D\left(\frac{10\,400}{T}\right)\frac{I_{\mathrm{a}}(\mathrm{Ar\,V})}{I_0}.$$

Potassium is represented by lines of [K IV], [K V], and [K VI]. Thus

$$\frac{N(\mathrm{K}^{+3})}{N(\mathrm{O}^{++})} \simeq 26{\cdot}2D\left(\frac{11\,300}{T}\right)\frac{I_{\mathrm{a}}(\mathrm{K\,IV})}{I_0}, \quad \frac{N(\mathrm{K}^{+3})}{N(\mathrm{O}^{++})} = 1{\cdot}54D\left(\frac{-2570}{T}\right)\frac{I_{\mathrm{neb}}(\mathrm{K\,IV})}{I_0}$$

$$\frac{N(\mathrm{K}^{+4})}{N(\mathrm{O}^{++})} \sim 1{\cdot}04D\left(\frac{2420}{T}\right)\frac{I_{\mathrm{neb}}(\mathrm{K\,V})}{I_0}, \quad \frac{N(\mathrm{K}^{+5})}{N(\mathrm{O}^{++})} = 3{\cdot}6D\left(\frac{-2120}{T}\right)\frac{I_{\mathrm{neb}}(\mathrm{K\,VI})}{I_0}.$$

The [Ca V] λ5309 transition is sometimes observed:

$$\frac{N(\mathrm{Ca}^{+4})}{N(\mathrm{O}^{++})} \simeq 28D\left(-\frac{1360}{T}\right)\frac{I_{\mathrm{neb}}(\mathrm{Ca\,V})}{I_0}.$$

The lines of potassium, calcium, and sodium appear in nebulae whose ionization level is appropriate. They yield abundances comparable with values expected for stars.

Before considering applications of these equations to specific nebulae, we must discuss the observational problems involved.

1. Observational Data; Requirements and Limitations

In the 'thirties when the theory of spectrum-line excitation in gaseous nebulae began to be developed in quantitative form, it became evident that very severe requirements would be placed on the observations. Inference of physical conditions in a nebula demands accurate line-intensity ratios, sometimes over large wavelength intervals. Often the critical spectral lines are very weak. Photoelectric photometry, which is indispensable for the stronger lines, is of limited usefulness for weak lines in a crowded region of the spectrum or for lines such as 3965 (He I), 3969 [Ne III] and 3970 Hg. In practice, one combines photoelectric and photographic spectrophotometry. The photoelectric scanner gives the integrated brightness of the nebula in the strongest lines and permits measurements of a few weaker ones if one has the patience and above all the telescope time. The calibrated photographic plate enables one to achieve both angular and spectral resolution.

Clearly, one would like to observe over as broad a spectral range as possible. The λ3000 – λ5000 region of the spectrum presents no particular difficulties with fast emulsions and modern nebular spectrographs. The region λ5000 – λ5900 can be handled nicely with an image converter and the λ6000 – λ6700 region again with red sensitive plates. From λ6700 to about λ9000 the hypersensitized photographic plate can be used, but in the infrared we can reach only the strongest line with appropriate photocells.

It has become increasingly evident that we also need angular scale; for some problems one must go to the Coudé focus of a large telescope to obtain carefully guided spectrograms.

One must mention the photometric difficulties inherent in measuring line intensities over a range of as much as 30000 (as e.g. in NGC 7662) and over a spectral range from 3000 Å to 10000 Å. The problem is exacerbated by intrinsic changes in intensity over short angular distances in the nebula.

The results reported here were obtained mainly at the Lick and Mount Wilson Observatories. Recently, we have collaborated with Kaler, who has obtained spectrophotometric observations at Kitt Peak. At Mount Wilson we have emphasized relatively high-dispersion Coudé spectroscopy of bright planetaries and photoelectric spectrophotometry with the Cassegrain scanner at the 60-inch. The effort at Lick Observatory has been centered on the nebular spectrograph at the prime focus of the 120-inch reflector, insofar as the blue spectral region is concerned. Thanks to Walker, who has promoted the use of the Lallemand electronic camera at the Coudé focus of the 120-inch, we now have data on the green-yellow region $\lambda 4700 - \lambda 5900$ in a number of bright planetaries.

Chopinet (1963) has pioneered in spectrophotometry of nebulae with an electronic camera. With suitable choice of developer and emulsion, and with due care in observing an appropriate comparison star, it is possible to attain an accuracy comparable with that obtained by good photographic photometry (Walker, 1963; Walker and Aller, 1965). Since photographic emulsions are notoriously slow in the region near $\lambda 5200$, the gain in speed offered by the image converter is well worth the extra effort involved.

The region longward of $\lambda 5900$ remains inadequately covered even for the brighter planetaries; there are many important lines in this region.

It will be of enormous importance to obtain observations from outside the Earth's atmosphere both in the infrared and in the heretofore inaccessible ultraviolet. Only a few planetaries are bright enough to be observed in this way even when space telescopes are fully operative (cf. Code, 1960; Aller, 1961; Osterbrock, 1963, 1967; Gould, 1966).

2. Effects of Stratification and Filamentary Structure

Analyses of planetary nebulae are complicated by effects of stratification – i.e. the concentration of more highly ionized atoms toward the center (cf. Bowen, 1928) and the tendency for the nebular gas to exhibit considerable structure (Minkowski, 1964).

Let us examine, first of all, some of the stratification effects as exhibited in the blue region of the spectrum. Figure 1 (upper) shows spectra of NGC 6778, IC 5217, J 900, NGC 6818, and NGC 6741 obtained with the Lick 120-inch with the prime-focus nebular spectrograph. The spectra of NGC 6741 and IC 5217 have been widened because these nebulae are very small. Pronounced stratification effects are evident in NGC 6818 and NGC 6778, both of which have faint central stars that appear to have a nearly purely continuous spectrum. In NGC 6818 the [Fe v] radiation seems strongly confined to the central region even though [Ne v] is spread throughout the nebula.

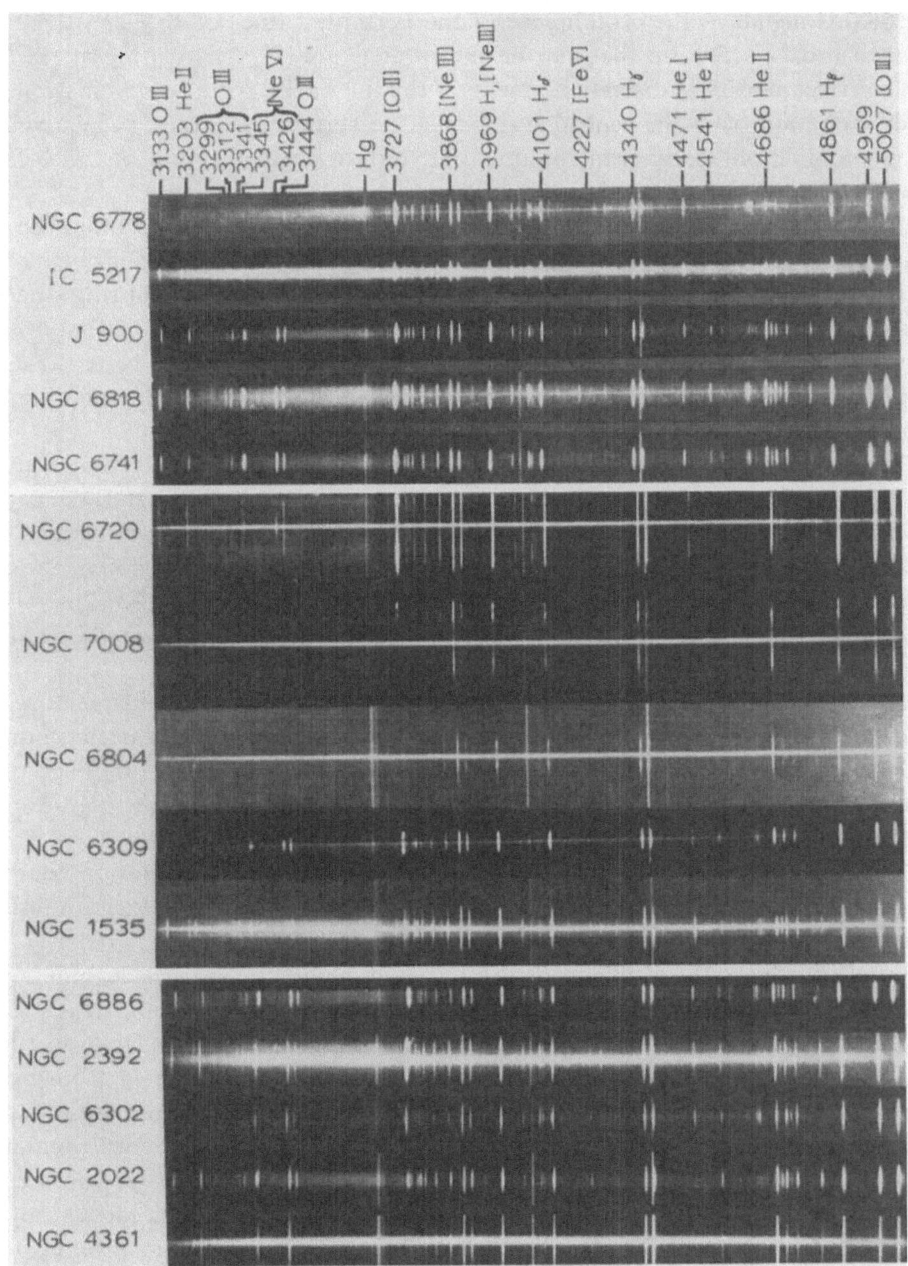

FIG. 1. *Spectra of planetary nebulae in the blue spectral region, taken with Lick Observatory prime-focus nebular spectrograph.*

Pronounced stratification effects in NGC 6778 are evident. An outstanding characteristic of this nebula is the prominence of the permitted lines of oxygen, carbon, and nitrogen ions – a feature that can be interpreted only by postulating an extremely high electron density in discrete filaments if the lines arise from recombination, or by direct excitation from the central star. Since the surface brightness is relatively low, the recombination mechanism requires that only a small fraction of the volume is actually filled with dense radiating matter.

The middle part of Figure 1 shows results for five nebulae with pronounced stratification effects. NGC 1535 is a classical two-ringed planetary of Vorontsov-Velyaminov's type IV. The spectrograph slit was placed through the bright ring; the bright streak is light scattered from the central star. NGC 1535 and NGC 6309 show pretty much the classical pattern of stratification as proposed by Bowen. Note the concentration of [Ne V] and 4541 He II towards the central star in NGC 6309 and the prominence of He I and 3727 [O II] in the outer layers.

The other three objects are of relatively low surface brightness. NGC 6804 displays a somewhat dim ring of rather uniform excitation. On the other hand, the centre of the ring in NGC 6720 is occupied by ions of high excitation (Minkowski and Osterbrock, 1960). Note especially the [Ne V] radiation which hugs the region of the central star. In the outer ring [S II] is enhanced as are also λ3727 [O II], He I λ3178, λ4471 and Mg I λ4571. Were it not for its low surface brightness, NGC 6720 would be an ideal object for a study of stratification effects.

NGC 7008 shows effects of both pronounced stratification and filamentary structure. The slit of the 120-inch nebular spectrograph was placed at a position angle of 25°. In the high-excitation blob, which is closer to the central star, lines of He II λ4686, [Ar IV], and O III are observed, whereas in the outer blob, the high-excitation lines are missing.

The bottom part of Figure 1 shows a number of very high-excitation nebulae, of which NGC 4361 is probably the most highly excited. Atmospheric excitation severely weakens the ultraviolet spectrum of NGC 4361.

Closely related to the problems of stratification, filamentary structure, and point-to-point excitation differences within gaseous nebulae is the question of the distribution of atoms among various stages of ionization. The ionization–distribution function introduced by Bowen and Wyse (1939) must fluctuate considerably from point to point in certain nebulae. The best way to solve this problem would be to observe the ions of each element in several stages of ionization. This is possible for helium, and to some extent for other elements such as neon, carbon, nitrogen, and potassium. Many elements are represented only by one or two ionization stages. For carbon, nitrogen, oxygen, and a few other elements represented by permitted lines it may yet be possible to separate the relative contributions of direct stellar excitation and recombination theories.

Some nebulae certainly contain condensations of greatly varying excitation and density. Suppose we take a series of nebulae of moderate to high surface brightness and arrange them in order of increasing excitation on the basis of lines of He II, [Ne V], etc. The [N I] and [N II] lines show a strikingly capricious behaviour (Wyse, 1942) which is well-exhibited in the Lick observations secured with the Lallemand electronic camera, as applied by Walker and his associates. The nitrogen lines tend to vary together. In the moderately low-excitation planetary NGC 6826, and in the high-excitation planetaries NGC 6818 and NGC 6826, the nitrogen lines are very weak. On the other hand they are strong in the Orion Nebula, in the low-excitation planetary IC 418 and in the high-excitation planetaries NGC 6741, Anon 21^h31^m and especially NGC 2440. Particularly marked is their great strength in the remarkable nebula NGC 6302.

Is this behaviour of the nitrogen lines an abundance effect or an excitation effect? We may answer this question partially at least by examining the spatial distribution of the nitrogen radiations within a given nebula (Walker and Aller, 1967).

In NGC 6720 the radiations of [Cl III], [N I], [N II], and He I all tend to be concentrated in the outer ring, while, as previously noted, He II, [Ne V], etc. are found in the inner parts. Even in the very small planetary NGC II 2003, [N II] is concentrated differently from other elements. In NGC 6543, [N I] and [N II] are concentrated in particular condensations, while [Fe VI] λ5676 and [Fe III] show quite different distributions. This object is an example of a nebula where there seem to be quite different excitations in different filaments. In the extremely inhomogeneous, broken-ring nebula, NGC 2392, the brighter, inner ring shows prominent lines of [Cl III], [Fe III], He I, He II, and [N II] while the fainter, outer blob shows [N I] as its strongest feature. The complicated nebula NGC 2440 (cf. Minkowski, 1964) shows considerably different blobs of nitrogen and ionized helium.

We are tempted to conclude that in objects such as NGC 2440, NGC 6302 and others mentioned above, filaments of both high and low excitation are involved. In some, the individual blobs are easily seen, but in others, such as NGC 7027, the blobs of different excitation lie below the resolution limits of our instruments.

In spite of these complexities, we must still try to find out what we can about the chemical composition of planetaries. Do field planetaries show composition abnormalities comparable to those found in M 15 by O'Dell *et al.* (1964)? That is, do there exist planetaries with chemical compositions differing greatly from the average?

Some years ago, many of us were inclined to believe that all planetaries probably had about the same chemical composition, and that apparent differences arose from difficulties in allowing for stratification effects, etc. One ratio which appears to be measurable with a reasonable degree of accuracy in most planetaries is the He/H ratio, and this seems to be nearly constant from one planetary to another (Mathis, 1957; Aller, 1964; O'Dell, 1963). In some nebulae, argon and metals such as sodium or potassium are observed in 2 or 3 ionization stages. Neon is sometimes observable as [Ne III], [Ne IV], and [Ne V], so that the neon/hydrogen ratio can be estimated.

Rather than present average results from a large number of nebulae (see e.g. Aller, 1964) we will illustrate some of the difficulties involved by comparing results gotten for three nebulae with rich bright-line spectra: NGC 7072, NGC 7662 and NGC 2022.

NGC 7027 is a tempting object for analysis because it exhibits such a rich spectrum. An earlier attempt (Aller, 1956) to analyze the spectrum of this object led to the conclusion that: "The results can be understood in terms of a nebula consisting of numerous filaments, knots, and tenuous regions such that the density may range from perhaps less than 10000 ions/cm^3 to perhaps something like 200000 ions/cm^3. The probable, but not directly observable, filamentary structure of NGC 7027 thus imposes a fundamental limitation on the accuracy with which the abundances of the ions can be found in this object." These conclusions were substantiated by Seaton and Osterbrock (1957) and by a direct photograph obtained by Minkowski. The nebula does have a distinct filamentary structure. With the availability of collisional line strengths for ions of the 3p^n row, additional relationships involving nebular and auroral transitions can be utilized. These all tend to indicate that the radiations of these ions, e.g. [S II], [Ar IV], [Cl III], originate in condensations whose densities are higher than the average density inferred from the angular diameter and surface brightness of the nebula.

If an attempt is made to use an average filament and an ionization-distribution curve deduced primarily from the ions of neon, the results obtained will be somewhat as given in Table 1. Successive columns give the element, the number of ions observed compared with the number of ions of O^{++} (we tabulate this for ions represented by forbidden lines), the estimated ratio of the total number of ions to the observed number of ions, and the logarithm of the number of ions on the scale $\log N(\mathrm{H}) = 12{\cdot}00$. The column headed 'Stellar Abundance' refers to an average deduced for our corner of the galaxy from solar system and stellar data (cf. Aller, 1961, p. 192). The

Table 1

Composition of NGC 7027

Element	$N/N(O^{++})$	$N_T/N_{obs.}$	$\log N$	Stellar Abundance	Neb-Star	(1956)
Hydrogen		1·0	12·00	12·00		12.00
Helium		1·0	11·23	11·21	+0·02	11·25
Nitrogen	3·1	32	8·5	8·05	+0·5	8·67
Oxygen	7·8	7·8	9·0	8·95	+0·05	8·93
Fluorine	0·0016	2·2	5·3	6·0	−0·7	5·63
Neon	1·25	1·05	8·2	8·70	−0·5	8·08
Sodium	0·002	1·86	6·4	6·30	+0·3	–
Sulphur	0·78	19·3	8·0	7·35	+0·65	8·03
Chlorine	0·061	6·5	6·9	6·25	+0·6	7·0
Argon	0·06	2·3	6·9	6·88	0·0	7·3
Potassium	0·0056	1·3	5·8	4·82	+1·0	6·1
Calcium	0·042	2·2	6·7	6·19	+0·5	6·1

next column gives the difference in the sense nebula minus star, and the last column gives the corresponding numbers deduced in 1956.

The nebula shows no stratification effects so we have assumed that the same electron temperature, $T_e = 15000$ °K, holds throughout although the density might be expected to change. We have assumed therefore that the same ionization–distribution curve holds throughout.

Glancing first at the differences between the stellar and nebular results, much of the discrepancy may represent the effects of bad estimates, but some may be real. The third column gives the factor by which the number of observed ions has been multiplied to give the total number of ions of that particular element. It is deduced from the distribution curve giving the relative number of ions in different stages of ionization; such a procedure can give only a rough estimate, but it can tell us whether or not we are observing a significant fraction of the ions of a particular element. Sulphur is observed only as [S II] and [S III], but most of the sulphur atoms must exist in more highly ionized stages. The correction factor of 19 is clearly very uncertain; it could easily be three times smaller. The largest factor is for nitrogen and here we are urgently in need of recombination rates so that we might deduce the numbers of missing ions from permitted lines of N II and N III.

Effects of filaments with differing densities and excitations are probably more serious for some ions than for others. Probably, argon, chlorine, and sulphur are the more seriously affected. In addition, further refinements are needed in the theory of collisional strengths for these ions. There is some hint that argon, potassium, and calcium may be more abundant than in a normal star, but the problem of the filaments of differing excitation and density must be solved before any such conclusions can be proclaimed.

Bowen and Wyse (1939) pointed out that the range in excitation in NGC 7662 was much smaller than in NGC 7027. In the bright ring of this nebula, at least, we are dealing with a more nearly homogeneous situation. Sulphur and nitrogen are represented only by low stages of ionization, and the vast bulk of atoms of these elements are not observable, at least by their forbidden lines. Stratification effects, which

Table 2

Abundances in NGC 7662

Element	N(total)/N(obs.)	log N	Element	N(total)/N(obs.)	log N
Hydrogen		12·00	Sodium	2·6	6·2
Helium	1·0	11.25	Sulphur	52	7·8:
Nitrogen	126	7·9	Chlorine	4·1	7·0
Oxygen	6·7	8·93	Argon	1·3	6·8
Fluorine	1·5	4·8	Potassium	1·3	5·8
Neon	1·0	8·1	Calcium	2·7	6·2

imply a possible change in electron temperature as well as ionization with distance from central star, are important in this nebula. The electron temperature deduced from the [O III] lines is in the neighbourhood of 13000 °K, but we have assumed that in the zone where the [Ne IV] lines are produced $T_e = 17000$ °K, while it rises as high as 20000 °K in the [Ne V] zone. The ionization curve, essentially based on neon, has been applied to all elements except helium. In particular, it seems unlikely that oxygen ions would show an appreciably different ionization distribution than neon. Unfortunately, large correction factors are required for nitrogen and sulphur – so the results quoted for these elements are little more than flimsy guesses. For O, Na, Cl, Ar, K the results agree reasonably well with NGC 7027. The abundances of F and Ca appear to be lower in NGC 7662.

As a third example, consider the high-excitation planetary NGC 2022. Here is a planetary for which composite estimate are difficult because of effects of high ionization. Because of its lower surface brightness, fewer lines are available and the number of observable ions is reduced. No information on the abundance of chlorine or sulphur can be deduced from available forbidden line data. The electron temperature is adopted as 17500 °K for all except the highest-excitation ions such as [Ne V], for

Table 3

Results for NGC 2022

Element	log N	Element	log N	Element	log N
Hydrogen	12·00	Oxygen	8·6	Argon	7·1
Helium	11·08	Nitrogen	7·9	Potassium	5·4
		Neon	7·6		

which we adopted $T_e = 20000°$. Helium and neon appear to define a good ionization curve; the intensity of O IV $\lambda 3411$ lines is consistent with a pure recombination origin. The nitrogen abundance is difficult to estimate. Argon and potassium seem to give reasonable results.

In other nebulae, such as NGC 7008 and NGC 6720 which have been studied by Seligman and Bohannan, respectively, attempts have been made to concentrate on individual knots and filaments, or particular zones of the nebula. By examining several regions of differing excitation it may be possible to put together a more definitive picture of a compositie nebula than is possible by considering an 'average' region or the integrated spectrum of the object.

There are some conclusions that would appear to emerge. Although the hydrogen/helium ratio seems to be nearly constant from one nebula to another – as various investigators have found, and although there do not appear offhand any confirmed examples as extreme as the M 15 nebula discussed by O'Dell, Peimbert, and Kinman (1964), there may be a number of real differences.

For the construction of models of planetary nebulae, and similar tasks where approximate abundances are needed, we propose the figures of Table 4 for rough guidelines.

Table 4

Logarithms of relative numbers of atoms

Hydrogen	12·00	Oxygen	8·9	Sodium	6·6	Argon	7·0
Helium	11·23	Fluorine	4·9	Sulphur	7·9	Potassium	5·7
Nitrogen	8·1:	Neon	7·9	Chlorine	6·9	Calcium	6·4:

Progress in this field requires the construction of detailed models of individual nebulae for which good observational data can be secured – radio-frequency measurements, isophotal contours and spectrophotometric measurements. For several nebulae the requisite information seems almost at hand.

Acknowledgments

This program was supported in part by the Air Force Office of Scientific Research, Grant No. 83-65 to the University of California, Los Angeles, by Aerospace Research Laboratories (ARP) at Wright-Patterson Air Force Base, Ohio, and by the National Science Foundation, Grant No. 6559 to the Ohio State University, Columbus, Ohio.

References

Aller, L.H. (1956) *Gaseous Nebulae*, Chapman and Hall, London.

Aller, L.H. (1961) *Les Spectres des Astres dans l'Ultraviolet Lointain*, Institut d'Astrophysique, Liège, p. 535.

Aller, L.H. (1961) *Abundances of the Elements*, Interscience Publishers, New York.

Aller, L.H. (1964) *Astrophys. Norw.*, **9**, 193.

Aller, L.H. (1964) *Pub. astr. Soc. Pacific*, **76**, 279.

Bowen, J.S. (1928) *Astrophys. J.*, **67**, 1.

Chopinet, M. (1963) *J. Observateurs*, **46**, 27.

Code, A.D. (1960) *Astr. J.*, **65**, 278.

Czyzak, S.J. (1967) in *Interstellar Medium*, Univ. of Chicago Press, Chicago (in press).

Czyzak, S. J., Krueger, T.K. (1967) *Proc. phys. Soc., London*, **90**, 623.

Gould, R.J. (1966) *Astrophys. J.*, **143**, 603.

Mathis, J.S. (1957) *Astrophys. J.*, **125**, 318.

Mathis, J.S. (1957) *Astrophys. J.*, **126**, 493.

Menzel, D.H., Aller, L.H., Hebb, M.H. (1941) *Astrophys. J.*, **93**, 230.

Minkowski, R. (1964) *Pub. astr. Soc. Pacific*, **76**, 197.

Minkowski, R., Osterbrock, D. E. (1960) *Astrophys. J.*, **131**, 537.

O'Dell, C.R. (1963) *Astrophys. J.*, **138**, 1018.

O'Dell, C.R., Peimbert, M., Kinman, T.D. (1964) *Astrophys. J.*, **140**, 119.

Osterbrock, D.E. (1963) *Planet. Space Sci.*, **11**, 621.

Osterbrock, D.E. (1967) *Phil. Trans. R. Soc. Lond. A* (in press).

Seaton, M.J. (1953) *Proc. phys. Soc., London*, **218**, 400.

Seaton, M.J. (1960) *Rep. Prog. Phys.*, **23**, 313.

Seaton, M.J., Osterbrock, D.E. (1957) *Astrophys. J.*, **125**, 66.

Seraph, H.E., Seaton, M.J., Shemming, J. (1966) *Proc. phys. Soc., London*, **89**, 27.
Walker, M.F. (1963) *Pub. astr. Soc. Pacific*, **75**, 430.
Walker, M.F., Aller, L.H. (1965) *Astrophys. J.*, **141**, 1318.
Walker, M.F., Aller, L.H. (1967) unpublished.
Wyse, A.B. (1942) *Astrophys. J.*, **95**, 356.

DISCUSSION

Flower: The weak intensity of the [N I] λ 5199 line in some nebulae is probably due to the absence of a region in which there is neutral nitrogen plus electrons.

Aller: The capricious behaviour of the [N I] and [N II] lines is interpreted as meaning that regions of very low excitation may be found in some nebulae, such as NGC 2440, but not in others, such as NGC 7662.

Reeves: The fact that the Ne/O ratio is smaller in nebulae than in stars is of great importance for nucleosynthesis. How certain are you of this result?

Aller: It is measured in B stars, and is good only to within a factor of 10.

Underhill: From my attempts to determine the Ne and O abundances in 10 Lacertae, I would reduce the ratio Ne/O by about a factor 5 from the usual 'cosmical abundance'.

Van Horn: Would you be willing to commit yourself to a general impression on the abundances? In particular, do the C and O abundances in the planetaries appear generally higher or lower than the cosmic abundances?

Aller: It is difficult to give firm numbers for the C/O ratio. The carbon abundance depends on the permitted lines of C II, C III and C IV. When a proper theory of recombination is worked out, I believe it will be possible to compare predicted and observed intensities for lines arising from levels of different excitation potential, in order to decide between recombination and stellar radiative excitation. The carbon abundances derived by our group seem rather high, but we thought this might be due to inadequacies in the recombination theory. As for O, we should compare abundances of O^{++} derived from O II recombination lines with those from the [O III] lines. So far, there is no evidence that the C/O ratio in planetaries actually differs from the local 'cosmic' abundance.

Gebbie: We have computed oxygen abundances for 38 planetary nebulae. The O^+ and O^{+2} abundances were obtained from forbidden lines, using the O II and O III collision strengths of Seaton and his collaborators, and using observed line intensities from all available data including recent photo-

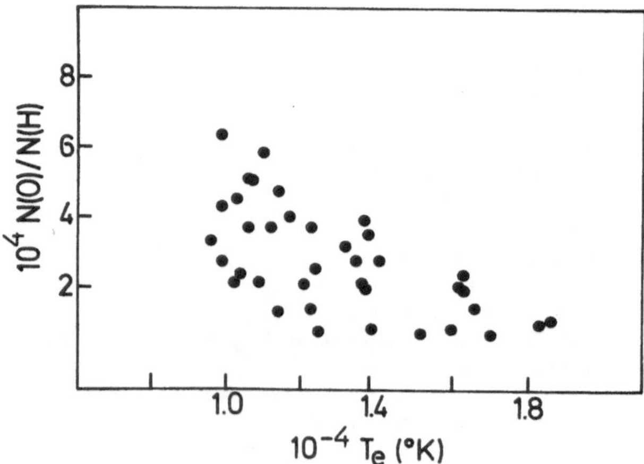

FIG. 2. *Abundances of oxygen relative to hydrogen in 38 planetary nebulae, plotted as a function of apparent electron temperature T_e.*

electric measurements of Aller and his collaborators, of O'Dell and of Vorontsov-Velyaminov and his collaborators. Allowance was made for the higher ions of oxygen by using Seaton's relation

$$\frac{N(\mathrm{O})}{N(\mathrm{H})} = \frac{N(\mathrm{O}^{+}) + N(\mathrm{O}^{+2})}{N(\mathrm{H})} \times \frac{N(\mathrm{He}^{+}) + N(\mathrm{He}^{+2})}{N(\mathrm{He}^{+})} .$$

Our results are plotted, for convenience, against electron temperature T_e (see Figure 2), although the apparent correlation with T_e is that which we would expect as a result of errors in the observed [O III] line intensities, and is probably not significant. Our results do, however, indicate that the oxygen abundance in planetary nebuale is lower by a factor of about 3 than the currently accepted cosmical value, $N(\mathrm{O})/N(\mathrm{H}) \sim 10^{-3}$.

Aller: This procedure differs from ours in that we have used the concentrations of trebly and quadruply ionized neon as derived from the [Ne IV] and [Ne V] lines to estimate the proportions of the more highly ionized oxygen atoms. Hence we derive a higher oxygen abundance. The problem should be re-examined with a more accurate ionization theory and models for the nebulae.

GENERAL DISCUSSION – THIRD SESSION

Faulkner: The general agreement between the interstellar extinction obtained from comparisons with radio observations, and that obtained from observed line ratios in the visible region argues for uniformity of the ratio of total absorption to selective absorption in the directions of the observed planetary nebulae. This contrasts with the variation of this ratio obtained from observations of some other objects, particularly H II regions.

Aller: It is quite possible to observe the spectra of many planetary nuclei and compare them with normal Population-I stars. In this way, if one believes the conventional interpretation of stellar spectra he can deduce the effective temperature of the observed stars. For example, Wilson and I found a T_e of about 35000 °K for the central star of NGC 2392.

Reeves: The O/H ratio of 3×10^{-4} reported by Gebbie would place the planetary nebulae between Population-II and Population-I stars, in agreement with the idea that they are Intermediate Population objects.

Feast: I would like to ask if anyone has made any abundance estimates for very distant planetaries in the direction of the centre, in the way that was done for the M 15 planetary?

Aller: Some distant faint planetaries can be observed but not many lines can be measured. Therefore our information on the abundances in these objects must be less precise.

O'Dell: Two very high-velocity objects in the direction of the galactic nucleus were photoelectrically measured in 1963 and found to have He/H ratios the same as nearby planetary nebulae.

Houziaux: Do the permitted O II or N I lines observed in planetaries show the same extent as the corresponding forbidden [O III] or [N II] lines?

Aller: The permitted lines of O II are observed in the brighter regions that show [O III] but the intensities are lower by a factor of 1000. The permitted N I lines fall in an unfavorable spectral region. For these reasons there are no observational measurements that can answer your question.

Weidemann: In view of Sargent's recent observation of a He deficiency in a hot star in M 15, the He/H measurement for the planetary nebula (K 648) is of special importance. What probable error would you assign to your result, which indicates a high He abundance of He/H = 0·18?

O'Dell: The scatter of the observations indicates an uncertainty of about 20%. The mean value found by O'Dell, Peimbert and Kinman has recently been confirmed by an independent photoelectric study at the Lick Observatory by Spinrad and Peimbert.

Session IV

STRUCTURE AND DYNAMICS

MORPHOLOGICAL STUDY OF PLANETARY NEBULAE

G. S. KHROMOV
(Sternberg State Astronomical Institute, Moscow, U.S.S.R.)

and

L. KOHOUTEK
(Astronomical Institute of the Czechoslovak Academy of Sciences, Czechoslovakia)

There are two approaches to the problem of the dynamics of planetary nebulae. The first is based on the investigation of the velocity fields in the nebulae. The second can be carried out by studying the spatial distribution of the matter in these objects. In an ideal case both approaches can be united. However, data on the velocity fields in planetary nebulae still are very scarce and fragmentary and their rate of accumulation is rather slow.

At the same time, there is sufficient observational basis for a morphological study of planetary nebulae. That is why the second approach has been chosen by the authors of the present paper; the morphology of planetary nebulae will be studied for the purposes of constructing spatial models of them, and of investigating the general character of the velocity fields in these objects.

In the first section of the present paper methods and results of analysis of the observational material will be presented. In the second part the data from the previous section will be used to derive a spatial model and to outline a dynamic picture.

1. A morphological study of objects of a given class should be started by developing an empirical classification of their observed forms. The variety of observed forms of planetary nebulae makes this first task rather difficult, and observational problems complicate it still more.

Probably for these reasons some earlier classification systems were entirely descriptive, contained no attempts at generalization and lacked a clear physical background. As a result the opinion was established that planetary nebulae are a collection of objects with a wide variety of different spatial structures.

We believe that this idea has no reasonable physical basis. It should be kept in mind that planetaries constitute a group of related objects united by a common genesis. It seems questionable that one mechanism which is responsible for the origin of planetary nebulae could create many entirely different spatial structures, or even a set of several different types of them. That is why in our morphological study primary attention has been paid to searching for common features inherent in all planetary nebulae.

The observational basis for the present study was provided by a collection of original

Osterbrock and O'Dell (eds.), Planetary Nebulae, 227–235. © *I.A.U.*

photographs and reproductions of planetary nebulae, which were collected by Perek and Kohoutek at the Astronomical Institute of the Czechoslovak Academy of Sciences in Prague. Originally this material was used in the preparation of the *Catalogue of Galactic Planetary Nebulae* (1967).

The total number of known planetaries is 1036; 295 objects were finally selected for the morphological study. The overwhelming majority of the others were stellar planetary nebulae and could not be studied from a morphological point of view. Some 80 objects showed overexposed images and could not be studied for this reason.

A basic principle of our classification of observed forms of planetary nebulae was the same as it was formulated by one of the authors some years ago (Khromov, 1962). This principle states that in each planetary nebula one can distinguish a bright and compact 'main structure' which is submerged into a fainter and elongated 'peripheral structure'. The average ratio between the surface brightnesses of the main and peripheral structures is about a factor of 10, which corresponds to at least a factor of 3 in density. It appears therefore that the main structure contains the largest portion of the observable mass of a planetary nebula.

The study of the observed forms of a large number of planetaries showed that there exist three types of observed forms of the main structure. These types – we have marked them by numerals 1, 2, and 3 – are clearly distinguishable and differ from one another by their degree of deviation from a regular ring-shaped form.

The peripheral structures are much more individual, but nevertheless one can trace certain common features. We were able to subdivide all observed forms of peripheral structures into three types – a, b, and d. Moreover, a set of simple geometrical characteristics of the main and peripheral structures was defined for further quantitative analysis.

The classification types of the main and peripheral structures can be seen in Figure 1, where all the principal cases are presented.

Table 1 shows the distribution of nebulae studied according to the types of their main and peripheral structures.

Table 1

Type of the nebula	1		2				3				
Characteristic of peripheral structure	–	d	–	a	b	d	–	a	b	ab	ad
Number of objects classified	66	14	49	27	2	6	26	34	25	11	3
Total number	80		84				99				

The total number of objects classified was 263, which is about 90% of the nebulae studied. Each nebula was classified by both of the authors of this paper independently. The cases of disagreement were rather rare. Peculiar objects which could not be properly classified within the limits of our scheme were mostly faint with no trace of regular structure.

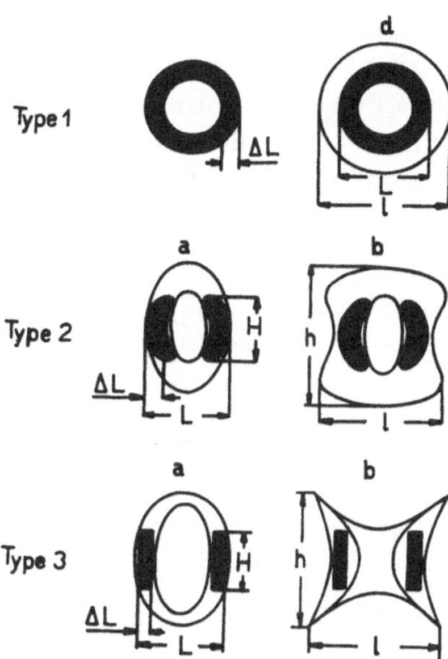

FIG. 1. *Schematic drawings of three classification types, with significant dimensions indicated.*

2. In the second part of our paper we shall consider the question of the spatial structure of planetary nebulae. Many authors have studied this problem, and among the papers which have been written on this subject one should particularly mention the classical study of Curtis (1918), containing many ideas developed later by others. Both Curtis' paper and the later papers are mostly devoted to individual cases and contain no attempts at generalisation of the results.

One of the most important studies in our opinion is the paper by Minkowski and Osterbrock (1960), in which two planetary nebulae – NGC 6720 and NGC 650-1 – were considered. In this paper it was definitely stated that many planetaries, in spite of their apparent forms, could have a toroidal structure.

Some years ago one of the authors of this report (Khromov, 1962) suggested that all observed forms of planetary nebulae can be explained by projecting upon the sky one common spatial structure. In the present paper an attempt is made to revise this old idea on the basis of more complete observational material.

The supposition that effects of projection can play a primary role in determining the observed structures of planetary nebulae forces us to review the question of the spatial orientation of the nebulae. We investigated the distribution of angles between the apparent axes of planetary nebulae and the galactic equator. The axis of a nebula of type 2 or type 3 was determined as the perpendicular to a line connecting the two maxima of brightness. The results of this study, which included 182 objects, definitely prove a random orientation of planetary nebulae.

Now we can analyse the morphological data provided by our classification. It is reasonable to begin by considering the main structure.

Remembering the three types of observed forms of the main structure of planetary nebulae, one can easily realize that such an empirical picture can be explained by the supposition that the main structure of the nebulae actually is a toroid. If this toroid

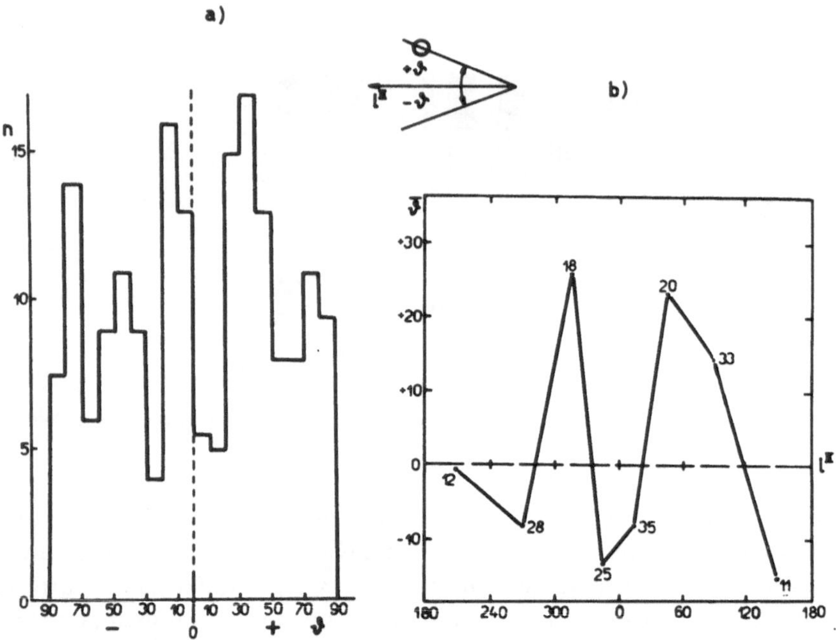

FIG. 2. *(a) Histogram of angle between axes of planetary nebulae and galactic equator. (b) Mean angles between axes and galactic equator as a function of galactic longitude. The figures at the points of the graph indicate numbers of objects contributing at a given point.*

is seen from the direction of its axis of rotation, we see a ring-shaped type-1 structure; looking at it from its equatorial plane we see a type-3 structure. Finally, the structures of type 2 correspond to an intermediate case of projection.

For simplicity we consider this toroid a right-angled one; its average geometric parameters can be determined from the observed classification data. The corresponding histograms show that the average value of the parameter $\Delta L/L = 0{\cdot}24$ is determined quite reliably by 211 objects. The second parameter, H/L, shows a considerable scatter, but an average estimate 0·65, determined by 183 objects, is reliable enough.

The mean spatial parameters of the main structure can be used to compute corresponding isophotal pictures that one would observe looking at such a structure at different angles between the line of sight and the axis of the toroid. It can be shown that this picture has a good qualitative agreement with the real one observed in the light of H and He lines.

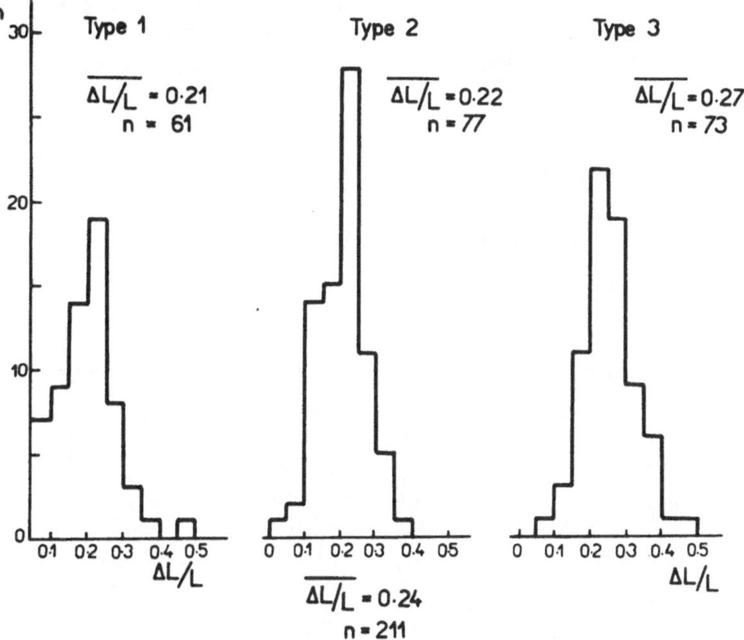

FIG. 3. *Histograms of relative ring thickness for each classification type separately.*

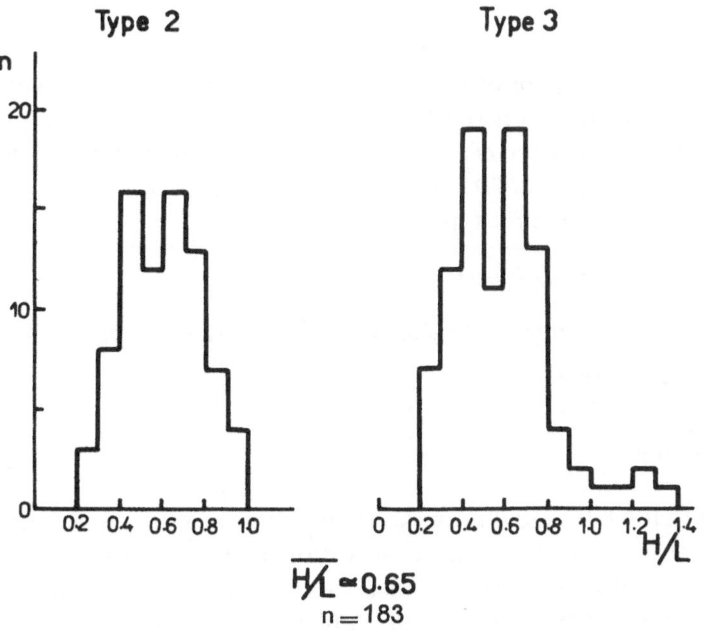

FIG. 4. *Histograms of relative ring heights for two classification types separately.*

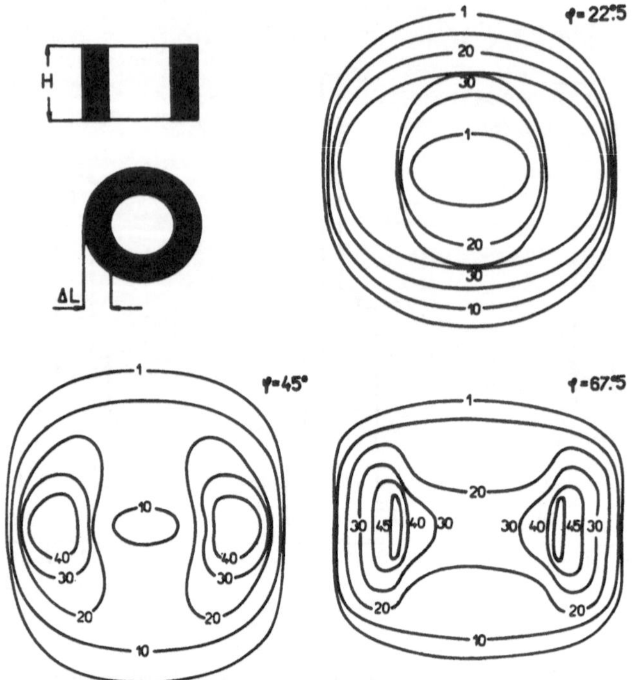

FIG. 5. *Computed isophotes for model structure with mean parameters derived from observed planetaries.*

It is interesting to compare observed numbers of nebulae of different classification types with those predicted on the basis of our isophotal pictures. Looking at the isophotes one can see that the critical angle which separates the ring-shaped nebulae from the other forms is about 30°; Table 2 contains the predicted and observed numbers of the nebulae of different types.

Table 2

Relative numbers of planetary nebulae of different types

Type of the main structure	1	2 + 3
Theory	1·0	6·1
Observations	1·0	2·3

It is evident that while a qualitative agreement between the theory and observations does exist, there is still a significant observational excess of the type-1 objects. This excess can be explained as a result of observational selection, in that very faint type-1 objects can be more easily distinguished on photographic plates than irregular type-3 objects. This interpretation can be proved by the dependence of the distribution of objects of different types on their surface brightness and on their galactic latitude. The first of the distributions clearly shows a decrease of the fraction of fainter nebulae from type 1 to type 3. The second one demonstrates a relatively higher degree of concentration of type-3 objects to the galactic plane; hence these objects are more distant and consequently brighter and more compact as well. One of conclusions following from this reasoning is that some dozens of extremely faint and nearby type 3 objects still remain undiscovered.

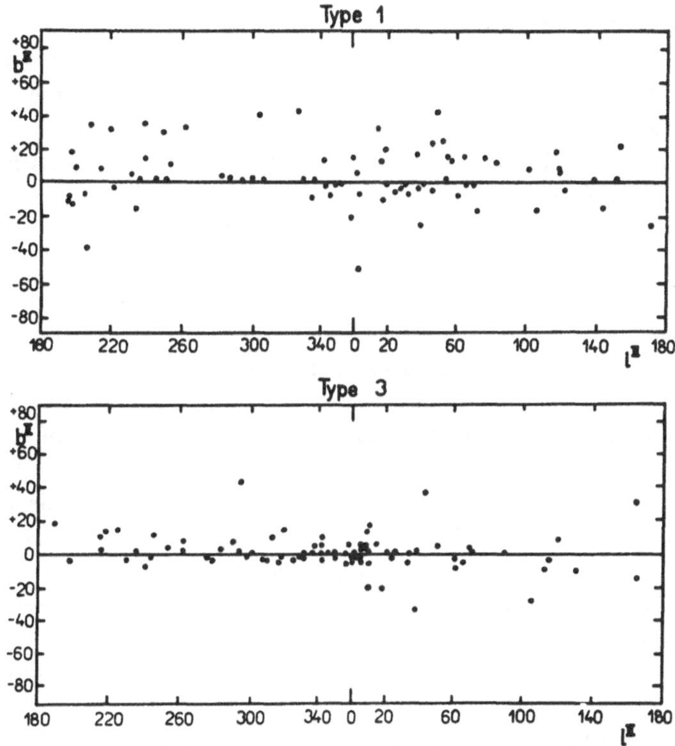

FIG. 6. *Galactic distribution of type-1 and type-3 planetaries separately.*

Let us consider now the peripheral structures of planetary nebulae. According to our classification there are three observed types of these structures. It is interesting to note that relatively regular, ring-like structures of type *d* are seen exclusively around regular ring-shaped type-1 main structures. It is seen in addition that type-*b* structures frequently appear about type-2 and type-3 nebulae, and that the outer diameter of a type-*b* structure usually exceeds that of its corresponding main structure. Considering the situation in terms of projection, one can conclude that regular type-*d* peripheral structures appear due to the projection of type-*b* structures, so the structures of type *d* do not represent an independent morphological case. An analogous idea has been expressed earlier by Minkowski and Osterbrock (1960) with respect to the nebulae NGC 650-1 and NGC 6720.

The latter conclusion can be supported by a comparison of numbers of planetary nebulae with the peripheral structures of *b* and *d* types. The ratio $N(2b+3b+3ab)/N(1d)=2{\cdot}7>1$, as it should be according to the general law of projection. It is worthwhile to mention as well, that the average diameter of type-*d* structures and the average *maximum* width of type-*b* and *ab* structures show a good deal of agreement: $l/L(1d)=1{\cdot}9$, $l/L(3b+3ab)=1{\cdot}5$.

It is interesting to note that, while the overwhelming majority of so-called double-shell planetary nebulae can be interpreted in the framework of the picture outlined above, some examples of real double-shell nebulae still exist. The fraction of them is rather small but not negligible. The typical samples of such objects have outer symmetrical condensations in the equatorial plane of the main structure. They are seen most distinctly among the type-3 nebulae, among which there are two definite cases of nebulae of this kind.

At the same time the difference between type *a* and type *b* is real. Trying to find any physical differences between the nebulae possessing these two kinds of peripheral structures we have studied in detail appropriate observational data. It is evident that type 3 nebulae offer the best opportunity to study peripheral structures. The mean observed galactic latitude of 3b nebulae is $|b^{II}|=3\overset{\circ}{.}7$, while that of 3a objects is $9\overset{\circ}{.}9$. It means that the mean distances of 3a and 3b objects differ by about the factor of 2·7. Taking into account the mean angular diameters of these objects ($\varphi(3a)=75''$, $\varphi(3b)=20''$), one can estimate that the ratio of their mean linear diameters is about of 1·4; this difference is thus not very significant.

The surface brightnesses of 3a objects are approximately by 1^m lower than that of 3b objects. The different character of the galactic concentration of type 3a and 3b nebulae may indicate that our observational data on these nebulae have been affected by a factor of observational selection.

There are perhaps two possible ways of further discussion. The first one states that the only small difference in linear diameters of type 3a and type 3b nebulae most likely indicates their age to be approximately the same. Then the whole variety of *a* and *b* structures could reflect different dynamical conditions in the respective nebulae – probably different shapes of the velocity field. (This point of view is shared by L.K.)

The second approach is based on the supposition that the small difference between the surface brightness of type 3a and 3b nebulae actually reflects the fact that the density in 3b objects is approximately 2·5 times that of 3a objects. And, vice-versa, we may suppose that planetary nebulae with an *a*-type periphery represent an older evolutionary stage of 3b objects. (This second point of view is shared by G.S.K.)

In conclusion we would like to consider briefly some dynamical consequences following from the foregoing spatial model of planetary nebulae. These consequences are as follows:

(1) The expansion velocity field in planetary nebulae must have axial symmetry.

(2) The expansion velocity must increase from the equatorial plane of the main structure to its poles.

(3) The direction of the maximum-velocity vector may not necessarily coincide with the polar axis. In fact, most elongated details of type-*b* peripheral structure are usually directed at an angle to the polar axis.

(4) The equatorial expansion velocities must be at least two times smaller than those in the polar region. (This conclusion is based on measurements of the relative extensions of these structures.)

(5) The character of the expansion velocity field probably is subject to evolutionary variations.

The expansion of a planetary nebula plays the role of the principal evolutionary factor. This expansion causes a density decrease, which in turn promotes the process of ionization of the nebula. An ionization front moving outwards through the nebula must significantly affect the original velocity field and the density distribution. One may suppose that nebulae with type-*b* peripheral structures are objects in which the process of ionization is almost finished, while nebulae with type-*a* structures are completely ionized objects.

To explain the origin of the toroidal structure of planetary nebulae, one has to search for a source of a primeval axially symmetric instability. We can suggest a mechanism which can probably explain the observed picture. If in a primeval planetary nebulae there is a slight concentration of density in the equatorial plane, it will focus the ionization front. As a result this inhomogeneity will be increased and something like the observed toroidal structure will appear. A similar result could be obtained on the supposition that the primeval velocity field originally had an axial symmetry. The proposed qualitative mechanism is very schematic, and further theoretical and observational investigations of the dynamics are needed.

In conclusion we would like to note that a detailed and more accurate account of this morphological study, together with discussion of peculiar observed forms of planetaries is to be presented in two or three papers in the *Bulletin of Astronomical Institutes of Czechoslovakia* in the beginning of next year.

References

Curtis, H.D. (1918) *Pub. Lick Obs.*, **13**, 55.
Khromov, G.S. (1962) *Soviet Astron. A–J.* (Engl. transl.), **6**, 370.
Minkowski, R., Osterbrock, D.E. (1960) *Astrophys. J.*, **131**, 537.
Perek, L., Kohoutek, L. (1967) *Catalogue of Galactic Planetary Nebulae*, Academia, Prague.

PROBLEMS OF GAS DYNAMICS IN PLANETARY NEBULAE

F. D. KAHN
(Dept. of Astronomy, University of Manchester, England)

ABSTRACT

The expansion of a typical planetary nebula is studied by means of a simple model, with spherical symmetry. It is shown that there is a build-up of radiation pressure, due to Lyman-α photons trapped in the nebula, and that this causes the initial acceleration outwards. After some $1{\cdot}5 \times 10^{11}$ sec the nebula will have a radius of the order of 10^{17} cm, and an expansion velocity of about 2×10^{6} cm/sec. At roughly this time the dynamical effects of radiation pressure begin to be superseded by those of the recoil pressure at the ionization front, which continues to dominate until the nebula is fully ionized.

Finally some reasons are considered why most nebulae do not have spherically symmetrical shapes, but why so often they appear to be ring-like or butterfly-shaped.

1. Introduction

The motion of the gas in a planetary nebula is influenced by the pressure of the hot, ionized gas, by the pressure of the Lyman-α radiation trapped in the nebula, and by the heating of the newly ionized gas at the front of the ionized region. In this paper no account will be taken of possible hydromagnetic effects, for there is little reason, from either observation or theory, to believe that magnetic fields of any consequence can exist in planetary nebulae

2. Some Simple Properties of H II Regions

The luminous part of a planetary nebula is an H II region, in which the hydrogen is kept almost fully ionized by the radiation of the central star. The work of Seaton and his group (Seaton, 1966; Harman and Seaton, 1966; and other references cited there) has led to a good description of the properties of such stars, and Seaton's data will be used here. During the evolution of a nebula the luminosity L_* of the typical central star varies between 2×10^{36} and 4×10^{37} erg/sec, and its surface temperature T_* between 4×10^{4} and 10^{5} °K.

The most important parameter is I_*, the rate of production of photons, by the central star, in the Lyman continuum. If the star radiates like a black body in the far ultraviolet (see, e.g., Gebbie and Seaton, 1963), then I_* is related to L_* by

$$I_* = i(T_*)\, L_* \tag{1}$$

and $i(T_*)$ varies only slowly with the temperature T_* of the star. For $T_* = 4 \times 10^{4}$ °K,

Osterbrock and O'Dell (eds.), Planetary Nebulae, 236–248. © *I.A.U.*

$i(T_*)=1{\cdot}2\times 10^{10}$ photons/erg; between $T_*=6\times 10^4\,°$K and $10^5\,°$K, $i(T_*)$ is virtually constant, at $1{\cdot}8\times 10^{10}$ photons/erg. A typical value for I_* is therefore 10^{47} photons/sec.

Now hydrogen is by far the most important constituent of the nebula. It may be assumed that, in the early phases of its evolution, the ionized portion of the nebula is entirely enclosed in a shell of non-ionized hydrogen. Almost all the Ly-c photons are absorbed in the H II region, and go to balance the recombinations of protons and electrons there. These statements will be checked later. For the present, let n_i equal the proton and the electron number density in the ionized region, which we take to be a sphere of radius R, and in which the ionized gas has a temperature $T_i=10^4\,°$K.

The recombination of a proton and an electron to the ground state produces another Ly-c photon. Thus Ly-c photons are lost only when recombinations occur to excited states, so that, in a first approximation,

$$I_* = \frac{4\pi}{3}\,\beta n_i^2 R^3\,. \tag{2}$$

At $10^4\,°$K the recombination coefficient to all excited states of the H atom is $\beta\approx 2\times 10^{-13}$ cm^3/sec. Equation (2) leads to an estimate of the total mass of ionized gas

$$M_i = \frac{4\pi}{3}\,n_i m_{\mathrm{H}} R^3 = \left(\frac{4\pi}{3}\frac{I_*}{\beta}\right)^{1/2} m_{\mathrm{H}} R^{3/2} \approx 2{\cdot}6\times 10^5 \left(\frac{L_*}{\beta}\right)^{1/2} m_{\mathrm{H}} R^{3/2}\,, \tag{3}$$

where m_{H} is the mass of the H atom. If L_* is held fixed, then M_i varies like $R^{3/2}$. As R increases, the mass of ionized gas increases, presumably at the expense of the non-ionized gas in the shell which surrounds it. If L_* and R increase together, M_i increases still faster.

Now a fraction ε of the available Ly-c photons is spent in ionizing the atomic hydrogen drawn into the H II region as the radius R increases. It will now be shown that ε remains small throughout. From Equation (3), with L_* fixed,

$$\frac{\dot M_i}{M_i} = \frac{3\dot R}{2R}\,. \tag{4}$$

The number of new protons entering per unit time is $\dot M_i/m_{\mathrm{H}}$, so that

$$\varepsilon I_* = \dot M_i/m_{\mathrm{H}} = 2\pi n_i R^2 \dot R\,. \tag{5}$$

From Equations (2) and (5), then

$$\varepsilon = (\beta n_i)^{-1}\,(\dot R/R)\,. \tag{6}$$

$(\beta n_i)^{-1}$ is the characteristic time for the recombination of a proton and an electron, about 10^9 sec with a reasonable value of n_i. $R/\dot R$ is the characteristic expansion time of the nebula, of the order of 10^{11} sec. Thus ε is of the order of 0·01. Alternatively, it may also be deduced from Equations (2) and (5) that

$$\varepsilon = \left(\frac{4\pi}{3\beta I_*}\right)^{1/2} \dot{R}R^{1/2} \approx 6 \times 10^{-18} \dot{R}R^{1/2},$$

with our typical value for I_*. Once again it is clear that, even though ε increases as the nebula expands, it always remains a small quantity. These conclusions will not be qualitatively changed if I_* varies with time.

A good approximation, usually made, is that a Ly-α photon will eventually be emitted in the H II region for every Ly-c photon absorbed there (see, e.g., Aller, 1956). The rate of production of radiant energy in the Ly-α line thus becomes

$$L_\alpha = I_* \chi_\alpha = i(T_*)\, \chi_\alpha L_*, \tag{7}$$

where $\chi_\alpha = 10{\cdot}15$ eV is the energy of the Ly-α photon. With the quoted value for $i(T_*)$

$$L_\alpha \approx 0{\cdot}3 L_* \tag{8}$$

as long as the temperature T_* lies between 5×10^4 °K and 10^5 °K.

3. The Transfer of Lyman-α Radiation

The Ly-α radiation cannot escape freely from the H II region, for there will always be a small proportion of atomic hydrogen present. In fact the ionization front is usually found at a distance from the central star such that

$$\tau_l = \int_0^R \kappa_l \rho_i (1 - x)\, dr \approx 10 \text{ or } 20,$$

where $\kappa_l \approx 3{\cdot}6 \times 10^6$ cm^2/gm is the opacity of atomic hydrogen at the Lyman limit, ρ_i is the density by mass of $(H^+ + H)$, and x is the fractional ionization within the H II region (see, e.g., Mathews, 1965). There should therefore be between 3×10^{-6} and 5×10^{-6} gm/cm^2 of atomic hydrogen along a line between the central star and the ionization front.

In the absence of thermal motion, atomic hydrogen has an opacity $\varpi/(u^2 + \eta^2)$ at frequency $\nu = \nu_\alpha - u$, where $\varpi = 10^{29}$ cm^2 gm^{-1} s^{-2}, $\eta = 2{\cdot}5 \times 10^7$ s^{-1}, and $\nu_\alpha = 2{\cdot}5 \times 10^{15}$ s^{-1}, the frequency of Ly-α. In the wings of Ly-α the H II region will therefore be opaque out to a frequency shift given by

$$u^2 \approx 5 \times 10^{-6}\, \varpi \approx 5 \times 10^{23} \text{ s}^{-2}, \quad \text{or}$$
$$u \approx \pm 7 \times 10^{11} \text{ s}^{-1}.$$

The non-ionized jacket around the H II region is even more opaque. It has a surface density

$$\sigma = M/4\pi R^2,$$

and if $M = 10^{33}$ gm and $R = 10^{17}$ cm, then $\sigma \approx 10^{-2}$ gm/cm^2, larger by some 3 orders of magnitude than the mass of atomic hydrogen along a unit column from the central star to the I-front. A photon with frequency close to ν_α therefore cannot escape from the nebula, but the indefinite build-up of radiation density is prevented by the gradual drift of the photons in frequency. There are two important mechanisms:

(i) The photons are essentially trapped in an expanding enclosure. By a classical physical result their frequency then varies inversely as the linear scale of the enclosure, at a rate given by

$$\dot{u} = \nu_\alpha \dot{R}/R\,.$$

The rate of change of J due to this red-shift (rs) mechanism is

$$\left(\frac{\partial J}{\partial t}\right)_{\mathrm{rs}} = -\,\nu_\alpha \frac{\dot{R}}{R}\frac{\partial J}{\partial u}, \tag{9}$$

where $J(u)\,\mathrm{d}u$ is the photon density between frequencies $\nu_\alpha - u$ and $\nu_\alpha - u - \mathrm{d}u$.

(ii) The photons are scattered by the H atoms in the H II region. The scattering coefficient depends on frequency and is determined by Doppler and by natural line broadening, according to the formula

$$p(u, u') = \frac{3\varpi\rho_{\mathrm{H}}c}{8\sqrt{2\pi}U}\int_0^\pi\int_{-\infty}^{\infty}\frac{1+\cos^2\theta}{u''^2+\eta^2}\exp\left[-\frac{1}{2U^2\sin^2\theta}\right.$$
$$\left.\times\left\{(u-u'')^2 - 2(u-u'')(u'-u'')\cos\theta + (u'-u'')^2\right\}\right]\mathrm{d}u''\,\mathrm{d}\theta\,, \tag{10}$$

which is easily proved. $p(u, u')\,\mathrm{d}u'$ is the probability per unit time, that a photon with frequency shift u should be scattered into the range of frequency shift u', $u' + \mathrm{d}u'$;

$$U = \frac{\nu_\alpha}{c}\sqrt{\frac{kT_i}{m_{\mathrm{H}}}} \approx 8\times 10^{10}\,\mathrm{s}^{-1}, \quad \text{when } T_i = 10^4\,{}^\circ\mathrm{K}$$

and ρ_{H} is the density, by mass, of atomic hydrogen.

The redistribution of photons in frequency occurs at a rate

$$\frac{\partial J(u)}{\partial t} = \int_{-\infty}^{\infty}\{p(u', u)\,J(u') - p(u, u')\,J(u)\}\,\mathrm{d}u'$$
$$= \int_{-\infty}^{\infty} p(u', u)\,\{J(u') - J(u)\}\,\mathrm{d}u'\,, \tag{11}$$

and the second step in these relations follows because p is symmetrical in u and u'.

When the spectrum is flat, $J(u')=J(u)$ for all pairs of u and u', and no net redistribution occurs.

This observation eases the calculation that follows. During the early phases of the evolution of a nebula, the Ly-α line is spread far out into the wings, and the spectrum close to ν_α will be rather flat. Thus no additional change is introduced into the calculated profile even when the transfer near ν_α is estimated by the correct formula, rather than by the approximate formula that really holds only in the wings of the line. The scattering in the wings is coherent, except for small Doppler shifts due to the thermal motion of the scattering atoms. It is easier to find suitable approximate formulae directly, rather than to derive them from (10).

Let a photon come from P, and let it be scattered at O into the direction OP', such that there is an angle θ between PO produced and OP'. Let OQ be the bisector of the angle POP'. Let the atom that scatters the photon have a velocity component v parallel to OQ. Then it may be shown that an incoming photon, having frequency ν, relative to the laboratory frame of reference, is scattered with frequency $\nu\{1+(2v/c)\sin(\theta/2)\}$ relative to that frame. The frequency change is

$$\delta u = -2\nu\frac{v}{c}\sin\frac{\theta}{2} \approx -2\nu_\alpha\frac{v}{c}\sin\frac{\theta}{2}. \tag{12}$$

Taking averages over all values of v

$$\langle \delta u \rangle_v = 0$$

and

$$\langle \delta u^2 \rangle_v = \frac{4\nu_\alpha^2}{c^2}\frac{kT_i}{m_{\mathrm{H}}}\sin^2\frac{\theta}{2}. \tag{13}$$

Scattering angles are distributed symmetrically with respect to $\theta=\pi/2$, so that

$$\langle\langle \delta u^2 \rangle_v\rangle_\theta = \frac{2\nu_\alpha^2}{c^2}\left(\frac{kT_i}{m_{\mathrm{H}}}\right) \equiv 2U^2, \tag{14}$$

and $U \approx 1{\cdot}1\times 10^{11}\ \mathrm{s}^{-1}$, with $T_i=10^4\ {}^\circ\mathrm{K}$.

Apparently then, in the wings of Ly-α,

$$p(u,u') = \frac{\varpi\rho_{\mathrm{H}}c}{u^2}\phi(u'-u), \tag{15}$$

where ϕ is a symmetrical probability distribution, normalized to unity, with variance U^2. Since $\eta^2 \ll U^2$ it is safe to neglect η^2 in the denominator. But this form for $p(u,u')$ is not symmetrical in u and u', as it ought to be. The correct approximation is therefore

$$p(u,u') = \frac{\varpi\rho_{\mathrm{H}}c}{2}\left(\frac{1}{u^2}+\frac{1}{u'^2}\right)\phi(u'-u), \tag{16}$$

and Equation (11) becomes

$$\frac{\partial J(u)}{\partial t} = \frac{\varpi \rho_{\mathrm{H}} c}{2} \int_{-\infty}^{\infty} \left(\frac{1}{u^2} + \frac{1}{u'^2} \right) \phi(u' - u) \left[J(u') - J(u) \right] du' . \tag{17}$$

In the wings individual steps in frequency are small compared with the frequency difference from the line centre, so that one may expand under the integral sign, and find

$$\frac{\partial J(u)}{\partial t} = \frac{\varpi \rho_{\mathrm{H}} c}{2} \int_{-\infty}^{\infty} \left[\frac{2}{u^2} - \frac{2(u' - u)}{u^3} \right] \phi(u' - u) \times \left[(u' - u) \frac{\partial J}{\partial u} + \frac{1}{2} (u' - u)^2 \frac{\partial^2 J}{\partial u^2} \right] du' , \tag{18}$$

to second order. No contributions arise from terms which are linear in $(u'-u)$, and to second order

$$\left(\frac{\partial J(u)}{\partial t} \right)_{\mathrm{sc}} = \frac{\varpi \rho_{\mathrm{H}} c}{2} \left[- \frac{2}{u^3} \frac{\partial J}{\partial u} + \frac{1}{u^2} \frac{\partial^2 J}{\partial u^2} \right] \int_{-\infty}^{\infty} (u' - u)^2 \phi(u' - u) \, du'$$

$$= \varpi \rho_{\mathrm{H}} c U^2 \frac{\partial}{\partial u} \left(\frac{1}{u^2} \frac{\partial J}{\partial u} \right). \tag{19}$$

This is the rate of change of photon density due to the scattering mechanism.

(iii) The photons leak through the atomic hydrogen in which they are trapped. The H II interior region of the nebula is much less opaque than the non-ionized shell. The interior therefore fills up to a uniform photon density $J(u)$ at any frequency. A gradient of photon density $J(u)/l$ will be set up across the thickness l of the shell. If ρ_{H} is the density of atomic hydrogen there, its optical depth at frequency $\nu_\alpha - u$ is $\varpi \rho_{\mathrm{H}} l / u^2$ and the photon flux per unit frequency interval is

$$F(u) = \frac{c J u^2}{3 \varpi \rho_{\mathrm{H}} l} \equiv \frac{c J u^2}{3 \varpi \sigma} . \tag{20}$$

A sphere of radius R contains $(4\pi/3)\, R^3 J(u)$ photons per unit frequency interval. The rate of leakage of photons is therefore

$$\frac{4\pi}{3} R^3 \left(\frac{\partial J}{\partial t} \right)_l = - 4\pi R^2 F(u)$$

and so

$$\left(\frac{\partial J}{\partial t} \right)_l = - \frac{c u^2}{\varpi \sigma R} J(u) . \tag{21}$$

Now combine Equations (9), (19) and (21), and in Equation (19) set $\rho_{\mathrm{H}}=\rho_i(1-x)$, the density of atomic hydrogen in the H II region. In a steady state

$$- v_\alpha \frac{\dot{R}}{R}\frac{\mathrm{d}J}{\mathrm{d}u} + \varpi\rho_i(1-x)\, cU^2 \frac{\mathrm{d}}{\mathrm{d}u}\left(\frac{1}{u^2}\frac{\mathrm{d}J}{\mathrm{d}u}\right) - \frac{cu^2 J}{\varpi\sigma R} + L_\alpha \delta(u) = 0 . \qquad (22)$$

The last term describes the rate at which Ly-α radiation is generated in the H II region. Equation (22) is solved by setting $u^2\,\mathrm{d}u \equiv \mathrm{d}z$, and then its solutions have the form $A_1 e^{\lambda_1 z} + A_2 e^{-\lambda_2 z}$, where λ_1 and λ_2 are both positive. Only a finite amount of energy can be trapped, so that both the solutions for positive and for negative z must tend to zero as $|z|$, and therefore as $|u|$, tend to infinity. The solutions are respectively proportional to $e^{\lambda_1 z}$ and $\mathrm{e}^{-\lambda_2 z}$ for $u<0$ (the blue side) and for $u>0$ (the red side). Values of λ_1 and λ_2 are, to a good approximation,

$$\lambda_1 = \frac{v_\alpha \dot{R}}{\varpi\rho_i(1-x)\, U^2 cR}, \qquad (23)$$

$$\lambda_2 = \frac{c}{\varpi\sigma\dot{R}v_\alpha}; \qquad (24)$$

and here $\lambda_1 \gg \lambda_2$. All this is true provided

$$\frac{\dot{R}^2}{kT_i/m_{\mathrm{H}}} \gg \frac{4\rho_i R(1-x)}{\sigma}. \qquad (25)$$

By an earlier argument the right-hand side of condition (25) is of order 10^{-3}; the condition is therefore obeyed if $\dot{R}$ is not less than 1/30 of the thermal velocity in the H II region.

It appears that the radiation field is much more sharply limited on the side of negative u, the blue side. Very few photons reach large enough shifts on that side to escape there; virtually all will eventually escape with red shifts. On the red side the frequency distribution of the trapped photons is given by

$$J(u) \propto \exp\left\{-\frac{cu^3}{3\varpi\sigma\dot{R}v_\alpha}\right\}. \qquad (26)$$

The line thus becomes quite flat near $u=0$; the same also is true on the side of negative u. The mean value of u at which a photon escapes (neglecting the few that escape on the blue side) is

$$\bar{u} = \frac{\int_0^\infty u\,F(u)\,\mathrm{d}u}{\int_0^\infty F(u)\,\mathrm{d}u}$$

$$= \frac{\int_0^\infty u^3 \exp\{-cu^3/3\varpi\sigma\dot{R}v_\alpha\}\,du}{\int_0^\infty u^2 \exp\{-cu^3/3\varpi\sigma\dot{R}v_\alpha\}\,du}$$

$$= 1{\cdot}8\left(\frac{\varpi\sigma\dot{R}v_\alpha}{c}\right)^{1/3} = 5{\cdot}7\times 10^{11}(\sigma\dot{R})^{1/3}; \tag{27}$$

with $\sigma = 10^{-2}$ gm/cm^2 and $\dot{R} = 10^6$ cm/sec as typical values

$$\bar{u} = 1{\cdot}2\times 10^{13}\,\mathrm{s}^{-1}. \tag{28}$$

Since $\bar{u}/v_\alpha$ is of order 0·5% the nebula expands about 0·5% during the period that a typical photon remains trapped. Note that $\bar{u}$ is more than 2 orders of magnitude larger than the thermal width U. This confirms that the Ly-α line spreads far out into the wings before its photons can escape from the nebula.

4. The Mechanical Effect of Radiation Pressure

It has now been shown that Ly-α radiation is generated in the ionized region of the nebula at a rate 0·3 L_* erg/sec, and that it loses a fraction $\bar{u}/v_\alpha$ of its original energy to the expanding shell of non-ionized gas. Thus the rate at which work is done on the shell is

$$M\dot{R}\ddot{R} = 0{\cdot}3L_*\bar{u}/v_\alpha \equiv 0{\cdot}3L_*\times 1{\cdot}8\left(\frac{\varpi M\dot{R}}{4\pi R^2 c v_\alpha^2}\right)^{1/3} \tag{29}$$

or

$$R^{2/3}\dot{R}^{2/3}\ddot{R} = 2\times 10^{-5}L_*/M^{2/3}. \tag{30}$$

To make a numerical estimate assume first that L_* is constant and R has the form at^s. Substitution shows that

$$R = 2{\cdot}2\times 10^{-2}\,\frac{L_*^{3/7}}{M^{2/7}}\,t^{8/7} \tag{31}$$

and

$$\dot{R} = 2{\cdot}5\times 10^{-2}\,\frac{L_*^{3/7}}{M^{2/7}}\,t^{1/7}. \tag{32}$$

In this case most of the acceleration occurs when t is small. Later the nebula expands at almost a constant rate. For example, if $L_* = 10^{37}$ erg/sec and $M = 10^{33}$ gm, then

$$\dot{R} = 2{\cdot}5\times 10^6\ \text{cm/sec} \quad\text{and}\quad R = 2{\cdot}2\times 10^{17}\ \text{cm}, \tag{33}$$

when $t = 10^{11}$ sec. For t larger or smaller by a factor 10, the estimate for R changes by only 45% upwards or downwards.

Solutions in which R varies like a power of t can also be obtained when the luminosity L_* is assumed to vary with time. This assumption is more reasonable, for observation seems to show that the typical central star evolves with increasing luminosity early in the life of a nebula (Seaton, 1966). To illustrate this case, let $L_* = l_* t^2$, and take $l_* = 10^{15}$ erg/s^3, so that $L_* = 10^{37}$ erg/sec when $t = 10^{11}$ sec. Equation (30) is now solved by

$$R = \alpha t^2 \quad \text{and} \quad \dot{R} = 2\alpha t\,, \tag{34}$$

where $\alpha = 6 \times 10^{-3} l_*{}^{3/7} / M^{2/7} \approx 6 \times 10^{-6}$ cm/s^2.

At time $t = 10^{11}$ sec the predicted values for R and $\dot{R}$ are somewhat smaller than for the case in which L_* is constant. This is as expected. For the time $t = 2 \times 10^{11}$ sec, Equations (34) predict $R = 2{\cdot}4 \times 10^{17}$ cm and $\dot{R} = 2{\cdot}4 \times 10^6$ cm/sec, but by then some of the approximations made earlier are beginning to fail. The luminosity L_* would equal 4×10^{37} erg/sec at that stage, and, according to Equation (3), the mass M_i in the H II region is about 6×10^{32} gm. An appreciable fraction of the shell has therefore been ionized, and the equations used here will cease to be valid. In fact the condition that M_i be small compared with M, the initial mass of the shell, is

$$2{\cdot}6 \times 10^5 \frac{l_* t^2}{\beta^{1/2}} m_{\mathrm{H}} R^{3/2} \ll M;$$

with R expressed in terms of t, by Equation (34), the inequality becomes

$$t^4 \ll \frac{M^{10/7} \beta^{1/2}}{123 m_{\mathrm{H}} l_*^{8/7}} \approx 2 \times 10^{45}\,, \tag{35}$$

or $t \ll 2{\cdot}2 \times 10^{11}$ sec.

The shell of atomic hydrogen ceases to exist after about $2{\cdot}2 \times 10^{11}$ sec. If our formulae were valid until then they would predict for that time a radius of 3×10^{17} cm and a velocity $2{\cdot}6 \times 10^6$ cm/sec, about twice the speed of sound in the H II region. The shell is then moving so fast that one cannot be sure that the ionized gas will fill the H II region uniformly. Nevertheless a straightforward comparison according to the present calculation shows that the radiation pressure and gas pressure become equally strong rather earlier than this time. At time t one finds that the radiation pressure is

$$p_R = \frac{M\ddot{R}}{4\pi R^2} \approx 27 \frac{M^{9/7}}{l_*^{3/7}} t^{-4}\,, \tag{36}$$

and the gas pressure

$$p_G = \frac{2(M_i/m_{\mathrm{H}})}{(4\pi/3) R^3} kT_i \approx 10^3 \frac{M^{3/7}}{l_*^{1/7}} t^{-2}\,, \tag{37}$$

so that p_R much exceeds p_G as long as

$$t \ll 0{\cdot}16 \left(\frac{M^3}{l_*}\right)^{1/7} \approx 1{\cdot}6 \times 10^{11}. \tag{38}$$

At $t = 1{\cdot}6 \times 10^{11}$ sec the estimated velocity of the shell is

$$\dot{R} = 2 \times 10^{-3} M^{1/7} l_*^{2/7} \approx 2 \times 10^6 \text{ cm/sec}.$$

Thus the radiation pressure p_R provides the initial acceleration, and makes the shell expand faster than the sonic speed in the ionized gas, before the gas pressure p_G can become as important as p_R. Soon after p_G begins to be significant the shell will disappear, in any case.

5. Pressure Exerted at the Ionization Front

There will also be dynamical effects at the ionization front which may influence the motion of the outer shell. Conditions at ionization fronts (or I-fronts) have often been discussed, in the context of the interstellar medium (Axford, 1961; Mathews, 1965; Hjellming, 1966). Only some order-of-magnitude estimates will be made in this section.

Briefly mass, momentum and energy must be conserved at the I-front, and these requirements are expressed by three conservation laws for

mass flux $$\Phi = \rho_0 v_0 = \rho v \tag{39}$$

momentum flux $$\Pi = p_0 + \rho_0 v_0^2 = p + \rho v^2 \tag{40}$$

and energy conservation $$2\mathscr{E} + \frac{5p_0}{\rho_0} + v_0^2 = \frac{5p}{\rho} + v^2. \tag{41}$$

Here p is the pressure, ρ the density and v the speed relative to the I-front. $\mathscr{E}$ is the net thermal energy gain, per unit mass, due to the ionization process (possibly allowing for subsequent cooling). The suffixes zero refer to the non-ionized gas ahead. If the front advances slowly relative to that gas and the temperature there is low, then $(5p_0/\rho_0) + v_0^2$ is negligible compared with $2\mathscr{E}$. To estimate a lower bound to the pressure ahead, one looks for the minimum value of Π, given Φ and $\mathscr{E}$. Provided v_0^2 is small compared with p_0/ρ_0, that is provided the I-front moves at low speed into the gas ahead, p_0 will approximately equal Π. With a little algebra one finds that

$$p_0 \approx \Pi = \frac{\Phi}{v}\left(\frac{2}{5}\mathscr{E} + \frac{4}{5}v^2\right) \geqslant \frac{4\sqrt{2}}{5}\Phi\,\mathscr{E}^{1/2}. \tag{42}$$

The rate of flow of mass into the H II region, per unit surface area, is

$$\Phi = \frac{\dot{M}_i}{4\pi R^2} \approx \frac{10^6 l_*^{2/7} M^{1/7} m_H}{\beta^{1/2} t}, \tag{43}$$

after substituting from Equations (3) and (34), and putting $L_* = l_* t$. Further $\mathscr{E} \approx kT_*/m_{\rm H}$, where T_* is the surface temperature of the central star, and it follows from Equations (42) and (43) that

$$p_0 \gtrsim \frac{10^6 l_*^{2/7} M^{1/7}}{\beta^{1/2}} (kT_* m_{\rm H})^{1/2} \, t^{-1} \, . \tag{44}$$

Putting $T_* = 10^5$ °K, and on substitution for β, the pressure p_I at the ionization front will be found to satisfy

$$p_I (\equiv p_0) \gtrsim 10^{-5} l_*^{2/7} M^{1/7} / t \, , \tag{45}$$

so that $p_I \gg p_R$ when

$$\frac{10^{-5} l_*^{2/7} M^{1/7}}{t} \gg 27 \, \frac{M^{9/7}}{l_*^{3/7}} \, t^{-4} \, , \quad \text{or}$$

$$\begin{aligned} t^3 &\gg 27 \times 10^6 \, \frac{M^{8/7}}{l_*^{5/7}} = 2 \cdot 5 \times 10^{33} \, , \quad \text{or} \\ t &\gg 1 \cdot 4 \times 10^{11} \, . \end{aligned} \tag{46}$$

Thus p_G and p_I begin to dominate p_R at about the same time. But note that p_G is then decreasing more rapidly than p_I.

6. Discussion

In the paper the description of the dynamics of a planetary nebula has been drastically simplified. Even so it seems clear that radiation pressure in Lyman-α will be very important, until the speed of expansion becomes too high. During this phase the nebula is expected to contain a large proportion of non-ionized hydrogen. After about $1 \cdot 5 \times 10^{11}$ sec the recoil pressure at the ionization front begins to be more important than radiation pressure and will dominate the motion, until the hydrogen has been ionized.

From a study of photographs it is quite obvious that spherical symmetry is not a common property among planetary nebulae (see, e.g., Henize, 1967; Westerlund and Henize, 1967; Khromov and Kohoutek, 1968; Perek and Kohoutek, 1967). Indeed the typical nebula is more likely to be ring-like, or, when seen sideways on, butterfly-shaped. There seem to be two ways in which this different symmetry might be introduced by the dynamical processes described in this paper.

One possibility is that the original, non-ionized nebula was an oblate spheroid. The atomic hydrogen shell formed from such a nebula would contain less mass per unit area over the poles, that is at the ends of the short axis of the spheroid. The polar parts of the shell would let through a greater flux of radiation and they would therefore be accelerated to a higher speed. Eventually there would be left behind a ring-like region, which expands more slowly, and has a higher density.

Alternatively, the primitive nebula might have been spherically symmetrical, with a central star that rotates fast. The stellar surface temperature is then a function of latitude, with hot regions near the poles, and cooler regions over the equator. In the early evolution of the nebula the uneven distribution of colour and luminosity over the star would probably have little influence on the manner in which radiation pressure repels the shell, since Ly-α photons undergo many scatterings, and presumably redistribute themselves in a symmetrical fashion. But later the ionization front would advance faster into the polar parts of the non-ionized shell, and the recoil pressure there would be greater. These parts of the shell would therefore be expelled with a higher speed, once again leaving a ring-like structure behind. It is possible that these rings would appear to be butterfly-shaped, when seen sideways on.

Finally a closer inspection of the photographs show that planetary nebulae also have a rather finely detailed filamentary structure. No doubt there is some dynamical mechanism, like a Rayleigh-Taylor instability, which gives rise to these features. Such problems need to be studied much more thoroughly.

References

Aller, L.H. (1956) *Gaseous Nebulae*, Chapman and Hall, London, p. 74.
Axford, W.I. (1961) *Phil. Trans. R. Soc. Lond.*, **A253**, 301.
Gebbie, K.B., Seaton, M.J. (1963) *Nature*, **199**, 580.
Harman, R.J., Seaton, M.J. (1966) *Mon. Not. R. astr. Soc.*, **132**, 15.
Henize, K.G. (1967) *Astrophys. J. Suppl. Ser.*, No. **126**.
Hjellming, R.M. (1966) *Astrophys. J.*, **143**, 420.
Khromov, G.S., Kohoutek, L. (1968) in the present volume, p. 227.
Mathews, W.G. (1965) *Astrophys. J.*, **142**, 1120.
Perek, L., Kohoutek, L. (1967) *Catalogue of Galactic Planetary Nebulae*, Academia, Prague.
Seaton, M.J. (1966) *Mon. Not. R. astr. Soc.*, **132**, 113.
Westerlund, B.E., Henize, K.G. (1967) *Astrophys. J. Suppl. Ser.*, **126**.

DISCUSSION

Underhill: How do you get the original neutral gas nebula around the star in order to start the process?

Menon: Is it implied that the neutral hydrogen shell was originally part of the star and not a part of the interstellar medium?

Kahn: One picture of the early evolution of a planetary nebula is to imagine that in the beginning a cloud of non-ionized gas surrounds the central star. The star then becomes hot and begins to create an H II region. The higher pressure in the H II region pushes back the H I region and forms it into a thin shell. According to this view it is almost necessary that an H I shell should surround the nebula at some stage.

Evans: This is a very attractive idea because of the frequency with which one observes a breakdown of spherical symmetry in the sense of a transition to a bipolar form. One example is Henize 150 where the ring has broken at opposite points and gas streams out, being re-excited by the central star as it crosses the beam.

Böhm: In the second phase of the central stars' evolution their luminosity decreases with time. How does this affect the development of the ionization front?

Kahn: While the non-ionized shell is still in existence, any reduction in the Lyman continuum flux at the I-front slows down the flow through the front. If the radiation from the star gets so weak that no ionizing radiation gets through the H II region to the I-front, then a recombination front will form.

Gurzadian: All these considerations are very interesting, but your theory should take into account the role of magnetic fields which might be present in the nebula.

Kahn: It is most doubtful whether one can reasonably assume that a magnetic field can exist which is strong enough to affect the fluid motions significantly.

Capriotti: The work of Kahn is quite pertinent to the study of the planetary nebulae. One way to test the validity of the assumption that the He I 2^3S state is depopulated through photo-ionization by Ly-α radiation is to estimate the length of time for which a sufficiently high Ly-α density can be sustained. The blanketing effect of a thin shell of neutral hydrogen has been proposed as the agent for the trapping of Ly-α radiation to the extent that it depopulates the 2^3S state at the observed rate. Kahn's result that this blanketing effect could last for only a very short time is a reason for rejecting this proposed mechanism.

Underhill: Kahn's remarks referred to the initial stages of development of a planetary nebula. Reeves has already suggested that the injection of low-energy protons might be a significant factor in causing the observed line spectrum of the nebula. Would not the original state of the nebula then be quite different from the state it would have if only radiation processes were occurring? In such a case the relaxation to the state which we observe might differ from that postulated by the types of theory discussed today (i.e. chiefly ionization by stellar ultraviolet and Ly-α and He II 303 lines). A continuous ejection of protons at 2000 km/sec from the central star is not unexpected if we extrapolate from the known behavior of Wolf-Rayet stars and O-type supergiants.

Münch: I wish to express the point of view held by Capriotti in somewhat different words. The need for a deactivation agent of the He I 2^3S level arose from the considerations regarding the nebular absorption of He I $\lambda 3889$ in stars imbedded in H II regions made by Wilson and myself. We found that considering two-photon decay, collisional deactivation, and photo-ionization by star light only, the observed absorption implied a very small emission measure, in disagreement with observation. On this basis I suggested that photo-ionization by Ly-α may play a role. The upward revision of the collisional deactivation rates reported at this conference by Seaton will help the situation, but will by no means completely explain the observed population of He I 2^3S. The lack of $\lambda 3889$ absorption in many stars imbedded in H II regions (and I have observed more than 100 of them) appears to me to be a serious problem.

Capriotti: The electron density in an H II region is probably low enough so that the $\lambda 10830/\lambda 5876$ line-intensity ratio is near the low-density limit. Therefore, one probably would not be able to use this particular ratio in order to estimate the relative population density of the 2^3S state and check on the results obtained by emission measures in $\lambda 3889$. On the other hand, because of the relatively low electron densities in the H II regions the deviation of the depopulation rate of the 2^3S state as determined by a $\lambda 3889$ emission measure from a theoretically predicted rate could hardly be blamed on underestimates of the collision cross-sections for $2^3S \rightarrow 2^1S$, 2^1P, and 1^1S transitions. It would be nice if the discrepancy between theory and observation as regards the depopulation of the 2^3S state of helium in both H II regions and the planetary nebulae could be explained on the same basis.

SUR LA STRUCTURE MORPHOLOGIQUE ET CINÉMATIQUE DE LA NÉBULEUSE 'HELIX'

G. CARRANZA, G. COURTÈS et R. LOUISE
(Observatoire de Cordoba, Argentine,
Observatoires de Marseille et de Haute-Provence, France)*
(Présenté par H. Andrillat, Faculté des Sciences, Montpellier)

La nébuleuse planétaire ayant le plus grand diamètre apparent est la nébuleuse Helix**; elle se prête très bien, en raison de ses larges dimensions ($15' \times 10'$), aux observations avec les techniques interférentielles courantes: la photographie monochromatique en lumière Hα ou [NII] λ6584 et l'interférométrie Perot-Fabry. On peut ainsi obtenir non seulement la morphologie de cette nébuleuse, mais aussi le champ des vitesses radiales au moyen des anneaux d'interférence des mêmes raies (Courtès, 1951, 1952, 1960).

Les photographies monochromatiques d'Helix effectuées à l'aide de filtres interférentiels à bandes passantes très étroites (4 à 6 Å), montrent les particularités suivantes (Figures 1 et 2) (Cruvellier, 1967):

La géométrie générale des images en lumière Hα est à peu près la même que celle des images en lumière [NII] 6584 Å.

Ces images diffèrent néanmoins en dimensions, l'image Hα étant légèrement plus petite (5 à 10%) que l'image en [NII].

Les filaments sont beaucoup moins nets, quelquefois plus larges, souvent moins intenses et généralement plus floculeux en lumière Hα qu'en [NII].

Le filament extérieur devient très faible en lumière Hα.

Sur le cliché Hα il y a une émission diffuse et 'floconneuse' à l'intérieur de la structure filamentaire qui est imperceptible ou absente en lumière [NII]. C'est, à notre connaissance, la première fois que l'on met en évidence cette structure très particulière de l'émission Hα interne d'Helix.

L'image Hα d'Helix ressemble assez à celle des planétaires ordinaires (sphères gazeuses) mais elle est difficile à accorder avec l'image [NII] qui suggère plutôt une structure filamentaire hélicoïdale et non une sphère gazeuse.

Pour se libérer autant que possible des effects subjectifs, on a étudié la distribution

* Les premiers interférogrammes étudiés dans cette publication ont été pris au télescope Crossley de l'Observatoire de Lick à la suite d'une invitation du Dr N. U. Mayall. Les suivants ont été obtenus à l'Observatoire de Haute-Provence au télescope de 120 cm.

** NGC 7293.

Osterbrock and O'Dell (eds.), Planetary Nebulae, 249–255. © *I.A.U.*

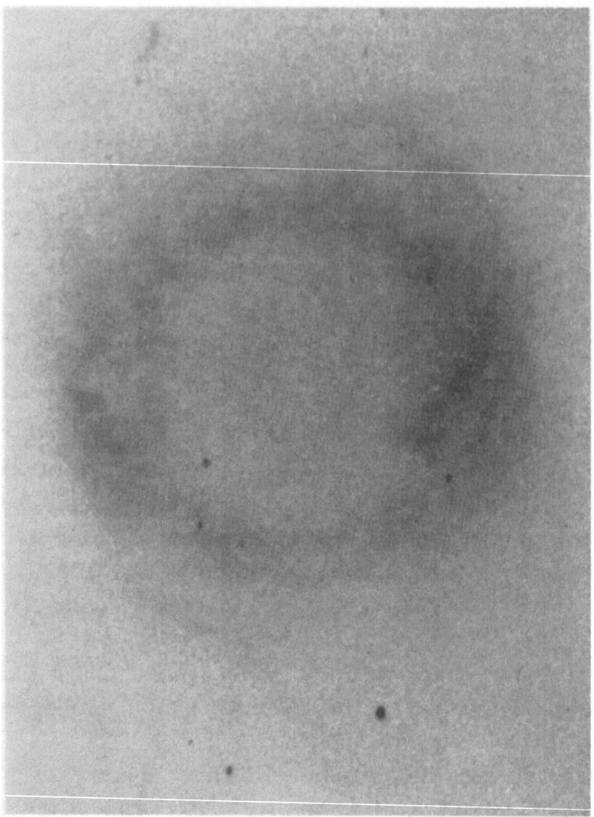

FIG. 1. *Photographie d'Helix prise avec un filtre interférentiel Hα. Télescope de 120 cm de l'Observatoire de Haute-Provence.*

par rapport à l'étoile centrale des régions les plus brillantes dans un cliché pris avec une bande passante très large (≈500 Å comprenant Hα λ6563 et [NII] λλ6548 et 6584). Ce cliché, obtenu au télescope de 5 m du Palomar, nous a été très aimablement communiqué par le Dr. Minkowski.

La morphologie d'Helix peut s'interpréter de deux façons différentes; la première, d'ailleurs suggérée par Minkowski, consiste à supposer que cette nébuleuse est la superposition de deux anneaux de gaz concentriques dont l'anneau externe serait en train de se dissocier (Figure 3). D'après une étude morphologique fondée sur la répartition des points les plus intenses de l'anneau interne, l'inclinaison du plan des anneaux par rapport au plan de projection est alors de $\alpha = 36°$ si l'on suppose que l'anneau est parfaitement circulaire. En raison de sa morphologie mal définie, l'anneau externe ne permet évidemment pas une telle analyse.

Si on admet une vitesse d'expansion radiale **V** à partir de l'étoile centrale des points situés sur l'anneau interne, la répartition de la projection de **V** sur la direction de

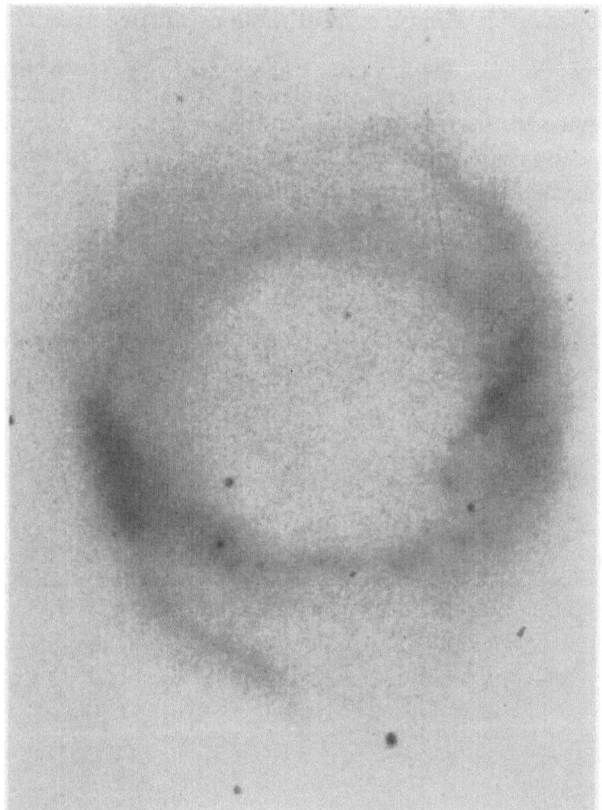

FIG. 2. *Photographie d'Helix prise avec un filtre interférentiel* [NII]. *Télescope de 120 cm de l'Observatoire de Haute-Provence.*

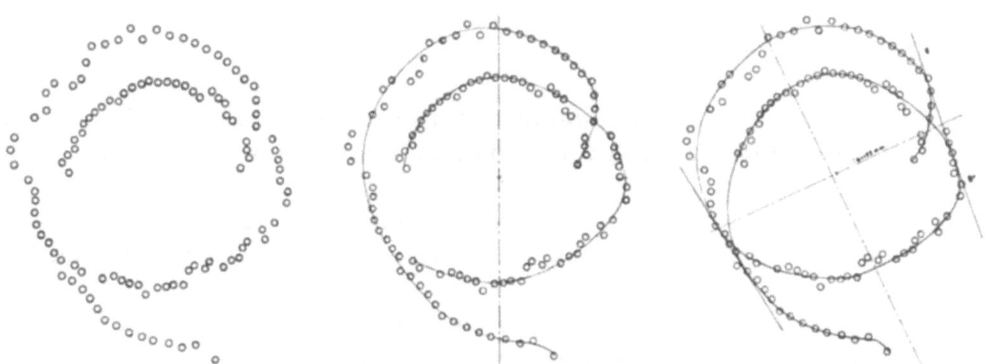

FIG. 3. *De gauche à droite: Points de densité photographique maximum; Nébuleuse Helix interprétée comme deux anneaux; Nébuleuse Helix interprétée comme une hélice.*

l'observateur en fonction de l'angle azimuthal φ est donnée par l'expression:

$$\text{Proj. } \mathbf{V}/\mathbf{u} = \mathrm{V} \sin\alpha \sin\varphi.$$

Quand on compare la distribution des vitesses obtenues avec la raie [N II] dans l'anneau interne à celle d'un anneau ayant la même géométrie en expansion radiale, on constate que l'accord est assez satisfaisant.

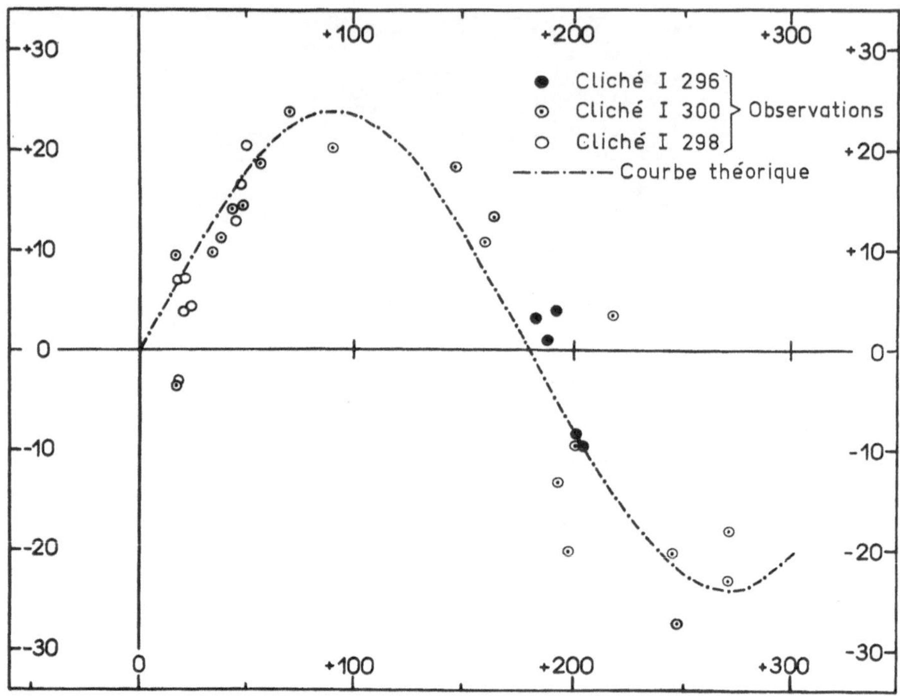

FIG. 4. *Répartition des vitesses de l'anneau interne.*

La Figure 4 montre cette comparaison. L'expansion qui permet le meilleur accord entre les vitesses observées et la courbe d'expansion théorique a pour module V= 39·1 km/sec.

Une difficulté, cependant, se présente: on ne peut rien espérer de la répartition des vitesses de l'anneau externe qui, selon l'hypothèse vraisemblable, serait en train de se dissocier.

Cette difficulté est levée si l'on suppose, comme le suggère l'image [N II] d'Helix, que le filament est unique et épouse une forme hélicoïdale. On supposera, pour simplifier (Figure 3), qu'il s'agit d'une hélice droite à pas constant. L'étude morphologique permet de déterminer les éléments de cette hélice.

(1) Inclinaison de l'axe de l'hélice par rapport au plan de projection $i = 54°$;

(2) Pas $= 0{\cdot}36$ en unité du diamètre de l'hélice.

Toujours dans l'hypothèse d'une vitesse radiale **V** constante à partir de l'étoile centrale, la répartition de la projection de **V** sur la direction de l'observateur est donnée par la relation:

$$\text{Proj. } \mathbf{V}/\mathbf{u} = \mathrm{V} \cos i \cos\theta \left[\operatorname{tg}\theta \operatorname{tg} i + \sin\varphi\right],$$

où $\mathbf{u}$ = vecteur unitaire porté par la direction de l'observateur.

Avec:
$$\begin{cases} \operatorname{tg}\theta = p\varphi/2\pi a \\ p = \text{pas de l'hélice.} \\ a = \text{rayon du cylindre dans lequel est inscrite l'hélice.} \end{cases}$$

La distribution des vitesses obtenues avec la raie [N II] a été comparée à celle qu'il y aurait dans une nébuleuse ayant la même géométrie et qui serait en expansion radiale à vitesse constante. Les résultats de cette comparaison sont indiqués sur la Figure 5. La vitesse radiale y est donnée en fonction de l'angle azimuthal φ dans le

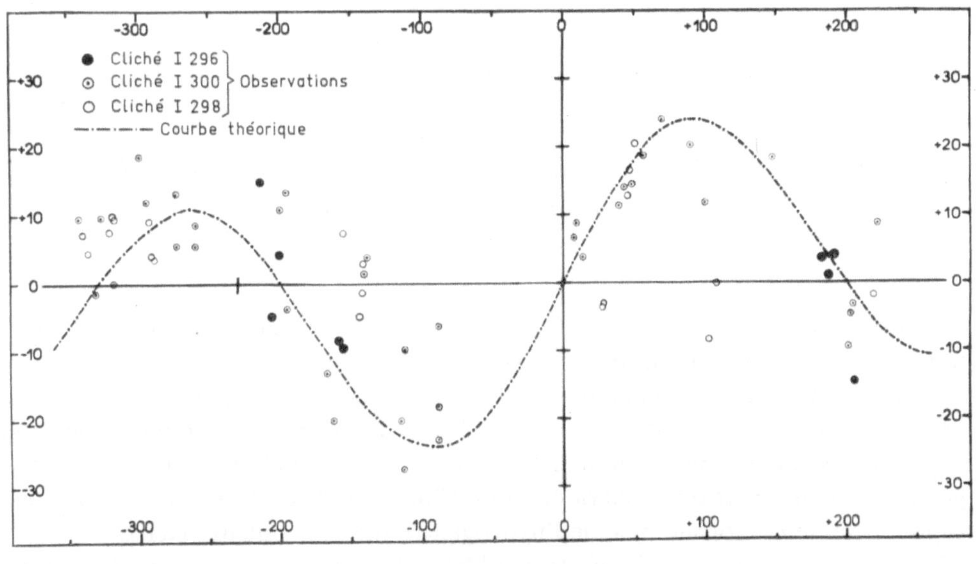

FIG. 5.

plan perpendiculaire à l'axe de l'Helix; cet angle est déterminé à partir des angles de position mesurés sur la photographie de la nébuleuse et des paramètres de l'hélice.

La répartition théorique des vitesses radiales sur l'hélice est en très bon accord avec les vitesses radiales mesurées sur la raie [N II]. La courbe théorique qui s'accorde le mieux avec les observations permet de déterminer la vitesse d'expansion radiale de la nébuleuse. Cette vitesse est de 37·4 km/sec.

Cette dernière hypothèse sur Helix a l'avantage de considérer toutes les vitesses radiales observées, ce qui n'est pas le cas dans la première hypothèse. Dans les deux cas, on retrouve à peu près la même vitesse d'expansion radiale de la nébuleuse par rapport à l'étoile centrale. Ceci tient au fait que l'anneau interne, assimilé à un cercle, a une géométrie très proche de celle de la spire centrale de l'hélice. En fait, les différences maximales des vitesses radiales entre ces deux modèles et pour la partie centrale sont de l'ordre de 2 km/sec. Il n'est donc pas possible de choisir entre les deux modèles à partir de la seule géométrie de la bande centrale.

On a également étudié la répartition des vitesses radiales mesurées sur la raie Hα. Cette étude a conduit aux résultats suivants:

(1) Les points expérimentaux se répartissent dans un intervalle de vitesses radiales −10 km/sec et −30 km/sec. (Figure 6).

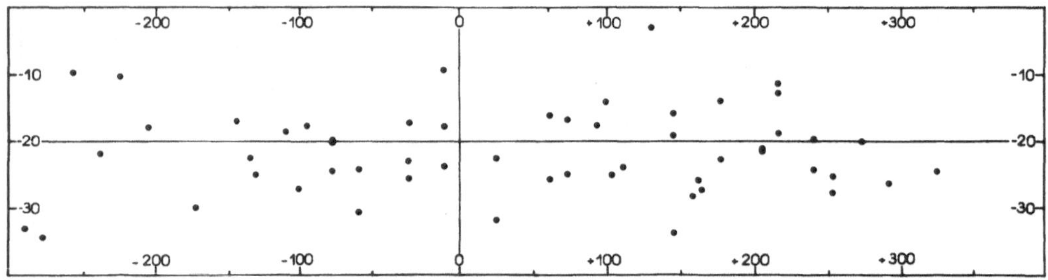

FIG. 6. *Répartition des vitesses radiales en Hα.*

(2) Il est impossible, quelle que soit l'hypothèse faite sur Helix, de tracer une courbe régulière qui passe entre les points expérimentaux.

(3) La vitesse radiale moyenne de l'ensemble de la nébuleuse par rapport au soleil est de −29 km/sec ±5, en bon accord avec les mesures spectroscopiques anciennes.

(4) Aucune structure cinématique nette ne se dégage de la répartition des vitesses de l'hydrogène; une répartition sphérique homogène est vraisemblablement la meilleure interprétation de la morphologie de l'hydrogène dans la nébuleuse Helix.

D'autre part, le rapport d'intensité Hα/[N II] dans la nébuleuse que l'on obtient aussi à l'aide des interférogrammes, montre qu'entre l'anneau externe et l'anneau interne, les conditions d'excitation sont différentes; dans l'anneau externe, il est en général plus favorable à [N II] que dans l'anneau interne; la valeur moyenne du rapport Hα/[N II] dans celui-ci est de 0·9; dans le filament externe, il est de 0·7. Si l'on a affaire à un seul filament hélicoïdal, cette variation ne peut être expliquée qu'au moyen d'hypothèses *ad hoc*.

Enfin, comme il a été dit, l'émission à l'intérieur des filaments n'est décelable que pour le rayonnement Hα; une telle diminution du rapport Hα/[N II] quand on s'éloigne de l'étoile centrale, a déjà été observée dans d'autres nébuleuses planétaires.

Les résultats précédents permettent de conclure qu'Helix est très probablement constituée par une hélice à pas constant dont la photographie donne l'image sur le plan de projection.

On n'a pas considéré la possibilité qu'Helix soit une planétaire sphéroïdale en expansion (combiné éventuellement avec rotation), ce qui pourrait cinématiquement être acceptable pour les raisons suivantes:

(1) L'extrême netteté de la structure filamenteuse en lumière [N II].

(2) L'absence de dédoublement des raies, fréquent dans les planétaires sphéroïdales en expansion.

Cette étude préliminaire ne nous permet pas d'obtenir un modèle définitif, mais seulement de délimiter les possibilités. Elle précise, néanmoins, certaines particularités structurales d'Helix que l'on ne peut tirer que de l'intégration des données morphologiques, cinématiques et physiques.

Les résultats sur l'expansion publiés par Liller (1966), fondés sur la mesure des mouvements propres, comparés aux nôtres, permettent d'estimer la distance entre 140 et 400 parsecs.

Bibliographie

Courtès, G. (1951) *C. r. Acad. Sci. Paris*, **232**, 1283.
Courtès, G. (1952) *C. r. Acad. Sci. Paris*, **234**, 506.
Courtès, G. (1960) *Ann. Astrophys.*, **23**, 155.
Cruvellier, P. (1967) *Ann. Astrophys.*, **30**, 1059.
Liller, W. (1960) *Astrophys. J.*, **144**, 280.
Minkowski, R.L. Communication privée.

Note bibliographique

NGC 7293 ou nébuleuse Helix a été découverte par Curtis sur une photographie prise au Crossley.
Premiers spectres: Campbell et Moore, *Publ. Lick. Obs.*, **13** (1918).
Vitesses mesurées par Wright, *Publ. Lick. Obs.*, **13** (1918).
Spectrogrammes par Minkowski, *Astrophys. J.*, **243** (1942).
Rapport des raies de [O II] par D.E. Osterbrock, *Astrophys. J.*, **131** (1960), 541.
Distances données par C.R. O'Dell (137 parsecs), *Astrophys. J.*, **135** (1962), 371.
Données photométriques par G.A. Gurzadian *et al.*, *Soobšč. Bjurak. Obs.*, **34** (1963) 59; *ibid.*, **35** (1964), 59.
Flux radio mesuré par D. Lynds, *Publ. nat. Radio Astr. Obs.*, **1** (1961), 85; Davis *et al.*, *Nature*, **206** (1965), 809; Oslee *et al.*, *Austr. J. Phys.*, **18** (1965), 187; M. Hughes, *Astrophys. J.*, **149** (1967), 377.

FILAMENTS OF THE HELIX NEBULA

B.A. VORONTSOV-VELYAMINOV
(Sternberg State Astronomical Institute, Moscow, U.S.S.R.)

Thirty years ago Baade discovered in the Helix Nebula extremely thin radial filaments, but since then their existence has been neglected. Data presented below were obtained by studying a copy of a plate obtained in 1954 in the light of Hα with the 200-inch telescope. Our goal is to draw attention to these filaments, which are of great importance for the understanding of planetary nebulae in general.

The thickness of the filaments is approximately 1″ of arc, and their extension is about 30″ of arc. The regular linear filaments point exactly at the nucleus of the nebula, and are best seen against the dark background inside the ring. Hundreds of filaments can be individually resolved, and thousands of fainter filaments must also be present. Apparently the bright ring of NGC 7293 is formed entirely of coalesced filaments. As a rule, each filament has a bright condensation at its inner end. The brightness of filaments can be from 19 mag to 20 mag. Adopting a distance for NGC 7293 of 100 parsec, the length of the filaments is 3000 AU and their width is 150 AU. Neither short filaments nor their condensations can be seen in the central part of the nebula. Thus the filaments and the ring they form actually are a kind of a thick toroid.

It is very improbable that the filaments represent gaseous streams directed from the envelope to the nucleus of the nebula. Instead, they must be formed by corpuscular currents ejected by the nucleus, or by tracks of an unknown agent. The absence of the filaments nearer to the nucleus than about 90″ even in the plane of the ring is enigmatic. The filaments are reminiscent of solar corpuscular streams, both being nearly cylindrical in cross-section. A thin column of ionized dense gas, ejected from the nucleus, probably first loses its luminescence and then becomes luminous again by the Zanstra mechanism just in the vicinity of the ring. It is difficult to explain the condensations in the gas, or the luminosity at the end of a filament, which faces the nucleus.

From the velocity of ejection of about 1000 km/sec, observed in Wolf-Rayet stars, one can easily calculate that the gas from the nucleus will reach the ring in some 70 years. Such motions are easy to detect from already existing plates. After penetrating into the expanding ring the corpuscular streams slow down and dissipate. The ejection from the nucleus can be recurrent and its duration can be variable. If the value of electron density in the filaments is of the same order of magnitude as that found in

Osterbrock and O'Dell (eds.), Planetary Nebulae, 256–258. © *I.A.U.*

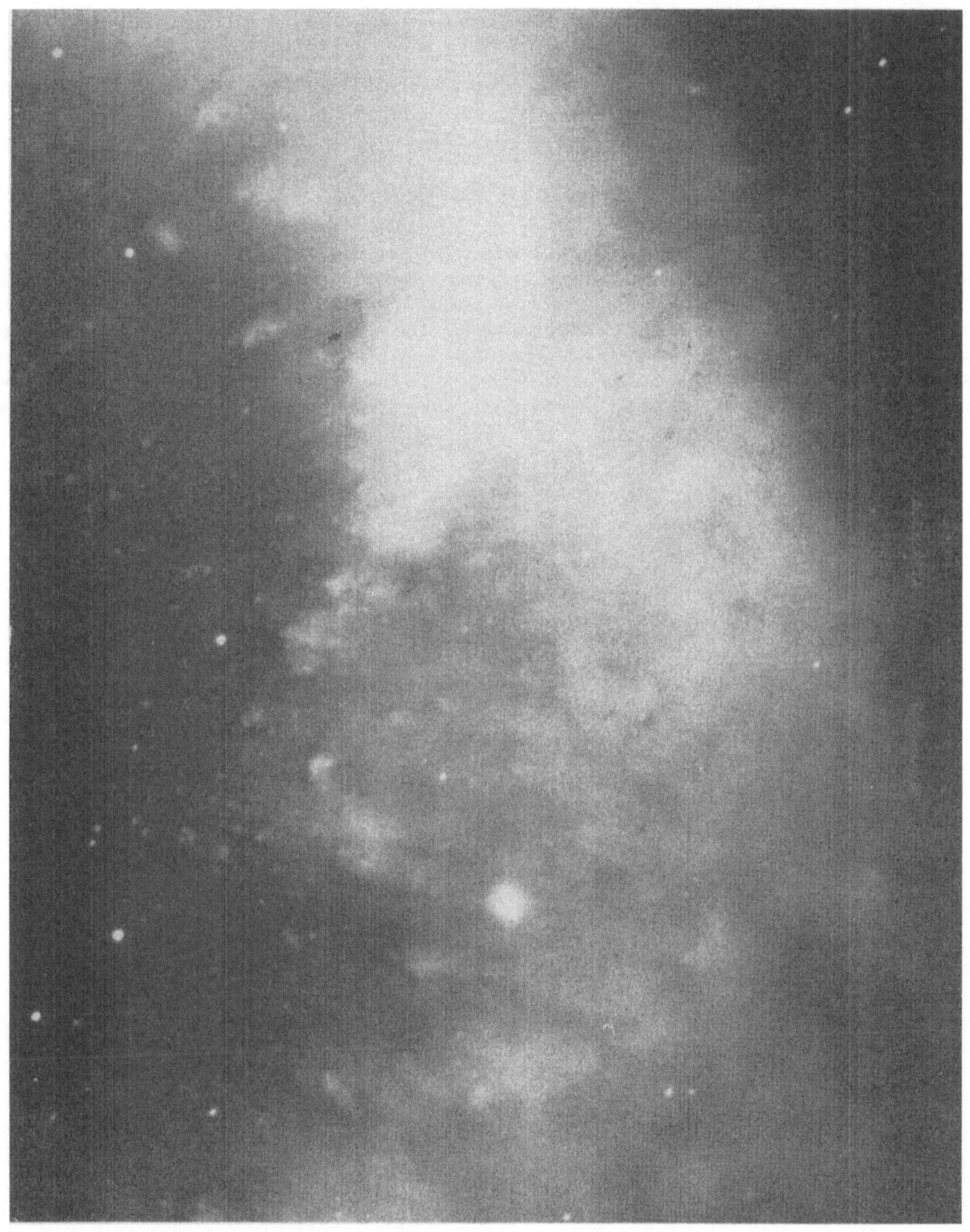

FIG. 1. *Part of NGC 7293, 200-inch plate taken in* $H\alpha$ + [*N* II], *showing filaments and condensations.*

other ring-shaped planetary nebulae, say $n_e = 500$ cm^{-3}, the mass of a single filament is 5×10^{25} gm; 1000 filaments have a mass 10^{-5} $M_{\odot}$, equal to the mass loss by a Wolf-Rayet star during 1 year. The creation of the nebular envelope would require 10^2–10^3 years of such activity by the nucleus.

The structure and origin of the nebular envelope of the Helix Nebula must be typical at least for all ring planetary nebulae.

DISCUSSION

Mathews: The instabilities in NGC 7293 could result from a Rayleigh-Taylor instability at the shock front where a stellar wind from the central star interacts with the bulk of the nebular matter.

Liller: We have measured the angular radial motion of three of the bright condensations within the ring of NGC 7293 on 100-inch Newtonian plates, the earliest of which was Baade's discovery plate. The mean radial motion of these condensations is $+1\overset{''}{.}0 \pm 0\overset{''}{.}4$ of arc per century, at a mean distance of 176″ of arc from the central star. If we assume the distance to be 200 parsec, this motion corresponds to a linear velocity of about 10 km/sec, considerably less than the velocities mentioned by Vorontsov-Velyaminov.

INTERNAL MOTIONS IN THE PLANETARY NEBULA NGC 6543

GUIDO MÜNCH
(Mount Wilson and Palomar Observatories, U.S.A.)

1. Introduction

The planetary nebula NGC 6543 has been described (Curtis, 1918) as "quite irregular, of helical form". Its high surface brightness and its dimensions ($16'' \times 22''$ of arc) make it an ideal object for observation with the multislit technique used by Wilson (1950) to study other planetaries. In fact, had NGC 6543 been accessible to the 100-inch Mount Wilson Coudé, undoubtedly it would have been included in Wilson's survey. The interest in NGC 6543 arises from the fact that no information is available regarding the motions in a planetary nebula with some sort of helical appearance. The obvious question that arises in this respect is whether the nebular material is actually arranged in space on an open helix, rather than in a flat spiral. The related problems of its lifetime and stability also may be elucidated by studying in detail the internal motions.

2. Observations

Two multislit spectrograms of NGC 6543 were obtained with the 72-inch camera of the Palomar Coudé spectrograph. With an exposure of 90 min, measurable images in the N1 and N2 lines of O^{++}, in Hβ, in $\lambda\lambda$3868 and 3967 of Ne^{++} and in $\lambda\lambda$3726–29 of O^{+} are reached. The multislit unit is the same as that used by Wilson *et al.* (1959) to study the Orion Nebula. The projected width of each slit is 0·015 mm or 0·068 Å, and the slit spacing (1·0 mm) corresponds to 1$''$.32 of arc in the plane of the sky. The slits were oriented by means of the field rotator, North–South in one plate and East–West in the other. The exciting star in both cases was centered on the central slit, by reference to a fiducial mark readily identifiable on the images of that slit.

The measurements of apparent wavelengths were carried out by referring micrometer settings at points separated by 0·2 mm along every slit to the comparison spectrum, in the manner described by Wilson *et al.* (1959). Thus, radial velocities for the lines of O^{++}, Ne^{++} and O^{+} have been obtained for a square grid of points covering the nebula with 1$''$.32 spacing. Tables 1, 2, and 3 give the radial velocities of these lines, the points to which they refer being identified by rectangular equatorial coordinates measured in integral units of the spacing and with origin at the central star.

The intrinsic widths of the nebular lines are appreciably larger than the instrumental profile and vary considerably from place to place. In certain areas the lines can be clearly

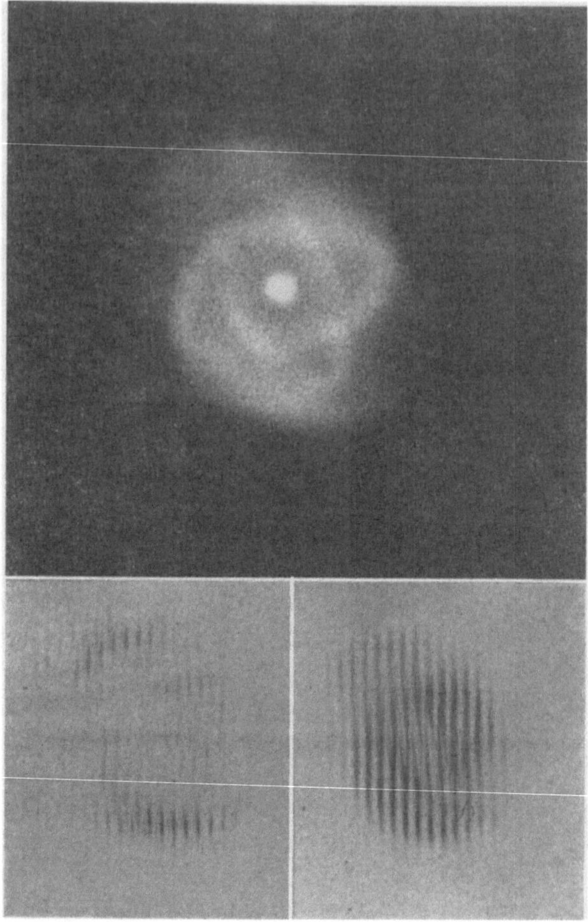

FIG. 1. *Image of NGC 6543 in radiation of the* [O III] *N1 and N2 lines at top. Multislit images in* [O II] *λλ3726–28 (left) and* [Ne III] *λ3867 (right) at bottom. In all images North is towards bottom and East to right.*

seen as double, and in a few points in the Ne^{++} images three components can be seen. This is why some entries in the tables contain more than one number. At those points where the lines appeared broad but unresolved, no attempt has been made to allow for blending effects in the radial velocities, but some idea of their importance can be gained by comparing the values with those of neighboring points. The general appearance of the O^{++} and Ne^{++} images is very similar and their radial velocities are nearly identical. The O^{+} image, unlike the images in lines of higher excitation, is densest in the outer regions. The radial velocities of O^{+} at those points where it is reasonably certain that there is no blending, nevertheless, agree with those of the

Table 1*

Radial velocities of [O III] λ5006·85 Å

	E_7	E_6	E_5	E_4	E_3	E_2	E_1	EOW	W_1	W_2	W_3	W_4	W_5	W_6	W_7
N7	–	94·	97·4	95·9	95·1	94·6	93·3	95·1	93·6	?	92·4	–	–	–	–
N6	98·0	98·2	97·0	96·8	90·6	85·3	85·7	87·4	88·2	87·4	86·8	89·	–	–	–
N5	88·2	88·2	91·0	87·5	84·9	84·4	77·6	80·2	83·3	85·4	86·6	83·2	82·0	–	–
N4	83·6 –	83·7 –	86·7 –	91·2 66·2	87·8 66·2	84·8 61·3	75·4	76·0	79·6	84·8	86·1	85·8	81·5	77·4	–
N3	78·2 –	78·1 –	81·5 –	81·8 60·8	82·2 58·5	80·7 61·3	75·0	76·0	79·4	82·5	85·8	84.8	81·8	80·3	81·0
N2	– 74·6	– 73·4	– 77·1	– 76·4	– 75·8	95·3 62·8	90·6 62·3	87·8 62·3	86·6 65·6	88· 67·2	87·9 67·6	81·5	80·7	78·6	79·8
N1	– 70·6	– 69·2	– 71·0	– 71·8	– 65·0	89·7	87·	85·6	81·8	80·4	86·6 66·4	78·0	74·0	72·2	74·8
NOS	66·0	65·4	66·0	69·0	87·6 64·0	100·2 58·2	99·	95·2	91·2	81·8	89·8 69·0	90·4 65·0	71·4	71·0	69·0
S1	62·6	61·4	62·2	63·5	66·4	79·4 51·7	87·4 45·8	86·4 43·2	82·6 –	78·2 –	76·2 –	86·1 62·0	88·2 61·6	– 63·4	– 68·8
S2	62·4	61·2	58·1	60·1	63·9	70·0 51·6	73·1 48·8	74·5 44·	74·9 –	73·0 –	71·1 –	79·9 60·9	80·2 54·7	– 58·4	– 61·8
S3	60·	58·6	56·2	54·4	56·0	59·8	61·6	69·4	70·2	75·8 56·6	73·1 53·	73·9 52·6	? 50·5	– 54·0	– 58·1
S4	–	58·6	58·4	55·2	56·4	58·2	59·4	69·6 55·8	71·9 54·2	73·2 49·2	74·2 46·7	72·6 48·0	? 48·5	– 53·4	– 56·1
S5		58·2	58·	53·6	55·5	55·8	56·4	65· 50·4	– 49·3	– 47·8	– 47·2	– 46·2	– 47·6	– 48·8	?
S6		–	–	?	47·2	47·9	48·0	44·8	46·0	44·4	42·5 31·	42·1 33·	43·3 33·	42·6	–
S7					?	41·	43·	41·	42·	38·	38·	36·4	37·1	?	–

* All entries have negative sign.

Table 2*

Radial velocities of [Ne III] λ3868·73 Å

	E_7	E_6	E_5	E_4	E_3	E_2	E_1	EOW	W_1	W_2	W_3	W_4	W_5	W_6	W_7
N6	–	–	?	?	?	95·	95·	93·2	94·	–	–	87·	–	–	–
N5	–	99·2	?	?	92·	86·	84·	87·0	86·6	89·8	88·2	?	–	–	–
N4	84·6	87·4	87·6	85·4	85·8	83·4	77·6	77·2	82·4	84·6	86·6	79·8	85·2	?	–
N3	81·2 –	78·4 –	83·6 –	87·0 –	86·6 –	83·6 62·4	84·4 63·1	83·3 65·6	78·8	81·6	83·5	85·4	84·	80·	–
N2	74·9 –	75·4 –	76·5 –	78·6 –	77·0 –	81·2 58·7	87·9 66·1	74·6	76·1	79·1	79·6	80·5	77·6	75·4	–
N1	– 70·0	– 69·8	– 73·6	– 71·0	– 68·3	– 62·2	87·9 60·3	86·3 58·4	81·9 58·2	83·8 62·2	81·6 65·8	83·8 67·9	83·7 69·0	?	–
NOS	– 65·0	– 63·1	– 65·2	– 67·0	– 59·9	– 55·2	87· 52·8	82·4 53·8	79·5 49·6	74·0 48·7	72·2	82·0 62·2	79·1 59·8	75·0 58·0	66·
S1	– 61·6	– 58·2	– 59·8	– 63·5	75·3 57·6	78·4 50·3	79·5 47·1	80·2 42·6	77·4 42·2	79·5 66·5 42·9	83·0 65·6 –	81·2 61·2 –	74·9 56·2 –	71·6 57·5 –	– 59·6 –
S2	56·3	54·7	54·6	57·5	58·	69·0 47·0	74·4 43·1	73·0 41·	70·7 37·	68·6 42·	76.0 59·8 –	77·7 52·8 –	– 54·4 –	– 59·2 –	– 59·8 –
S3	52·	50·8	50·0	52·2	55·4	54·9	62·8 45·9	64·8 36·6	65·2 –	63·6 –	73· 53·4 –	76· 50·7 –	– 48·2 –	– 61·0 46·6 –	– 62·0 50·1 –
S4	49·	48·8	48·6	46·8	48·0	48·7	51·0	50·8	52·0	65·2 45·9	– 47·0	– 47·2	– 45·0	– 47·0 36·9	– 50·8 40·0
S5	–	47·	48·0	42·8	44·2	48·4	51·2	53·3	54·5	48·8	– 40·0	– 40·1	– 41·4	– 44·3	– 47·0
S6			–	42·	44·	45·4	47·8	46·8	– 39·0	– 36·4	– 35·2	– 35·7	– 36·6	– 36·2	?
S7						?	37·	38·	43·	?	33·	31·7	27·8	28·8	–

* Alletries have negative sign.

Table 3*

Radial velocities of [O II] $\lambda 3726{\cdot}50$ Å

	E_7	E_6	E_5	E_4	E_3	E_2	E_1	EOW	W_1	W_2	W_3	W_4	W_5	W_6	W_7	W_8
N7																
N6																
N5	–	97·3	–	–	–	–	–	–	82·9	83·6	–	–	–	–	–	–
N4	88·9	90·9	–	–	–	–	70·1	72·8	79·2	88·7	88·9	91·4	–	–	–	–
N3	82·7	84·7	–	–	–	61·4	64·2	65·6	–	88·	89·8	89·6	–	–	–	–
N2	75·8	74·5	–	–	–	55·8	56·8	–	–	–	95·0	92·7	–	–	–	–
N1	71·0	69·0	–	–	–	51·5	57.	–	–	–	90·	90·9	–	–	–	–
NOS	66·1	64·1	–	–	46·4	47·	49·	–	–	88·	91·4	89·2	–	–	69·	68·0
S1	61·2	59·5	53·8	–	–	–	–	–	–	–	87·2	86·9	–	–	67·0	69·
S2	55·1	53·7	53·0	–	–	–	–	–	–	–	82.	–	–	–	62·6	64·8
S3	53·0	52·0	49·3	47·0	–	–	–	–	–	–	78·	–	–	–	56·2	57·2
S4	52·4	49·0	44·4	41·3	–	–	–	–	–	–	78·?	–	–	–	50·8	54·6
S5	–	50·	46·	45·8	44·6	49·8	49·8	–	–	–	–	–	–	45·8	44·4	–
S6	–	–	–	44·0	45·6	51·7	53·1	–	–	–	–	–	–	40·9	39·6	–
S7	–	–	–	–	–	–	–	–	–	–	–	–	26·1	30·2	–	–
S8	–	–	–	–	–	–	–	–	–	–	–	–	25·8	28·	–	–

* All entries have negative sign.

higher excitation lines. The Hβ image differs from the others mainly because the line widths are noticeably broader, indicating that the broadening is to some extent thermal in origin.

3. Interpretation

Although on a first impression the radial velocity field appears very complex, it is not difficult to discern a relatively simple trend in those points with single unblended lines. The most obvious feature is that the extreme radial velocities, around −100 and −25 km/sec, appear at the NE and SW corners of the nebula, the 'ansae' referred to by Curtis (1918). This situation is quite different from that found in the planetary nebulae with nearly circular symmetry (Wilson, 1950), where the extreme radial velocities are observed at the centre (two components). The whole set of radial velocities for the high-excitation lines, disregarding obvious blends, can be divided into two groups according to their magnitude. The largest velocities (in absolute value), together with the violet-displaced components at points where the lines appear double, very nearly define an area very nearly elliptical with axial ratio 7:4 and major axis in position angle 23°. In the same fashion, the smaller velocities, together with red-shifted components, define another ellipse, similar and parallel to the former. These two ellipses are partially overlapping in space but not in velocity. This general arrangement of the points according to their radial velocities is sketched in Figure 2, where the helical arms seen in the O^{++} images have been outlined. Along these

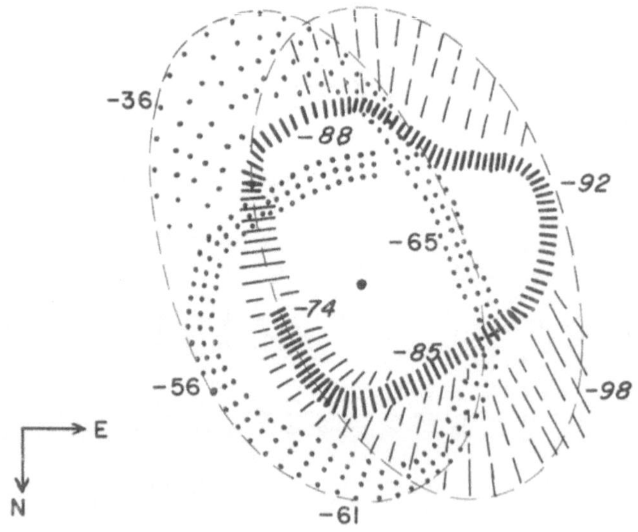

FIG. 2. *Regions of the planetary nebula NGC 6543 where the radial velocities of the* [O III] *lines are in absolute value small (dotted) and large (hatched). A few representative values of the velocities (km/sec) are entered for the approaching (italic figures) and receding (roman figures) loops.*

structural features the radial velocities vary gradually, indicating continuity in their motion as well as in their density. It is noticed that the line joining the centres of the ellipses is nearly parallel to the minor axes of the ellipses and contains the exciting star at the centre. This fact, together with the similarity of the envelope ellipses, suggests that their true shape is nearly circular and that in space the nebular material is arranged on two helical surfaces with generatrix passing through the central star and forming an angle of $\cos^{-1}(4/7)=55°$ with the line of sight. A sketch of the suggested spatial arrangement of the nebular arms is shown in Figure 3. These arms are not, as

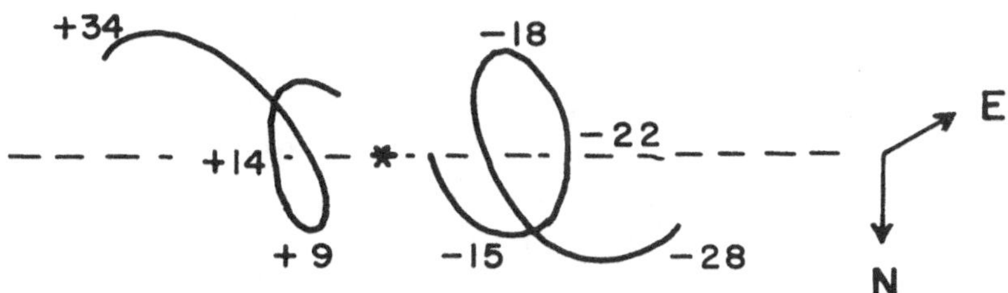

FIG. 3. *Side view perspective of NGC 6543. Only the densest parts of the nebular arms, as seen in the O++ radiation, are outlined. The representative radial velocities given in km/sec are relative to the central star and are uncorrected for foreshortening.*

depicted, isolated features, but rather are condensations on helical sheets which contain sufficient matter to shield the outer regions from the ultraviolet radiation of the central star, as required by the observed ionization stratification.

The relative radial velocity between the inner parts of the approaching and receding loops (from the star) is approximately $V_0=24$ km/sec, and if it is the result of a uniform symmetrical expansion from the central star, the time T since the expansion began is obtained from the distance D of the nebula and the angular separation (5″.5) between the centre of the two ellipses expressed in radians. If the distance is 1000 parsecs (O'Dell, 1963), the lifetime of the nebula is thus 1000 years, about 1 order of magnitude shorter than that generally considered 'typical' for a planetary nebula (Abell and Goldreich, 1966). It should be noticed that the dimensions in cross-section of the nebular arms are at present a few seconds of arc. Since the velocity of sound in the nebula is around 20 km/sec, the thermal diffusion of the arms in 1000 years would not be sufficient to make them lose their identity. That is to say, if the ejection took place from two localized regions diametrically opposite to each other on the surface of the star parent to the present nucleus, and, thus, was confined to a tube of diameter small compared to its length, not enough time has elapsed for the arms to diffuse completely. In fact, this argument can be inverted to fix the time since ejection and hence derive the distance, once it is supposed that no *ad hoc* forces (e.g. of a magnetic nature) are stabilizing the nebular arms. It should be pointed out that to stabilize the arms at present magnetically, a field strength of the order of 5×10^{-3} Oersted is required. This would imply an impossibly large field in the early phases of the expansion. Under the plausible assumption, then, that the loops will continue their present axial expansion and thermal diffusion uniformly, one can see that when the 'typical' lifetime of a planetary has elapsed the helical arms seen today in NGC 6543 will have disappeared and the nebula will be a low surface brightness amorphous planetary.

References

Abell, G.O., Goldreich, P. (1966) *Publ. astr. Soc. Pacific*, **78**, 232.
Curtis, H.D. (1918) *Publ. Lick Obs.*, **13**, 55.
O'Dell, C.R. (1963) *Astrophys. J.*, **138**, 67.
Wilson, O.C. (1950) *Astrophys. J.*, **111**, 279.
Wilson, O.C., Münch, G., Flather, E., Coffeen, M.F. (1959) *Astrophys. J. Suppl. Ser.*, **4**, 199.

DISCUSSION

Liller: We have measured the angular radial motion for several filaments in NGC 6543 on 60-inch Cassegrain plates. The mean motion we have derived is $+0''\!.5 \pm 0''\!.1$ of arc per century.

Terzian: I would like to mention that the observed optically thin radio flux of NGC 6543 is 0·9 flux units and that predicted from its Hβ flux is 1·6 flux units. This large difference probably is not due to observational errors.

EMISSION-LINE PROFILES IN PLANETARY NEBULAE

DONALD E. OSTERBROCK
(Washburn Observatory, University of Wisconsin, U.S.A.)

This research was undertaken with the idea of measuring as accurately as possible the internal-velocity distribution in planetary nebulae, in order to compare the observational measurements with hydrodynamical models of expanding nebulae. Much of the work was done in collaboration with J. S. Miller and D.W. Weedman. All the observational data were obtained photographically with the Coudé spectrograph of the 100-inch telescope at Mt. Wilson, using an image rotator, a 900 line/mm grating, and an *F*/5·2 camera, giving a dispersion of about 4 Å/mm in the blue and about 6 Å/mm in the red. The measured velocity resolution is approximately 5–6 km/sec. The data for five nebulae have been published (Osterbrock *et al.*, 1966) while data for three more, NGC 2392, NGC 3242, and IC 418 are discussed here for the first time.

All the spectrograms I will discuss here were taken with the slit through the central star, and all the tracings I will discuss were made at the centre of the nebular image, so they refer to the distribution in radial velocity of material along the line of sight through the centre of the nebula. The measurements published in the first paper showed that the nebular emission lines are not only double, as is well known from the work of Wilson (1948), but also have widths which vary systematically with atomic weight in the sense that the lightest element, H, has the widest lines, and the heaviest elements (O, N) have the sharpest lines. The widths are partly due to thermal Doppler broadening, but in addition there must be considerable range in mass motion, for the profiles of the lines and their widths cannot be interpreted as due to thermal broadening alone. In typical planetaries such as NGC 7662 the observations show that along a line through the centre of the nebula there is material with velocity ranging between approximately +10 and +40 km/sec, and also material with velocity ranging between −10 and −40 km/sec. This range of velocity is probably due to the outward gradient of expansion velocity discovered by Wilson (1948) from the observed correlation between ionization potential and measured velocity of expansion, though from the tracings of the spectrograms taken at the centre of the nebula alone it is impossible to separate this effect from any possible turbulent velocity that might be present.

The new measurements show that NGC 3242 is a typical nebula with double lines, and I will not discuss it further here. IC 418 is interesting in that though the [N II] lines are double, the [O III] lines are single (Figure 1). This is due to the fact that O^{++} is concentrated to the centre of the nebula where the expansion velocity is lower, while N^{+} occurs in the outer part of the nebula where the expansion velocity is larger.

Osterbrock and O'Dell (eds.), Planetary Nebulae, 267–269. © *I.A.U.*

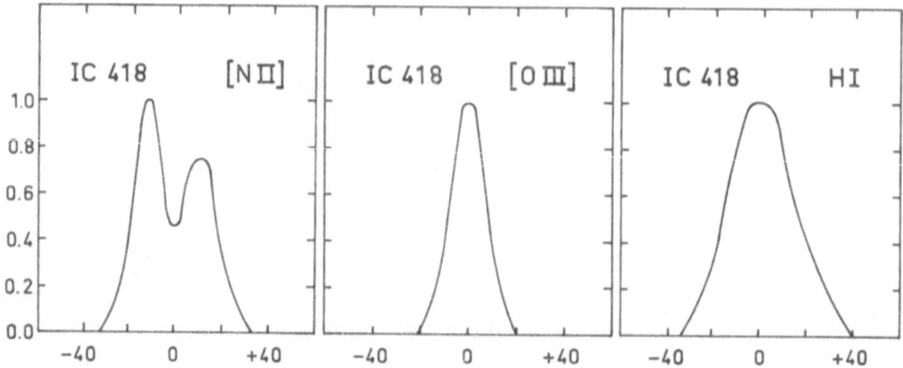

FIG. 1. *Measured line profiles of* [*N* II] *(average of 4 separate exposures),* [*O* III] *(average of 6), and H* I *(average of 5) at the centre of IC 418.*

The observed H I lines are also single, partly as a result of the fact that H has a higher thermal Doppler width, and partly due to the fact that the H lines are formed throughout the nebula. At the present time we are carrying out calculations to see if the H I profile can be quantitatively understood on the basis of velocity distributions obtained from the [N II] and [O III] lines.

The most interesting emission lines are those observed in NGC 2392, the planetary nebula with the largest expansion velocity measured by Wilson (1948). The [O III] images reproduced in Figure 2 show the complications very well. In the bright ring

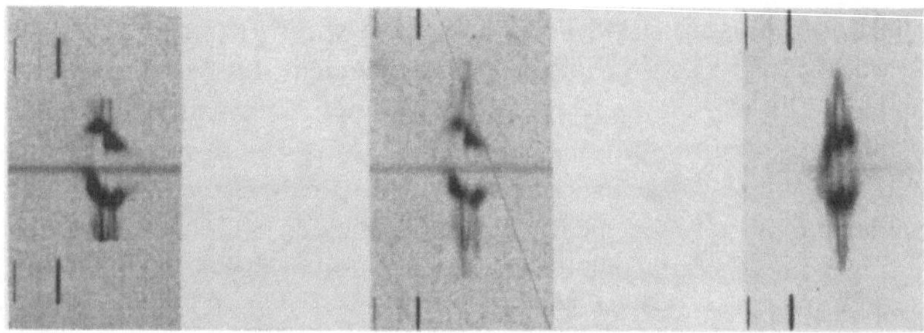

FIG. 2. *Reproductions of* [O III] *λ5007 emission line of NGC 2392. Left and centre are two different exposures with slit oriented North–South through centre of nebula; right, slit East–West through centre.*

there is a range in radial velocity of over 100 km per second. The outer structure seems to belong to a different kinematic system from the inner structure, as is also the case in NGC 7662. The spectrum of NGC 2392 taken with the slit East–West shows particularly clearly that in addition to the system expanding with approximately 50 km/sec radial velocity at the centre of this nebula, there is another component with approximately 0 km/sec radial velocity, perhaps connected with the outer structure of this nebula. The schematized theoretical picture of a spherically symmetrically expanding nebula is evidently far too drastic a simplification to apply to NGC 2392.

Further analysis of the spectrograms all along the slit length (that is to say all along a line that projected on the sky goes through the centre of the nebula and that therefore is the trace of a plane through the centre of the nebula) by Weedman tends to confirm the picture of Wilson (1958) that many planetary nebulae can be understood as prolate spheroids with axial ratio approximately 3:2. Measurements of the line profiles at the ends of the slit image (that is looking tangentially through the nebula) show that the turbulence is small, and that any turbulent velocity that does exist is less than 5 km/sec. In NGC 7009, examination of the slit images show that there are many abrupt near-discontinuities in velocity (of the order of 3–5 km/sec), that velocity changes are correlated with density changes, and that there is a symmetry to the patchy inhomogeneous structure of the nebula, in the sense that regions of high density tend to be diametrically opposite regions of high density, and regions of low density tend to be diametrically opposite other regions of low density. Theoretical work on the interpretation of these measured line profiles is continuing.

This research was based on observational material obtained as a guest investigator at Mount Wilson and Palomar Observatories, and was supported by the National Science Foundation.

References

Osterbrock, D.E., Miller, J.S., Weedman, D.W. (1966) *Astrophys. J.*, **145**, 697.
Wilson, O.C. (1948) *Astrophys. J.*, **108**, 201.
Wilson, O.C. (1958) *Rev. mod. Phys.*, **30**, 1025.

DISCUSSION

Liller: Does NGC 2392 appear to have an outward velocity proportional to radius, or constant with radius as Wilson's data and our data seem to indicate?

Osterbrock: We have not measured the plates of this nebula yet, so I cannot answer the question.

Münch: On the basis of your material can you say definitely, for one selected planetary nebula at least, whether the microturbulence is less than the thermal root-mean-square velocity? This point is important, because the galactic H II regions have microturbulence that is just transonic, and therefore there must be a fundamental difference in this respect between the two classes of objects.

Osterbrock: Weedman's measurements show that in at least two planetaries, NGC 7009 and NGC 7662, if there is any microturbulence its root-mean-square velocity is less than 5 km/sec.

O'Dell: Can the [O I] λ6300 line profile be measured? This would be especially interesting as it should arise in the outermost regions of the nebula.

Osterbrock: No, it was too faint to be photographed at this high dispersion with the equipment I had. An image tube would help a lot.

Aller: By combining multislit spectra and monochromatic isophotal contours, it is possible to get information not only on the velocity pattern in a planetary nebula, but also on the three-dimensional distribution of the emitting matter. Such observations have been made for NGC 7662, using multislit spectrograms taken by Wilson. The bright inner shell contains a number of intensity fluctuations, indicating knots and condensations, as one would expect. The illustration is published in *Progress in Fluid Dynamics*, edited by Temple and Seeger.

INTERNAL MOTIONS IN THE PLANETARY NEBULA NGC 6853

P. V. SHEGLOV
(Sternberg State Astronomical Institute, Moscow, U.S.S.R.)

Je voudrai bien parler la langue classique de l'interférometrie, mais le temps nous est cher et je passerai à l'Anglais.

I have made an interferometric study of the planetary nebula NGC 6853. A system with a Fabry-Perot étalon was used, but not a common one. In the classical arrangement known from Fabry's time (Figure 1), the étalon is placed in the exit pupil, formed by a

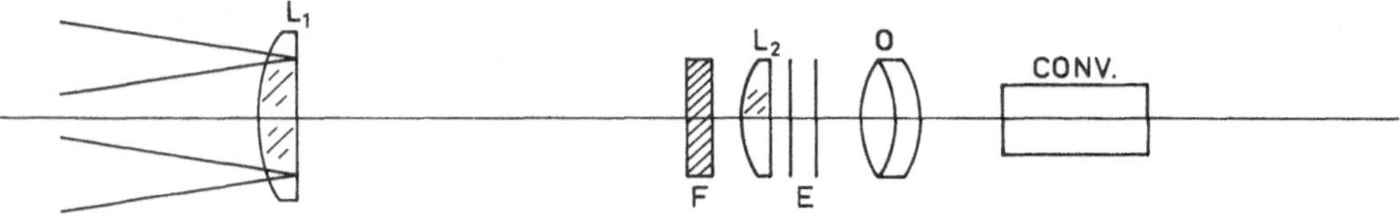

FIG. 1. *A classical arrangement of the Fabry-Pérot étalon.*

field lens, i.e. in the plane where the image of the telescope mirror is formed by this lens, while in our system the étalon is in the focal plane of the telescope and the image of the nebula is formed in the plane of the étalon (Figure 2).

In the classical system the fringes are superimposed on the image of the nebula; we

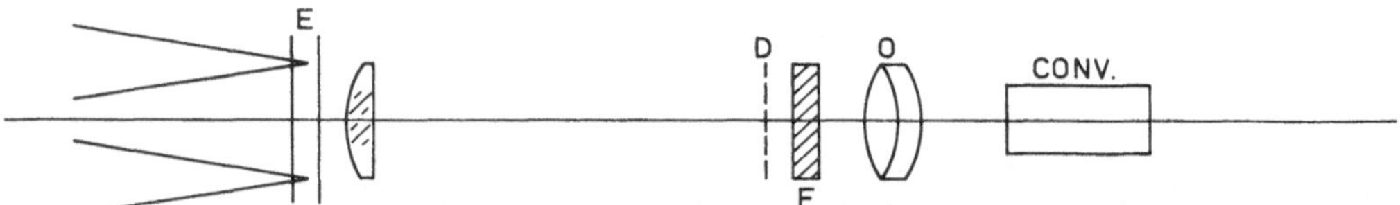

FIG. 2. *A narrow-band filter with Fabry-Pérot étalon.*

see it only at the points where the conditions of maximum transmission are fulfilled. The II system acts as a narrow ($\delta\lambda \approx 0{\cdot}7$ Å) filter; we obtain a picture of the whole nebula in a chosen interval of wavelength. This interval can be changed by changing the annular diaphragm in the exit pupil of the system. The equivalent bandwidth of such a system is 1·1 Å, and the transparency is about 50%. The dimensions of the

Osterbrock and O'Dell (eds.), Planetary Nebulae, 270–272. © *I.A.U.*

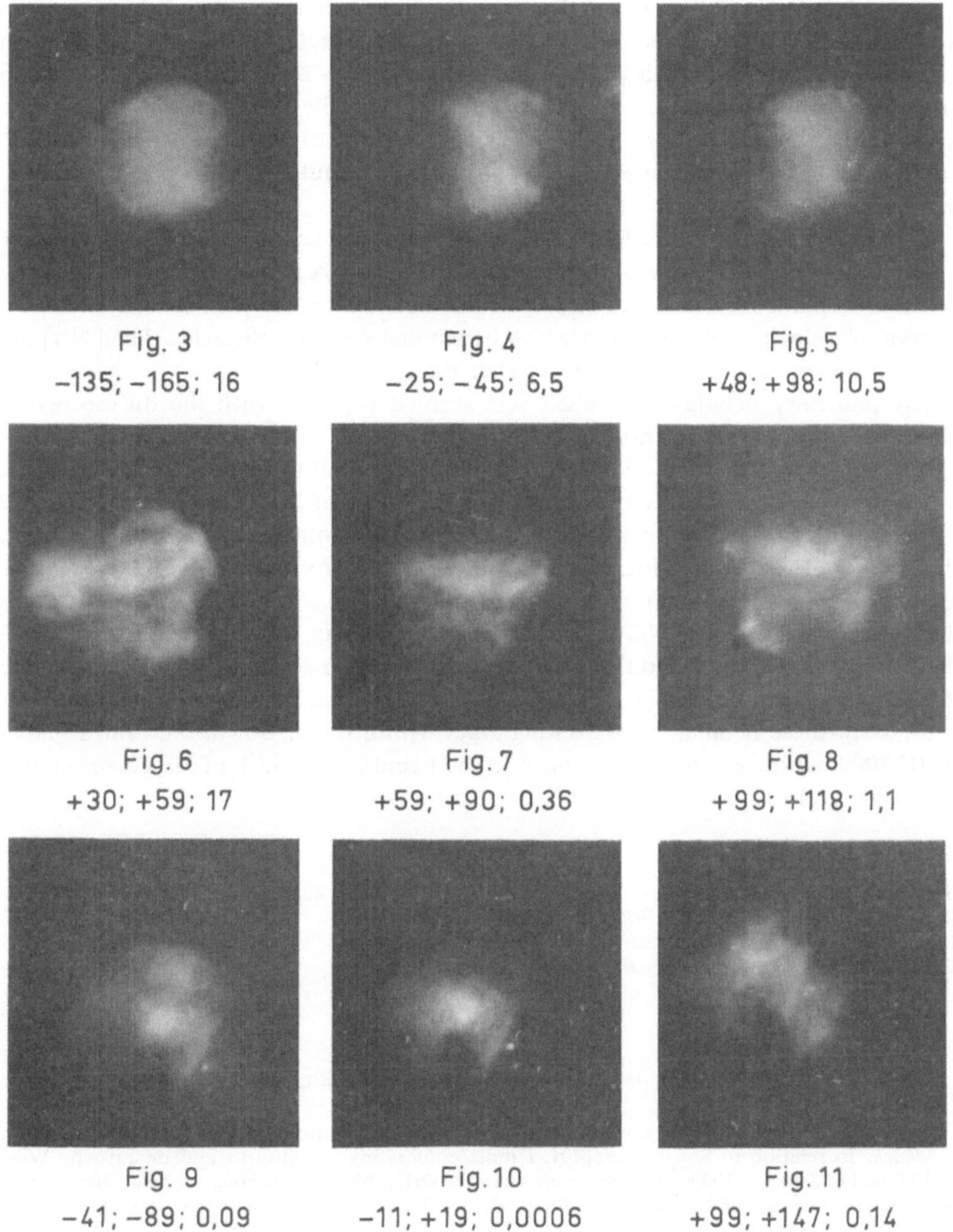

FIG. 3–11. *Three nebulae, each in three different velocity intervals as indicated by first two numbers (in km/sec); third number is relative exposure. Figures 3–5 are NGC 6853; 6–8, NGC 6618; 9–11, NGC 1976.*

fringe isolated are, however, smaller than the image of the mirror, which results in stopping down the telescope. The loss of light is by a factor of $D_0^4/(D_1-D_2)^4$; we cut down the aperture of the telescope in the ratio $D_0^2/(D_1-D_2)^2$, and use only a fraction of the line profile. In the case of a rectangular line profile filling all the mirror image we loose another factor $D_0^2/(D_1-D_2)^2$, where D_0 is the diameter of exit pupil, and D_1 and D_2 are the inner and the outer diameters of the annular diaphragm. The image is therefore photographed by a contact image converter.

The overlapping of the orders of the étalon can be a source of serious trouble; the spacing of the étalon used was chosen to make the 6584 Å and 6548 Å [N II] lines coincide and to put them at a distance of ½ order, i.e. 750 km/sec from Hα. It allowed an interval of velocities of ±400 km/sec to be studied without obstacles. The [N II] lines were partly transmitted by the interference filter.

The planetary nebula NGC 6853 was studied together with the diffuse nebulae NGC 6618 and 1976. All three nebulae are seen in an interval of radial velocities of ±120 km/sec relative to the Sun. The bright parts of all three nebulae are seen in all the velocity intervals studied. The changes of shape of NGC 6618 and NGC 1976 when the velocity interval is shifted are much more pronounced than the changes of shape of NGC 6853. The brightness of the high-velocity parts is about 2–5% of the brightness of the nebula in the central part of its emission line. In Figures 3–11 we see the nebulae NGC 6618, NGC 1976 and NGC 6853 in different velocity intervals; these intervals in km/sec and the relative exposures in an arbitrary system are given on the figures. It must be noted that high velocities are found also in NGC 6611 and NGC 6853. Both these nebulae have exciting stars within them. But in the central part of NGC 7000, whose exciting star is outside the nebula, the width of Hα corresponds to the thermal velocities of gas motion.

References

Gershberg, R. E., Sheglov, P. V. (1964) *Astr. I. USSR*, **41**, 3, 425.
Sheglov, P. V. (1963) *Astr. Cirk. USSR*, **266**.
Sheglov, P. V. (1967*a*) *Astr. Cirk. USSR*, **414**.
Sheglov, P. V. (1967*b*) *Astr. Cirk. USSR*, **423**.

DISCUSSION

Andrillat: Quel est le temps de pose nécessaire pour obtenir la photographie de la Nébuleuse Dumb-bell?

Sheglov: Il est 10 min pour les parties les plus intenses, et 1 heure pour les régions les plus faibles.

Münch: In relation to Sheglov's report, I must confess my inability to understand why Wilson and I failed to detect ±100 km/sec components in the Orion Nebula. Our instrumental characteristics and exposures were such that we should have detected such components if they were 5% as strong as the main component.

Vaughan: Is it certain that we have no contamination by very faint lines of other ions in the bandpass of the isolating filter, and that no étalon ghosts are present?

Sheglov: A single modern Fabry-Perot étalon allows us to obtain a contrast of about 100 for close spectral lines; for a spectrograph this is more difficult.

DYNAMICAL EVOLUTION OF A MODEL PLANETARY NEBULA

W. G. MATHEWS
(University of California, San Diego, U.S.A.)

The gas-dynamical equations for conservation of mass, momentum and energy, the ionization equation, and the equation of transfer for the ionizing radiation have been solved in a manner which approximates conditions in planetary nebulae. The ionized and neutral gas are assumed to heat and cool according to a linearized rate proportional to

$$\tfrac{3}{2}k(T - T_{eq}), \quad \text{where} \quad T_{eq}(\text{H II}) = 10^4\,°\text{K} \quad \text{and} \quad T_{eq}(\text{H I}) = 500\,°\text{K}.$$

The initial configuration is a spherical shell of mass $M_n = 0{\cdot}307\, M_\odot$ with outer radius 0·027 parsec, thickness 0·002 parsec, temperature 500 °K, density $7{\cdot}5 \times 10^5$ cm^{-3}, and a uniform outward velocity of 20 km/sec. The luminosity, temperature, and surface gravity of the central star are assumed to vary as functions of the projected nebular radius according to the evolutionary scheme developed by Seaton and Harman. At each step in the calculation the projected radius of the nebula is determined and the ionizing photon luminosity at the inner edge of the nebula is therefore known. As in earlier models (Mathews, 1966), a dynamical pressure is found to be required to maintain the shell appearance of the nebula. The stellar wind which produces this dynamical pressure at the inner edge of the nebula ($R = R_{in}$) is proportional to $R_{in}^{-2}\,[u_0^2 - 2(GM_* g)^{1/2}]^{1/2}$. Here a simple ballistic theory is used in which the gas leaves the stellar surface with a velocity $u_0 = 1000$ km/sec. The central star of mass $M_* = 1{\cdot}0\, M_\odot$ has a surface gravity g which is calculated as a function of time from the projected radius. The dynamical pressure eventually drops to zero when $g \geqslant u_0^4/(4\,GM_*)$.

As the gas at the inner edge begins to ionize, the pressure throughout the nebula is equalized by a shock which moves outward through the neutral gas. Later, when about 1/10 of the nebular mass is ionized, a second shock is released from the ionization front, and this shock moves through the neutral shell reaching the outer edge after about 1400 years. The density in the H I gas just behind this shock is quite large ($10^6 - 10^5$ cm^{-3}), and the outward gas velocity increases from within until it reaches a maximum of 40–80 km/sec just behind the shock front. The projected appearance of the nebula during this stage has a double-ring structure similar to many observed planetaries. The most interesting development is an apparent *shrinking* of the (projected) nebula which occurs from 1200 to 1300 years, i.e. just after the nebula is completely ionized. This is caused by the large gas velocities in the outer parts of the shell which, when ionized, expand so fast into the external vacuum that an apparent

Osterbrock and O'Dell (eds.), Planetary Nebulae, 273–274. © *I.A.U.*

inward-moving rarefaction wave results. During this same period the outward expansion velocity which would be observed at the centre of the nebula would be only ~ 22 km/sec. The recent proper-motion observations of the Lillers, which show apparent still-stands or shrinkages, might well be caused by this effect. The projected radius at the instant when the nebula first becomes thin to the Lyman continuum is 0·068 parsec, in excellent agreement with the value given by Seaton and Harman. The subsequent evolution of the completely ionized nebula is very similar to the models calculated earlier (Mathews, 1966). A massive ejection followed by a stellar wind thus appears to explain many of the observed features of planetary nebulae.

Reference

Mathews, W. G. (1966) *Astrophys. J.*, **143**, 173.

DISCUSSION

Seaton: Is it possible to say what fraction of the energy absorbed from the central star goes into dynamical motion (the remainder being emitted as radiation of the nebula)?

Kahn: The fraction of the energy output of the star which goes into kinetic energy of motion is of the order of one or a few percent.

Savedoff: Is not the stellar wind introduced in your models just to maintain the central holes? It seems to me that if the condition of constant initial velocity were relaxed, one could construct other initial velocity fields which would maintain the central depression.

Williams: In a previous paper, you gave parameters for the stellar wind which were consistent with the values, $N \approx 1$ cm^{-3}, and $U_{+} \approx 550$ km/sec. In view of the sharp peak in the density distribution in the present models, would you now change these numbers?

Mathews: The velocity of mass loss cannot be determined from these models. The dynamical pressure of the stellar wind depends on both the gas density and the square of the gas velocity just before shocking against the inner edge of the planetary shell. The earlier figures which I mentioned were only representative possibilities.

EXPLANATION OF DIFFERENT SHAPES OF PLANETARIES IN TERMS OF MOVEMENT IN MAGNETIC AND GRAVITATIONAL FIELDS

E. WOYK (ELIŠKA CHVOJKOVÁ)
(Astronomical Institute, Czechoslovak Academy of Sciences, Czechoslovakia)

ABSTRACT

Many features observed in solar prominences, planetary nebulae and the terrestrial magnetosphere (e.g. the filamentary structure, suspended clouds and arcs or toroids hovering above the equator) lead to the conviction that magnetic fields play a significant role in these objects. The magnetic mechanism (Woyk, 1967) described briefly below may be able to explain the shapes of all known planetaries. The field intensity sufficient for supporting the mechanism described need not exceed about 10^{-6}–10^{-4} G.

1. Theory of Particle Paths in Magnetic and Gravitational Fields

A. SPIRALLING IN A COLLISIONLESS PLASMA

Besides the well-known magnetic-mirror level r_H, an upper reflecting level r_g can also often arise due to gravitation (see Figure 1). The level r_g is the highest altitude that a particle with a given initial path-radius and gyrofrequency can attain in a gravitational field. Thus in a collisionless plasma some particles should continually spiral up and down along the same field line, perpetually reflected between r_H and r_g. At r_H and r_g the path is perpendicular to the magnetic field H; its deviation from H is there $i=90°$. The angle i is a minimum at the level r_W (characteristic for each path) at which the particle attains the critical velocity v_W. For a perfect dipole $v_W = v_{esc}/3^{1/2}$ (v_W and the escape velocity vary with the distance r from the centre). A particle reflected just when its velocity coincides with v_W merely circles around the field line without any drift along or across it (dashed circles in Figure 1). Otherwise, below r_W the particle velocity v is higher than v_W (higher even than the local v_W corresponding to the investigated point), above r_W the particle is slower.

B. EFFECT OF ELASTIC COLLISIONS

Let us deal with a plasma cloud emerging from below the magnetic surface (i.e. from below the level where the magnetic pressure p_H exceeds the kinetic pressure,

Osterbrock and O'Dell (eds.), Planetary Nebulae, 275–278. © *I.A.U.*

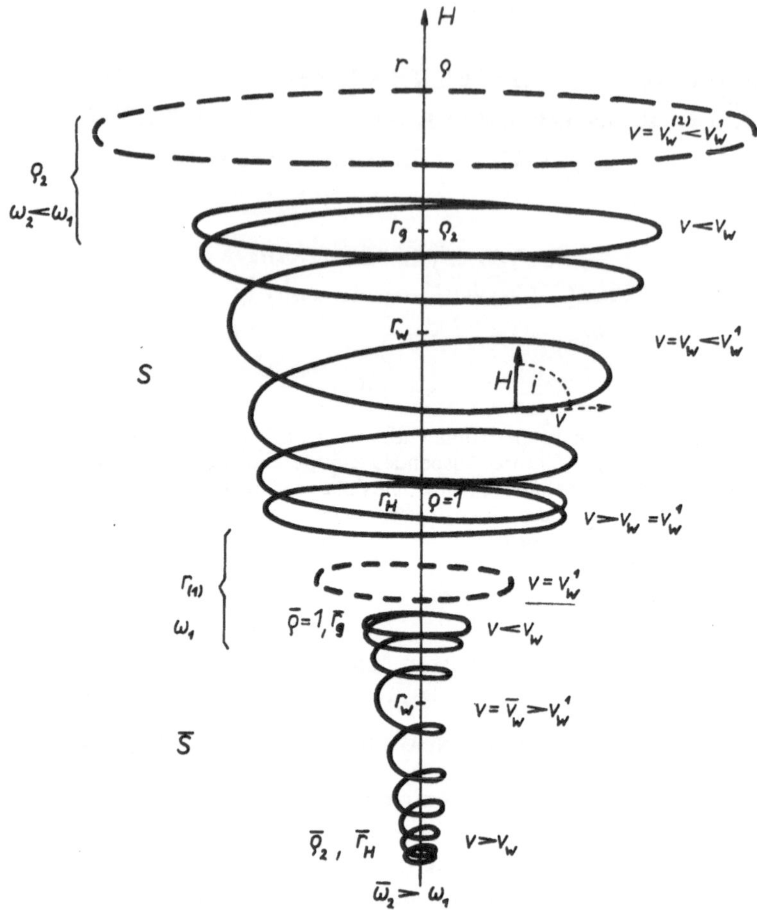

FIG. 1. *Spiralling of charged particles along magnetic-field lines in an extended gravitational field, neglecting collisions. Particles reflected with the critical velocity* v_W *can merely circle around H (dashed lines); faster particles spiral below* r_W, *slower above* r_W. *The necessary path formulae are summarized by Woyk (1967) or Chvojková (1965).*

p_{kin}, and where the gyrofrequency ω_H is higher than the collision frequency). With regard to the velocity distribution the plasma cloud should spread out along the field line and disappear below the magnetic surface again (either at the same point or at the magnetically conjugate point). At high altitudes, however, where particles slide rather than spiral along the field line (where $i \approx 0$), elastic collisions contribute to an increase of i, and therefore the mean $\bar{r}_H$ and $\bar{r}_g$ shift closer to each other. When the lower reflecting level $\bar{r}_H$ has emerged above the surface the remaining plasma can no longer escape below the surface. Apparently immobile clouds, arcs or bunches are thus created in spite of an extremely high inner particle velocity, comparable with the escape velocity. Clouds in which the mean particle velocity $\bar{v} > v_W$ tend to accumulate closer to the top of the magnetic loops, while clouds with lower mean particle velocity $\bar{v}$ tend to accumulate nearer their bottom.

2. Application to Planetary Nebulae

The shapes of Figure 2 assume that a uniform burst of plasma occurs from the whole surface. The expanding plasma tears magnetic-field lines from the central star, so that the resulting field $H = Cr^{-n}$ with $n \leqslant 3$ is far different from a magnetic-dipole

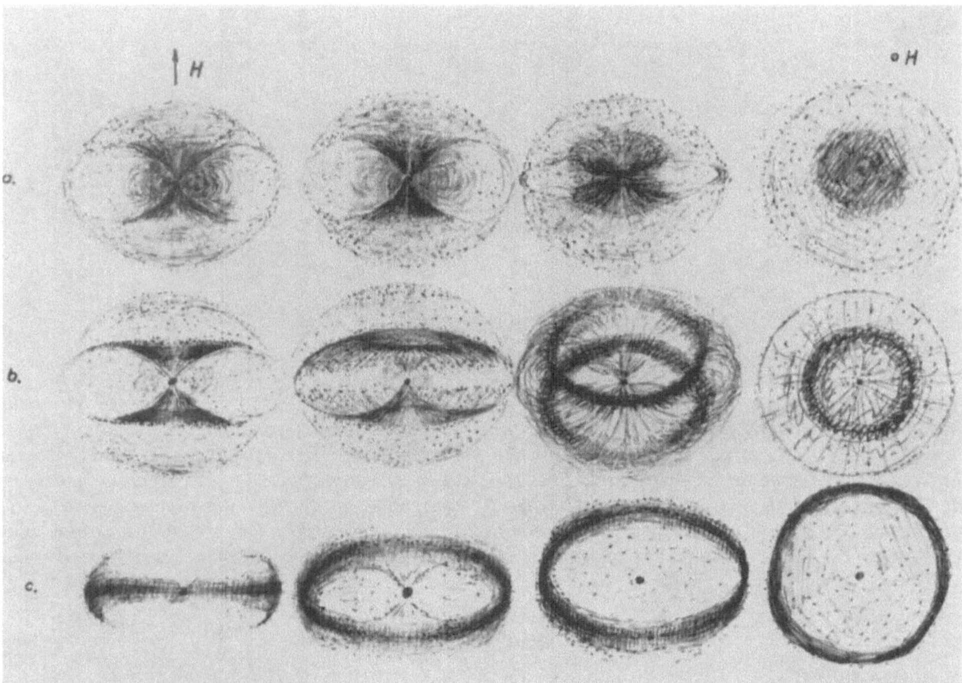

FIG. 2. *Shapes arising from uniform plasma bursts ejected from the whole stellar surface. (a) The mean particle velocity is slow (then the mean $\bar{r}_W$ is close to the bottom of the field lines, and also the plasma of the polar caps has been trapped between $\bar{r}_H$ and $\bar{r}_g$). – (c) High (over escape) velocity of spiralling particles (accumulation at the top of magnetic loops; polar caps have escaped into the interstellar space). Left: equatorial view; right: polar view.*

field. Plasma streams bursting from equatorial regions disappear at the conjugate points, producing a relatively hollow cave within the more dense outer-plasma toroid in which – according to collisions and $\bar{v}$ – a considerable remainder of the initial plasma jet becomes trapped, either at the top or at the bottom of the magnetic-field lines. (The diffusion along or across H produced by elastic collisions is very slow).

Figure 3 deals with some non-uniform effects. When the increase of density has led to $p_{\text{kin}} > p_H$, the magnetic-field system collapses. Also a very violent burst can produce the most irregular shapes with distorted broken filaments (Crab Nebula).

The upper model is not regarded as final. The aim of this contribution is only to demonstrate that the magnetic field should not be regarded as insignificant as has been done up to now.

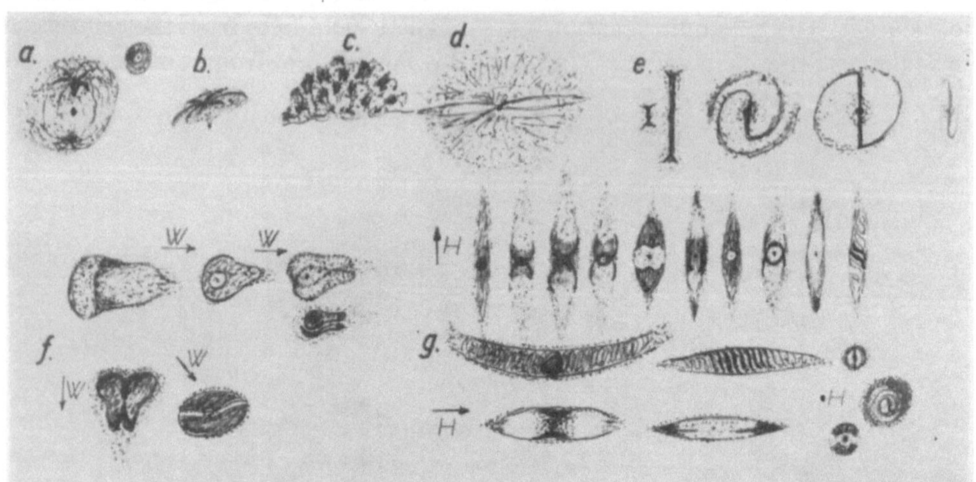

FIG. 3. *Secondary effects: (a) Ejection from preferred active centres; the velocity is high enough to attain the other hemisphere. – (b) The same effect but slower; the stream cannot reach the top of the magnetic loop. – (c) Plasma bunches arising when the mean particle velocity is v_W; $\bar{r}_H$ and $\bar{r}_g$ are now relatively close to each other. – (d) Field lines torn out by a violent burst. – (e) Streams ejected from polar regions only (the distortion can be created by precession or by the outer magnetic field). – (f) Pear- or egg-like shapes probably due to the interstellar wind observed by Oort. – (g) Cigar-like shapes created when the outermost intensity of H has decreased so much as to equal the outer, interstellar field.*

References

Chvojková, E. (1965) *Bull. Astr. Inst. Čsl.*, **16**, 63.
Chvojková, E., Klepešta, J. (1965) *Bull. Astr. Inst. Čsl.*, **16**, 70.
Woyk, E. (1967) *J. atmos. terr. Phys.*, **29**, 465.

DISCUSSION

Flower: Would you expect to see synchrotron radiation from these objects?

Woyk: Yes, I think so.

Aller: Could a field of the expected magnitude be detected from the Zeeman effect of the 109α line?

Terzian: I do not think that we can observe the Zeeman effect in the H 109α line if the magnetic fields are as low as 10^{-5} gauss.

THE ROLE OF MAGNETIC FIELDS IN THE ORIGIN AND STRUCTURE OF PLANETARY NEBULAE

DONALD H. MENZEL
(Harvard College Observatory, Smithsonian Astrophysical Observatory, U.S.A.)

That the characteristic, quasi-symmetrical structure of planetary nebulae may result from the presence of imbedded magnetic fields has undoubtedly occurred to many astronomers. Gurzadian (1962), for example, employed the widely used equation

$$p + H^2/8\pi = \text{const}, \tag{1}$$

where p is the pressure of the ionized gas and H the magnetic-field intensity. This equation specifies that the sum of the gas and magnetic pressures should be constant for a given value of the radius.

As often applied, however, Equation (1) is wrong. In particular it is wrong when the assumed magnetic field is a simple dipole field. For magnetic fields do not act directly on the gas. They act only on electric currents flowing in the medium. In a dipole, all the current is concentrated at or near the origin. To study this phenomenon, let us turn to the basic equation from which (1) results, under special circumstances.

The force equation of magnetohydrodynamics is:

$$\rho \frac{\mathrm{d}\mathbf{v}}{\mathrm{d}t} = - \nabla p - \rho \nabla V + \frac{1}{4\pi} (\nabla \times \mathbf{H}) \times \mathbf{H}, \tag{2}$$

where ρ is the density, $\mathbf{v}$ the velocity, V the scalar gravitational potential, and $\mathbf{H}$ the vector magnetic field. For a dipole field the **curl,** $(\nabla \times \mathbf{H})$, vanishes and thus no magnetic force exists, as previously noted. For electromagnetic forces to occur, the current density $\mathbf{J}$,

$$\mathbf{J} = \frac{1}{4\pi} \nabla \times \mathbf{H}, \tag{3}$$

must not vanish. The force is a vector, $\mathbf{J} \times \mathbf{H}$, perpendicular to both $\mathbf{J}$ and $\mathbf{H}$.

Consider a field with the following characteristics. From the vector potential,

$$\mathbf{A} = \mu \mathbf{e}_\theta \frac{r \sin\theta}{(a^2 + r^2)^{3/2}}, \tag{4}$$

derives the field,

$$\mathbf{H} = \nabla \times \mathbf{A} = \mu \left[\mathbf{e}_r \frac{2\cos\theta}{(a^2 + r^2)^{3/2}} + \mathbf{e}_\varphi \frac{(r^2 - 2a^2)\sin\theta}{(a^2 + r^2)^{5/2}} \right]. \tag{5}$$

Osterbrock and O'Dell (eds.), Planetary Nebulae, 279–281. © *I.A.U.*

Here, r, θ, ϕ are the usual spherical coordinates, $\mathbf{e}_r$, $\mathbf{e}_\theta$, and $\mathbf{e}_\phi$ the associated unit vectors, and μ the magnetic moment. The disposable parameter, a, determines the general shape of the field. When $a=0$, we obtain the characteristic dipole field with its singularity at the origin.

$$\nabla \times \mathbf{H} = \mu \mathbf{e}_\phi \frac{15a^2 r \sin\theta}{(a^2 + r^2)^{7/2}}. \tag{6}$$

Note that the current vanishes, as it should, for the dipole condition. For $r \gg a$, Equation (5) approximates to the standard dipole field.

Equation (2) separates, for the hydrostatic case, into two partial differential equations:

$$\frac{\partial p}{\partial r} = -\frac{GM\rho}{r^2} - \frac{15\mu^2 a^2 r (r^2 - 2a^2)\sin^2\theta}{4\pi(a^2 + r^2)^6}, \tag{7}$$

$$\frac{1}{r}\frac{\partial p}{\partial \theta} = \frac{30\mu^2 a^2 r \sin\theta\cos\theta}{4\pi(a^2 + r^2)^5}. \tag{8}$$

The latter integrates immediately to

$$p = p_0(r) + \frac{15\mu^2 a^2 r^2 \sin^2\theta}{4\pi(a^2 + r^2)^5}, \tag{9}$$

wherein $p_0(r)$ is a 'constant of integration', a function of r only. Clearly, it represents the pressure distribution along the polar axis, $\theta=0$ or π.

Differentiate (9) with respect to r, and substitute into (7). The result is:

$$\rho = \rho_0 + \frac{105\mu^2 a^2 r^5 \sin^2\theta}{4\pi GM(a^2 + r^2)^6}. \tag{10}$$

The second terms of (9) and (10) represent the increase of pressure and density produced by the magnetic field. The phenomenon is a true 'pinch effect', resulting from the equatorial current.

If magnetic fields are to contribute appreciably to the pressure and density, the second terms of these equations should at least be of the same order as the first. Take the extreme case where p_0 and ρ_0 are zero. The temperature distribution will then be independent of μ. From the equation of state, we get

$$T = pm/k\rho = GMm(a^2 + r^2)/7kr^3, \tag{11}$$

where k is Boltzmann's constant and m the mean molecular mass.

Now let $T=10^4$ and $M=10^{33}$ g. We get

$$r^3/(a^2 + r^2) = 7 \times 10^{12}\ \text{cm}. \tag{12}$$

For a planetary, we must set $r \sim 10^{17}$. Solving for a, we get the unreasonably large figure of 10^{19} cm. In other words, the electric currents responsible for the field would have to be located in a region 100 times the radius of the luminous gas, an unacceptable condition.

The difficulty arises from the low gravitational field at large distances from the central star. No equilibrium solution is possible under such conditions, whatever the field may be, unless the temperature of the gas is reduced to the order of 1 °K, when it would be neither ionized nor luminous. Introducing non-zero values for p_0 and ρ_0 does not substantially alter the problem.

This study clearly indicates that self-contained, static magnetic fields do not contribute directly to the form and structure of a planetary. Inclusion of mass motion, such as radial expansion, only serves to make matters worse, though the temperature problem no longer remains. Instead of (9), we obtain an equation something like the following:

$$\rho v^2/2 = \frac{15\mu^2 a^2 r^2 \sin^2\theta}{4\pi(a^2 + r^2)^5}, \tag{13}$$

in which the density of kinetic energy, with v the velocity of expansion, replaces the gas pressure, p. But the same problem arises as before. No satisfactory simultaneous solution of (10) and (13) exists, except for physically insignificant values of v. Although the assumed magnetic field is somewhat arbitrary, no reasonable variation is likely to represent the problem.

Hence, although the model fails for planetary nebulae, it does hold some promise for application to magnetic stars and stars with distended atmospheres. This application will be the subject of a second paper, in which I shall develop another possible relationship between such fields and the observed structure of planetaries.

In brief, the shells that comprise a planetary nebula must be considered objects totally disconnected gravitationally or magnetically from the central star. Internal magnetic fields may still contribute to nebular structure, but no static equilibrium solution exists. For example, magnetic fields could have dictated the shell-like or ring-like structure when the nebula was very small, perhaps part of the atmosphere of the central star. After ejection, the slowly expanding shell has retained some semblance of its original form.

Reference

Gurzadian, G. A. (1962) The Magnetic Fields in Planetary Nebulae, in *Vistas in Astronomy*, 5, Ed. by A. Beer, Pergamon Press, New York, p. 40.

DISCUSSION

Khromov: May I ask you about the origin of the postulated magnetic field?

Menzel: If you can tell me what causes the electric currents in sunspots I can answer that question. I have postulated the existence of currents, which must occur, not just in the central star, but throughout the volume of the nebula.

THE OBSERVATION OF MONOCHROMATIC INTENSITY DISTRIBUTIONS IN PLANETARY NEBULAE BY MEANS OF ELECTRONOGRAPHY*

MERLE F. WALKER
(Lick Observatory, University of California, U.S.A.)

and

GERALD E. KRON
(U.S. Naval Observatory, U.S.A.)

ABSTRACT

The linear response and high information gain of the electronographic process make it an important new technique for the observation of monochromatic intensity distributions in planetary nebulae. Preliminary results for NGC 6720, obtained with the U.S. Navy electronic camera and the 61-inch astrometric reflector, are presented.

Work in recent years (Frieser and Klein, 1958; Frieser *et al.*, 1959; Vernier, 1959; Méallet, 1961; Duchesne and Méallet, 1962; Valentine, 1966) has demonstrated that a linear relationship exists in electronography between the number of incident electrons and the photographic specular density in the silver-halide recording emulsion, to very high values of the density (Kron and Papiashvili, 1967; Kron, 1967; Walker, 1967). Emulsions exposed in an electronographic-type image-converter (electronic camera) will thus have a photographic density proportional to the intensity of the light incident on the photocathode over a large range of intensities. In addition, the fine grain and high quantum efficiency of the electronographic process permits the recording of the same information in $\leqslant 1/15$ the time required photographically, depending upon the wavelength (Walker, 1966*b*).

Electronic cameras have been used in astronomical spectrophotometry (see e.g. Walker and Popper, 1964; Aller and Walker, 1965; Walker, 1966*a*), in measuring the surface brightnesses of galaxies (Kron, 1967) and globular clusters (Kron and Papiashvili, 1967), and it has been shown that they can be used for the determination of stellar magnitudes (Walker and Kron, 1967). However, the small sizes of the photocathodes of presently available electronic cameras restrict their application to studies of objects or fields of relatively small extent.

One problem for which present electronic cameras are well suited is that of studying monochromatic intensity distributions in planetary nebulae. Using the electronic camera, direct-intensity pictures can be made of the nebulae in the light of particular emission lines by observing through narrow-band interference filters. Equal-interval

* Contributions from the Lick Observatory, No. 269.

iso-intensity contour maps of the distribution of intensity or surface-brightness in the nebulae may then be made from these pictures by scanning them in a Joyce, Loebl recording microdensitometer, operated as an isodensitracer. This microdensitometer has been carefully designed to record specular densities up to a photographic density of 6. The action of the densitometer depends ultimately on matching the plate transmission with that of an optical wedge. The wedges supplied by the manufacturer have a linear variation in density with distance along their length to an accuracy of $\pm 1\%$. Contours corresponding to (equal) density increments on the plate as small as 0·005 may be traced. Contours of successively higher density (or intensity) are indicated by the three different modes of operation of the recording pen, in the following sequence: blank, dash, dot.

A preliminary study of this technique has been made using the U.S. Navy Electronic Camera, developed by one of us (Kron), attached to the focus of the 61-inch astrometric reflector of the U.S. Naval Observatory Flagstaff Station. The scale in the focal plane of this telescope is 13″.5/mm, which is reduced electronically to 21″.1/mm on the recording emulsion in the electronic camera. The telescope is geometrically $f/10$, but owing to the large size of the flat secondary mirror, operates optically at $f/12$. Tests indicate that despite the slow optical speed of the telescope, images of planetary nebulae of moderate surface brightness can be obtained in the light of the stronger emission lines with reasonable exposure times. Thus, using a telescope having an optical speed of about $f/5$, it should be possible to obtain direct-intensity pictures of a number of planetaries in the light of relatively faint high- and low-excitation lines.

Figure 1 reproduces the electronographic picture obtained in the light of λ4861 Hβ, while other pictures were also obtained in λ3727–29 [O II] and λ5007 [O III]. The scale of the reproduction is 1″.1/mm, and the exposure times and filter data for all the pictures are given in Table 1.

Table 1

Filter and exposure data for electronographic observations of NGC 6720

λ	Filter		Exposure Time
	Half-Intensity Width (Å)	Peak Transmission (%)	
3727–29	40	42	$1^h 00^m$
4861	19	56	1 31
5007	21	66	0 30

The sensitivity of the (Sb-Cs) cathode of the electronic camera was about 20 μa/lu, and the images were recorded on Ilford K5 nuclear research emulsion. The plates were developed 5 min in D-19 at 65 °F. The slightly granular appearance of the pictures results from the fact that the image tube used was an experimental one

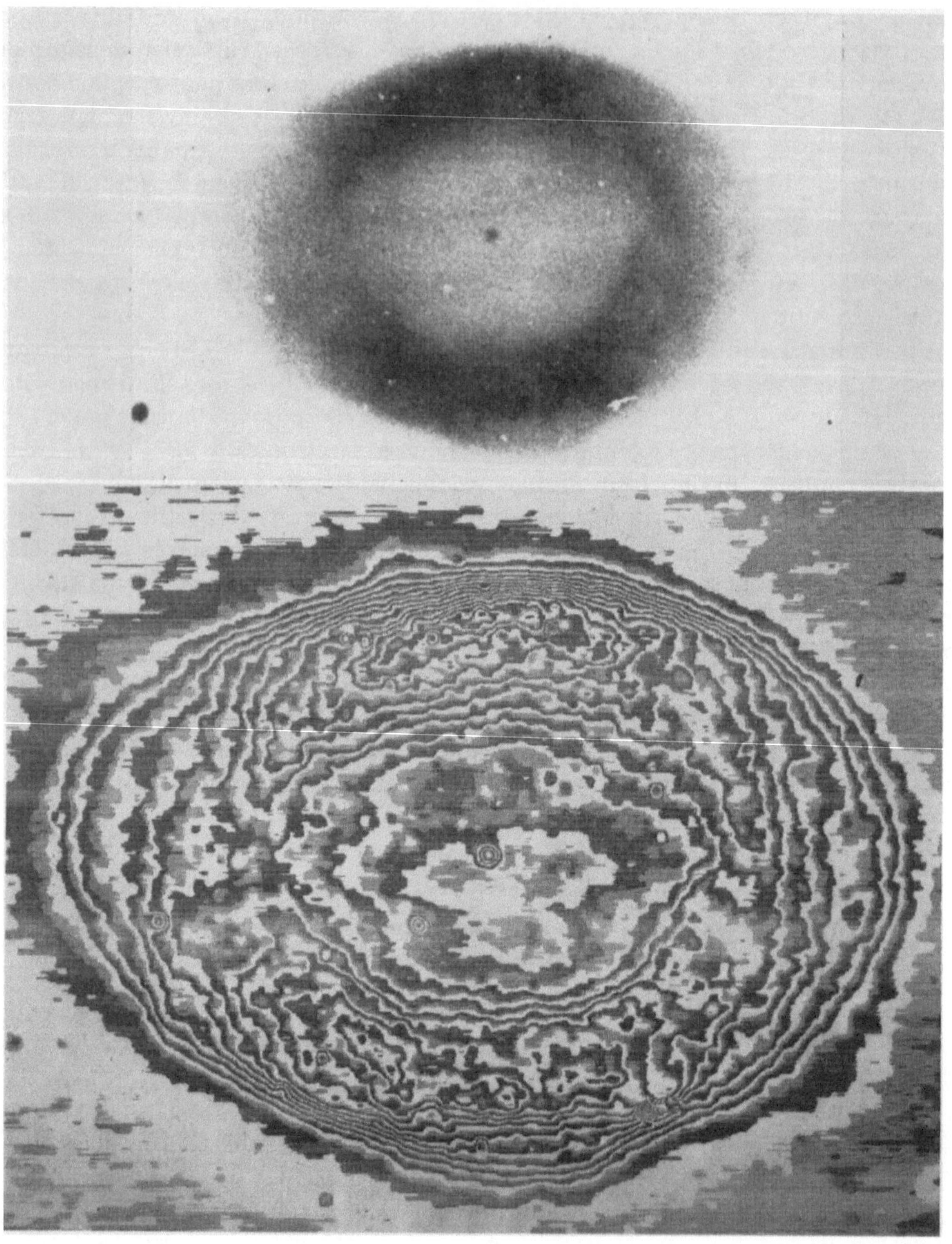

FIG. 1. *Upper: Electronographic picture of NGC 6720 in the light of λ 4861 Hβ; lower: Intensity distribution of NGC 6720 in the light of λ 4861 Hβ.*

with a ground cathode surface to eliminate the effects of light reflected to the photocathode from the elements of the electron-optical system. The resolution of the pictures, however, is set by the seeing at the times of observation; the best seeing occurred during the exposure on $\lambda 4861$, where the image of the central star has a diameter of $1''.8$.

The lower part of Figure 1 shows the isodensity or iso-intensity tracing obtained from the picture taken in Hβ above. In this tracing the size of the scanning aperture, projected on the emulsion, was $50\ \mu \times 50\ \mu$, the interval between successive scans was $5\ \mu$, and the density increment between successive scans was 0·02. The enlargement of the contour map over the original plate is 27·8 times. Comparing these iso-intensity contours with those published by Aller (1956), the large increase in information recorded by the electronographic method is apparent. It is to be noted that these particular Ilford plates contained a number of defects. These are readily identifiable on the contour maps by comparing these with the original pictures. Eventually, the two-dimensional maps of a given nebula in different emission lines might be put on a homogeneous intensity system for that nebula by means of the electronographic spectrophotometry of the planetary nebulae recently carried out by Aller and Walker (1965, 1967), which gives the relative intensities of the emission lines at a few particular points in the nebula. Such material could then be used for two-dimensional spectrophotometric investigations such as the study of the detailed ionization structure of the nebula.

The present results indicate that the application of electronography to the photometric study of the planetary nebulae constitutes an important new observational technique. It is also clear that in order to extract the maximum amount of information by this method, the electronographic pictures must be taken under conditions of excellent seeing, and with a telescope giving sufficient scale that the definition of the pictures is always seeing-limited.

Acknowledgements

It is a pleasure to thank Mr. K. Janes for his assistance in making both the observations and the contour maps of NGC 6720. Travel support for one of us (Walker) to visit Flagstaff was provided by a grant from the National Science Foundation.

References

Aller, L.H. (1956) *Gaseous Nebulae*, Chapman and Hall, London, p. 247.
Aller, L.H., Walker, M.F. (1965) *Astrophys. J.*, **141**, 1318.
Aller, L.H., Walker, M.F. (1967) in preparation.
Duchesne, M., Méallet, M. (1962) *C.r. Soc. sav. Paris*, **254**, 1400.
Frieser, H., Klein, E. (1958) *Z. angew. Phys.*, **10**, 337.
Frieser, H., Klein, E., Zeitler, E. (1959) *Z. angew. Phys.*, **11**, 190.
Kron, G.E. (1967) unpublished.

Kron, G.E., Papiashvili, I.I. (1967) *Publ. astr. Soc. Pacific*, **79**, 9.
Méallet, M. (1961) Mémoire présenté à la Faculté des Sciences de l'Université de Paris pour obtenir le Diplôme d'Études Supérieures de Sciences Physiques.
Valentine, R. (1966) *Advances in Microscopy*, **1**, 180.
Vernier, P. (1959) *Bull. astr., Paris*, **22**, 83.
Walker, M.F. (1966*a*) in *Stellar Evolution*, Ed. by R.F. Stein and A.G.W. Cameron, Plenum Press, New York, p. 405.
Walker, M.F. (1966*b*) *Advances Electronics, Electron Phys.* **22B**, 761.
Walker, M.F. (1967) unpublished.
Walker, M.F., Kron, G.E. (1967) *Publ. astr. Soc. Pacific,* **79**, 551.
Walker, M.F., Popper, D.M. (1964) *Astrophys. J.*, **139**, 168.

ON THE ORIENTATION OF THE MAJOR AXES OF PLANETARY NEBULAE

V. P. GRININ and A. M. ZVEREVA
(Crimean Astrophyscial Observatory, U.S.S.R.)

The orientation of the major axes of planetary nebulae was investigated and the results are given in Figure 1. The solid line gives the distribution of 132 planetary nebulae as a function of the angle $\Delta\phi$ between the major axis of the nebula and the galactic equator. It can be seen that the distribution is not uniform, and is not symmetrical about $\Delta\phi = 0$. The dashed line gives the distribution of the same nebulae as a function of the angle between major axis of the nebula and the direction of the interstellar magnetic field, which is determined from the stellar polarization measurements of stars close to the individual nebulae. This distribution is less uniform than the first one but is symmetrical about $\Delta\phi = 0$. These results lead to the conclusion that the orientation of planetary nebulae is partly determined by the interstellar magnetic field.

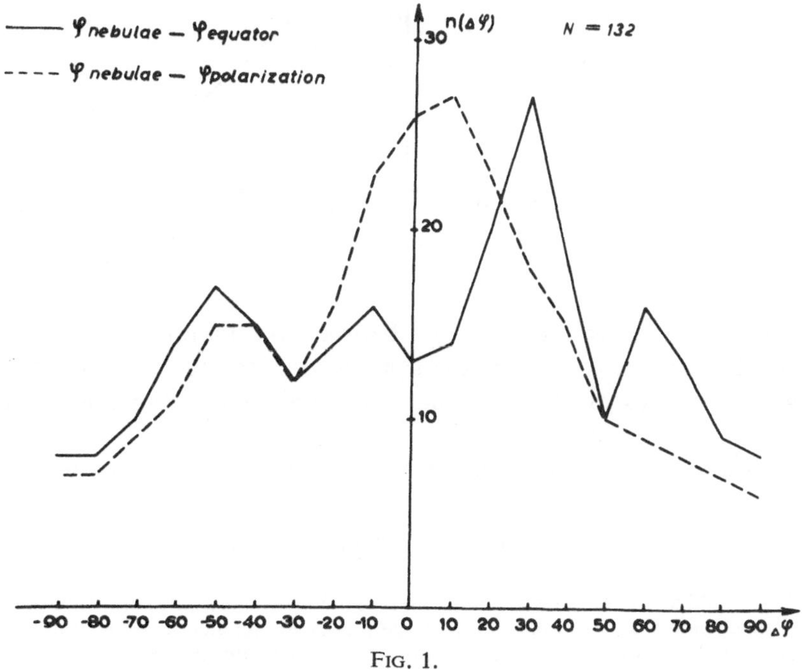

FIG. 1.

Osterbrock and O'Dell (eds.), Planetary Nebulae, 287–289. © *I.A.U.*

In order to check this conclusion, the planetary nebulae were divided into two groups, with large and small galactic latitude ($b \gtrless 10°$). Then the above procedure was repeated, with the results given in Figure 2. It can be seen that the effect is stronger for nebulae with $b < 10°$, and is almost absent for nebulae which are far from the galactic equator. The conclusion that a correlation exists between the orientation of planetary nebulae and the interstellar magnetic field is thus confirmed.

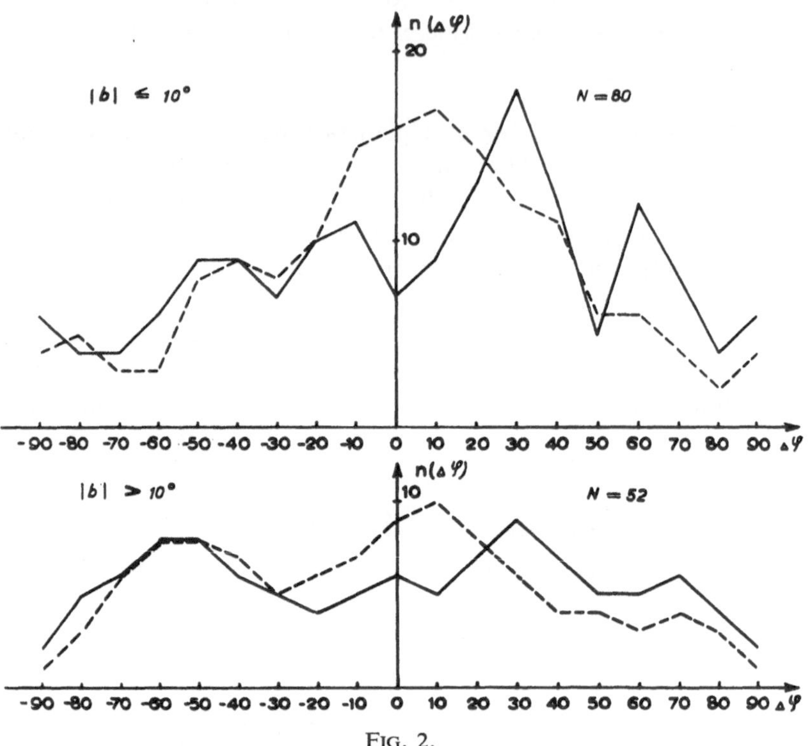

FIG. 2.

It is noted that the mean angle $\langle \Delta\phi \rangle$ between the major axis of the planetary nebulae and the galactic equator is the same as was found by Shain for the mean angle between the axes of elongated dark and emission nebulae and the galactic equator.

This contribution is to be published in full in *Astrophysica* (U.S.S.R.).

DISCUSSION

Gershberg: I should like to emphasize that Grinin and Zvereva's results show that there is no way to avoid the conclusion that the expansion of large planetary nebulae is influenced by the galactic magnetic field. It is known that the same phenomenon exists in diffuse nebulae. In diffuse nebulae, however, the density is lower than in planetary nebulae. However, the similarity in orientation implies that the ratio of magnetic pressure to gas pressure must be about the same in the two types of nebulae. Therefore the magnetic fields near planetary nebulae must be larger than the mean galactic field by a factor 10–30. It seems that such an intensified magnetic field can result from the compression of the galactic field by an expanding nebula.

GENERAL DISCUSSION – FOURTH SESSION

Evans: It is essential to have a practical definition for a planetary nebula. Many brighter cases are disputed. The statistics of fainter objects (for which the possible observations are very limited) will be severely affected by the definition adopted. The practical definition now in use seems to be that a planetary nebula is an object which occurs in a catalogue of planetary nebulae.

Minkowski: In judging whether an object should be classed as a planetary nebula, the internal velocities should be taken into account. Unlike the planetary nebulae which show maximum velocities of expansion at their centres, NGC 6302 has zero velocity of expansion at the centre, and the North preceding loop seems to have been ejected with maximum velocity in the direction of its axis (Minkowski and Johnson, *Astrophys. J.*, **148**, 1967, 659).

As to the question of how to define a planetary nebula, there is no better way than to accept any object in a catalogue of planetary nebulae if nobody has serious objections.

Mathews: In view of the fact that stellar winds appear to be necessary in dynamical models of the nebula, it would seem that the appearance of NGC 6302 in the sky and its velocity field could both result from a stellar wind accelerating a model similar to that suggested by Khromov. If the initial ejection of nebular mass contained more mass in the equatorial region of a spherical shell, then the less massive polar regions would be more easily accelerated outward, expanding faster and reaching greater distances. Magnetic fields are therefore not needed to explain these apparent loops. Other nebulae such as NGC 6537, NGC 2346, NGC 650-1, and CD $-29°$ 13998 are similar to NGC 6302 while nebulae like NGC 40 may represent a less advanced stage of this process. It would seem wrong, therefore, to suggest that NGC 6302 is uniquely unlike any other planetary and that it should be put into a class of its own.

Underhill: Arguments that planetary nebulae are the result of strong stellar winds are attractive. However, even if these winds do exist, they do not force us to conclude that all the central stars of planetary nebulae (odd-shaped gaseous bodies with nebular spectra) are at late evolutionary stages. Some Wolf-Rayet stars and some O stars are known to have stellar winds and to be surrounded by gaseous nebulae. These objects clearly cannot be placed at a very late stage of evolution. In other words, the characteristics of the surrounding nebula are not sufficient to determine the basic characteristic of the exciting star needed for stellar evolution studies.

Mathews: For mass-loss, the age of the star is not as important as the temperature and surface gravity. The observations of mass-loss by Morton refer, of course, to

Osterbrock and O'Dell (eds.), Planetary Nebulae, 290–293. © *I.A.U.*

supergiants, but the same mechanism may produce mass-loss in planetary-nebula nuclei.

Abell: If we attempt to derive evolutionary tracks for central stars, and do so by comparing different nebulae, which are presumably in different stages of evolution, it is important to consider objects for which the stars, before ejection, all had nearly the same mass. This condition is satisfied by the majority of planetaries, which have a disk distribution, and thus probably have been ejected from stars of about 1·0 or 1·2 solar masses. But we also class as 'planetaries' objects surrounding, e.g., Wolf-Rayet stars, which quite probably have a history different from that of the more common disk-population planetaries. Ideally, I think a definition of a planetary nebula should include a specification of the kind of star that produced it.

Rose: Most planetary nebulae are observed to have low velocities of expansion (10–100 km/sec). Perhaps the presence of low expansion velocities should be used in deciding if a peculiar object is a planetary nebula. In general, the expansion velocities for gas surrounding stars that are blue and not planetary nebulae (e.g. Wolf-Rayet stars) show considerably higher velocities of expansion.

Capriotti: Among the low surface-brightness planetary nebulae discovered by Abell, how many appear to be spherically symmetric?

Abell: About 56 out of 86 seem to be regular.

Capriotti: The implication has been made several times in this discussion that all planetaries evolve toward a more chaotic state, or at least toward a state of axial symmetry. However, from the Abell survey it appears that at least some of the planetary nebulae remain spherically symmetric, or evolve toward a state of spherical symmetry.

Abell: Most of the 86 planetaries that were found on the Palomar Sky Survey appear as fairly homogeneous disks or rings, but a large minority show irregularities, or are even highly chaotic. I agree with you that the morphological appearance should probably not be considered as fundamental in the definition of planetary nebulae, since it may merely represent different evolutionary stages, or perhaps only minor differences in the original density distributions within the nebulae. Far more fundamental is the kind (i.e. mass) of the star that produced a nebula, but of course this is not easy to know. Nevertheless, nebulae surrounding Wolf-Rayet stars should certainly not be included with other planetaries in studying the evolution of the involved stars.

Khromov: I would like to make a comment on the study of the structure of planetary nebulae. Due to stratification effects, the photographic images of a given planetary nebula can be very different in different spectral lines. If the images resulting from the light from several different ions in a planetary nebula are intermixed, due to the properties of the filter-plate combination that is used, the resulting structure of the nebula may appear highly complicated.

So, in studying the morphology, one should be sure to treat his observational material with a proper critical spirit.

Menzel: I have the impression that every star might like to be a planetary nebula, but most of them simply do not make the grade. Our Sun, for example, is trying very hard, but it has succeeded in growing only feeble prominences, an inconspicuous corona, and an even less conspicuous solar wind. And yet, as I have pointed out elsewhere, our corona possesses many excitation features resembling those of planetaries: forbidden lines, filamentary structure, stratification, and a form suggestive of magnetic fields.

In my opinion only two main types of gaseous nebulae exist: diffuse and planetary. The former owe their luminosity to stars accidentally and temporarily wandering through them. The latter are closely associated generically with a central star. Such a star probably exists (or has existed) even when we cannot see it. The nebula owes its existence to the star. Therefore I think we should be generous in our definition of planetary nebulae. We should not tie it to some theoretical idea of the nature of the nebula or its nuclear star. If a star has an extended luminous envelope and if it appears in a catalogue of planetary nebulae, that is good enough for me.

Woltjer: I wish to comment briefly on what can be expected of magnetic fields in planetaries. For a field to be dynamically important it is first of all required that the magnetic energy be comparable to the kinetic and thermal energies. This implies a field of about 5×10^{-4} *G*. As to the nature of the field, it can be tied to the central star, it can have arisen from a stellar field by expansion, or it can have been produced by a field stretching mechanism, as in the solar wind. The latter can only give a field of small scale.

When we consider a field-containing region (with inner radius R) anchored in or close to the central star (mass M), we can show (Woltjer, L., *Astrophys. J.*, **148**, 1967, 291) that for a steady system we must have the magnetic energy $E_{\mathrm{mag}} < \zeta GM^2/R$, with ζ a factor close to unity, provided the gravitational potential energy in the region outside R is negligible and provided no pressure is applied from the outside. From this we find that a field that originates primarily in or close to the central star cannot have a strength of more than 3×10^{-6} *G* in a typical planetary shell. The same result can be demonstrated for a field resulting from the expansion of a stellar field. Our conclusion thus is that magnetic fields in planetaries, if present, are mostly of small scale. In this case they cannot significantly influence the large scale structure of the planetaries, which more probably is to be understood in hydrodynamic terms. Of course we do not exclude the idea that the ejection of matter from the star can be influenced by a stellar field.

In the outermost tenuous parts, the interstellar magnetic field can perhaps be of some importance. However, no large effects are expected because this field will not be compressed strongly around the planetary, since the expansion velocities are at most mildly supersonic with respect to the interstellar Alfvén velocity, in all but a few cases.

Finally we note that if a small scale field of 5×10^{-4} *G* were present the diffusion of galactic high-energy electrons would lead to a non-thermal radio source of less than

0·01 flux units at 400 Mc/sec. Thus the eventual detection of non-thermal radio emission from planetaries would imply the local production of electrons with energy of several hundred MeV.

Gurzadian: I wish to remind you that all observational data show that at the present time it is impossible to build a theory of the form of planetary nebulae without taking into account the role of magnetic fields, perhaps of the dipole type, connected with the shape of the nebulae.

Session V

CENTRAL STARS

THE ATMOSPHERES AND SPECTRA OF CENTRAL STARS

K. H. BÖHM*
(University of Heidelberg, Germany)

ABSTRACT

After a brief discussion of the determination of the effective temperature, the surface gravity and the chemical composition of the central stars, we describe the typical problems arising in the computation of static and dynamical models of the very hot atmospheres of these objects. Special attention is paid (1) to the strong non-greyness of the atmosphere, (2) the electron-scattering contribution to the source function, (3) the contribution of the higher ions of C, N, O and Ne to the absorption coefficient, (4) the importance of radiation-pressure effects. The application of the computed energy distribution to the determination of the ionization stratification of the nebula and to the calculation of the bolometric correction is discussed.

A formation of emission edges due to the Schuster effect does not seem to occur in any of the static *non-grey* model atmospheres which have been computed so far. The reason is that the steep temperature gradient in the uppermost layers of these models strongly favors the formation of absorption edges. Using very simple dynamical models by Schmid-Burgk (1967) we discuss the possibility of a considerable flattening of this temperature gradient by a hydrodynamic outflow driven by radiative acceleration. We argue that the Schuster effect may be much more important, if these atmospheres are not in hydrostatic equilibrium.

The position of central stars of different spectral type in the T_{eff}-g plane is discussed.

1. Introduction

A study of the atmospheres of central stars is of great interest for the following reasons:

(1) It enables us to calculate the energy distribution of the true surface fluxes of these stars, thus permitting:

(a) a theoretical determination of the ionization stratification of the nebulae,

(b) the definition of an improved scale of Zanstra temperatures,

(c) the computation of improved bolometric corrections.

(2) It must finally offer a possibility to understand the rather unusual line spectra (e.g. of type WR, Of and NGC 246, see below) of many of these objects, thus permitting a direct determination of chemical abundances.

(3) We have an opportunity to study the physics of very hot stellar atmospheres in the range $4\times10^4\,°K \lesssim T_{eff} \lesssim 2\times10^5\,°K$.

Let us assume that the structure of the atmosphere of a central star is determined by the effective temperature T_{eff}, the surface gravity g and the chemical composition

* Now at the University of Washington, Seattle, Wash.

Osterbrock and O'Dell (eds.), Planetary Nebulae, 297–316. © *I.A.U.*

X_i of the atmosphere. This excludes (tentatively) the possibility of magnetic fields or rotation playing an important role in determining the structure of these atmospheres, but it does not exclude the possible instability of the atmosphere due to a strong 'radiative acceleration' (see below).

It is a very fortunate circumstance that in the case of central stars T_{eff} and g can be determined without understanding the rather complex line spectra of these objects. In this respect the study of central star atmospheres is even simpler than the investigation of other stellar atmospheres. Consequently we shall discuss in this paper:

(1) the determination of T_{eff} and g (without using line spectra),

(2) the computation of static and dynamic model atmospheres for these stars, the calculation of the frequency distribution of the emergent fluxes and application to the determination of the ionization structure of the nebula, and only finally

(3) we shall briefly consider the spectra of central stars (which are not yet understood in detail).

2. The Determination of T_{eff}, g and the Chemical Composition

The knowledge of the Zanstra temperatures and the luminosity L (determined either by the $R(M)$- or the $R(N_e)$-method, see below) together with a fairly crude estimate of the masses of the central stars may be used to determine T_{eff} and g in the following way. Tentatively, T_{eff} may be identified with the Zanstra temperature*, g may be determined using the well-known relations:

$$L = 4\pi R_s^2 \sigma T_{\text{eff}}^4 \tag{1}$$

$$g = \frac{GM_s}{R_s^2} \tag{2}$$

where L and the stellar mass M_s are assumed to be known. R_s is the stellar radius, G the gravitational constant and σ the Stefan-Boltzmann constant.

Consequently the necessary empirical data are the Zanstra temperature, the luminosity and the mass of the central star. Let us discuss briefly the determination of these data:

(A) It is well known that in Zanstra's method one counts the number of photons beyond the Lyman limit (of HI, HeI or HeII) by measuring the number of photons in one or several recombination lines of these ions and compares this to the number of photons in the visual region. It is very important that the Zanstra temperature be identified with the effective temperature only for a nebula which is optically thick in the Lyman continuum – otherwise the Zanstra temperature gives only a lower limit

* Strictly speaking a new scale of Zanstra temperatures based on the wavelength distribution of the emergent flux for model atmospheres should be used. Such a Zanstra-temperature scale has been determined by Capriotti and Kovach (1968), using the model atmospheres by Böhm and Deinzer (1966).

for T_{eff}. Since we are only interested in definite values for T_{eff}, some criteria for the optical thickness of the nebula have to be applied. Such criteria have been worked out by Harman and Seaton (1966). They are based on comparison of the recombination lines He II 4686 and Hβ (optical thickness in He II), He I 4471 and Hβ (optical thickness in He I, provided He II is absent) and on the presence of O I lines (optical thickness in H I). The importance of using the He II lines for a study of the optical thickness of a nebula has been emphasized earlier, especially by Wurm and Singer (1952). Harman and Seaton (1966) have given a list of 42 central stars, for which the nebula is optically thick in at least one Lyman continuum and for which, consequently, fairly reliable effective temperatures can be determined.

(B) The determination of the luminosity of central stars requires a knowledge of the distance of the planetary, the apparent brightness of the central star and the bolometric correction for such a high-temperature object. The distance is equal to the ratio (R/ϕ) of the linear and the angular radius. Two methods have been used for the determination of R:

(1) Often one uses the so-called Shklovsky method (see Minkowski and Aller, 1954; Shklovsky, 1956), in which it is assumed that all planetary nebulae have approximately the same mass. The method makes use of the relation between the surface brightness $S_{H\beta}$ in Hβ and the electron density N_e

$$S_{H\beta} = \frac{\varepsilon}{3} R N_e^2 h\nu_{H\beta} \alpha_{H\beta} . \tag{3}$$

ε is the filling factor, R the radius of the nebula, $\alpha_{H\beta}$ the recombination coefficient for hydrogen leading to the emission of Hβ. In addition one applies the obvious relation between the ionized mass of the nebula and N_e:

$$M = \frac{4\pi}{3} R^3 \varepsilon N_e m_H . \tag{4}$$

If M is assumed to be known and $S_{H\beta}$ has been measured, R can be determined from (3) and (4). For obvious reasons this procedure has been called the $R(M)$-method by Seaton (1966). Osterbrock (1960), O'Dell (1962), Khromov (1962) and Seaton (1966) have tried to check the assumptions of the $R(M)$-method. The assumption of constant mass for all nebulae seems to be justified as a crude approximation. Obviously the mass given by Equation (4) is only identical with the total nebular mass for an optically thin nebula, so that this method can be applied only in this case.

Seaton (1966) has developed an approximate theory of $R(M)$ in the optically thick case. He finds the following relation between $R(M)$ and the actual radius R of an optically thick nebula

$$\frac{R(M)}{R_0} = \left(\frac{R}{R_0}\right)^{2/5} \left(\frac{L_0}{L}\right)^{1/5} , \tag{5}$$

showing that $R(M)$ increases monotonically with increasing R, but decreases with increasing L of the central star. Since in the early stages of a star following the Harman-Seaton sequence the luminosity of the central star is rather low, the $L^{-1/5}$-dependence overcompensates the $R^{2/5}$-dependence of $R(M)$. This leads to an explanation of the fact that there exists a minimum of the observed $R(M)$-values as found by O'Dell (1963). This possibility gives us additional confidence in the applicability of the $R(M)$-method.

(2) The second method is the so-called $R(N_e)$-method (Harman and Seaton, 1964; Seaton, 1966). In this procedure one determines N_e from the ratio of forbidden-line intensities (e.g. from the ratio [OII] 3726/[OII] 3729), and then uses Equation (3). The disadvantage of this method is its strong dependence on an exact knowledge of the filling factor and the details of the density distribution within the nebula. This method has been applied by Seaton (1966) to those nebulae which are optically thick in hydrogen and to which consequently the $R(M)$-method cannot be applied.

(C) The mass can be estimated only in a rather crude way. We may follow the argument by O'Dell (1963) and Osterbrock (1966) that the distribution of the planetary nebulae vertical to the galactic plane indicates $M \approx 1{\cdot}2\ M_\odot$ for the original star (according to the distributions for stars of different mass given by Schmidt, 1963). This original mass of $1{\cdot}2\ M_\odot$ is compatible with our present ideas about the evolution of these objects. Furthermore, if the nebula contains $0{\cdot}6\ M_\odot$ (as suggested by Seaton, 1966), we get a mass of $0{\cdot}6\ M_\odot$ for the central star in satisfactory agreement with the masses of normal D A white dwarfs (Weidemann, 1963).

Using the methods described under (A) and (B), $T_{\rm eff}$ and L can be determined and the HR-diagram can be drawn. This can be most easily done using nebulae which are optically thick in HeII (thus permitting the derivation of a reliable HeII Zanstra temperature) and optically thin in HI (so that the $R(M)$-method can be applied). The main part of the Harman-Seaton sequence (see especially Seaton, 1966) is defined by these objects. Only the L of the very early (low-luminosity) part of the Harman-Seaton sequence has to be determined by the $R(N_e)$-method. O'Dell (1968) has pointed out that the results for the Magellanic Clouds (by Miss Webster) indicate that the luminosities for the early phases of the Harman-Seaton sequence might be larger than the values derived by Seaton (1966). From our point of view it is also important to note that the central stars of the nebulae which are optically thick in HeII and HI seem to have considerably higher $T_{\rm eff}$ (up to $2{\cdot} \times 10^5$ °K) than the maximum value reached by the Harman-Seaton sequence ($T_{\rm eff} \approx 1{\cdot} \times 10^5$ °K).

Having available the HR-diagram of the central stars and the mass estimates described in paragraph (C), we can calculate the position of the central stars in the $T_{\rm eff}$-g diagram, which is an extremely useful starting-point for model-atmosphere calculations. Such a diagram for the Harman-Seaton sequence and for the original sequence of O'Dell (1963) is shown in Figure 1. We have also drawn the curve (a straight line in the doubly logarithmic diagram) which separates the region permitting hydrostatic solutions from the region in which such solutions are impossible (because the radiative acceleration g_{rad} is larger than the surface gravity g). This curve is defined by the relation

$$g = \frac{\sigma T_{\text{eff}}^4}{c} \sigma_{\text{el}}, \tag{6}$$

where σ_{el} is the cross-section for Thomson scattering. The simple relation (6) is only valid if Thomson scattering is the dominant absorption mechanism close to the instability limit. This seems to be the case (Böhm and Deinzer, 1966).

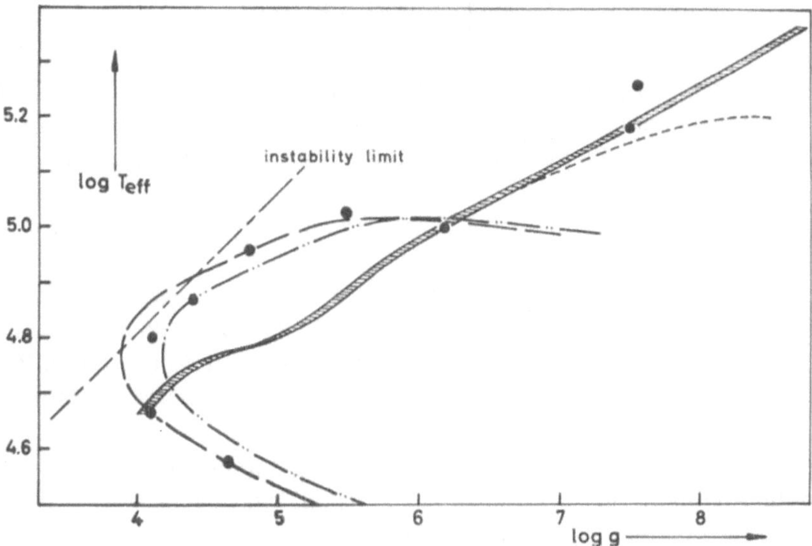

FIG. 1. *The schematic Harman-Seaton sequence for $M = 1\ M_\odot$ (–. .–) and $M = 0{\cdot}5\ M_\odot$ (– –) and the (old) O'Dell sequence (widened line) in the $\log g$-$\log T_{\text{eff}}$-plane. The dots correspond to the position of computed non-grey model atmospheres. The instability limit is defined by Equation (6).*

Finally we have to consider the determination of chemical abundances. Since there is little hope of gaining direct information without first understanding the structure of these atmospheres, the best possible working hypothesis is to assume that the central star atmospheres have the same chemical composition as the nebulae. We have used the nebular chemical composition given by Aller (1965).

3. Model Atmospheres and Energy Distribution of Surface Fluxes

Having T_{eff}, g and the chemical composition available, we can now proceed to the computation of model atmospheres.

It is useful to first give a brief list of the typical difficulties encountered in the calculation of very hot stellar atmospheres:

(1) Because of the strong absorption edges of He II and the higher ions of C, N, O and Ne these atmospheres are strongly *non-grey*.

(2) Especially in the intermediate part of the Harman-Seaton sequence, the electron-scattering contribution to the source function becomes rather large. This fact leads to a considerable increase in the complexity of radiative equilibrium computations.

(3) The 'radiative acceleration' has to be taken into account in the calculation of the hydrostatic equilibrium, i.e. the surface gravity g has to be replaced by

$$g - \frac{\pi}{c} \int_0^\infty (\kappa_\nu + \sigma_{el}) F_\nu \, d\nu, \tag{7}$$

where F_ν is the monochromatic radiative flux*. Since the middle part of the Harman-Seaton sequence lies either very close to or actually crosses the instability line (see Figure 1), we have to keep in mind the possibility that (7) may be negative in large parts of the atmosphere and that therefore no hydrostatic solution is possible.

(4) The radiative equilibrium is probably influenced considerably by strong lines in the far ultraviolet.

(5) Even the 'classical' model atmospheres in this T_{eff}-g-range are rather extended. We find e.g. for a non-grey atmosphere of $T_{eff} = 7{\cdot}4 \times 10^4\,°K$ and $g = 2{\cdot}51 \times 10^4$ cm sec^{-2} a thickness of $1{\cdot}2 \times 10^{10}$ cm. This has to be compared to the radius of 7×10^{10} cm for such a star. It is obvious that treating such an atmosphere as plane-parallel is only a crude approximation.

Model atmospheres of central stars have been computed by Gebbie and Seaton (1963), Gebbie (1967), Böhm and Deinzer (1965, 1966) and Böhm (1965, 1967). Gebbie and Seaton have paid special attention to point 2 (influence of electron scattering on S_ν) but have assumed a grey temperature stratification and have neglected the contribution of the high ions of C, N, O, and Ne to the absorption coefficient. Böhm and Deinzer, on the other hand, have constructed non-grey models with a high degree of flux constancy. They have included the higher ions of C, N, O and Ne in the computation of the absorption coefficient, but they have used an iteration method for the determination of the electron-scattering part of the source function. This procedure converges only sufficiently fast if at least in the deeper parts of the atmosphere electron scattering is not too important. An ideal method would be a combination of both procedures (Gebbie-Seaton and Böhm-Deinzer). At the moment, however, this would require a very large amount of computing time, since the Gebbie-Seaton method would have to be applied for every frequency in every iteration step of the flux-iteration method.

Both groups have taken into account point 3 (radiative acceleration) and both have neglected points 4 and 5.

* In the following we have neglected the surface effects. These are due to the very rapid increase of F very close to the surface in wavelength regions in which κ_ν is very large compared to $\bar{\kappa}$.

These authors have restricted themselves to the consideration of hydrostatic models. Presently Schmid-Burgk (1967) is carrying out an investigation of hydrodynamic models for central stars in which (7) is negative in a large part of the atmosphere (see below).

Let us first consider the construction of model atmospheres in hydrostatic and radiative equilibrium. In the usual form of the hydrostatic equation, g has to be replaced by Equation (7). The condition of radiative equilibrium can be expressed as

$$F = \Sigma F_\nu = \Sigma \phi_\nu (S_\nu) = \text{const}. \tag{8}$$

(independent of depth), where the integral operator ϕ_ν is defined as

$$\phi_\nu (S_\nu) = 2 \left\{ \int_{\tau_\nu}^{\infty} S_\nu (t_\nu)\, E_2 (t_\nu - \tau_\nu)\, \mathrm{d}t_\nu - \int_0^{\tau_\nu} S_\nu (t_\nu)\, E_2 (\tau_\nu - t_\nu)\, \mathrm{d}t_\nu \right\}. \tag{9}$$

The source function S_ν is

$$S_\nu = \frac{\kappa_\nu}{\kappa_\nu + \sigma_{\mathrm{el}}} B_\nu + \frac{\sigma_{\mathrm{el}}}{\kappa_\nu + \sigma_{\mathrm{el}}} J_\nu . \tag{10}$$

J_ν is the monochromatic mean intensity and has to be determined for every frequency separately by solving the Schwarzschild-Milne integral equation

$$J_\nu = \Lambda_\nu \left(\frac{\kappa_\nu}{\kappa_\nu + \sigma_{\mathrm{el}}} B_\nu \right) + \Lambda_\nu \left(\frac{\sigma_{\mathrm{el}}}{\kappa_\nu + \sigma_{\mathrm{el}}} J_\nu \right), \tag{11}$$

where

$$\Lambda_\nu (S_\nu) = \frac{1}{2} \int_0^{\infty} S_\nu (t_\nu)\, E_1 (|t_\nu - \tau_\nu|)\, \mathrm{d}t_\nu . \tag{12}$$

The most effective way to solve (8) is the use of a 'temperature correction procedure' (cf. Unsöld, 1951; Böhm, 1954; Avrett and Krook, 1963; Lucy, 1964; Henyey, 1967) in which one calculates ΣF_ν for a given approximation of the temperature stratification and then converts the difference ΔF between the computed total flux and the required constant flux ($\sigma/\pi\ T_{\mathrm{eff}}^4$) into a correction ΔB of the approximate stratification of the Planck function. Böhm and Deinzer have used Lucy's (1964) method which is a modification of the Unsöld (1951) flux-iteration procedure. The following relation between ΔB and ΔF has been used:

$$\Delta B - \frac{\kappa_J}{\kappa_P} J = B - \frac{\kappa_J}{\kappa_P} \left\{ \frac{\Delta F(0)}{2} + \frac{3}{4} \int_0^{\tau} \frac{\kappa_F}{\kappa_P} \Delta F\, \mathrm{d}\tau \right\}. \tag{13}$$

Here the three mean values κ_J, κ_P and κ_F of the absorption coefficient are defined by:

$$\kappa_J = \frac{1}{J}\int_0^\infty \kappa_\nu J_\nu \, d\nu; \quad \kappa_P = \frac{1}{B}\int_0^\infty \kappa_\nu B_\nu \, d\nu; \quad \kappa_F = \frac{1}{F}\int_0^\infty (\kappa_\nu + \sigma_{el}) F_\nu \, d\nu. \tag{14}$$

The method of Böhm and Deinzer is an iteration scheme, in which one calculates $\Sigma\phi_\nu(S_\nu)$ (see Equations (8) and (9)) and then applies (13). Between two successive 'flux-iteration' steps one iterates Equation (11) twice for every frequency using the B_ν found by the flux iteration.

Gebbie and Seaton assume that the stratification of B_ν is known in advance (grey atmosphere). Consequently we only have to solve Equation (11). This is done by rewriting (11) as an integral equation for:

$$\psi_\nu = \frac{\sigma_{el}}{\kappa_\nu + \sigma_{el}} \{J_\nu - B_\nu\} \tag{15}$$

subtracting out the singularity of the integrand and replacing the integral by a gaussian sum. This leads to a system of linear inhomogeneous equations which can be solved.

Böhm and Deinzer (1966), and Böhm (1967) have calculated 9 models in the temperature range $3{\cdot}8 \times 10^4\,°\mathrm{K} \leqslant T_{eff} \leqslant 1{\cdot}8 \times 10^5\,°\mathrm{K}$. These models lie partly on the Harman-Seaton and partly on the O'Dell sequence (see Figure 1). Gebbie has computed 9 model atmospheres for the range $1{\cdot}0 \times 10^5\,°\mathrm{K} \leqslant T_{eff} \leqslant 2{\cdot}5 \times 10^5\,°\mathrm{K}$; $4{\cdot}8 \leqslant \log g \leqslant 8{\cdot}0$ using the Gebbie-Seaton method.

It is important to note that the non-grey models show (like non-grey models of cooler stars) a very steep temperature gradient close to the surface (Böhm and Deinzer, 1965). This is a reaction of the atmosphere to the rapid increase of F_ν in the ν-region with large κ_ν when one approaches the surface.

With one exception ($T_{eff} = 4{\cdot}625 \times 10^4$; $g = 1{\cdot}26 \times 10^4$) it has been possible to construct model atmospheres with a flux constancy of better than 1%.

What kind of emergent flux distributions do we get for these models? Here we have to distinguish between the results of Gebbie and Seaton (1963) and Gebbie (1967) on the one hand, and Böhm and Deinzer (1966) and Böhm (1967) on the other. For the reason discussed above the results of Böhm and Deinzer show effects of the steep ('non-grey') temperature decrease at the surface, whereas such effects do not appear in the results of Gebbie and Seaton. Furthermore Gebbie and Seaton neglect absorption by the higher ions of C, N, O and Ne. This has the consequence that for wavelengths which are sufficiently small compared to 228 Å (Lyman edge of He II), electron scattering becomes the only absorption mechanism. Therefore they get a rather high F_ν (larger than the black-body radiation for $T = T_{eff}$) in this frequency range. On the other hand the absorption by high ions of C, N, O and Ne may have been somewhat overestimated by Böhm and Deinzer, who used the hydrogen-like approximation for these ions.*

A possibly very important effect in these high-temperature atmospheres is the Schuster-effect as pointed out by Gebbie and Seaton (1963). It is well known (cf. Kourganoff, 1963) that J_ν is smaller than B_ν close to the surface in those ν-regions in which B_ν increases only slowly with τ_ν. This is the case e.g. in the 'red' part of the spectrum. ('Red' means: at a wavelength considerably longer than the λ corresponding to the maximum of the Planck function.) In this case S_ν increases abruptly as we pass through the absorption edge going from longer to shorter wavelengths: As shown by Equation (10), the jump in κ_ν has the effect that S_ν jumps from a lower value (close to J_ν) to a higher value (close to B_ν): it shows an emission edge. If S_ν behaves in this way in a large part of the atmosphere, one might expect to find an *emission edge* also in the emergent flux $F_\nu(0)$. On the other hand, one has to keep in mind that there is also present the well-known effect which usually leads to the formation of an *absorption edge*: One looks deeper into the atmosphere on the long wavelength side of the edge

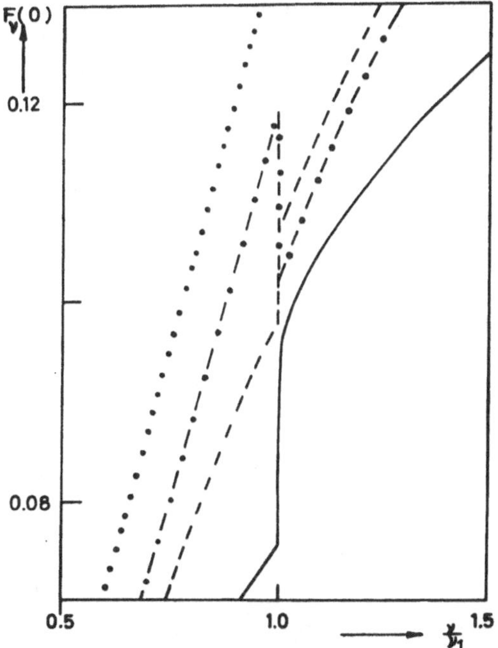

FIG. 2. *The frequency distribution of $F_\nu(0)$ at the H I-Lyman edge for grey model atmospheres of $T_{eff} = 1{\cdot}0 \times 10^5$ and $\log g = 4{\cdot}8$ (——); $\log g = 5{\cdot}0$ (— — —) and $\log g = 5{\cdot}2$ (—·—). The dotted curve shows the black-body distribution. This figure has been taken from Gebbie (1967).*

* A computation using absorption coefficients determined by the quantum-defect method (Flower, 1968) has just been started.

than on the short wavelength side. Consequently one might expect that there will be a certain amount of cancellation between both effects and that a fairly accurate knowledge of the atmospheric structure will be required in order to decide whether an emission or absorption edge will be present.

As pointed out by Mrs. Gebbie (1967), the formation of an emission edge will be favored in atmospheres of relatively low g (see Figure 2). This is due to the great importance of electron scattering in atmospheres of low surface gravity.

It is also important to note that, according to the statement made above, the formation of an emission edge will be favored by a small temperature gradient and will be suppressed by a steep T-gradient. Consequently in a non-grey atmosphere (having a rather steep T-gradient close to the surface; see above) there is a strong tendency towards the suppression of emission edges. So it is understandable that Gebbie and Seaton (using grey models) often find emission edges for the H I Lyman continuum, whereas the same edge appears in absorption for all non-grey models (Böhm and Deinzer, 1966; Böhm, 1967) which have been computed. Typical diagrams of $F_\nu(0)$ for two atmospheric models of different T_{eff} and g are presented in Figure 3. Besides a rather strong absorption edge of He II and a fainter H I edge, one sees also rather pronounced absorption edges of ions like Ne IV, O V, Ne V etc. One important aspect is the strong depression of the F_ν (0) (as compared to the Planck function) shortward of

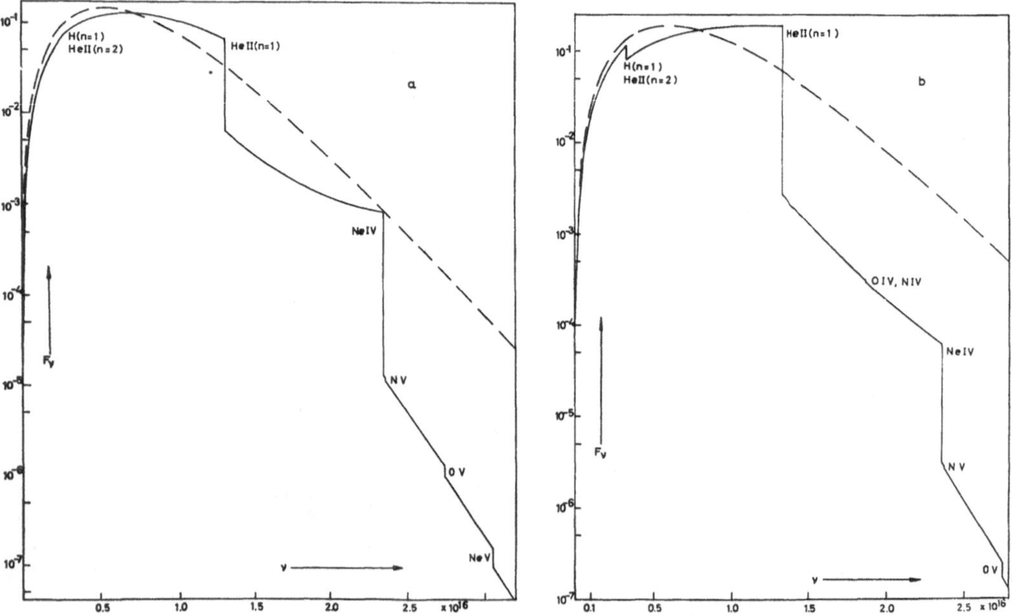

FIG. 3. *$F_\nu(0)$ for model atmospheres with $T_{\text{eff}} = 9{\cdot}12 \times 10^4\,°K$, $g = 6{\cdot}3 \times 10^4$ cm sec^{-2} (a), and $T_{\text{eff}} = 1{\cdot}0 \times 10^5\,°K$, $g = 1{\cdot}5 \times 10^6$ cm sec^{-2} (b). The broken curve shows the black-body distribution for the same temperature.*

$\lambda 228$ in general and especially shortward of the Ne IV edge. It is obvious that this depression must have a strong influence on the ionization stratification of the nebula. Since we have used only a crude approximation for the computation of the absorption coefficients of the C-, N-, O- and Ne- ions, it is important to ask: How strongly do these results depend on an accurate knowledge of the absorption coefficient for these ions? We have made one test calculation for a star of $T_{\rm eff} = 7{\cdot}4 \times 10^4$ °K; $g = 2{\cdot}51 \times 10^4$ cm sec^{-2}, in which we have arbitrarily reduced the absorption coefficient for all C-, N-, O- and Ne-ions by a factor of 6·67 as compared to our earlier calculations. Surprisingly the general qualitative picture is changed rather little by these large changes in the absorption coefficient. The main reason is that the absorption coefficient immediately shortward of the two most important absorption edges (O IV – N IV and N V) is so high that one essentially sees only the 'surface' of the atmosphere in these wavelength regions. Even after an artificial reduction of κ_ν in this ν-range by a factor 6·67 one still essentially sees only the surface. These ideas can be applied only to $T_{\rm eff} \approx 7 \cdot \times 10^4$ °K; in other $T_{\rm eff}$ ranges the influence of an error in κ_ν may be much larger.

The models can be used to determine a better scale of Zanstra temperatures and improved bolometric corrections for central stars. Improved Zanstra temperatures have been determined by Capriotti and Kovach (1968). They find (for a fixed ratio of the measured Hβ intensity to the continuum intensity) higher Zanstra temperatures for non-grey model atmospheres than for black-body radiation. Using the models by Böhm and Deinzer (1966) they find e.g. a Zanstra temperature of $1{\cdot}18 \times 10^5$ °K for a star which would have a Zanstra temperature of $1{\cdot}0 \times 10^5$ on the black-body scale. Bolometric corrections for model atmospheres have been computed by Böhm (1967*b*) and have been compared to those calculated for black-body radiation. The results are given in Table 1.

Table 1

Bolometric corrections for central stars of planetary nebulae. (non-grey model atmospheres)[a]

Model No.	$T_{\rm eff}$(°K)	g(cm sec^{-2})	References[b]	B.C.$_{\rm N.GR}$[c]	B.C.$_{\rm B.}$[d]
1	$3{\cdot}800 \times 10^4$	$4{\cdot}5 \times 10^4$	II	−3·65	−3·36
2	$4{\cdot}625 \times 10^4$	$1{\cdot}26 \times 10^4$	I	−4·27	−3·93
3	$6{\cdot}310 \times 10^4$	$1{\cdot}26 \times 10^4$	I	−5·26	−4·85
4	$7{\cdot}400 \times 10^4$	$2{\cdot}51 \times 10^4$	II	−5·73	−5·34
5	$9{\cdot}100 \times 10^4$	$6{\cdot}3 \times 10^4$	I	−6·36	−5·97
6	$1{\cdot}000 \times 10^5$	$1{\cdot}5 \times 10^6$	I	−6·81	−6·27
7	$1{\cdot}060 \times 10^5$	$3{\cdot}16 \times 10^5$	I	−6·89	−6·44
8	$1{\cdot}500 \times 10^5$	$3{\cdot}06 \times 10^7$	I	−7·93	−7·54
9	$1{\cdot}800 \times 10^5$	$8{\cdot}6 \times 10^7$	II	−8·25	−8·11

[a] All bolometric corrections given here refer to the colour V of the U-B-V system.
[b] Paper from which the $F_\nu(0)$ has been taken: (I) = Böhm and Deinzer (1966), (II) = Böhm (1967).
[c] B.C.$_{\rm N.GR.}$ = bolometric correction for the non-grey model atmosphere.
[d] B.C.$_{\rm B.}$ = bolometric correction for black-body radiation with $T = T_{\rm eff}$.

The relatively large deviations of $F_\nu(0)$ from black-body radiation in some ν-regions has important consequences for the ionization stratification of the nebula. These have been studied by Goodson (1967). Since all central stars which have been investigated have a higher $F_\nu(0)$ than the black-body curve between the He I and He II Lyman edges and a lower $F_\nu(0)$ shortward of λ228, one finds e.g. much more O III and less O II in a nebula excited by a 'star' with a non-grey model atmosphere than in a nebula excited by black-body radiation of the same T_{eff}. The reason is that the O II ionization edge lies between the He I and He II edges, in the ν-region where the radiation field is rather strong. On the other hand the O III ionization edge occurs in a ν-region in which the radiation field is strongly depressed. Consequently there will be relatively few O II and O IV ions, but relatively many O III ions. A similar effect occurs in the ionization of N. This is clearly shown in Figures 4 and 5 in which we compare the ionization stratification which one gets for the computed $F_\nu(0)$ of a non-grey model atmosphere of $T_{\text{eff}}=46250\,°\text{K}$ and the stratification following from the black-body energy distribution for the same T_{eff}. It is important to note that the black-body model leads to a pronounced Ne V zone, whereas one gets much less Ne V from the $F_\nu(0)$ of a non-grey atmosphere. As pointed out by Williams (1967), this leads to difficulties in the interpretation of the observed spectra of planetaries. There is still some hope that these can be avoided when a calibration of the Zanstra-temperature scale is used which is based on the computed $F_\nu(0)$ for non-grey atmospheres (Capriotti and Kovach, 1968). If this should not be the case we have to ask: How can the strong ultraviolet (for λ228 Å) deficiency be avoided? From what has been stated above it is

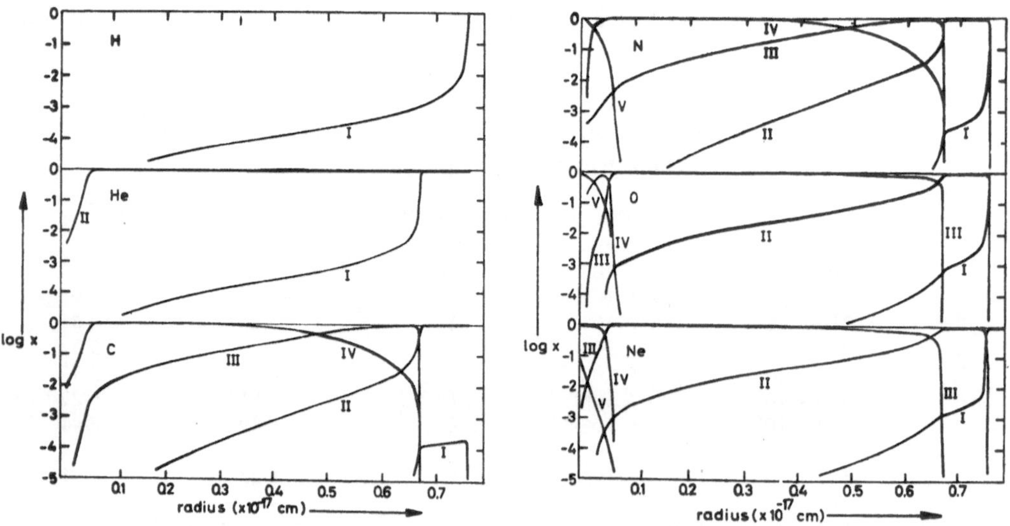

Fig. 4. *The logarithm of the abundance x of the different ions (designated by Roman numbers) of H, He, C, N, O and Ne as a function of the distance from the central star in a planetary nebula of constant density excited by stellar black-body radiation of $T=46250\,°K$. After Goodson (1967).*

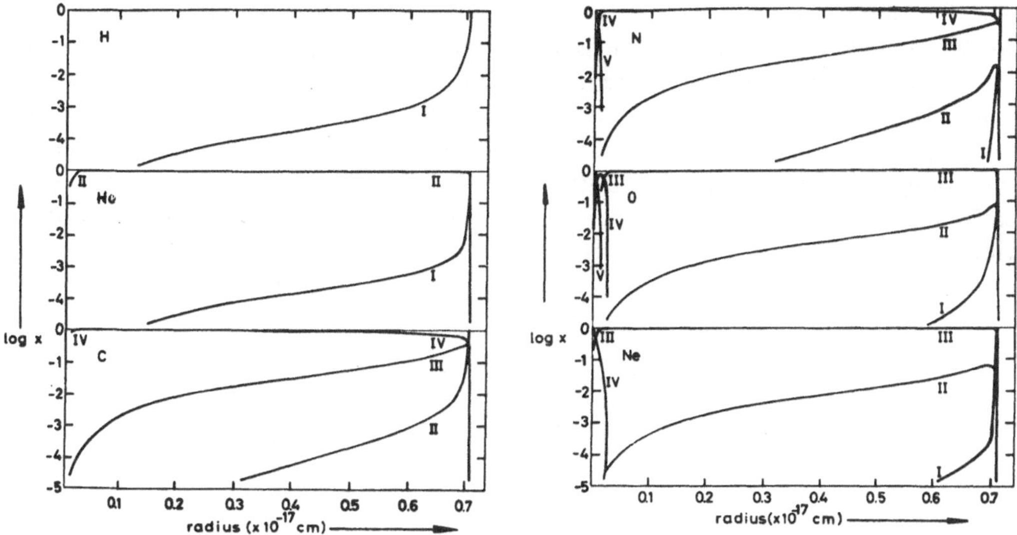

FIG. 5. *The same as Figure 7, but with excitation by the radiative flux $F_\nu(0)$ of a non-grey model atmosphere of the same temperature and $g = 1{\cdot}24 \times 10^4$ cm sec^{-2}. After Goodson (1967).*

obvious that we get a smaller deficiency if either the surface g is low or the temperature gradient is sufficiently flat. Whether a lower surface g alone would be sufficient to convert the ultraviolet deficiency into an ultraviolet excess as required by Williams is not known at present. How could we get a flatter temperature gradient in the high layers of the atmosphere?* One important possibility is that a considerable fraction of all central star atmospheres are not in complete hydrostatic and radiative equilibrium, but are continuously losing mass, e.g. due to the instability discussed in connection with Equation (7). The possible evidence for such a process is partly empirical (central stars having Wolf-Rayet spectra, large emission line width in central stars of the NGC 246 type; see Greenstein and Minkowski, 1964) and partly theoretical (a hydrodynamical explanation of planetary nebulae probably requires a stellar wind from the central star; Mathews, 1966). Schmid-Burgk (1967) has started a study of the hydrodynamics of a hot atmosphere in which Equation (7) is negative in a large part of the atmosphere. At the moment his calculations are not yet finished and he has available only certain similarity solutions (Sedov, 1959) of the problem. In order to find such similarity solutions the original depth-dependence of the temperature and the ρ- and T-dependence of κ has to be simplified in such a drastic manner that they are no longer realistic from a physical point of view. The problem is considered as a time-dependent one in the sense that one starts out from a stable situation and then

* It would not make sense to argue that the observations show that a grey stratification is 'better' than the non-grey one, unless physical arguments could be given for this peculiar result.

lets the total radiative flux F grow with time while keeping g fixed. Schmid-Burgk's solutions show that the $T(\tau)$ becomes much flatter as time goes on (and consequently the outflow of mass becomes more important). We may suspect that this will be also true for models which are more realistic from a physical point of view.* Consequently one expects that the especially steep temperature gradient in the upper layers of the non-grey atmospheres will be flattened and that therefore the $F_\nu(0)$ will more easily be influenced by the Schuster-effect. Moreover, the depression in the far ultraviolet will be reduced considerably. In Figure 6 the $T(\tau)$-relation is shown as a function

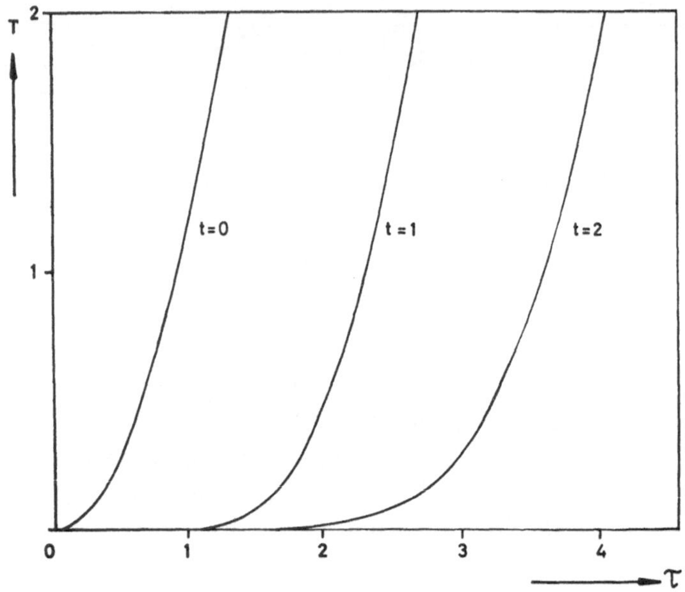

FIG. 6. *The $T(\tau)$ relation as a function of time for Schmid-Burgk's similarity solutions.*

of time for one of Schmid-Burgk's similarity solutions. The initial stratification is a polytropic atmosphere with $\Gamma = 11$, which means that

$$\rho \propto z^{1/10} \tag{16a}$$

$$T \propto z, \tag{16b}$$

where z is the geometrical depth. The opacity is assumed to be proportional to $\rho T^{-3/5}$. The time-dependence of F is assumed to be:

$$F \propto t. \tag{17}$$

* Schmid-Burgk is now doing such calculations, in which the system of partial differential equations is solved numerically. The main difficulty is of course a correct treatment of the radiative terms.

The diagram clearly shows the increasing flattening of the $T(\tau)$-relation with increasing time.

4. Spectra of Central Stars

We are still very far from understanding the spectra of central stars in any detail. Since all classical methods of quantitative spectral analysis do not seem to work for most of these rather peculiar objects, the only promising approach seems to be the following:

(1) Use the available information on T_{eff} and g to construct model atmospheres (as described in the preceding sections);

(2) Predict the line spectra for these models;

(3) Compare the predicted and the observed spectra and improve the models by a trial and error method.

Steps 2 and 3 have not yet been carried out. (But, as we have seen in the preceding section, even the use of the empirical information on the far *UV* continuous spectra already leads to better model atmospheres.) As has been emphasized by Aller and Liller (1967), a direct spectroscopic analysis should be possible for the absorption-line O stars among the central stars.* But even when considering these objects one has to keep in mind that they probably have a considerably higher g than normal O stars.

Very useful surveys of the empirical data on the spectra of central stars have been given by Vorontsov-Velyaminov (1953), Aller (1956), Aller and Liller (1967), and Perek and Kohoutek (1967).

Six different types of spectra seem to occur, namely:

(1) Wolf-Rayet spectra, mainly of type WC or intermediate between WC and WN. (Only recently Bertola (1964) has found a central star of spectral type WN.)

(2) Of-spectra (e.g. NGC 2392; see Wilson, 1948) often showing some deviations from the spectra of population-I Of-stars (cf. Oke, 1954).

(3) Spectra intermediate between Of- and Wolf-Rayet type.

(4) Absorption-line O spectra.

(5) Purely continuous spectra (within the presently obtainable spectral resolution).

(6) Spectra of the NGC 246 type (which are high-excitation spectra with strong O VI emission lines). Very high temperature objects of this type have been studied by Greenstein and Minkowski (1964).

Among the brightest planetary nuclei there is a fairly large number of Wolf-Rayet stars. There is some hope that some fundamental questions about the nature of the Wolf-Rayet phenomenon may be answered using empirical information about planetary nuclei. One may ask: Can the instability of the atmospheres of Wolf-Rayet stars be due to the action of the 'radiative acceleration' discussed above? If that should be the case all Wolf-Rayet stars should lie beyond (or at least close to) the instability line drawn in Figure 1. From this point of view it is interesting to find out whether stars of

* A determination of ionization temperatures in these objects has been carried out by Aller (1948) following the work of Petrie (1947) for normal O stars.

the six different spectral groups cover different regions of the g-T_{eff}-plane. It is obvious that such a study can be especially useful from our point of view, if we restrict ourselves to objects for which reasonably accurate values of g and T_{eff} can be given. Consequently we shall limit our study to the objects studied by Harman and Seaton (1966) and Seaton (1966), since these authors have paid special attention to the determination of the optical thickness of the nebulae which are used for the determination of T_{eff} and L. Even so, in view of the uncertainties in the determination of T_{eff} and g (see above), we have to expect a large scatter of the points. Spectral classification is available (see Perek and Kohoutek, 1967; Aller, 1956; Aller and Liller, 1967) for 28 of the 42 objects with known values of T_{eff} and g. These objects have been listed in Table 2.

Table 2

Relation between spectral type, effective temperature and surface gravity for 28 central stars

Object	Adopted Spectral Type for Central Star [a]	T_{eff}(°K)	g(cm sec^{-2})
NGC 40	WC 8	$3{\cdot}39 \times 10^4$	$1{\cdot}66 \times 10^5$
IC 351	W	$9{\cdot}12 \times 10^4$	$5{\cdot}75 \times 10^4$
NGC 1501	WC 6	$7{\cdot}25 \times 10^4$	$7{\cdot}25 \times 10^4$
NGC 1535	cont.	$7{\cdot}41 \times 10^4$	$1{\cdot}86 \times 10^4$
IC 418	WC 7	$4{\cdot}27 \times 10^4$	$2{\cdot}88 \times 10^4$
NGC 2022	cont.	$9{\cdot}12 \times 10^4$	$5{\cdot}13 \times 10^4$
IC 2149	O 7	$4{\cdot}90 \times 10^4$	$8{\cdot}13 \times 10^2$
NGC 2371–2	cont.	$1{\cdot}00 \times 10^5$	$1{\cdot}91 \times 10^5$
NGC 2392	Of	$6{\cdot}76 \times 10^4$	$9{\cdot}33 \times 10^3$
NGC 3242	cont.	$9{\cdot}33 \times 10^4$	$7{\cdot}41 \times 10^4$
NGC 6058	cont.	$7{\cdot}25 \times 10^4$	$2{\cdot}51 \times 10^4$
NGC 6210	Of	$5{\cdot}01 \times 10^4$	$3{\cdot}16 \times 10^6$
NGC 6309	cont.	$9{\cdot}55 \times 10^4$	$3{\cdot}98 \times 10^4$
NGC 6543	W	$6{\cdot}61 \times 10^4$	$1{\cdot}48 \times 10^4$
NGC 6572	W	$6{\cdot}17 \times 10^4$	$8{\cdot}71 \times 10^3$
NGC 6751	WC 6	$7{\cdot}59 \times 10^4$	$1{\cdot}70 \times 10^4$
NGC 6804	cont.	$7{\cdot}25 \times 10^4$	$6{\cdot}61 \times 10^3$
BD +30°3639	WC 8	$4{\cdot}57 \times 10^4$	$9{\cdot}12 \times 10^1$
NGC 6826	Of+W	$6{\cdot}92 \times 10^4$	$1{\cdot}32 \times 10^4$
NGC 6853	cont.+He I 5876	$1{\cdot}32 \times 10^5$	$6{\cdot}76 \times 10^6$
NGC 6891	O 7	$5{\cdot}50 \times 10^4$	$4{\cdot}37 \times 10^3$
NGC 6905	WC 6	$1{\cdot}02 \times 10^5$	$1{\cdot}48 \times 10^5$
NGC 7008	cont.	$9{\cdot}77 \times 10^4$	$3{\cdot}16 \times 10^5$
NGC 7009	cont.	$8{\cdot}13 \times 10^4$	$5{\cdot}75 \times 10^4$
NGC 7026	WC 6	$9{\cdot}77 \times 10^4$	$2{\cdot}63 \times 10^4$
IC 5217	W	$7{\cdot}41 \times 10^4$	$2{\cdot}82 \times 10^4$
NGC 7662	cont.	$1{\cdot}00 \times 10^5$	$3{\cdot}72 \times 10^4$
HD 138403	Of	$4{\cdot}27 \times 10^4$	$2{\cdot}69 \times 10^3$

[a] From Perek and Kohoutek (1967), Aller (1956), Aller and Liller (1967).

Of these 28 central stars 11 are WR-stars, 3 Of-stars; 2 absorption-line O stars; 1 object is intermediate between Of and WR, 11 show a continuous spectrum (without detectable absorption or emission lines). Unfortunately there is only one star of the NGC 246 type in Seaton's list, namely the central star of NGC 246 itself. But NGC 246 is probably optically thin for H I and He II. Consequently the Zanstra temperature defines only a lower limit for T_{eff} and the central star cannot be located in the g-T_{eff}-plane. The position of central stars of different spectral type in the g-T_{eff}-plane is shown in Figure 7.

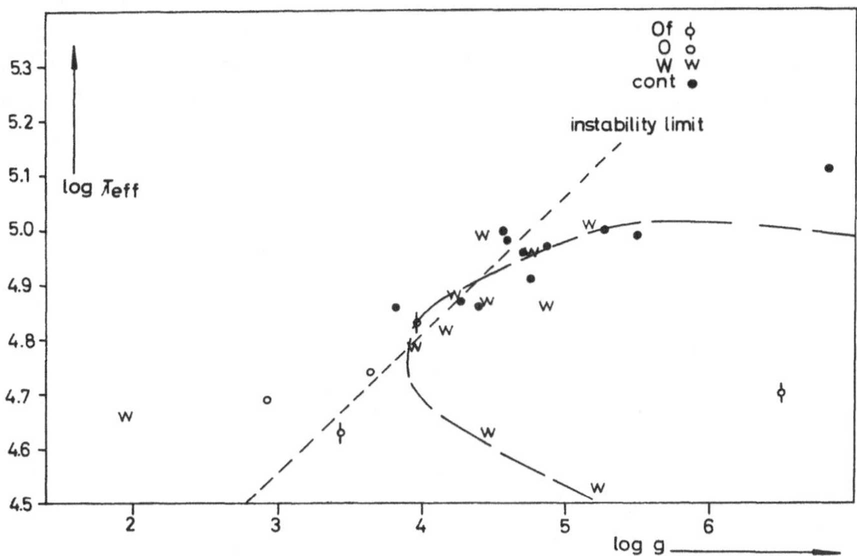

FIG. 7. *The position of central stars of different spectral types in the g-T_{eff}-plane.*

Though the scatter is considerable, certain trends can easily be recognized. The stars with 'purely continuous' spectra occur only in the high-temperature part of the Harman-Seaton sequence (starting at about $T_{\text{eff}} \approx 7{\cdot}5 \times 10^4\,°\text{K}$). Among the stars available in our list all objects in advanced evolutionary stages (up to $\log g \approx 7$, $T_{\text{eff}} \approx 10^5\,°\text{K}$) seem to belong to this spectral class. (It would be interesting to know, how the very hot objects of NGC 246 type fit into this picture.) The Wolf-Rayet stars occur over a wide range of temperatures ($3{\cdot}3 \times 10^4\,°\text{K} \lesssim T_{\text{eff}} \lesssim 1{\cdot} \times 10^5\,°\text{K}$), but they are all rather close to the instability limit, except for the rather low-temperature objects for which the scatter is enormous. (These are the objects for which the $R(N_e)$-method has been used.) It is difficult to decide, whether the value of g could be wrong by more than 2 orders of magnitude in an object like the central star of NGC 40. If that should be the case this Wolf-Rayet star could be also shifted into the neighbourhood of the instability limit and it would be very convincing to assume that the Wolf-

Rayet phenomenon has to do with the radiative acceleration as discussed above in connection with Equations (6) and (7). One difficulty with this hypothesis is that the Of-stars and even the absorption-line O stars seem to lie as close to the instability limit as the WR stars. With the very limited accuracy of giving a star's position in T_{eff}-g-diagram we do not yet understand why a number of objects do become O and Of-stars and do not show WR spectra.

One star, namely BD+30°3639 lies so far to the left in the T_{eff}-g-diagram (i.e. it occurs in such an extremely unstable region) that one has the feeling that something must be wrong with the determination of g for this object. It seems probable that the luminosity of this object has been considerably overestimated. The difficulty could be avoided by assuming that the mass of the nebula surrounding BD+30°3639 is considerably smaller than that of other planetary nebulae. This would lead to a smaller R(M) and consequently to a smaller distance.

Note added in proof

Computations using quantum defect absorption coefficients for the higher ions of C, N, O and Ne have been carried out in the meantime by Böhm (*Z. Astrophys.*, in press). Hidalgo, Hummer and Mihalas (private communication) have calculated model atmospheres for central stars using new absorption coefficients for the high ions of C, N, O and Ne calculated by M. B. Hidalgo (*Astrophys. J.*, in press). Both calculations show that for stars with $T_{eff} > 10^5$ °K the absorption jumps in $F_\nu(0)$ due to the high ions of C, N, O and Ne are smaller than originally predicted by Böhm and Deinzer (1966). On the other hand, it is correct that the ultraviolet flux is really strongly suppressed by these ions.

I am most grateful to Dr. D.G. Hummer for informing me about the work by Hidalgo, Hummer and Mihalas in advance of publication.

References

Aller, L.H. (1948) *Astrophys. J.*, **108**, 462.
Aller, L.H. (1956) *Gaseous Nebulae*, Chapman and Hall, London.
Aller, L.H. (1965) in *Astronomie und Astrophysik*, Ed. by H. H. Voigt, Springer-Verlag, Berlin, p. 571.
Aller, L.H., Liller, W. (1967) in *Stars and Stellar Systems*, vol. VII, Ed. by G.P. Kuiper and B.M. Middlehurst, University of Chicago Press (in press).
Avrett, E.H., Krook, M. (1963) *Astrophys. J.* **137**, 874.
Bertola, F. (1964) *Publ. astr. Soc. Pacific*, **76**, 241.
Böhm, K.H. (1954) *Z. Astrophys.*, **34**, 182.
Böhm, K.H. (1965) *Z. Astrophys.*, **62**, 167.
Böhm, K.H. (1967) Unpublished.
Böhm, K.H. (1967*b*) *Z. Astrophys.*, **67**, 219.
Böhm, K.H., Deinzer, W. (1965) *Z. Astrophys.*, **61**, 1.
Böhm, K.H., Deinzer, W. (1966) *Z. Astrophys.*, **63**, 177.
Capriotti, E.R., Kovach, W.S. (1968) *Astrophys. J.*, **151**, 991.

Flower, D.R. (1968) in the present volume, p. 77.
Gebbie, K.B., Seaton, M.J. (1963) *Nature*, **199**, 580.
Gebbie, K.B. (1967) *Mon. Not. R. astr. Soc.*, **135**, 181.
Goodson, W.L. (1967) *Z. Astrophys.*, **66**, 118.
Greenstein, J.L., Minkowski, R. (1964) *Astrophys. J.*, **140**, 1601.
Harman, R.J., Seaton, M.J. (1964) *Astrophys. J.*, **140**, 824.
Harman, R.J., Seaton, M.J. (1966) *Mon. Not. R. astr. Soc.*, **132**, 15.
Henyey, L.G. (1967) *Astrophys. J.*, **148**, 207.
Khromov, G.S. (1962) *Astr. Zu.*, **39**, 468.
Kourganoff, V. (1963) *Basic Methods in Transfer Problems*, Dover Publ., New York.
Lucy, L.B. (1964) in *Proc. 1st Harvard-Smithson. Conf. on Stellar Atmospheres*, p. 93.
Mathews, W.G. (1966) *Astrophys. J.*, **143**, 173.
Minkowski, R., Aller, L.H. (1954) *Astrophys. J.*, **120**, 261.
O'Dell, C.R. (1962) *Astrophys. J.*, **135**, 371.
O'Dell, C.R. (1963) *Astrophys. J.*, **138**, 67.
O'Dell, C.R. (1968) in the present volume, p. 361.
Oke, J.B. (1954) *Astrophys. J.*, **120**, 22.
Osterbrock, D.E. (1960) *Astrophys. J.*, **131**, 541.
Osterbrock, D.E. (1966) in *Stellar Evolution*, Ed. by R.F. Stein and A.G.W. Cameron, Plenum Press, New York, p. 381.
Perek, L., Kohoutek, L. (1967) *Catalogue of Galactic Planetary Nebulae*, Academia Publishing House of the Czechoslovak Academy of Sciences, Prague.
Petrie, R.M. (1947) *Pub. Dom. astrophys. Obs., Victoria*, **7**, 321.
Schmid-Burgk, J. (1967) Unpublished.
Schmidt, M. (1963) *Astrophys. J.*, **137**, 758.
Seaton, M.J. (1966) *Mon. Not. R. astr. Soc.*, **132**, 113.
Sedov, L.I. (1959) *Similarity and Dimensional Methods in Mechanics*, Infosearch Ltd., London.
Shklovsky, I.S. (1956) *Astr. Zu.*, **33**, 222, 315.
Unsöld, A. (1951) *Naturwissenschaften*, **38**, 525.
Vorontsov-Velyaminov, B.A. (1953) *Gasnebel und Neue Sterne*, Verlag Kultur und Fortschritt, Berlin.
Weidemann, V. (1963) *Z. Astrophys.*, **57**, 87.
Williams, R.E. (1967) *The Ionization of Planetary Nebulae*, Preprint.
Wilson, O.C. (1948) *Astrophys. J.*, **108**, 201.
Wurm, K., Singer, O. (1952) *Z. Astrophys.*, **30**, 387.

DISCUSSION

Underhill: There are plausible reasons for believing that the spectra of classical Wolf-Rayet stars are the result of collisional excitation, probably from streams of protons or α-particles, rather than of purely radiative processes. This is probably true also for the WR spectra of central stars of planetary nebulae, especially where O VI emission is present as well as C III and O III emission. A few classical Wolf-Rayet stars are known with O VI emission, but the appearance of O VI emission does not correlate uniquely with WC or WN type.

In 1958 I pointed out, using rather rough theory, that the precise shape of the continuous spectrum at 228 Å resulting from the C, O, and N continua would strongly affect the ionization ratio in the outer atmosphere and the relative intensities of lines from the C, N and O ions.

The widths and strengths of the stellar emission lines of WR central stars are sufficient to show that this part of the spectrum (WR lines) are formed in a fairly dense region ($N \simeq 10^{13}$–10^{14}) where chaotic motions may be large.

Some early work of mine on the effects of allowing for the curvature of a stellar atmosphere indicate that absorption features become sharper and a tendency to emission occurs. Probably this effect on the predicted spectra from models available at present is not negligible.

Böhm: It is true that the computations indicate a considerable extension of some atmospheres, but the extension does not seem large enough to lead to considerable changes if the effects of spherical symmetry are included.

Underhill: In the work reported, the electron scattering assumes non-coherence. This may make a significant change in the shape of the predicted UV spectrum, especially in the case of the C, N, O continua which are sufficiently close together to be handled, in some cases, as broad lines, rather than as continuous features. Taking account of the non-coherence of electron scattering is an improvement in the theory which is greatly to be desired.

UBV PHOTOMETRY OF THE BRIGHT NUCLEI OF THE PLANETARY NEBULAE

E. B. KOSTJAKOVA, M. V. SAVEL'EVA, O. D. DOKUCHAEVA, R. I. NOSKOVA
(Sternberg State Astronomical Institute, Moscow, U.S.S.R.)

The data, which we have at present for the planetary nebulae central stars, are in general insufficient and uncertain. Even the photographic magnitudes of the brightest nuclei are usually given in the known catalogues with accuracies of 2^m or 3^m. Therefore, the determination of central stars' magnitudes by the photoelectric method seems to be of great importance.

In the spring of 1966 we began photoelectric observations of the bright planetary nuclei at the Crimean Station of the Sternberg State Astronomical Institute. We used the 48-cm parabolic reflector at the Cassegrain focus with a photometer employing an EMI-photomultiplier. The filters used gave an instrumental photometric system close to that of the UBV system.

For the first series of observations we selected nebulae of large angular diameters and relatively low surface brightness, but with central stars brighter than $13^m.5$. The radiation of the nebula itself was taken into account by measuring the nebular background in several points in the vicinity of the nucleus.

Correction for the atmospheric extinction was carried out by means of Nikonov's method. For this purpose the polar star HD 80354 (A0; $V = 9^m.76 \pm 0.02$, $B-V = +0.22 \pm 0.01$, $U-B = +0.11 \pm 0.02$, according to our determinations) was selected as an atmospheric reference and observed several times every night.

Twelve members of the galactic cluster NGC 6633 were used as standards to reduce the instrumental photometric system to the UBV one. The colors B-V and U-B and the magnitudes V were taken from the work of Hiltner *et al.* (1958). The whole observation program was carried out from May until December of 1966.

Table 1 gives the results for the nuclei of 24 planetary nebulae: the colors B-V and U-B and the magnitudes V, with their root-mean-square deviations; the last column of the table shows the number of observations for each object.

The central star magnitudes of three nebulae – NGC 1501, 2371-2 and 6804 – are close to the limiting magnitude of the instrument used. Therefore, the results for these objects are of lower accuracy than for other nuclei.

Four nebulae – NGC 6543, 6572, 6853 and IC 3568 – have a relatively high surface brightness in comparison with the other nebulae of our list. In these cases the correction for the nebula radiation could not be done quite correctly.

Osterbrock and O'Dell (eds.), Planetary Nebulae, 317–319. © *I.A.U.*

Table 1

PlN	V	B-V	U-B	*n*
NGC 40	11·64 ± 0·02	− 0·05 ± 0·00	− 0·60 ± 0·02	3
NGC 246	11·98 ± 0·04	− 0·35 ± 0·02	− 1·18 ± 0·04	6
NGC 1501	14·75 ± 0·22	+ 0·33 ± 0·28	− 0·86 ± 0·14	3
NGC 1514	9·64 ± 0·01	+ 0·53 ± 0·00	+ 0·07 ± 0·02	5
NGC 1535	11·59 ± 0·06	0·00 ± 0·05	− 0·94 ± 0·02	4
IC 418	9·57 ± 0·07	+ 0·12 ± 0·05	− 0·91 ± 0·02	3
NGC 2371-2	14·82 ± 0·08	− 0·28 ± 0·12	− 1·14 ± 0·02	4
NGC 2392	10·54 ± 0·02	− 0·12 ± 0·04	− 1·08 ± 0·02	4
VV 68	8·31 ± 0·01	+ 0·10 ± 0·04	− 0·21 ± 0·07	4
IC 3568	11·39 ± 0·01	+ 0·67 ± 0·03	− 0·56 ± 0·09	3
NGC 6058	13·83 ± 0·04	− 0·48 ± 0·09	− 0·76 ± 0·10	3
IC 4593	10·80 ± 0·06	+ 0·07 ± 0·02	− 0·87 ± 0·09	4
NGC 6210	9·87 ± 0·10	+ 0·66 ± 0·02	− 0·39 ± 0·05	3
NGC 6543	9·74 ± 0·02	+ 0·36 ± 0·03	− 0·73 ± 0·02	4
NGC 6572	9·06 ± 0·11	+ 1·10 ± 0·00	+ 0·15 ± 0·34	2
NGC 6751	12·87 ± 0·19	+ 0·41 ± 0·07	− 0·45 ± 0·05	3
NGC 6804	14·42 ± 0·31	+ 0·06 ± 0·44	− 0·19 ± 0·28	3
NGC 6826	10·22 ± 0·04	0·00 ± 0·03	− 0·97 ± 0·02	8
NGC 6853	14·06 ± 0·09	− 0·27 ± 0·04	− 1·30 ± 0·04	3
NGC 7008	13·33 ± 0·03	− 0·37 ± 0·05	− 0·58 ± 0·07	5
A 64	13·27 ± 0·02	− 0·29 ± 0·01	− 1·07 ± 0·06	3
NGC 7293	11·90 ± 0·12	+ 0·75 ± 0·04	+ 0·23 ± 0·10	2
IC 1470	12·67 ± 0·06	+ 0·92 ± 0·08	− 0·21 ± 0·01	3
A 68	13·26 ± 0·02	+ 0·26 ± 0·02	− 0·10 ± 0·04	4

In the case of NGC 6210, due to its small dimensions and high surface brightness, the correction for the nebular radiation was not done. The data concern the entire object and not its nucleus alone.

For the nebula NGC 7293 the identification of the central star seems to be doubtful. The results obtained can be demonstrated in Figure 1, where the color-color diagram is drawn for the central stars studied. The nuclei with photoelectric colors, determined by Abell (1966), are plotted in the figure as crosses for comparison. The figure also shows the curve of the main-sequence stars, the locus of unreddened black bodies and the reddening lines. We can see in this figure that the majority of nuclei are situated in the region of the hottest stars.

We can make an approximate correction for the interstellar extinction for many planetaries. After such a procedure many of their nuclei are also located in the region of the hottest stars. However, we can see that some planetaries remain in the region of yellow stars, even after an interstellar extinction correction. These mainly concern the above-mentioned doubtful planetaries and nuclei with possible duplicity. The most interesting objects in that aspect are: V-V 68, A-68, NGC 7293 and NGC 1514. We can point out that A-10, studied by Abell, also showed a yellow color. Kohoutek's (1968) recent work dealt with NGC 1514 in detail.

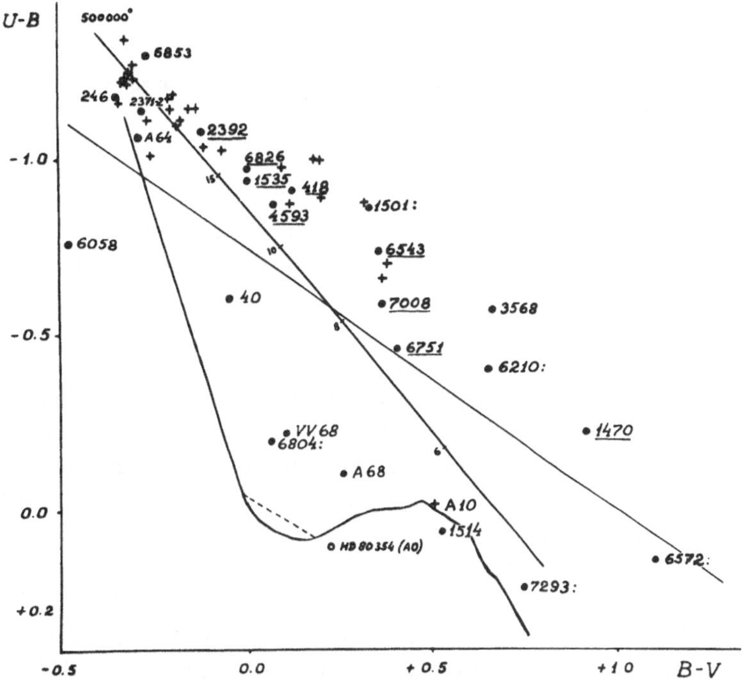

FIG. 1. *Color-color diagram for central stars of planetary nebulae. Observations by Abell are shown by crosses, results of this investigation by filled circles. The fine lines show the locus of blackbody colors and the reddening slope. The underlined numbers show the nuclei which can reach the region of the hottest stars after an interstellar extinction correction.*

Evidently, each of the above-mentioned planetaries needs additional and more detailed study; at any rate, this problem may be considered open to question. We intend to carry on with this work; the second series of observations have already begun.

References

Abell, G.O. (1966) *Astrophys. J.*, **144**, 259.
Hiltner, W.A., Iriarte, B., Johnson, H.L. (1958) *Astrophys. J.*, **127**, 539.
Kohoutek, L. (1968) in the present volume, p. 324.

DISCUSSION

W. Liller: I would like to ask Mrs. Kostjakova if she used special filters to suppress nebular radiations.

Kostjakova: We excluded the luminosity of the nebula in this way: We observed the background of the nebula with the same filters in several points, in the close vicinity of the nucleus. So we could exclude the luminosity of the nebula itself. In the majority of cases the background of the nebula was of the same order as the luminosity of the night sky. Only in a few cases (about three) was the luminosity of the nebulae larger.

PHOTOMETRIC OBSERVATIONS OF THE CENTRAL STARS OF PLANETARY NEBULAE

WILLIAM LILLER and CHENG-YUAN SHAO
(Harvard College Observatory, Cambridge, Mass., U.S.A.)

Currently at Harvard, we are engaged in a program for measuring multi-color magnitudes of the central stars of planetary nebulae. We plan to determine magnitudes of all favourable planetaries having central stars brighter than 16th photographic magnitude. The results, when completed, will give us improved information on (1) the amount of interstellar extinction for the individual objects, (2) the luminosities and colors of the central stars, (3) interstellar reddening corrections for spectrophotometric studies, and (4) the existence of variability among the central stars.

The nebular stars are measured both photoelectrically and photographically through special filters (with effective wavelengths close to those of the *U*, *B*, *V* system) to exclude the stronger nebular radiations, and also with standard *U*, *B*, *V* filters. While our observations are made primarily with the 155-cm reflector at Harvard's Agassiz Station, we also plan to use telescopes at the Kitt Peak National Observatory and at Cerro Tololo in Chile. During the 1966–67 season, we obtained data for the central stars of more than 30 planetaries, mostly NGC objects. At present, we have under observation about an equal number of stars selected from the Perek-Kohoutek Catalogue (1967).

In addition to determining the magnitudes of the nebular nuclei, we also measure some nearby stars around each nebula to check any light variation of the nucleus and to estimate the amount of interstellar reddening and extinction in the direction of the nebula. We have scheduled extensive observations of some central stars of suspected or known variability. This report summarizes some preliminary results of our program.

Table 1 lists the stars for which we have photoelectric data. The column headings *V*, *B-V* and *U-B* have their conventional meanings. The letter *W* represents the yellow observations made with a Wratten-15 filter (transmission λ5100 and redwards), which is used to eliminate the strong [O III] emission from the nebulae. The *W* magnitudes and the *B-W* colors, transformed to the *U*, *B*, *V* system, are given in parenthesis.

Because of the high surface brightness of most NGC objects, it is often difficult to separate entirely the central star from the surrounding nebulosity of complex structure and non-uniform intensity distribution. We always attempt to observe with the smallest diaphragms (8″ on the Newtonian and 3″ on the Cassegrain photometers) to

Osterbrock and O'Dell (eds.), Planetary Nebulae, 320–323. © *I.A.U.*

Table 1

Preliminary magnitudes and colors of the central stars of planetary nebulae

Nebula	*V* (or *W*)	*B-V* (or *B-W*)	*U-B*
246	11·95	−0·37	−1·20
1501	(14·2)	(−0·2)	−0·7
1514	9·42	+0·55	−0·06
1418	(9·7)	−0·2	−1·0
2022	(14·9)	+0·6	−0·6
2392	(10·4)	+0·4	−0·7
A 30	14·36	−0·06	−1·02
4361	12·9	−0·3	−1·3
A 36	11·43	−0·32	−1·22
6058	(13·7)	−0·3	−1·0
6543	(10·8)	+0·6	−0·5
6720	14·7	−0·4	−1·3
M 1–67	11·05	+1·07	−0·09
6804	(13·9)	+0·6	−0·7
+30° 3639	10·10	−0·06	−0·71
6853	13·82	−0·29	−1·16
6857	(13·30)	+1·04	−0·15
7094	13·60	−0·23	−1·20
7293	13·50	−0·34	−1·30

Remarks: 246, 1514, A 30 and A 36 – Nebula faint, contamination negligible.
6720, 6857, 7094 and 7293 – Ring excluded.
+30° 3639 – Nebular disk included.
6853 – Corrected for nebula.

exclude the background nebula, but sky conditions do not always make it possible. The standard *U*, *B*, *V* filters all transmit a number of bright nebular emissions. The *B* and *V* filters are 30 and 40% transparent at the wavelengths of the very strong green [O III] lines; the [O II] pair at λ3727 can often seriously affect the *U* measurements, although the greater relative intensity of the central star tends to minimize this influence. The *W* measurements, which are least affected by nebular emissions, best represent the yellow magnitudes of the stars. Therefore, except for a few objects noted in the remarks of the table, the listed *B-W* values are too small while the *B-V* and *U-B* colors are too large. We are now experimenting with two blue filters, one (glass) which eliminates the green nebular lines and the other (interference) which transmits only the spectrum between Hγ and Hδ lines.

For the objects not suitable for photoelectric measurements, the data have been supplemented by photographic observations with 'anti-nebula' filters. On moderately exposed plates the nebular material can be seen more easily than at the photometer eyepiece through diaphragms. Furthermore, it is possible to use defining apertures (such as an iris photometer provides) that are no larger than the star image itself. For this work, we have established photoelectric secondary standards in the immediate neighbourhood of each central star.

A few objects of special interest have received particular attention and we discuss them in turn.

NGC 1514: Because the nebula of this object is much fainter than its central star, the nebula did not appreciably contaminate the photoelectric observations of the star. The colors we observed for the star (*B-V* = +0·55, *U-B* = −0·06) and for other stars in its vicinity indicate a *B-V* color excess ~$0^{m}_{.}5$, in agreement with the results of Kohoutek (1966), who suggests that the central star is a reddened A0 star with appreciable ultraviolet excess attributed to an invisible hotter companion.

If the true nucleus of NGC 1514 is actually a double system with an O-type star fainter than its apparent A0 nucleus, the presence of the hotter component might be detectable by observations at very short wavelengths. In order to do so, we observed the star with a photoelectric spectral scanner at Agassiz Station on October 13, 1966. Two scans (Figure 1) from 3100 Å to 5400 Å showed that the central star is indeed a reddened star of the expected spectral type with detectable excess ultraviolet radiation.

Kohoutek 1-2: The nucleus of this planetary was found (Kohoutek, 1964) to be

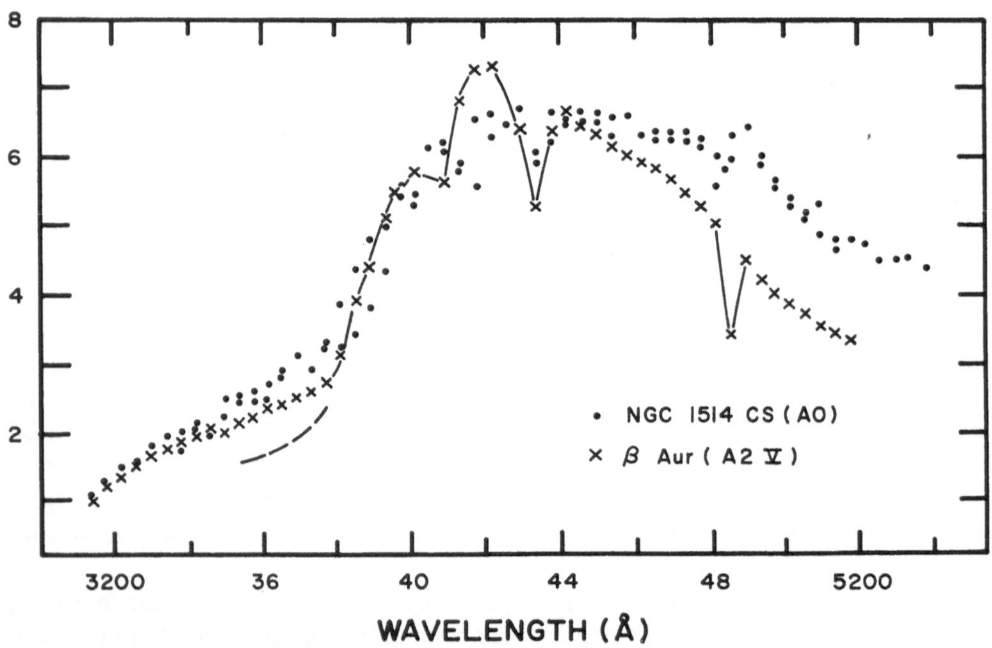

FIG. 1. *Spectrophotometric scans of the central star of NGC 1514. The two scans of the central star from 3100 Å to 5400 Å, Δλ = 50 Å, are plotted together with the single scan of β Aur. Each dot or cross represents one observation. At the red end of the spectrum, the intensity of the reddened nucleus is higher than that of β Aur. The spectrum of an A0 star reddened by ~ $0^{m}_{.}5$ in B-V is estimated to fall along the dashed line at the UV end, whereas the observed central-star continuum is higher (color bluer), confirming the existence of an UV excess. The ordinate is relative intensity.*

conspicuously variable on the overlap portion of two Palomar Sky Survey prints for two successive nights. From a search of the Harvard photographic plate collection, we confirmed the reported variability of the star. All available data appear in Figure 2, which shows that the star's photographic magnitude varies from 15·7 to at least 17·4. While the star spends most of its time at an intermediate magnitude (16·6 average), brightness changes of $1^m.5$ are possible in about 1 day. These facts seem to suggest irregular fluctuations. We plan to make more observations from Kitt Peak in October and November 1967 to further the study of its interesting light variation.

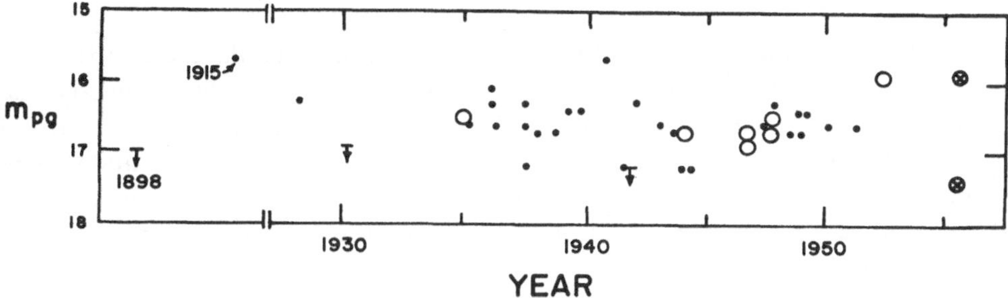

FIG. 2. *Photographic light variation of the central star of Kohoutek 1-2. Circles represent observations with an accuracy of* $\pm 0^m.1$*, and dots* $\pm 0^m.3$*. The two crossed circles indicate magnitudes from the Palomar Sky Survey prints. Arrows show when the star was fainter than the plate limit.*

Abell 36: This nebula possesses the brightest central star ($V = 11{\cdot}5$) in Abell's (1966) list of objects. The colors we measured agree well with those found by him, but the V magnitude differs by 0·1 magnitude. Because he suggested the possible variability of this star, we examined some 200 Harvard plates taken between 1931 and 1952. No light variation ($\pm 0^m.2$) was found from our visual estimates.

Acknowledgements

We thank Dr. Martha Liller for her help and the National Science Foundation for its support.

References

Abell, G.O. (1966) *Astrophys. J.*, **144**, 259.
Kohoutek, L. (1964) *Bull. astr. Inst. Csl.*, **15**, 161.
Kohoutek, L. (1966) *Bull. astr. Inst. Csl.*, **18**, 103.
Perek, L., Kohoutek, L. (1967) *Catalogue of Galactic Planetary Nebulae*, Academia, Prague.

THE BINARY-STAR HYPOTHESIS FOR THE NUCLEUS OF NGC 1514

L. Kohoutek

(Astronomical Institute, Czechoslovak Academy of Sciences, Czechoslovakia)

It is well known that the central star of the planetary nebula NGC 1514 ($\alpha_{50}=$ $4^h06^m\!.1$, $\delta_{50}=+30°39'$; $l^{II}=165{\cdot}5$, $b^{II}=-15{\cdot}3$) differs from the other planetary nuclei by its high brightness (relative to the nebula) and late-type spectrum. The difference $B_* - B_n = 1^m\!.9$, and especially the A0-spectral type (Chopinet, 1963) are quite atypical for central stars. For these reasons we began a complex study of this object (Kohoutek, 1967; Kohoutek and Hekela, 1967) on the basis of the following sources of observational material:

(1) Palomar Sky-Survey prints (Schmidt-camera, Palomar Observatory);

(2) direct photographs in UBV system with the 2-m Schmidt-camera in Tautenburg (1962);

(3) photoelectric UBV photometry with the 65-cm reflector at Ondřejov (1964–65),

(4) spectrograms of the central star using the 122-cm reflector in Asiago (1965), and relatively infrequent observational data from the older literature.

Direct photographs show the nebula as almost circular and of dimensions 135″ × 121″ with a trace of an outer envelope (~175″), with a non-homogeneous brightness distribution and with condensations forming an irregular ring. Detailed investigations of isophotes (in the B region) disclosed two condensations at a 55° position angle symmetric to the central star (Figure 1). According to our concept of the morphology of planetary nebulae (Khromov and Kohoutek, 1968) the nebula may perhaps be classified as type 3 and the condensations may be considered to be the projection of a ring on the celestial sphere (the main axis is at a 145° position angle).

The distance of the nebula, calculated according to the statistical method from the mean surface brightness and from the angular diameter, was estimated as $r \simeq 480$ parsec; thus the linear diameter corresponds to $D=0{\cdot}30$ parsec. The applied value of interstellar extinction, $A_V=2{\cdot}0$ mag, was derived from photoelectric and spectrographic observations. The electron density $N_e=290$ cm^{-3} ($N_e=140$ cm^{-3}) of the principle and outer envelope, respectively, and hydrogen mass of the nebula $0{\cdot}10_{\odot}$ ($0{\cdot}07_{\odot}$) indicate that this is a typical planetary nebula occurring rather at a late evolutionary stage.

For the central star let us first assume the spectral type to be A0. Then, however, the following contradictions with observational data appear:

Osterbrock and O'Dell (eds.), Planetary Nebulae, 324–328. © *I.A.U.*

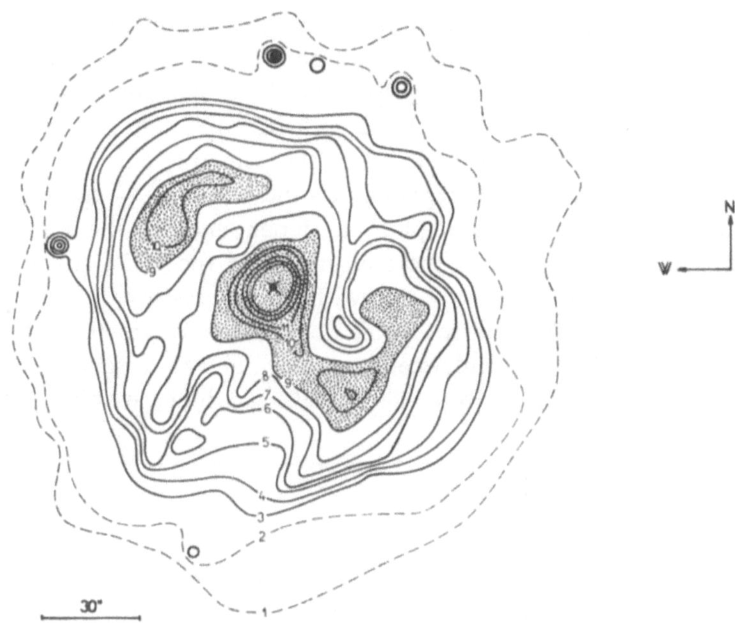

FIG. 1. *Isophotic contours of NGC 1514 in the B region* [*density scale:* $D = 0{\cdot}03$ *(1), 0·06 (2), 0·10 (3), 0·14 (4), 0·18 (5), 0·22 (6), 0·27 (7), 0·33 (8), 0·40 (9), 0·50 (10), 0·60 (11)*].

(a) Relative to common stars of A0 type the central star of NGC 1514 shows an extraordinary U-B excess of about $-0{\cdot}4$ mag, determined photoelectrically and spectroscopically.

(b) The theoretical Strömgren zone of completely ionized hydrogen is about five times smaller than the observed radius of the nebula.

(c) The temperature of the exciting star, derived from the intensity of the Hβ and HeII 4686 lines (Chopinet, 1963), is about six times higher than the corresponding value of the A0 spectral type.

We are convinced that the results mentioned are best accounted for by the existence of a second, much hotter component of the central star. A preliminary model of the nucleus of NGC 1514 gives the following parameters:

The physical double-star consists of

A0 III (A-component) + blue sd (X-component):

$$M_{\mathrm{V,A}} = -1 \qquad M_{\mathrm{V,X}} = +1{\cdot}2$$
$$T_{*,\mathrm{A}} = 10800\,°\mathrm{K} \qquad T_{*,\mathrm{X}} = 60000\,°\mathrm{K}$$
$$R_{*,\mathrm{A}} = 4{\cdot}1\ R_{\odot} \qquad R_{*,\mathrm{X}} = 0{\cdot}45\ R_{\odot}.$$

The luminosity class of the A-component was determined from the equivalent width of the Hγ line and from the electron density in the atmosphere of this star. The temperature of the X-component was estimated by means of the Harman-Seaton (1966) method from the magnitude of the central star and from the Hβ and He II 4686 intensities. There is no discrepancy between the dimension of the Strömgren zone of completely ionized hydrogen and the observed nebular radius at the temperature $T_{*,X}=60000\,°K$; the optical thickness of the principle envelope is $\tau=0{\cdot}085$.

The observed UV-excess is explained by the existence of the second component of the planetary nucleus. We can plot the dependence of the difference between the monochromatic magnitude of the continuum of the NGC 1514 central star and a common A0 star (in this case γ Tri) with the expression $1+f(\lambda)$, where $f(\lambda)$ is the reddening function. The course of this dependence ought to be linear and its slope should be a measure of the interstellar extinction. The observed dependence $\Delta m[1+f(\lambda)]$ (Figure 2) shows a drop at $\lambda=3880$ Å, which, within observational errors,

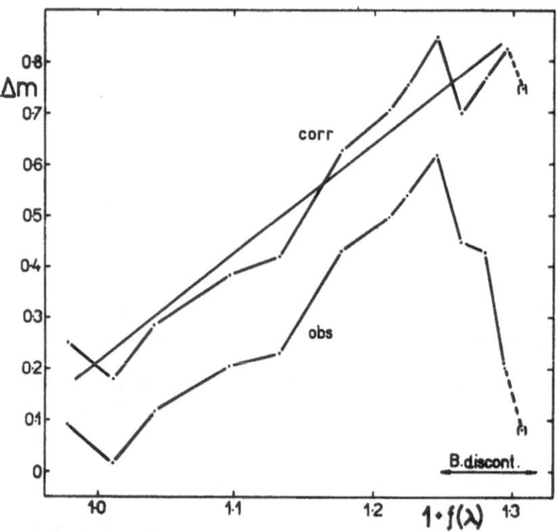

Fig. 2. *The dependence of difference Δm between the monochromatic magnitude of the continuum of NGC 1514 star and of γ Tri on the expression $1+f(\lambda)$: obs = observed; corr = corrected by introducing the X-component.*

disappears by introducing the X-component of the planetary nucleus having the above parameters.

Figure 3 demonstrates the position of the stars of NGC 1514 on an H-R. diagram, specifically the position of the undivided nucleus (black circle) and that of the A and X-components. For comparison, the figure also contains the positions of 39 planetary nuclei (Harman and Seaton, 1966) and the evolutionary track of the nuclei of planetary nebulae determined by Seaton (1966).

An attempt at direct verification of the binary-star hypothesis has been made. First, the radial velocity of the central star was measured for six Balmer lines on three

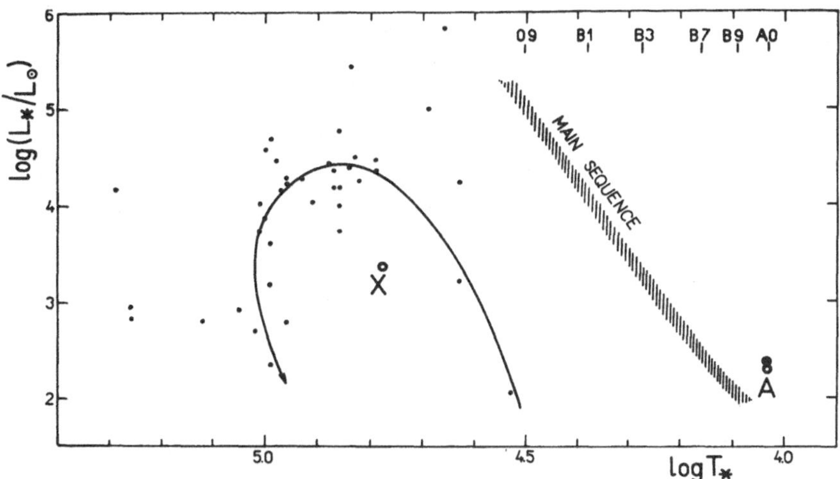

FIG. 3. *The position of NGC 1514 on the H-R. diagram: black circle = the undivided nucleus; A, X = the A- and X-components.*

Table 1

Date	V_{cor}
Jan. 8, 1954	+59·7 km/sec
Sep. 19, 1965	+86·5
Oct. 5, 1965	+66·2
Arithmetic mean	+70·8 km/sec

spectra made in Asiago (dispersion 74 Å/mm at Hγ) with the results given in Table 1, the mean error being about ± 8 km/sec.

It is of interest that the mean V_{cor} of the central star agrees with the radial velocity of the nebula, reported by Chopinet (1963), +70·6 km/sec. This agreement is good evidence supporting strongly the hypothesis that the observed A-component is actually connected with the nebula and that it forms a physical pair with the hot subluminous star. The probability of a random occultation of two early-type stars would have been very small. The next paper deals with further measurements of the radial velocity of this central star which have been carried out in Asiago.

Photoelectric observations of the central star with the 65 cm reflector at Ondřejov have not revealed any light changes greater than the observational errors. Recently Lawrence *et al.* (1967) found evidence suggesting periodic oscillations of 855 and 138 sec with amplitudes of 0·01–0·02 mag. More detailed photoelectric and spectrographic observations of this object are very desirable and they are part of the program for our 2-m reflector at Ondřejov. Similarly, additional evidence on the possible binary nature of other central stars of planetary nebulae would be of great value, particularly for the study of the evolution of central stars.

References

Chopinet, M. (1963) *Publ. Obs. Haute Provence*, **6**, No. 34.
Harman, R.J., Seaton, M.J. (1966) *Mon. Not. R. astr. Soc.*, **132**, 15.
Khromov, G.S., Kohoutek, L. (1968) in the present volume, p. 227.
Kohoutek, L. (1967) *Bull. astr. Inst. Csl.*, **18**, 103.
Kohoutek, L., Hekela, J. (1967) *Bull. astr. Inst. Csl.*, **18**, 203.
Lawrence, G.M., Ostriker, J.P., Hesser, J.E. (1967) *Astrophys. J.*, **148**, L161.
Seaton, M.J. (1966) *Mon. Not. R. astr. Soc.*, **132**, 113.

RADIAL VELOCITY OF THE CENTRAL STAR OF NGC 1514

A. MAMMANO, R. MARGONI, M. PERINOTTO
(Asiago Astrophysical Observatory, Italy)

We would like to report on some spectroscopic observations made of the central star of the planetary nebula NGC 1514, which we think can support the binary hypothesis now advanced by Kohoutek.

Between 1965 and 1967, 21 spectra (dispersions of 72 and 42 Å/mm at Hγ) have been obtained with the 122-cm reflector at Asiago. The radial velocity of the central star of NGC 1514 varies, according to our measurements, from $+35 \pm 10$ km/sec to $+95 \pm 10$ km/sec. Three maxima and three minima can be recognized among the 17 spectra measured so far. On the other hand, the physical association between the star and the nebula is confirmed by the agreement between the radial velocity of the absorption lines of the star and that of the nebular lines which is according to Chopinet (private communication), 70 ± 18 km/sec, obtained with a dispersion of 380 Å/mm.

These preliminary results appear to confirm the binary nature for the nucleus of NGC 1514, although further analysis is required to test the possibility that we are confronted with a close binary, and this work is in progress.

DISCUSSION

Miller: If the radial velocity variation of the star is interpreted as binary motion, what would be the best estimate of the orbital period?

Perinotto: It is not yet possible for us to give an estimate of the period. We need a larger number of spectra than we have.

A METHOD OF DETERMINATION OF THE TEMPERATURE OF NUCLEI OF THE PLANETARY NEBULAE

G. S. KHROMOV
(Sternberg State Astronomical Institute, Moscow, U.S.S.R.)

I would like to make a few comments on the problem of the determination of the temperatures of nuclei of planetary nebulae.

It can be shown that in a high-excitation planetary nebula two independent Strömgren zones of heavy elements can be found. The first is a He III zone; the second corresponds to a zone of the luminescence of Ne V forbidden lines and originates due to combined absorption by the heavy ions Ne IV, O V and N V. If so, a relatively independent method of the determination of the color temperature of nuclei can be suggested.

This method follows from a straightforward speculation, that in a stationary ionization zone the number of ionizations equals the number of recombinations and therefore, that the number of quanta in a corresponding interval of the stellar spectrum is proportional to the volume of this ionization zone.

If one knows the relative radii of some ionization zones in a given nebula one can calculate the relative intensities of the appropriate spectral intervals and then calculate the colour temperature of the star.

This method has been applied to 6 planetary nebulae with known angular diameters of the luminosity zones of He II and Ne V, and the results are presented in Table 1. All of the colour temperatures lie in an interval from 100000° to 200000°; their

Table 1

Temperatures of central stars

Designation	T_c	T_z	T_{HeII}
IC 2165	140000°	49000°	108000°
NGC 2392	160000°	25000°	
NGC 2440	150000°	42000°	100000°
NGC 6818	140000°	38000°	200000°
NGC 6886	160000°	52000°	
NGC 7662	130000°	78000°	

T_c = determinations by the author;
T_z = Zanstra temperatures (Zanstra, 1960; Gurzadjan, 1962);
T_{HeII} = He II temperatures (Gurzadian, 1962; Aller, 1960; Zanstra, 1960).

Osterbrock and O'Dell (eds.), Planetary Nebulae, 330–331. © *I.A.U.*

individual scatter being moderate. The approximate temperatures obtained are much higher than the Zanstra temperatures (Zanstra, 1960; Gurzadian, 1962), but are of the same order of magnitudes as those determined by the He II lines intensities (Gurzadian, 1962; Aller, 1960; Zanstra, 1960).

This result – independent of any distance or density estimates – probably proves that there is a significant ultraviolet excess in the spectra of nuclei of some planetary nebulae.

References

Aller, L. (1960) *Bull. astr. Inst. Netherl.*, **15**, 249.
Gurzadian, G. A. (1962) *Planetary Nebulae*, Moscow.
Zanstra, H. (1960) *Bull. astr. Inst. Netherl.*, **15**, 237.

DISCUSSION

Flower: I mentioned in my paper on Tuesday that I had made estimates of the combined opacities of ions such as Ne IV, O V, and N V at about 100 Å. I find that the combined opacity rises to 0·5–1·0 depending on the abundance values used. However, it is not clear that absorption of radiation by He II can be neglected in the Ne V ionization zone and if absorption by He II is significant, it will change the central star temperature estimates of Dr. Khromov.

Khromov: The calculations mentioned were very qualitative, and there is a danger that something has been missed. But I would like to emphasize, that an independent luminosity zone of [Ne V] still exists and I do not consider any other. It proves that there is an independent, corresponding luminosity zone. The contribution I have presented is a very brief account of a paper, which is in press in *Mon. Not. R. astr. Soc.* There are some facts in this original paper, which show that an additional source of opacity, if it still exists, does not affect the results significantly. (This paper has now appeared: Khromov, G.S. (1967) *Mon. Not. R. astr. Soc.*, **137**, 181.)

THE POSSIBILITY OF BETA DECAY IN THE ATMOSPHERES OF NUCLEI OF PLANETARY NEBULAE

G. A. Gurzadian
(Branch of Byurakan Astrophysical Observatory, U.S.S.R.)

At present there is some basis for questioning the usual presentation of the nuclei of planetary nebulae as being ordinary stars. If the quantity of mass released by a star during its formation or at some stage of its evolution serves as a criterion for its not being stationary, then the nuclei of planetary nebulae are hardly the most non-stationary objects. A strong deviation from Planck's radiation law may also serve as a criterion for non-stationarity when the deviation occurs in the far short-wave region of the spectrum in nuclei and similar objects (e.g. stars of Wolf-Rayet which are not nuclei of planetary nebulae). Finally, the existence of synchrotron radio radiation in some planetary nebulae must be explained by the exceptional activity of their nuclei. It would seem that the nuclei of these nebulae are suppliers if not of typical relativistic electrons then at least of electrons with high energies.

In the present report an attempt will be made to show that between these phenomena, there probably exists some common trait and these phenomena may be understood if the existence or appearance of so-called fast electrons is assumed (i.e. electrons with energies of the order 10^6–10^7 eV) in the outer regions of the atmosphere of the nucleus.

Primary attention in this article will, however, be devoted to the deviation from Planck's law of the radiation in the short-wavelength region of the spectrum of the nucleus.

One must consider as one of the important results of recent color observations of the nuclei of planetary nebulae the establishment of positive color indexes of a number of nuclei. Thus, according to data by Abell (1966), the color index of the nucleus of the nebula A 32 is $+0\overset{m}{.}62$, and that of A 37 is $+0\overset{m}{.}56$. In these cases it is not necessary to consider the influence of interstellar reddening because the Galactic declination of the first nebula is $+33\overset{\circ}{.}8$, and of the second $+42\overset{\circ}{.}2$. Among the remaining objects of Abell's list, even after introducing corrections for the effect of interstellar reddening, the color index of at least 10 nuclei appeared within the range of $+0\overset{m}{.}3$ to $+0\overset{m}{.}8$. The existence of red nuclei was also determined by Kazarian (1967) in a number of nebulae (e.g. for the nucleus of nebula V-V 421 the color index, corrected for interstellar extinction, is equal to $+0\overset{m}{.}32$). Stars with such color indexes cannot cause nebular luminescence. If we do not take into account the rather trivial assumption that the

Osterbrock and O'Dell (eds.), Planetary Nebulae, 332–338.

luminescence of the indicated nebulae may be caused by invisible hot components in the nuclei, then one must conclude that there is a great excess of luminescence in the short-wave region of the spectrum of these cold nuclei.

Let us consider another example. Among the nuclei of planetary nebulae we meet very many stars of the Wolf-Rayet type which are apparently different from the field Wolf-Rayet stars only in their brightness, being identical with them in other respects. However, it is known that the excitation temperatures found with the help of emission lines of various ions greatly vary from one another even within one particular star. In this case one observes a rather well-pronounced effect in that the excitation temperature rises with increased ionization potential of the atom. Figure 1 illustrates the

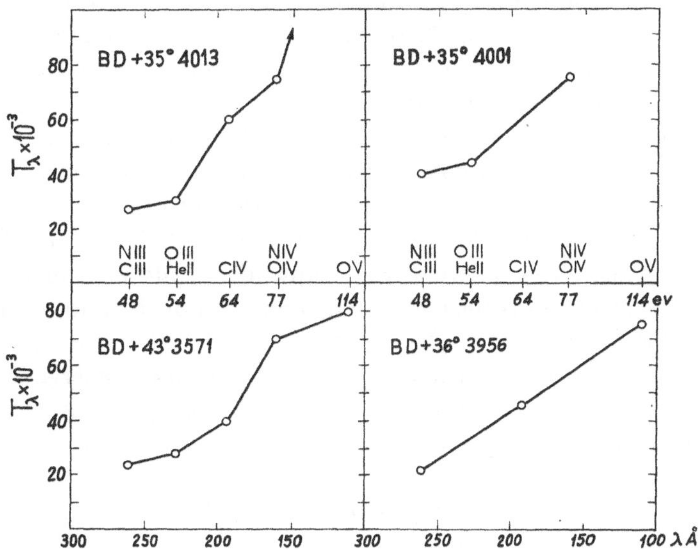

FIG. 1. *Color temperatures of central stars determined for several ions, plotted against the wavelength at the ionization boundary. The ionization potential is also shown.*

dependence of excitation temperature T_* on ionization potential (in this case as a function of the boundary wavelength λ of the ionization) for a few Wolf-Rayet stars. These diagrams are derived using Aller's (1943) data. Such a result is qualitatively obtained also when the temperature is determined by the method of Zanstra.

A number of attempts have been made to show the causes or factors which, if not considered when determining the temperature of Wolf-Rayet stars, might lead to such a result (e.g. increase of temperature with the increase of ionization potential). A detailed survey of the latter point has been made by Solobev. However, it is unlikely that future improvements of methods of temperature determination of Wolf-

Rayet stars, taking into account all possible factors, could introduce any basic change in the general character of the above regularity.

Therefore, by admitting that the increase of temperature in the short-wave region of the spectrum really takes place, one must conclude that a deviation from the perfect radiator exists in these stars and that the effect is greater at increasingly higher frequencies.

The inverse Compton-effect might be a possibility as a mechanism leading to such deviations, i.e. the scattering of light quanta on fast electrons with energies E, of the order of 10^6–10^7 eV. A formal examination of this process as applied to stellar atmospheres has been made by Gurzadian (1965).

Let us assume that the release or outflow of gaseous matter from the interior of Wolf-Rayet stars is accompanied by the appearance of fast electrons in its atmosphere. The scattering of photons by fast electrons takes place with an increase in frequency according to the correlation $\nu \approx \nu' (E/mc^2)^2$, where ν and ν' are the frequencies of the photon before and after collision with an electron.

Let us designate through τ the effective optical depth in fast electrons of the layer: $\tau = \sigma_e N$, where $\sigma_e = 0{\cdot}665 \times 10^{-24}$ cm^2, N is the total number of fast electrons in the column with the base 1 cm^2. The layer of fast electrons may be immediately on the photosphere or far from it. Let us assume that the intensity of the thermal radiation released from the photosphere at the frequency ν follows Planck's formula and is equal to $B_\nu(T_0)$, where T_0 is the effective temperature of the photosphere: $B_\nu(T_0)$ is simultaneously the intensity at the base of the layer of fast electrons, where $\tau = 0$.

Further assuming that the indicated layer consists of mono-energetic electrons, i.e. the energy of all electrons is equal to E, we shall have for the intensity $I_\nu(\mu, \tau)$ emitted from such a layer in some frequency ν (see details in Gurzadian, 1965):

$$I_\nu(\mu, \tau) = B_\nu(T_0)\, C_\nu(\mu, \tau), \tag{1}$$

where

$$C_\nu(\mu, \tau) = \left(1 + \frac{1}{4\pi} \frac{1}{\mu^4} \frac{\exp(x_0) - 1}{\exp(x_0/\mu^2) - 1}\, \tau\right) e^{-\tau}, \tag{2}$$

and where

$$\mu = E/mc^2; \quad x_0 = h\nu/\kappa T_0 .$$

From (1) and (2) follows, that as a result of the scattering of fast electrons on light quanta issuing from the photosphere (or from the atmosphere of the star) an increase in the intensity of the short-wave region of the spectrum takes place at the expense of its weakening in the long-wave region. The excess of the energy in the short-wave region of the spectrum, naturally, leads to a rise of the effective temperature in that region.

In order to obtain a quantitative estimate of the growth of the temperature in the various regions of the spectrum, let us substitute the function of $I_\nu(\mu, \tau)$ in (1) by using Planck's function at some effective temperature T_ν at a frequency ν. Putting also the expression $B_\nu(T_0)$ in (1) we obtain:

$$e^x = 1 + \left[1 + \frac{1}{4\pi\mu^4}\frac{\exp(x_0) - 1}{\exp(x_0/\mu^2) - 1}\tau\right]^{-1} [\exp(x_0) - 1]\, e^\tau,$$

where $x = h\nu/\kappa T_\nu$, and T_ν is the new temperature. In the regions of short waves (essentially those at $\lambda < 1000$ Å) we shall have from (3) to a sufficient degree of approximation:

$$T_\lambda = T_0 \frac{\mu^2}{1 + (\lambda\kappa/hc)(\ln 4\pi\mu^4 + \ln e^\tau\tau^{-1})\,\mu^2 T_0}. \tag{4}$$

Thus, if T_0, τ and μ are known, then with the help of (3) or (4) we can determine the effective temperature T_λ as it changes with wavelength.

In a circuitous manner the dependence of T_λ on λ was determined at three values of T_0 (10000°, 20000°, and 30000°) in the limits of τ from 0.1 to 0·00001 and at $\mu^2 = 10$, which corresponds to the energy of fast electrons of the order of $1{\cdot}5 \times 10^6$ eV. The results of the calculations are shown in Figure 2. The ionization boundaries of various ions are also indicated there.

As shown in the figures, the effective temperature in the short-wavelength regions - shorter than 300 Å – increases several times over the original temperature. In addition, the effective temperature increases monotonically with increasing frequency, i.e. with the growth of the ionization potential. It is characteristic that the effective temperature in the short-wave region can, according to observations (see Figure 1), increase the temperature of a star by a factor of 2 or more in the long-wavelength region even at very small values of the optical thickness of the layer of fast electrons, i.e. when $\tau \sim 10^{-4}$–10^{-5}. This ultraviolet excess is not of thermal origin. Therefore, it is seen that excitation, typical of gaseous nebulae and expanded envelopes of stars, is expected from a comparatively cold star if it is surrounded by a cloud of fast electrons.

It seems to us that the appearance of fast electrons on the outer regions of the atmosphere of a star is a general phenomenon appearing in all or almost all stars. For example, in the case of dwarfs of late spectral types, the presence of fast electrons may lead to a sharp increase of the brightness of a star in the visible ultraviolet region of the spectrum, to excitation of emission lines, etc. (these questions are examined in detail by Gurzadian, 1966). In the case of the Sun the hypothesis of fast electrons explains the observed spectra of X-rays of non-thermal origin (X-ray flares), etc. In the case of those hot stars which are the nuclei of planetary nebulae, the usual Wolf-Rayet stars and those of type O, the fast electrons lead to a sharp increase of the far ultraviolet end of their spectrum.

The truly remarkable thing is that in all of the above cases the best accordance, both qualitatively and quantitatively, of the theory with observations is obtained with $\mu^2 \simeq 10$, i.e. in the presence of fast electron energies of the order of 10^6–10^7 eV. This

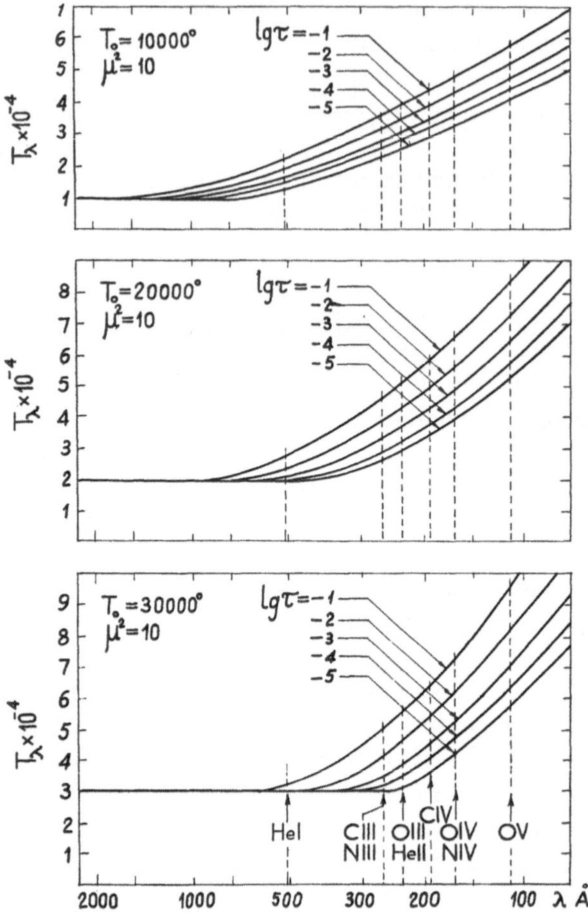

FIG. 2. *Theoretical color temperatures for atmospheres optically thick to fast electron scattering.*

energy is the order of the electron energy of β-decay of some unstable nuclei.

Ambartsumian (1954) showed that continual emission is connected with the wide range of phenomena observed in various non-stable objects, and pointed out the non-thermal nature of this emission. Refraining from suggesting a concrete physical mechanism, which might have led to the generation of this emission, Ambartsumian did, however, indicate the possibility of transporting energy in definite amounts from the internal parts of the stars to the outside layers by means of some mechanism unknown to us.

The above considerations allow us to draw the following conclusion: the carriers of stellar interior energy may be fast electrons which appear on the outer regions of the atmospheres of stars as a result of the β-decay of non-stable nuclei or nucleons. The latter, in turn, are ejected out to the external surface.

It is difficult to indicate a type of nucleus responsible for β-decay in stellar atmospheres. The decay of a neutron, during which protons and anti-neutrinos are discharged along with electrons, cannot explain the very quick generation of energy from flaring stars; the period of half-decay of neutrons is of the order of 12 min, i.e. comparatively great. In some respects β-decay of He^6 nuclei is of interest. First, the period of the decay of He^6 is not large – less than 1 sec. Secondly, the maximum of the energy of the decay electron is at $\mu \sim 3{\cdot}5$, i.e. $\mu^2 \approx 10$. Third, the product of the decay of He^6 is lithium, the anomalous content of which is established in many non-stationary stars. For example, in the atmospheres of T Tauri stars lithium is 100 times more abundant than in the Sun (Bonsak and Greenstein, 1960, Wallerstein *et al.*, 1965). We can parenthetically note that lithium cannot be discovered in the atmospheres of the nuclei of planetary nuclei. Because of the high temperature, lithium will be ionized there, and all of the lines connected with the basic energy level of Li II lie in the region 180 Å. Finally, helium is the second most abundant element in stars. That fact, however, does not mean that we admit the possibility of the short-lived isotope of helium He^6 issuing directly from the interior of the stars. The nuclei of He^6 are in all probability formed in the atmosphere of stars after the issue of interstellar substances.

Now back to the non-thermal radio emission of planetary nebulae. We do not yet possess strong proofs in favor of the synchrotron origin of this radiation; but it is possible to affirm that it must be synchrotron radiation. The analysis of the given observations indicates that the parameter (γ) of the energy spectrum of relativistic electrons is unexpectedly large, of the order of 4–5, and even larger. This must be contrasted with that of the cosmic radiation of non-thermal origin where it is not greater than three. This means that the spectrum of the relativistic electrons, responsible for radiation in planetary nebulae, is nearly mono-energetic, or rather similar to a gaussian, i.e. close to the energy spectrum of the electrons of β-decay. The generation of synchrotron radiation under the conditions of planetary nebulae ($H \simeq 10^{-3}$–10^{-4} gauss) requires that the energy of electrons should be 10^8–10^9 eV, which is 2–3 orders of magnitude more energy than electrons from β-decay. Such electrons may appear in a nebula either as a result of β-decay of nuclei unknown to us or by means of acceleration of electrons coming from β-decay already within the bounds of a nebula until the magnitude of 10^8–10^9 eV is reached. In the latter case the nucleus of a planetary nebula plays the part of injector of earlier accelerated particles breathing into the nebula electrons with energies of the order of 10^6–10^7 eV, and the nebula plays the part of an accelerator, where the electrons are driven until the energy of 10^8–10^9 eV is reached. How and in what regions of the nebula this acceleration takes place is not clear to us, particularly if we take into consideration the fact that according to ideas developed by Gurzadian (1962) magnetic fields in the basic envelopes of planetary nebulae must be of the bipolar type.

The difficulties connected with the means by which fast and relativistic electrons appear in planetary nebulae are great. Under such circumstances the possibility of highly energetic electrons generated in nebulae by other means should not be excluded. However this may occur, it is quite evident that the source of electrons must be the nuclei of nebulae. In that sense the classification of nuclei as objects with maximum non-stationarity is valid. In the present stage of the investigation of this matter we must, apparently, be satisfied with that conclusion according to which the appearance of fast electrons on the external regions of nuclei of nebulae and then their acceleration within the nebulae may explain the emission-line luminescence of planetary nebulae of comparatively low-temperature nuclei, as well as the fact of radio radiation having a synchrotron origin. The β-decay of some non-stable nuclei may be one of the possible mechanisms of the appearance of such electrons in the external regions of nuclei of nebulae.

References

Abell, G. (1966) *Astrophys. J.*, **144**, 259.
Aller, L.H. (1943) *Astrophys. J.*, **97**, 135.
Ambartsumian, V., Mustel, E., Severny, A., Sobolev, V. (1952) *Theoretical Astrophysics*, Moscow.
Ambartsumian, V. (1954) *Comm. Burakan Obs.*, **13**.
Bonsak, W.K., Greenstein, J.L. (1960) *Astrophys. J.*, **131**, 83.
Gurzadian, G.A. (1962) *Planetary Nebulae*, Moscow.
Gurzadian, G.A. (1965) *Astrophys.*, **1**, 319. (Also in *Doklady Acad. Sci. U.S.S.R.*, **166**, 1966, 53).
Gurzadian, G.A. (1966) *Astrophys.*, **2**, 217.
Kazarian, M. (1967) *Comm. Burakan Obs.*, **38**, 25.
Wallerstein, G., Herbig, G., Conti, P. (1965) *Astrophys. J.*, **141**, 610.

DISCUSSION

Reeves: Why do you want to get your high-energy electrons from β-decay? I cannot think of any way of making He^6 in reasonable amounts in stellar atmosphere or nebulae. Electromagnetic acceleration of electrons would seem much more reasonable.

Gurzadian: It is a very complex question and I cannot answer now. He^6 may be one of the possible, but not only source for the existence of the fast electrons in the outer regions of the star.

Salpeter: What was the assumed ratio of the rate of feeding energy into the β-decay electrons to the optical luminosity?

Gurzadian: Nearly 1000 if the optical thickness of the layer of fast electrons on the processes of Thomson scattering is 10^{-3}.

Thompson: The subject of non-thermal radio emission from planetary nebulae has been mentioned several times both today and in previous discussions. In case any misunderstanding remains, I should like to emphasize that there is really no significant evidence that non-thermal emission occurs. All cases of deviation from a thermal spectrum which have been examined in detail have been found to result from source confusion, and there is no reason to suppose that any of the remaining cases will be different.

SPECTRA OF THE CENTRAL STARS OF PLANETARY NEBULAE

LAWRENCE H. ALLER
(University of California, Los Angeles, Calif., U.S.A.)

1. Introduction

The central stars of planetary nebulae represent seemingly well-defined late stages of stellar evolution. Theoretical investigations with predictions of evolutionary tracks impose difficult requirements for observational data. Measurements of spectral energy distributions, of colors, and of magnitudes, and spectroscopic observations are all urgently needed.

Here one is confronted not only with the usual difficulties attending observations of faint stars, but also with interference from the radiation of the nebulae themselves. The nebular hydrogen and helium lines usually mask the corresponding absorption features of the stellar spectrum. If the nebular spectrum is of low excitation, stellar He II lines such as λ4686 or λ4541 may sometimes be observed in absorption. Incipient emission in λ4686 often fills in the corresponding stellar absorption line. Only for a few stars, such as the nucleus of NGC 246, NGC 1514, and NGC 3132 and the Abell objects (1966) observed by Greenstein and Minkowski (1964) are the nebulae so faint that interference from the nebulae spectrum is not significant.

Observations of the broad-emission-line spectra of central stars of the Wolf-Rayet type are less affected by confusion with nebular lines unless the star is very faint. Many of these stars show lines near λ4640 which coincide with radiations of nebular origin, the confusion being particularly serious for stellar planetaries.

Table 1 lists those nebulae whose nuclear spectra have been observed at Lick Observatory. A few are sufficiently bright to be photographed with the coudé spectrograph at the 120-inch. Most of them require observations with the air-Schmidt camera and nebular spectograph at the prime focus. The dispersion at the coudé is 16 Å/mm; at the prime focus the dispersion is 97 Å/mm. Table 2 lists those nebulae whose spectra display a strong continuum and in which no central stars have been observed. It is impossible to tell whether the continuum in some 'stellar' planetaries originates from the nebula or from a central star, but it is certain that in objects such as NGC 2440 or NGC 7027 the continuous spectrum is of nebular origin.

2. Planetary Nebulae with Continuous and Absorption-Line Spectra

Many objects listed as exhibiting purely continuous spectra may have weak ab-

Osterbrock and O'Dell (eds.), Planetary Nebulae, 339–354. © *I.A.U.*

Table 1

Planetary Nebulae Whose Central Stars have been Observed

Nebula	Description	Notes
NGC 40	WR	1, 2, 6
NGC 246	Of	1, 6
NGC 650	contin	4
IC 1747	WR	2
IC 351	contin	
NGC 1501	WR	2
IC 289	contin	
NGC 1514	A	1, strong K line
NGC 1535	O, abs	6
J 320		5
IC 418	Of	1, 2
NGC 2022	contin	
IC 2149	O, abs	1
NGC 2371-2	WR	
NGC 2346	A	
NGC 2392	Of	1, 2, 6
NGC 2438	contin	
NGC 2452	WR	2, 4
NGC 3132	A	strong K
NGC 3242	contin	6
NGC 4361	O, abs	3
IC 3568	O, abs	3
NGC 6058	O, abs	3
IC 4593	Of	1, 6
NGC 6210	Of	1, 6
IC 4634	contin?	
NGC 6309	contin	
NGC 6445	contin	4
NGC 6543	Of	1, 2, 6
NGC 6572	Of?	1, 5
NGC 6567	contin	
NGC 6629	O, abs	2
NGC 6720	contin	
NGC 6751	WR	1, 2
NGC 6778	contin	
NGC 6790	contin	5
NGC 6803	contin	
NGC 6804	O, abs	3
BD + 30°3639	WR	1, 2, 6
NGC 6818	contin	
NGC 6826	Of	1, 2, 6
NGC 6853	O, abs	3
NGC 6884	contin?	5
NGC 6886	contin	4
NGC 6891	Of	1, 2
IC 4997	Of	1, 5, 6
NGC 6905	WR	1, 2
NGC 7008	O, abs	3
NGC 7009	contin	6
NGC 7026	WR	2
IC 5217	contin?	5
NGC 7354	contin	4
NGC 7662	contin	6
CD−32°14673	O, abs	
VV 286	contin?	5

WR = Wolf-Rayet type.
contin = continuous with no absorption or emission.
O, abs = continuous with absorption lines.
Of = continuous with narrow emission lines and possibly also absorption features.

Notes:

[1] Photometric measurements and/or detailed descriptions have been published (Aller, 1948, 1956; Andrillat, 1959; Kohoutek, 1966; Swings, 1940; Swings and Struve, 1940; Swings and Swensson, 1952; Wilson, 1948; Wilson and Aller, 1954).

[2] Photometric measurements and descriptions are in preparation.

[3] Absorption-line intensities are listed in Table 3.

[4] Spectrum is very weak.

[5] Stellar origin not certain or nebular continuous spectrum may seriously influence stellar spectrum.

[6] Observed with coudé dispersion, 16 Å/mm.

Table 2

Nebulae with Strong Continuous Spectra for which no Central Stars have been Observed

Anon 0^h26^m	NGC 6537	NGC 6741
IC 2165	NGC 6644[1]	NGC 6833[1]
NGC 2440	IC 4732[1]	NGC 7027
J 900	IC 4776[1]	IC 5117[1]

[1] Stellar origin not certain or nebular continuous spectrum may seriously influence stellar spectrum.

sorption lines which, at the dispersions employed, are masked by the overlying nebular emission. Emission lines, characteristic of Of stars may also be missed if they coincide with strong nebular lines. We can assert with some confidence that bright stars such as the nuclei of NGC 3242, NGC 7009, and NGC 7662 have purely continuous spectra with no detectable trace of emission or absorption features.

Observations of the spectra of the brighter absorption-line stars, such as the nucleus of NGC 6891, 6826, IC 4593, and NGC 6210 substantiate earlier studies (Aller, 1948; Wilson and Aller, 1954) and are being analysed now by Miss Sarah Heap. Table 3

Table 3

Absorption-Line Spectra of Several Planetary Nuclei
Intensities expressed in equivalent widths W_λ

λ	NGC 4361	II 3568	NGC 6058	NGC 6804	NGC 6853	NGC 7008
4861		2·1				1·8
4686	1·5			3·	2·0	1·4
4541	0·9	0·8				
4340	1·7	1·0	2·7	2·7	1·7	1·8
4200		0·5				
4101·7	1·6	0·6	2·6	2·7	0·7	2·1
3970	1·2	0·8	2·0	1·2		1·7
3933						0·4
3889	1·1	1·0	1·7	1·0	0·9	1·1
3835	1·0	0·6	1·5			0·8
4797		0·4	1·2			0·7
3770		0·2	1·0			0·6
3750			0·3			0·4
Suggested Spectral Class	06	05	09?	09	07	07

lists approximate intensities expressed as equivalent widths for stars observed at the prime focus. At a dispersion of 97 Å/mm with an overlapping nebular line spectrum, only rough estimates are possible. Furthermore, note that the presence of an overlying nebular continuum causes all of the tabulated equivalent widths to be lower limits. Therefore, although the tabulated intensities may be useful for estimating spectral classes, they should be regarded as little better than eye estimates. Generally, I widened the spectra by drifting, but only a small degree of widening is possible for the spectrum of NGC 4361 and the tabulated intensities are, therefore, particularly uncertain.

In CD $-32°$ 14673 ($\alpha = 18^h 53^m 24^s$, $\delta = -32°17'$), the nebular Balmer lines completely mask any stellar hydrogen features, while the Pickering λ4541 and λ4200 He II lines have equivalent widths of 0·6 Å and 0·5 Å, respectively.

Further studies of these objects should be carried out with high dispersion and an image converter. Walker and Aller have carried out a few preliminary tests at the coudé focus of the 120-inch.

3. The Spectrum of the Nucleus of NGC 6543

Descriptions of the spectrum of the central star of NGC 6543 have been given by Swings (1940); for quantitative estimates of line intensities see Aller (1943, 1956, p. 211). With the coudé spectrograph of the 120-inch it is possible to measure not only the total intensities of the nebular lines but also their profiles.

With the higher dispersion and decreased contamination from the nebular spectrum gained at the coudé focus, Hγ and Hδ show profiles suggestive of the P Cygni type. The higher members of the series appear in absorption. The Balmer-line profiles are complicated because they are blended with the lines of ionized helium. Hence the centers appear displaced from the nebular emission lines.

Table 4 compares the spectrum observed at three different dates: In 1945 with a Cassegrain spectrograph at McDonald Observatory (Aller, 1956) and on June 18 and August 16, 1965, at Lick Observatory. In 1945 the NV λ4603 and λ4619 lines seemed to show P Cygni characteristics; later they seemed to be mostly in emission. Some change was noted in the appearance of complex structures between λ4630 and λ4659. The O III, O IV, N III, and N IV lines observed shortward of λ3760 show comparable intensities, although there is some hint that the $3s^3S - 3p^3P$ transitions were stronger on plate ES 4325 than on either other date of observation. The λ4686 He II line shows definite intensity and profile variations.

Line intensity and profile changes are well known in classical Wolf-Rayet stars where they may be of very short duration, Oke (1954). Probably the variations in planetary nebulae nuclei are also of very short duration (Wilson and Aller, 1954), but we have no easy means of proving this – unless an image converter can be used.

Thus as has been suggested often, it seems likely that the Of nuclei are stars with unstable atmospheres which are ejecting material into the surrounding nebular shells at the present time.

Table 4

The Spectrum of the Nucleus of NGC 6543

λ	Ion	1945	June 1965 EC 4329	Aug 1965 EC 4458
4685·8	He II	13·8	9·1	12·6
4658·6	C IV	2·5	0·6	1·6
4648	C III	1·6	pr	0·6:
4641	N III		0·74	0·74
4632	N III		0·5	0·6
		0·4	0·35	0·3?
4619·4	N V	0·2a	pr	pr
		0·7	0·99	0·79
4603·2	N V	0·3a	–	pr
4541·6	He II	0·8a		
			0·3:	0·3:
4340	Hγ		0·4a:	0·4a:
4116	Si IV		0·20	0·2
4101	Hδ		0·3:	0·3:
			0·5a:	0·5a:
4088	Si IV	0·8	0·6	0·5
		1·0		
4058	N IV	0·16a	1·2	1·6
3970	H, He		1·24a	1·6a
3889	H, He	1·3a	1·1a	0·7a
3835	H	0·6a	0·7a	0·7a
3797	H	0·7a	0·6a	0·7a
3771			0·5	0·5a
3760	O III		0·3	0·3
3757	O III		0·2	0·4
3755	O III/N III		0·3	0·4
3737	O IV		0·3	0·4
3714	N IV		0·4	0·4
3563·4	O IV	0·6	0·6	0·6
3560·9	O IV	0·7	0·6	0·8
3555	He I	–	0·3	0·2
3484·9	N IV	0·4	0·9	0·5
3483·0	N IV	0·5	0·8	0·5
3478·7	N IV	0·5	1·6:	0·6
3409 ± 3411·8	O IV	2·1	2·8	2·4
3403·6	O IV	0·8	1·0	1·0
3396·8	O IV	0·2	0·6	0·5
3385·5	O IV	0·3	0·7	0·4
3381·3	O IV	0·3		
3375	O IV	0·4	0·3:	

An absorption feature is denoted by 'a'.

4. Spectra of the Wolf-Rayet Type

We summarize here only some general features of the spectra of the Wolf-Rayet-like nuclei. Since the pioneer studies of the nuclei of NGC 40 (Paddock, 1915) and of BD

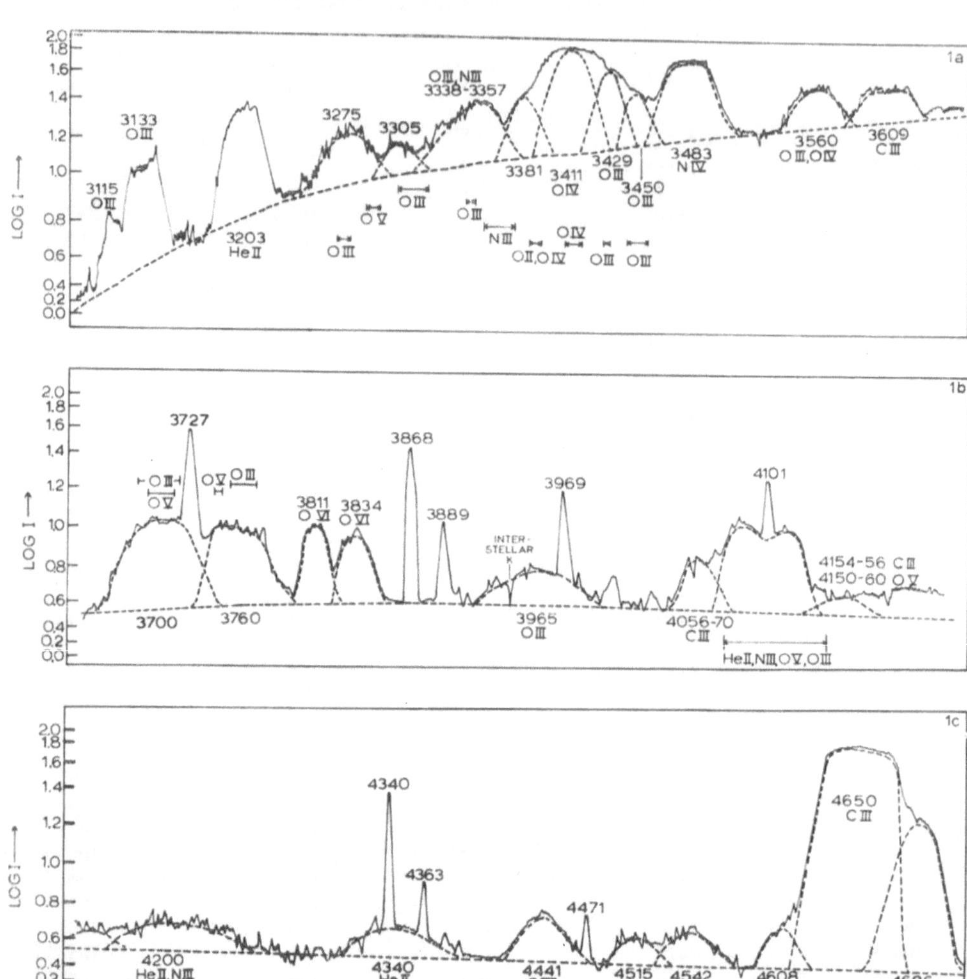

FIG. 1. *Tracings of the spectra of the Nuclei of NGC 6751, λ3100 – λ4400. In this and following figures the position of the continuum and suggested resolution of some complex emission structures are indicated by dotted lines. Lines of nebular origin appear as sharp spikes. A logarithmic scale of intensities is indicated at the left.*

FIG. 2. *Comparison of the λ4400 – λ4700 region in the central stars of NGC 6905, NGC 7026, IC 1747, IC 2003 and NGC 2371-72. Note the great strength of the emission features in NGC 6905, NGC 7026 and IC 1747. In IC 2003, the emission features of the nuclear spectrum are overwhelmed by the strong lines in the nebular spectrum.*

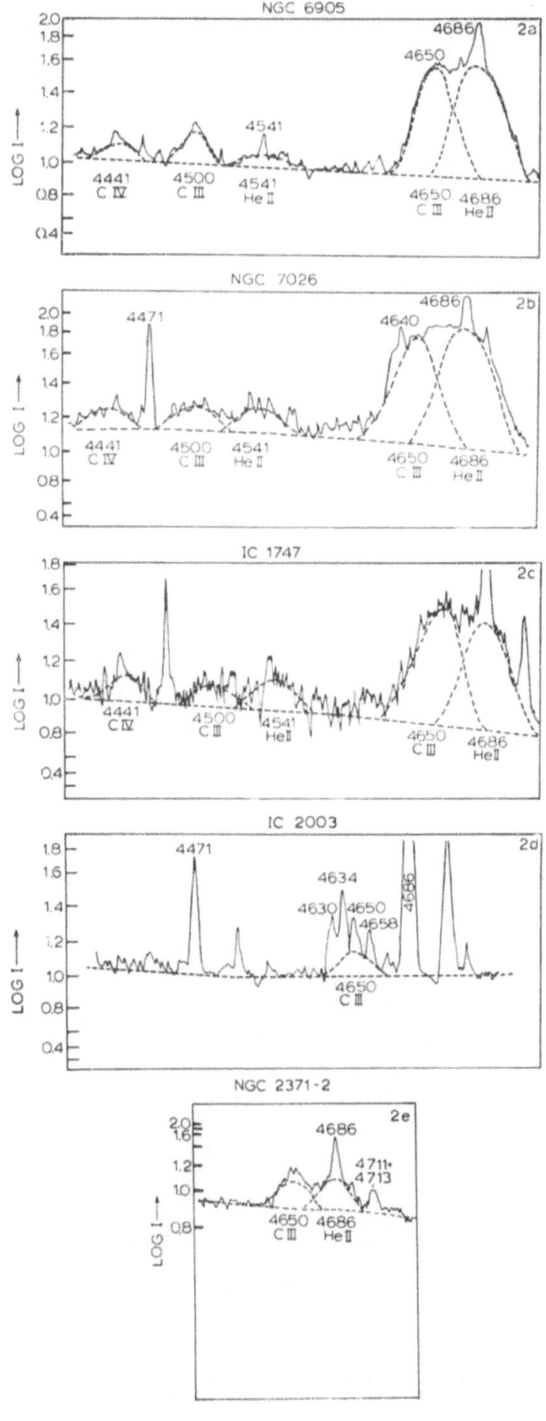
NGC 6905
2a
4686
4650
4541
4441
C IV
4500
C III
4541
He II
4650
C III
4686
He II
LOG I
NGC 7026
2b
4471
4640
4686
4441
C IV
4500
C III
4541
He II
4650
C III
4686
He II
IC 1747
2c
4441
C IV
4500
C III
4541
He II
4650
C III
4686
He II
IC 2003
2d
4471
4634
4630
4650
4658
4686
4650
C III
NGC 2371-2
2e
4686
4711+
4713
4650
C III
4686
He II

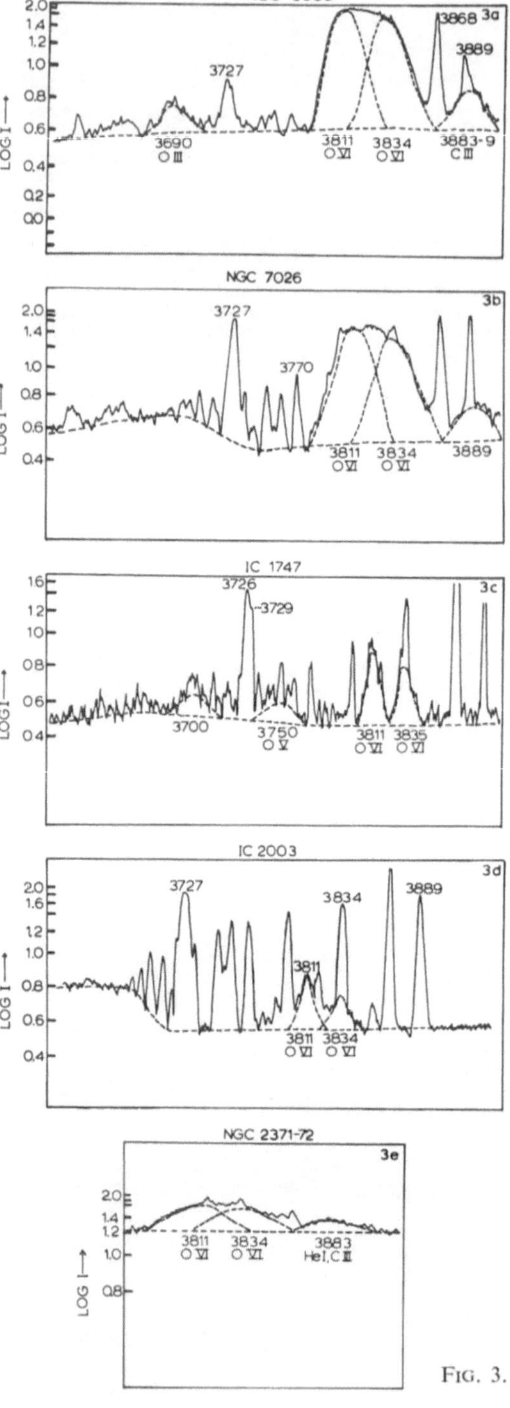

NGC 6905
3a
3727
3868
3889
3690
O III
3811
O VI
3834
O VI
3883-9
C III
LOG I
NGC 7026
3b
3727
3770
3811
O VI
3834
O VI
3889
IC 1747
3c
3726
3729
3700
3750
O V
3811
O VI
3835
O VI
IC 2003
3d
3727
3834
3889
3811
3811
O VI
3834
O VI
NGC 2371-72
3e
3811
O VI
3834
O VI
3883
HeI, C III

FIG. 3.

$+30°3639$ (Campbell, 1894) many investigations of these objects have been published (for a summary of earlier work see e.g. Aller, 1956, p. 205–209).

A full discussion of these Wolf-Rayet spectra must await the completion of a study now being conducted jointly by the author and Miss Lindsey Smith at Lick Observatory. In this program we compare the spectra of the planetary nuclei with those of classical Wolf-Rayet stars. Some general remarks can be supplied at this time.

Figure 7 compares the spectra of several planetary nuclei of the Wolf-Rayet type as photographed with the Lick prime focus nebular spectrograph. The spectrum of a B9 star, θ Crateris, is given for comparison. In Figures 1–4 we reproduce tracings of

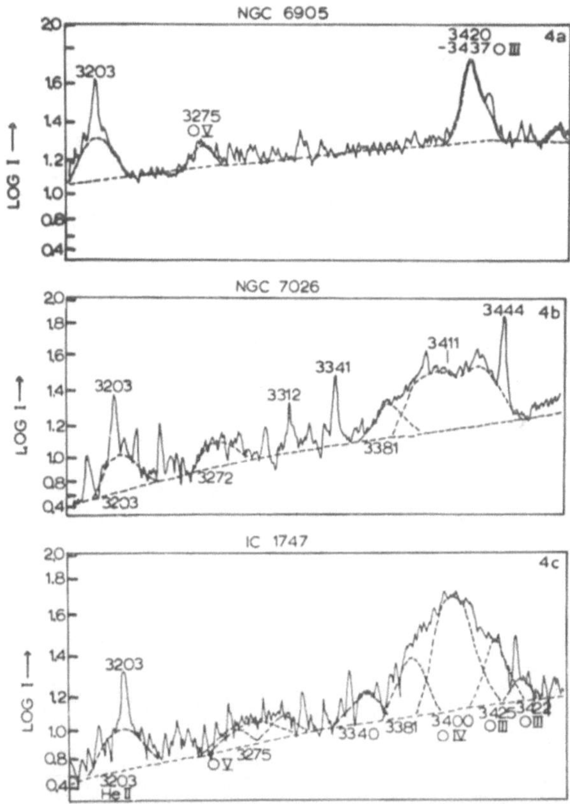

FIG. 4. *The ultraviolet region of the spectra of the central stars of NGC 6905, NGC 7026 and IC 1747. The O*III *lines (λ3420–3437) in NGC 6905 are narrow compared with the complex structures in NGC 7026 and IC 1747. The dotted lines indicate a possible resolution of this structure into separate components.*

FIG. 3. *The O*VI *λ3811, 3834 features in the central stars of NGC 6905, NGC 7026, IC 1747, IC 2003, and NGC 2371-72. The broad O*VI *lines attain great strength in the nuclei of NGC 6905 and NGC 7026, but are weaker in IC 1747, and in IC 2003 where the continuum is largely of nebular origin. In the central star of NGC 2371-72, the O*VI *lines have broad profiles upon which are superposed narrow emission lines.*

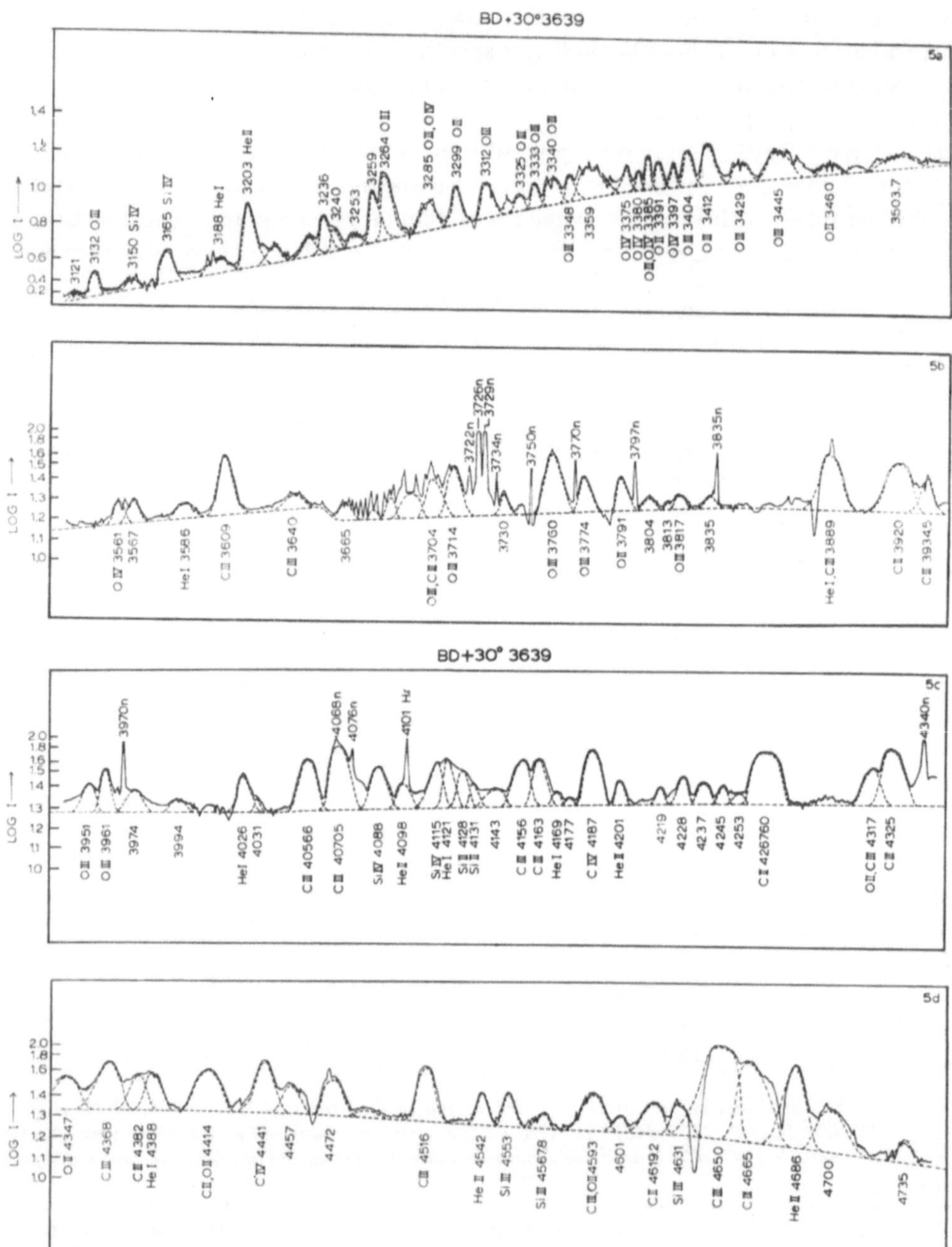
BD+30°3639
5a
LOG I
3121
3132 O III
3150 Si IV
3165 Si IV
3188 He I
3203 He II
3236
3240
3253
3259
3264 O II
3285 O II, O IV
3299 O III
3312 O III
3325 O III
3333 O III
3340 O III
O III 3348
3359
3412
3429
3445
3460
3503.7
5b
3722n
3726n
3729n
3734n
3750n
3770n
3797n
3835n
He I 3586
C III 3609
C III 3640
3665
3730
O III 3760
O III 3774
3804
3835
He I, C III 3889
C II 3920
BD+30° 3639
5c
3970n
4068n
4076n
4101 Hδ
4340n
3974
3994
He I 4026
4031
C III 40566
C III 40705
Si IV 4088
He II 4098
Si IV 4115
He I 4121
4143
C III 4156
C III 4163
He I 4169
4177
C IV 4187
He II 4201
4219
4228
4237
4245
4253
C II 426760
C III 4325
5d
C III 4368
4457
4472
C III 4516
He II 4542
Si III 4553
Si III 45678
4601
C II 46192
Si III 4631
C III 4650
C III 4665
He II 4686
4700
4735

FIG. 5a.

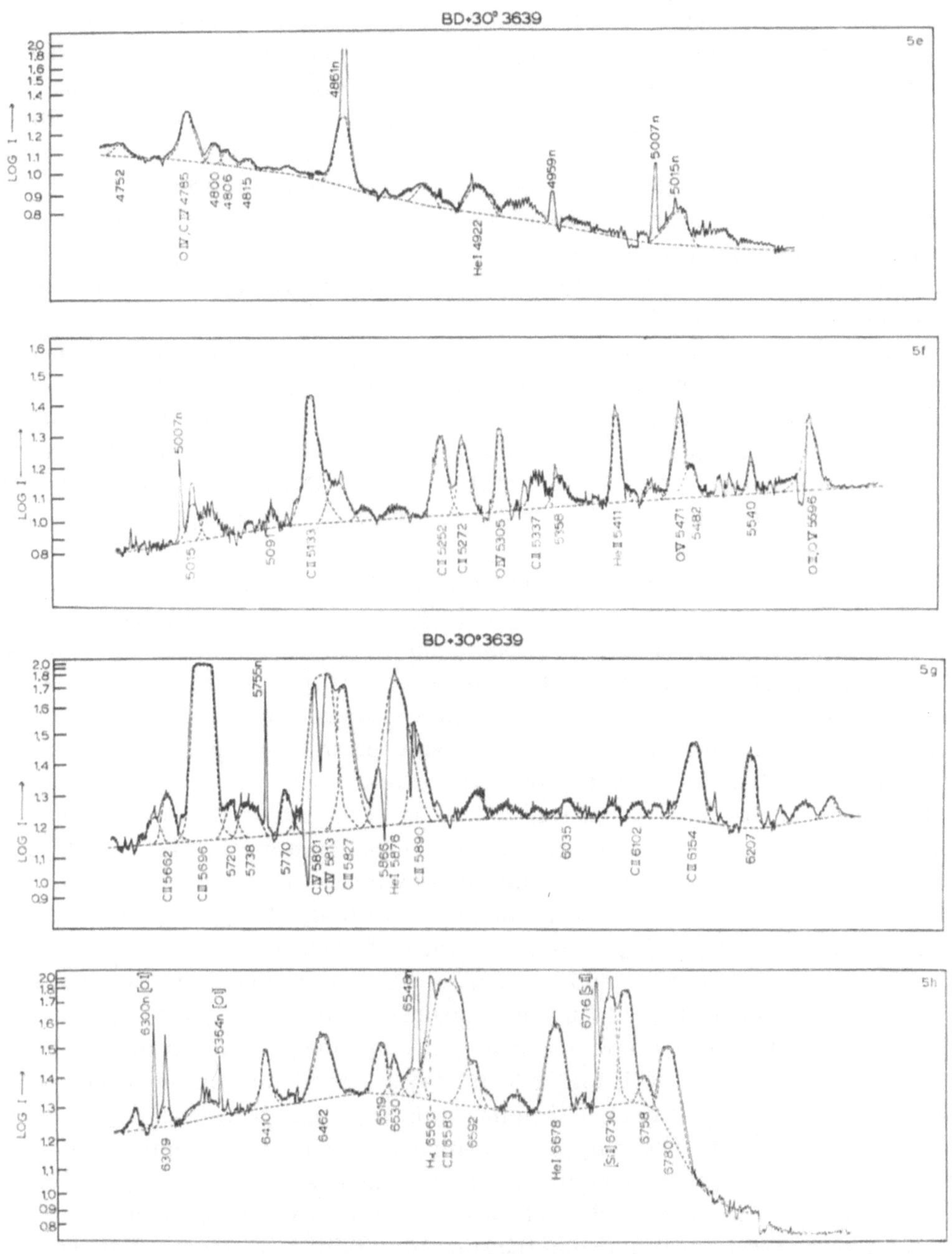

FIG. 5b.

FIG. 5. *A tracing of the spectrum of the central star of BD + 30° 3639, λ3100–λ6800. The spectrum is characterized by the sharpness of its lines. Note the absorption features on the violet edges of the lines of He λ3889, C III λ4650, C IV λ5801, and He I λ5876. Tracings a–e were made from plates secured with an Eastman IIaO emulsion; f, g, and h on plates secured with an Eastman IIaF emulsion.*

the more interesting regions of the spectra. Figure 1 shows a tracing of the spectrum of the nucleus of NGC 6751; this strongly resembles spectra of classical Wolf-Rayet stars of the carbon sequence, although N IV $\lambda 3483$ is present and strong. The dotted lines indicate the estimated position of the continuum and the suggested resolution of several strong lines into their respective components. The (relative) log I scale is indicated on the left, but no attempt has been made to allow for the wavelength variations of the transmission of the atmosphere and the spectrograph, or for the wavelength variation of the system sensitivity.

Figures 2a–e compare the $\lambda 4430$ – $\lambda 4700$ regions of the spectra of the nuclei of NGC 6905, NGC 7026, IC 1747, IC 2003, and NGC 2371-2. In Figures 3a–e a similar comparison is made of the $\lambda 3650$ – $\lambda 3900$ regions, while Figures 4a–c compare the ultraviolet regions of the spectra of the nuclei of NGC 6905, NGC 7026, and IC 1747. The spectra have been arranged in order of decreasing strength of the O VI doublet – $\lambda 3811$, $\lambda 3834$. In the spectrum of the nucleus of NGC 2371-2 these lines display sharp cores and broad wings. The spectrum of the nucleus of NGC 246 represents a further step in this sequence; in this spectrum the O VI lines are very sharp and weak with no underlying broad feature.

Quantitative comparison is rendered difficult by the fact that the continua are strongly affected by contributions from the nebulae as well as by the presence of overlying emission lines. The great strength and dominance of the O VI lines in some of these spectra is not paralleled among the spectra of the classical Wolf-Rayet stars.

The nuclei of NGC 1501 and NGC 2452 are also objects of this peculiar Wolf-Rayet type, resembling the nucleus of NGC 6905, with spectra displaying strong broad O VI lines and broad C III $\lambda 4650$ as the dominant features.

Campbell's hydrogen envelope star is, however, characterized by relatively narrow emission lines. Figure 5 displays tracings of the spectrum from $\lambda 3121$ to $\lambda 6800$. The presumed background continuum is indicated by the dotted line; it seems to be fairly well defined, except in the red where the plate sensitivity changes rapidly with wavelength. Notice the absorption lines associated with metastable levels of He I, $\lambda 3889$, C III $\lambda 4650$, C IV $\lambda 5801$, 5813, and also $\lambda 5876$. The carbon lines, especially C II $\lambda 5696$ and 6580 and C III $\lambda 4650$, dominate the spectrum. A detailed comparison of this stellar spectrum with that of its nearest classical WR analogue is now in progress.

The nucleus of NGC 40 is also a Wolf-Rayet star of the carbon sequence, with numerous broad, strong lines of helium, carbon, and oxygen ions. Except for the wavelength-dependent effects of plate sensitivity, transmission of optics, and atmospheric transmission, the tracings are on a true intensity scale. For the region $\lambda 4650$ to $\lambda 5900$ I am indebted to Merle Walker for observations with the Lallemand electronic camera; in the green spectral region the star is too faint to be recorded at the coudé on the usual emulsions in reasonable exposure times.

Czyzak and Aller have obtained photoelectric scans of the spectra of both BD + 30° 3639 and NGC 40 with the Cassegrain spectrum scanner at the 60-inch telescope on

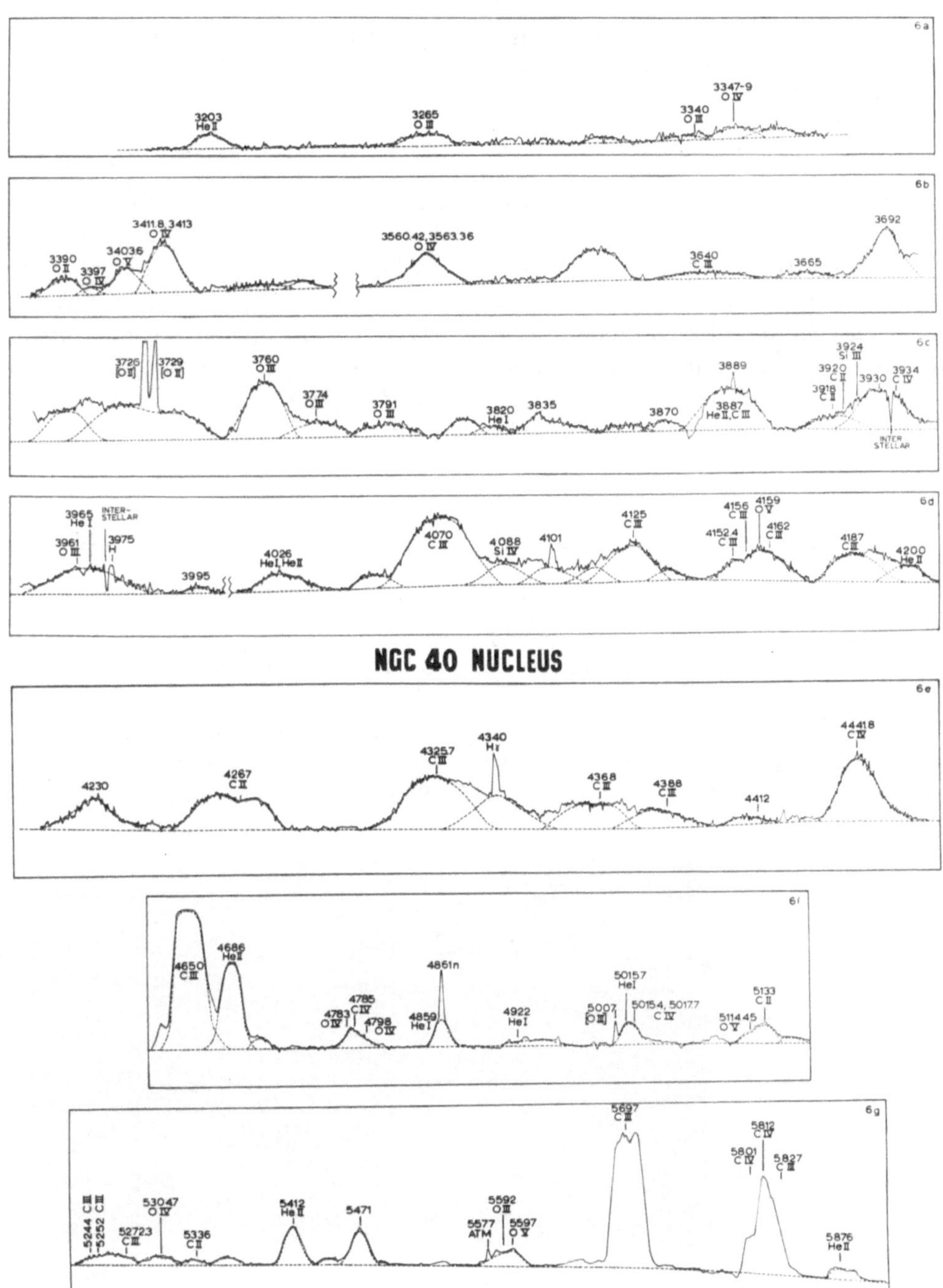

FIG. 6. *Tracing of the nuclear spectrum of NGC 40. The tracings are on a direct intensity scale. The region λ4650–λ5876 was observed with the Lallemand cell in collaboration with Merle F. Walker.*

Mt. Wilson. These data will make it possible to put all the photographic relative intensity measurements on an absolute scale.

5. Comments on the Hertzsprung Russell Diagram

A number of workers have prepared $\log L/L_{\odot}$ vs. $\log T_{\text{eff}}$ plots for the central stars of planetary nebulae. Uncertainties in the bolometric corrections make such plots highly subjective. The well-known discordances between the temperatures appropriate to the spectra of absorption line stars and the 'Zanstra temperatures' of these same objects as obtained by Harman and Seaton (1966) point up some of the difficulties involved. One may interpret the H and He II line intensities with the aid of Mihalas' (1965) model atmospheres with $\log g = 4{\cdot}0$ and $N(\text{He})/N(\text{H}) = 0{\cdot}15$. We find the following comparison with Seaton's (1966) results:

Central star temperatures (thousands of degrees)

Object	spec.	LHA	MJS	Object	spec.	LHA	MJS	Object	spec.	LHA	MJS
IC 2149	07·5	35 ± 2	49	NGC 6058	09?	33 ± 1	72	NGC 6826	07	39 ± 2	69
NGC 2392	06	43 ± 5	67	IC 4593	06	41 ± 2		NGC 6853	07	40 ± 2	132
NGC 4361	06	40	–	NGC 6210	06	42 ± 2	50	NGC 6891	07	37 ± 2	55
IC 3568	05	50	–	NGC 6804	09	33 ± 1	72	NGC 7008	07	37 ± 1	98

Miss Heap, who is examining this problem in detail, expresses the hope that discrepancies may be eliminated for some objects such as the nuclei of IC 2149, NGC 6210, and NGC 6891; those of NGC 6853 and 7008 present problems.

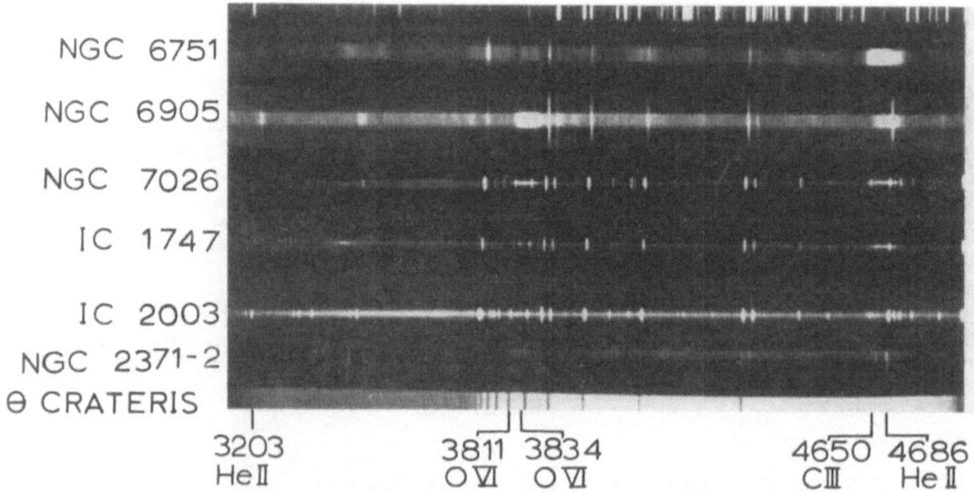

FIG. 7. *Comparison of the spectra of several stars of the Wolf-Rayet type.*

If one uses Minkowski's distance scale (quoted by Aller, 1965), O'Dell's (1963) bolometric corrections, central star magnitudes, and temperatures when spectral data are not available, the stars listed in Table 1 are found to yield an HR diagram that shows considerable scatter, neither confirming nor contradicting the sequence proposed by Seaton (1966). Very real uncertainties in the distance scale in addition to those in the temperature and bolometric corrections obliterate any well-defined evolutionary sequences that may exist. Also, if many of these central stars are binaries – as seems indeed true for NGC 1514 (Kohoutek, 1966), it would be very difficult to construct a meaningful HR diagram.

O'Dell and others have tried to segregate the various types of stars in the HR diagram and it may be that if lower temperatures and luminosities are accepted for the absorption-line stars, the different types of spectra may correspond to different domains in an HR diagram.

7. Summary

Spectra of planetary nuclei obtained with the Lick 120-inch telescope give additional data for a number of stars heretofore inadequately observed and give more detailed information on a number of the brighter stars. Characteristics of nuclei with Wolf-Rayet-like spectra are shown in detail.

The contradiction between stellar temperatures deduced from Zanstra methods and those appropriate to their spectral classes is found to hold for additional objects. Thus radii and luminosities deduced for these stars must be taken with reserve pending resolution of this difficulty. It is suggested that uncertainties in our present data on distances, temperatures and luminosities, preclude the deduction of precise evolutionary tracks for these stars.

Acknowledgements

Expenses for necessary travel to Lick Observatory to secure observations were supplied from University of California, Los Angeles and NASA Grant NsG 237-62. Reductions of all observational data were supported by AFOSR Grant 83-65 to UCLA from Air Force Office of Scientific Research. Bruce Bohanan assisted in the reductions of the data for NGC 6543. I am grateful to Merle Walker for permission to reproduce the tracing of the spectrum of the NGC 40 nucleus obtained with the Lallemand cell. Particular gratitude is due to Miss Lindsey Smith for her careful review of the manuscript and for many helpful discussions of problems of Wolf-Rayet stars.

References

Abell, G.O. (1966) *Astrophys. J.*, **144**, 259.
Aller, L.H. (1943) *Astrophys. J.*, **97**, 135.
Aller, L.H. (1948) *Astrophys. J.*, **108**, 462.
Aller, L.H. (1956) *Gaseous Nebulae*, Chapman & Hall, London.

Aller, L.H. (1965) *Landolt-Bornstein Tables, Group VI*, Vol. I in the series: Springer, Berlin, p. 566.
Andrillat, Y., Andrillat, H. (1959) *Ann. Astrophys.*, **22**, 104.
Campbell, W.W. (1894) *Astronomy and Astrophysics*, **13**, 461.
Harman, R.J., Seaton, M.J. (1966) *Mon. Not. R. astr. Soc.*, **132**, 15.
Kohoutek, L. (1967) *Bull. astr. Inst. Csl.*, **18**, 103.
Mihalas, D. (1965) *Astrophys. J. Suppl.*, **9**, 321.
O'Dell, C.R. (1963) *Astrophys. J.*, **138**, 67.
Minkowski, R., Greenstein, J. (1964) *Astrophys. J.*, **140**, 1601.
Oke, J.B. (1954) *Astrophys. J.*, **120**, 22.
Paddock, G.F. (1915) *Lick Obs. Bull.* **9**, 30.
Seaton, M.J. (1966) *Mon. Not. R. astr. Soc.*, **132**, 113.
Swings, P. (1940) *Astrophys. J.*, **92**, 289.
Swings, P., Struve, O. (1940) *Proc. nat. Acad. Sci. Am.*, **26**, 454, 548.
Swings, P., Swensson, J.W. (1952) *Ann. Astrophys.*, **15**, 290.
Wilson, O.C. (1948) *Astrophys. J.*, **108**, 201.
Wilson, O.C., Aller, L.H. (1954) *Astrophys. J.*, **119**, 243.

GENERAL DISCUSSION - FIFTH SESSION

Münch: All of us who have attempted to construct high-temperature model atmospheres are aware of the difficulties introduced by the large values of the radiation pressure and its gradient. The problem is not purely computational but it may actually happen that the outer layers are not in hydrostatic equilibrium and continuous mass loss is taking place. On the observational side, I may remark that in the bluest stars known (those at the extreme blue sequence of the globular clusters), although possessing extremely blue B-V colors, the line spectrum is definitely cooler, and I suggest that there is a false photosphere which is optically thick in the Balmer lines but not in the Paschen continuum.

Underhill: When an O-type model atmosphere is constructed ($T_{eff} \simeq 35000°$) including the opacity due to the resonance lines between 912 and 1500 Å, the radiation-pressure gradient term becomes very large in the outer atmosphere. A hydrostatic equilibrium atmosphere could only be obtained by empirically reducing the radiation-pressure gradient to an acceptably small value in the outer part of the atmosphere.

No difficulty with standard-model atmosphere methods is encountered if the line blanketing is left out of consideration. Models without line blanketing in the far UV, however, are not very real from a physical viewpoint. Considerable doubt exists about their value in interpreting the spectra of stars of type B I and earlier.

Aller: The profiles of the hydrogen and helium lines seem to be similar to those in main-sequence stars (see e.g. the paper by Wilson and Aller (1954), *Astrophys. J.*, **119**, 243). The lines are evidently broadened by Stark effects rather than by rotation. While shells are certainly present in some instances, e.g. the nuclei with Of stars and NGC 6543 which may resemble P Cygni, I believe that we are seeing the 'normal' photospheres of the stars in many instances, whatever constitutes the normal photosphere of an O star.

Seaton: I would like to emphasize again that, if one trusts the results of model atmosphere calculations, then one should use the calculated fluxes (instead of black-body fluxes) in deducing temperatures by Zanstra-type methods.

Capriotti: I would like to report on some recent work done by William Kovach and me. We have used the Böhm and Deinzer models in order to estimate the effective temperatures of the central stars. We have used the same method that Harman and Seaton employed. One has

$$\frac{I(\lambda 4686)}{I_{pg}(\text{star})} \sim \int_{\nu_0}^{\infty} \frac{F_\nu(T_{eff})}{h\nu} \, d\nu \Big/ \int_0^{\infty} F_\nu(T_{eff}) \, S_{pg} \, d\nu . \qquad (1)$$

Osterbrock and O'Dell (eds.), Planetary Nebulae, 355–358. © *I.A.U.*

$I(\lambda 4686)$ is the observed intensity in the $\lambda 4686$ line of He^+ and I_{pg} (star) is the observed intensity in the continuum of the central star in the photographic region of the spectrum. ν_0 is the frequency that corresponds to the ionization potential of He^+, $F_\nu(T_{eff})$ is the emergent flux of a Böhm and Deinzer model star with effective temperature T_{eff}, and S_{pg} is a filter function. We can re-write Equation (1) as

$$\frac{I\,(4686)}{I_{pg}\,(\text{star})} \sim \left[\int_{\nu_0}^{\infty} \frac{B_\nu(T_{eff})}{h\nu}\,d\nu \Big/ \int_{0}^{\infty} B_\nu(T_{eff})\,S_{pg}\,d\nu\right] \times \left[\int_{\nu_0}^{\infty} \frac{F_\nu(T_{eff})}{h\nu}\,d\nu \Big/ \int_{\nu_0}^{\infty} \frac{B_\nu(T_{eff})}{h\nu}\,d\nu\right]. \tag{2}$$

$B_\nu(T_{eff})$ is the Planck function. We have left $F_\nu(T_{eff})$ in integrand in the denominator of Equation (1) because in the frequency region where the product $F_\nu(T_{eff})\,S_{pg}$ is large, the Böhm and Deinzer model for a given value of T_{eff} is just like a black-body model. The Böhm and Deinzer models show flux deficits at frequencies greater than ν_0.

Therefore

$$\int_{\nu_0}^{\infty} \frac{F_\nu(T_{eff})}{h\nu}\,d\nu \Big/ \int_{\nu_0}^{\infty} \frac{B_\nu(T_{eff})}{h\nu}\,d\nu < 1\,. \tag{3}$$

Since the product on the right-hand side of Equation (2) is equal to some observationally determined quantity, then the quantity $[\int_{\nu_0}^{\infty} \{B_\nu(T_{eff})/h\nu\}\,d\nu / \int_{\nu_0}^{\infty} B_\nu(T_{eff})\,S_{pg}\,d\nu]$ has to be as big or bigger when one uses a Böhm and Deinzer model than it is when one uses a black-body spectral distribution. Consequently, the effective temperature that one obtains using a Böhm and Deinzer model is as big or bigger than the effective temperature obtained through the use of a black-body model. We obtained temperature differences as large as 24000 °K. The highest temperature considered by Böhm and Deinzer was 150000 °K. Therefore, for temperatures larger than that we had to extrapolate.

At any rate, as said before, it is a bit premature to conclude that the central stars of planetary nebulae have flux excesses at frequencies greater than ν_0 on the grounds that then the predicted $\lambda 4686/H\beta$ intensity ratios would agree with the observed intensity ratios. One would first have to construct a model nebula with a central star having a temperature that is consistent with the type of model star used. A more detailed report on this study has now been published: Capriotti, E.R., Kovach, W.S. (1968), *Astrophys. J.*, **151**, 991 (*Ed.*)

Osterbrock: Are the abundances indicated by line spectra of planetary nuclear stars consistent with the assumption that their atmospheres have the same composition as nebulae?

Aller: I know of no observational data inconsistent with this assumption. The spectra of the nuclei of planetaries do not appear to be inconsistent with the suggestion that they have a chemical composition similar to that of the surrounding nebula. I think that we must regard their compositions as normal.

Kohoutek: One comment on the distance of NGC 1514. I derived the mean distance from the following individual values applying our photographic and photoelectric data:

Author	*Method* pg parsec	pe parsec
Shklovsky (1956)	416	359
O'Dell (1962)	720	621
Abell (1966)	522	488
Kohoutek (1962)	488	383
Perek (1963)	442	397

Mean distance 484 ± (m.e.) parsec

In this case the accuracy of the mean value is less than 10%.

Underhill: The object NGC 1514 seems to be at a distance of 480 parsec where the radial velocity of the A0 III star and nebula is about +70 km/sec. This velocity is much larger than that expected for an A-type star close to the Sun; thus, one finds reason to conclude that the system may indeed belong to an old fast-moving population.

Perek: The radial velocity of NGC 1514 is not normal for an A-type star but it is not normal for a planetary nebula at that galactic longitude either. It is rather an outside value.

Evans: I would urge further observations of the radial velocity of this central star to determine the orbital elements. In the ordinary way the mass function of a single-lined binary is not very interesting but in this case a good estimate of the minimum mass of the hypothetical blue star could be made. The mean radial velocity is very high for a population-I star of this type.

There is a considerable resemblance to the A star in NGC 3132 where the star and nebular velocities did not come out quite the same but the possibility of velocity variation did not occur to us. It will be worthwhile referring to this. Incidentally, it seems probable that there is no considerable relative proper motion of the A star because Sir John Herschel – a very accurate draftsman – made a drawing in the period 1834–38 which closely resembles the modern picture.

Westerlund: Number 36 in Henize's Catalogue of Southern Planetary Nebulae has a star of A type centered in it. The velocity difference between this star and nebula appears similar to that of the star in NGC 3132 and the nebula itself.

Münch: When we observe faint blue stars at high galactic latitude we find mostly spectra which look like late B type, indistinguishable at low dispersion, from 'normal' main-sequence stars of the same type. I do not see why we have to insist on making the central star of NGC 1514 a normal main-sequence A0 type. More likely it is a star of population II such as those found in globular clusters immediately outside the RR Lyrae gap.

Shao: The existence of a planetary nebula in a binary system is very interesting and important in studying the evolution of the physical properties of the planetary component. Therefore, I should like to call attention to another object, VV 68. It was first discovered by Vorontsov-Velyaminov in 1960 and listed as a probable planetary. The star is $V_{\mathrm{m}}=8{\cdot}5$ and the spectral type is as late as B 9. The surface brightness of the nebulosity is very faint (about 15th mag/arc-minute^2). If this star is the real nucleus of the planetary, it would make it the brightest central star of all planetary nuclei. Therefore, we need both photometric and spectroscopic studies of this object.

Abell: There was a preliminary finding list of planetary nebulae discovered on the Palomar Sky Survey in *Publ. astr. Soc. Pacific*, **67** (1955), 258. Later a more complete list of new planetaries, and a discussion of them, appeared in *Astrophys. J.*, **144**, (1966), 259. The latter list not only includes the objects in the former list, but also additional ones, and the catalogue numbers are different. To avoid confusion I urge investigators to refer to their numbers in the *Astrophys. J.*, not the earlier *Publ. astr. Soc. Pacific* note.

Session VI

ORIGIN AND EVOLUTION

OBSERVATIONAL ASPECTS OF THE EVOLUTION OF PLANETARY NEBULAE AND THEIR CENTRAL STARS

C. R. O'DELL
(The Yerkes Observatory, Williams Bay, Wisc., U.S.A.)

The greatest activity in the area of evolution of the central stars of the planetary nebulae has occurred in the last two decades, although there have been significant contributions by many workers of considerable insight ever since the early studies at the Lick Observatory demonstrated the physical nature of this class of nebula. We shall try to draw together here the many facets of the present picture of this rapidly developing and highly significant phase of stellar evolution. Unfortunately, the picture is not at all as complete as one would like, and there remain several points of scientific contention; however, we should like to present the results of several investigators. A certain amount of redundancy with previous papers given prior will be necessary, but we shall endeavor to treat these topics briefly and in the context of the evolution of the nebulae and their stars.

Although this is a very rapid state of stellar evolution, the study of the changes in the nebulae and their stars is singularly favored. This is true for simple discovery because of their emission-line nature, which yields to surveys at relatively high spectral purity by filters and objective-prism spectroscopy techniques, that has resulted in the discovery and identification of a relatively complete sample in the solar neighborhood and a significant fraction of those objects in as far as the Galactic centre. The astrophysical process of changing the abundant high-energy ultraviolet central star photons into optical emission lines through photo-ionization followed by recombination and collisional excitation of metastable levels means that we have tools for the determination of the stellar and nebular temperatures, masses, luminosities and distances. This feature of conversion of photons from the spectral region of highest luminosity into readily observable photons is the greatest advantage of these systems and has made it possible to say more about the planetary nebula systems than any other advanced evolutionary system, except perhaps for the white dwarfs.

The discussion of the evolution of these systems depends primarily upon two features: (1) the determination of central star temperatures (since temperatures and the bolometric corrections are large), and (2) the determination of the distances to the nebulae (in order to find the apparent luminosities). We would like to briefly discuss these two areas and how they affect the present evolutionary picture.

Osterbrock and O'Dell (eds.), Planetary Nebulae, 361–375. © *I.A.U.*

1. Stellar Temperatures

Stellar temperatures may be inferred from spectroscopic characteristics and from the astrophysical Zanstra (1926) method. We shall first consider the former.

With a very few exceptions the assigned spectral types are quite early, falling into the O classification. There is a broad range of subdivisions, however; these are listed in Table 1.

Table 1

Stellar Types	Spectral Characteristics
Wolf-Rayet	Broad emission lines Of, He II, C III, etc.
O types	Hydrogen, He I and He II in absorption
Of types	Narrow emission of C III, N III, He II, etc.
Continuum	No emission or absorption features seen at the low dispersions employed

The division into types is by no means as clear cut and simple as this table would indicate. The absorption lines in the O types range from strong, deep, and narrow lines to those verging upon the continuum classification because of the low central depth and great width of their lines. The selectively brighter (apparently and intrinsically) Wolf-Rayet, strong line O's and Of types were studied early, and it was noted by Aller (1956) and others that a large range of line widths exists both in emission and absorption. Greenstein and Minkowski (1964) published the results of a study selectively oriented to the central stars associated with the lower surface brightness nebulae. In that investigation it was seen that generally the hydrogen lines were broad and weak and that the degree of excitation was quite high, even finding strong O V in absorption and O VI in emission in NGC 246. The crude selection effects of first studying the high surface brightness, classical nebulae and then those of lower surface brightness give some indication of different spectral characteristics associated with various nebulae. This tendency is much more clearly delineated when one considers the final absolute luminosity-temperature tracks that are derived. However, there is no strong evidence that there exists a single valued, monotonic sequence of spectral types, although there does exist the general feature of the most luminous central stars usually having stronger and deeper absorption lines – often associated with broad emission lines similar to the Of and Wolf-Rayet stars. The intrinsically less luminous stars have selectively broader and shallower absorption lines (or even none at all) than do the earlier spectral types and show emission and absorption features of high states of ionization. The observational difficulties in obtaining adequate spectra of these stars are considerable, arising from the superimposed strong nebular emission and from the faintness of the stars themselves. There remains much to be done in this area for observers possessing adequate equipment.

More quantitative measures of the stellar temperatures can be derived from the original method developed by Zanstra (1926) and its variations. In its simplest form, this theory derives from the supposition that all stellar photons capable of the ionization of a hydrogen atom will do so and that upon recombination and cascade of the ionized and excited hydrogen atoms, a sufficient optical depth in Lyman-line radiation prevails so that each original photo-ionization is ultimately followed by the emission of a Balmer line or continuum photon. This condition is stated in the first equation.

$$N(\text{Stellar Lyman Continuum Photons}) \geq N(\text{Nebular Balmer Photons}),$$

where the inequality arises if the nebula is not optically thick to the Lyman continuum. Under the assumption of knowledge of the continuum flux distribution of the central star, one can then derive the functional relation between the stellar and nebular brightness

$$m_{\text{star}} + 2{\cdot}5 \log F(\text{H}\beta) \leqslant 2{\cdot}5 \log F(T_s) + \text{Constant}.$$

This method is in essence a very broad base-line color index, where the emission-line flux in Hβ is used as a quantitative measure of the amount of ultraviolet stellar radiation. From the resultant temperatures and continuing the assumption that the flux distribution is known (usually a blackbody), it is a simple matter to calculate bolometric corrections (O'Dell, 1963*a*).

This same basic method can also be extended to other abundant elements, such as He I and He II, as was originally done by Würm and Singer (1952) and more recently by Harman and Seaton (1966). The primary advantage lies with He II since a nebula becomes optically thin to it at a much later stage in its expansion than does hydrogen. This means that He II Zanstra temperatures can be used rigorously after the hydrogen temperatures have become lower limits. The question of the validity of the usual assumption of a stellar atmosphere emitting as a blackbody must, of course, arise; since it is well known that in the population-I early-type stars very severe depressions of the flux at the Lyman discontinuity can occur. This can be easily tested by reference to the detailed model atmospheres calculated by Gebbie and Seaton (1963) and by Böhm and Deinzer (1965). The latter models, which were calculated for temperatures and surface gravities relevant to the early evolutionary picture indicate that the deviations from the black-body assumption in derived temperatures and absolute luminosities are not extremely large compared to the other uncertainties involved. The necessary corrections are in the sense that the effective temperatures should be larger than derived and will be most important for the coolest central stars.

2. Distances to the Nebulae

The distance to the planetary nebulae has been the subject of considerable discussion during the last several years (Shklovsky, 1956; O'Dell, 1962; Seaton, 1966; Abell, 1966) owing to an increased appreciation of the importance of this parameter and the existence of the accurate emission-line brightnesses coming from the work of Liller (1955) and of Osterbrock and his students (Capriotti and Daub, 1960; Collins *et al.*, 1961; O'Dell, 1962, 1963*b*). Space does not permit a discussion of the difficulties of the calibration of the distance scales, but we shall briefly review the basic considerations of this problem. It is now widely recognized that planetary nebulae exist in both the completely ionized and partially ionized states, depending upon their ultraviolet luminosity and nebular size and density. The method of distance determination to be applied depends upon the condition of ionization. The basic equations describing the observable characteristics of a planetary nebula are given below:

$$\mathscr{F}(\mathrm{H}\beta) = 1{\cdot}33\pi S^3 \varepsilon N_e^2 \alpha E(\mathrm{H}\beta)$$

$$\Phi = S/D$$

$$F(\mathrm{H}\beta) = F(\mathrm{H}\beta)/4\pi D^2 ,$$

where $\mathscr{F}$ is the total Hβ luminosity of a nebula of linear radius S, density N_e, and fraction of volume filled by material ε. $N_e^2 \alpha E(\mathrm{H}\beta)$ is the volume emissivity in Hβ. The observed angular radius (Φ) and observed flux $F(\mathrm{H}\beta)$ are, of course, related to the distance D, so that one can write an expression for the linear radius

$$S = 3F(\mathrm{H}\beta)/\Phi^2 \varepsilon \alpha E(\mathrm{H}\beta) N_e^2 .$$

To apply this method one needs to know the flux in Hβ, the angular size, the approximate electron temperature, the filling factor (estimated from photographs) and the density (usually found from forbidden-line data). This simple expression is rigorous and universally applicable, and has most widely been used by Seaton (1966) and his collaborators. There are no assumptions made of the Lyman-continuum optical depth, this being a great advantage of the method. However, this method has in practice been applied only to the highest surface brightness nebulae because of the difficulty of obtaining for the lower surface brightness nebulae the emission-line ratios necessary for the determination of the electron densities. The basic limitation of this method is the strong effects of filamentary structure on the resultant nebular distance and hence stellar luminosity. The filling factor used by Seaton (1966) was based upon examination of direct photographs and primarily considered the large-scale geometric distribution. This filling factor may be a considerable overestimate, however, because of the possible presence of unresolved filamentary structure, something often seen in very good photographs and indicated from analysis of forbidden-line ratios (Seaton and Osterbrock, 1957). The derived luminosity by this method is also very sensitive to each observable parameter:

$$L_{\mathrm{star}} = [36\pi/\alpha^2 E^2(\mathrm{H}\beta)^2]\,[l_{\mathrm{star}}\, F(\mathrm{H}\beta)^2/\phi^6 \varepsilon^2 N_e^4] .$$

For conditions of density not so large that collisional de-excitation is important, the density derived from forbidden-line ratios (usually the [O III] λ3727 doublet) are weighted towards being representative of the denser filaments since the emissivity increases as the square of the density. This would mean that errors would arise in the use of the luminosity relation above since a large part of the nebular emission might come from the low-density regions. The magnitude of such effects can be seen from considering a two-state nebula where $(N_{\text{high}}/N_{\text{low}})^2 \gg 1$ and $\varepsilon_{\text{low}}/\varepsilon_{\text{high}} > 1$, a model suggested by Seaton and Osterbrock's measurements of NGC 7027. In this case the correction factor in the luminosity expression becomes $(\varepsilon_{\text{low}}/\varepsilon_{\text{high}})^2$, e.g., in a nebula where three times as much volume is filled by the low-density material as for the high density, the true luminosity will be nine times larger than that derived from the elementary application of this method. The postulation of such filaments as general features of the nebulae is somewhat *ad hoc* in nature and may not generally be applicable; however, the luminosities derived by this method should probably be considered as lower limits.

It is possible (under much more restrictive conditions) to determine the distances of planetary nebulae for which the densities are not known. If the nebulae are completely ionized, the ionized (visible) radius will equal that of the gas itself ($S = R$) and the mass of the nebula is given by

$$M = 1{\cdot}33\pi R^3 \varepsilon N_e m_H .$$

Combining this expression with those given before and eliminating the electron density one can derive the generalized expression for optically thin nebulae:

$$R = \left[\frac{3\alpha E(\mathrm{H}\beta)}{16\pi^2 m_{\mathrm{H}}^2} \cdot \frac{M}{\varepsilon} \cdot \frac{\phi^2}{F(\mathrm{H}\beta)}\right]^{1/5},$$

which was first used extensively in only slightly different form by Šklovsky (1956). This method requires a determination of the factor $(M^2/\varepsilon)^{1/5}$. The usual assumption has been that the variations in this quantity are sufficiently small that an average value can be determined from calibration procedures on a few objects, and this value can then be applied to all other optically thin nebulae. O'Dell (1962) has tested this assumption of constancy by examination of the density-surface brightness relation expected, while Seaton (1966) has noted a leveling-off of a mass function as one passes from optically thick to optically thin nebulae. Seaton has employed this apparent upper limit for the mass derived from a few optically thin planetaries of known electron densities for calibration of his distance scale, while O'Dell has relied upon the average calibration found from the expansion of NGC 6720, the binary nature of the central star of NGC 246 and statistical parallaxes. Seaton's published distances are 45% larger than O'Dell's for optically thin nebulae. If one tries to apply the optically thin method to an optically thick nebula, an overestimate of the distance will result – a fact applied by Minkowski by using both methods on a single object to discriminate thick and thin nebulae. The most useful criterion for discrimination between types is the appearance of neutral oxygen forbidden lines that arise in the interface of ionization while relative strengths of He III, He II and H II lines can be helpful. Although by no means precise, it is possible to obtain distances good to a factor of 2 for most individual planetary nebulae at this time.

3. Resultant Picture of the Evolution

The evolution is seen in the regions that are populated in a logarithmic plot of stellar temperature and luminosity. Actual evolution through this populated region is indicated by the fact that the mean size of the nebular shell changes from sub-region to sub-region. Since it is reasonable to assume that the nebulae are all rather similar, and it is known that the material is in expansion, one can trace the sense of the evolution and determine its time-scale. In order to tie the entire picture together it is necessary to use the results of several approaches, depending upon the conditions of optical depth; we shall, therefore, discuss three sub-regions of stellar luminosity separately.

A. The *earliest stages* are characterized by a relatively high luminosity, temperatures not as high as found later and very dense surrounding nebulae. The latter characteristics dominate to make this a region where the nebulae are selectively optically thick to ionizing radiation, a feature brought out in Galactic planetaries by the presence of the [O I] lines. There are two main sources of material for this region – the distances of Seaton (1966) on the optically thick scale plus hydrogen Zanstra temperatures for Galactic objects, and the study in the Magellanic Clouds by Miss Webster (1967). Unfortunately, there are relatively few optically thick Galactic planetaries with low central star temperatures so that it is possible to easily misinterpret the form of the early evolution. The data by Miss Webster in the Magellanic Clouds is quite homogeneous due to the constant and known distance to the nebulae and she has derived Zanstra temperatures in the usual manner. In this case the cool end of the sequence is relatively well populated and it is possible to obtain a clearer picture. A final object falling into this region is the planetary nebula K 648 in the globular cluster M 15 (O'Dell *et al.*, 1964) whose distance is accurately known (for these purposes).

There seems to be little question about a general rise in stellar luminosity with increasing temperature in this phase, even after allowance for the many sources of uncertainty. The only remaining controversy is the degree to which the planetaries hook upwards. Seaton has argued that his two points at $\log L$ about 2 indicates a very strong hook while the Magellanic Cloud data of Miss Webster and K 648 would indicate that the rise occurs, but that there is essentially no other evidence for its occurring before a logarithmic luminosity of $+3{\cdot}5$. A very strong hook might be more compatible with the idealized models of Vila (1966), Shaviv (1967), L'Ecuyer (1966), Rose (1966) and Bautz (1968), although models such as some of those of Deinzer (1967) are flatter in this region. Although there has not been a thorough, homogeneous statistical approach to the spectral study of these central stars, there is obviously a very large, if not nearly complete, fraction that display Wolf-Rayet features.

The existence of W-R features in the spectra of stars has often been associated with mass loss, and it is relevant to ask if there is evidence of continued mass loss from the central stars to the nebulae during this period. This question is very hard to settle since one can determine a mass only for the ionized region, which is continuously changing as the mean density decreases. The disappearance of the W-R features at hotter, later stages argues that mass loss cannot continue much beyond the time when a typical object is identified as a planetary nebula system. The radius range in this earliest stage is about 0·04–0·12 parsecs.

B. What we shall designate here as the *intermediate stages* those nebulae characterized by a rather well-defined low optical depth in the hydrogen Lyman continuum. The neutral oxygen lines have disappeared and the constant mass method can be employed to find the distances. The absorption lines in the stars have become weaker and broader and the broad emission features are usually absent. The very fact that allows us to obtain the relatively good distances means that the hydrogen Zanstra temperatures will only be lower limits. Seaton (1966) has avoided this difficulty by adopting criteria based upon relative line strengths to pick out those objects that remain optically thick to HeI, so that the helium Zanstra temperature can provide a rigorous value. We show in the illustration the results of inclusion of these stars, where one can see that the range of temperatures is not extremely large. Across this region there appears to be a general decrease in luminosity as the nebular radius increases from 0·12 to 0·25 parsecs. There are many more nebulae that have already been identified and probably fall into this class, but they do not appear here because the test for the validity of the helium Zanstra temperature requires emission-line ratios, which are not known for many of these objects, since this is a region where the surface brightness of the nebula is decreasing as the fifth power of the nebular radius. In actuality this region is probably populated as heavily as the earliest stage.

C. In the *late stages* the radii of the gas shells are over 0·25 parsecs due to continued expansion of the material. Since the stars seem to have dropped significantly in luminosity, but have not changed greatly in temperature, the effective stellar radii are significantly smaller than in the prior two stages; thus, the radiation dilution has increased. At the same time the mean nebular gas density is continuously decreasing, which would increase the volume ionized; therefore, we have competing mechanisms for determining the optical depth in the Lyman continuum, a fact pointed out by Seaton (1966). The total number of ionizing photons emitted by the star per unit time can easily be related to the ionization radius, as can the luminosity of the star and its temperature and radius. Since there exists an average relation between the nebular density and the total gas radius, we can derive a general expression for the ratio of the ionization radius and the gas radius,

$$\frac{S}{R} \simeq 48 \frac{R}{\varepsilon^{1/3}} \left(\frac{L}{T_{\text{star}}}\right)^{1/3},$$

which will be a function of the shell size, and the stellar temperature and luminosity. This means that the transition between complete and partial ionization will be described by a straight line of small slope in our logarithmic temperature-luminosity plot, with the ordinate shifted according to the total gas radius. That is, at a given point on the diagram all nebulae smaller than the transition size will be optically thick. By inserting numerical values for the derived sizes one can see that a few of the very low surface brightness nebulae are probably optically thick and many more fall into a marginal zone.

We plot in Figure 1 the hydrogen Zanstra temperatures and the luminosities based on the optically thin distance scale – which involves a logical inconsistency. However, the temperatures are probably rather close while the luminosities are not critically dependent upon the exact ionized mass due to the low power of the mass dependence. Actually, there is a very interesting problem all along the optically thin sequence. This is the fact that the hydrogen Zanstra temperatures never become greatly different from the true values while one would naively theoretically expect that the lower limit would become much lower than the true value as the transition to being completely ionized occurs. Capriotti has derived quantitative values for the optical depth in the Lyman continuum by reference to the He I Zanstra temperatures and has found values of the optical depth of less than one in many cases although an abrupt drop of several orders of magnitude as might be expected is not present.

The effect of underestimating the Zanstra temperatures in this region would move the stellar points up and to the left while the application of the wrong method of distance determination would lower the points in the ordinate. Definitely, there are uncertainties in the positions of the individual points in this region, but these are probably small compared with the intrinsic range of positions that are observed. The occupied region is reasonably well defined in our luminosity-temperature plot. The points shown on the figure correspond only to those known systems that, from the sizes of their nebular envelopes, fall close to regaining the condition of significant hydrogen Lyman continuum optical depth and hence populates the lower region to the exclusion of the upper – within the classification of the late stage of development.

It is very important to note that there seems to be a rather continuous sequence of conditions present here, carrying from the high-luminosity stars down through those close to the solar luminosity. The lowest-luminosity stars are comparable in luminosity, temperature, and spectrum to some of the hottest white dwarfs that have been studied. It is altogether reasonable to think that a progression of nebular sizes represents a progression in age since the appearance and nature of the objects are very similar. This enables one to derive an approximate time-scale for passing through the various stages discussed here. Taking as a typical maximum size radius the value 0·7 parsecs and an average expansion velocity of 20 km/sec, we find a time-scale of about 35000 years. This low value, indicating very rapid evolution of the illuminating stars is

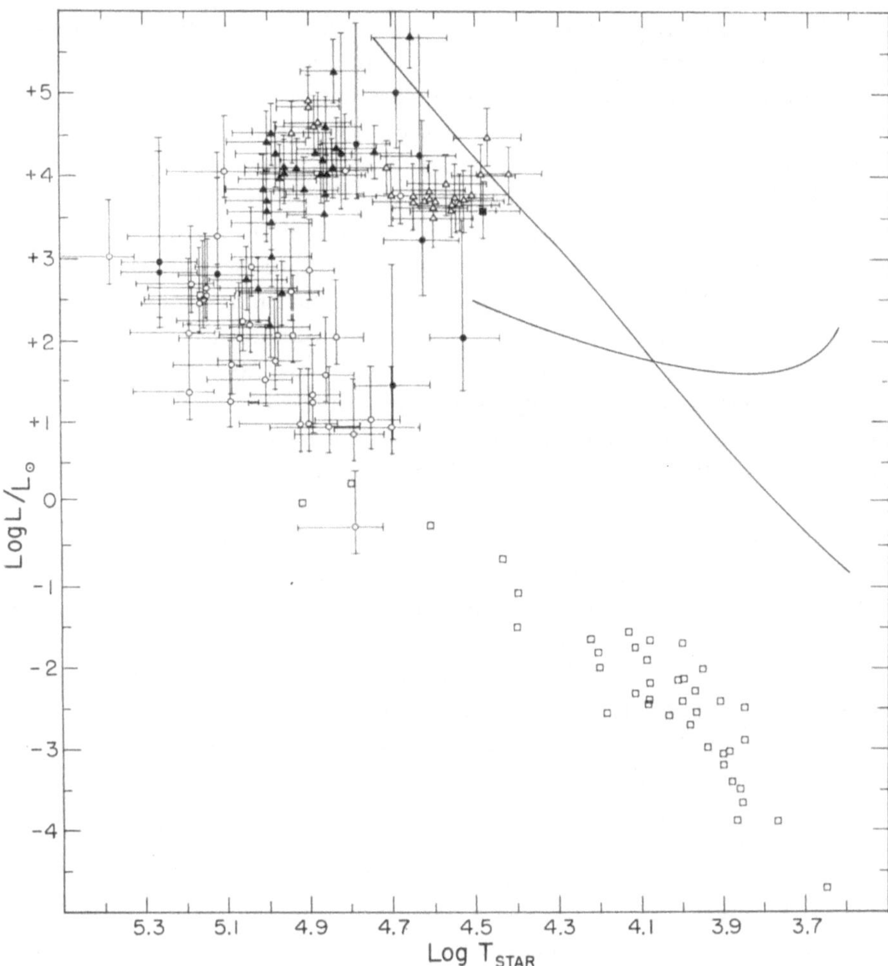

FIG. 1. *The composite luminosity-temperature diagram for the nuclei of planetary nebulae and white dwarfs. Open triangles are nuclei in the Magellanic Clouds; filled triangles are optically thin objects with He*II *Zanstra temperatures; filled circles are optically thick nebulae; open circles are marginally optically thick; and the filled square is K 648 in M 15. Luminosities for the optically thin objects are an average of Seaton's and O'Dell's values. The probable errors are derived from reasonable assumptions of errors of parameters entering the calculations as described in the text.*

probably good to a factor of 2–3 and hence represents a rigorous, although sometimes difficult, feature to match in the calculation of theoretical central stars.

It is very interesting that the higher luminosity stars fall along the limit ($\log L \simeq +4{\cdot}2$) where radiation and gas pressure are the same. One would expect the outer parts of the star to be more liable to ejection near this condition, a fact which is even more interesting when one notes that many of the stars of highest luminosity show

W-R and Of spectral features. These features have disappeared by the time (or perhaps because) the stars have dropped to lower luminosities.

The population along this evolutionary sequence, in terms of total number of known planetary nebula systems is a continuously decreasing function as the star and nebula becomes fainter. This is because of the rapid ($S \sim R_{\mathrm{gas}}^{-5}$) decrease of nebular surface brightness. This lower population is contrary to what one would expect from theoretical models for such stars, where the rate of change of stellar characteristics decreases with time and hence the population in our diagram should increase. The rapidly increasing difficulty of identification as a planetary nebula on present surveys probably more than covers other effects and makes it especially dangerous to infer conclusions about physical processes occurring in this lower region from the low population. Very narrow band-pass emission-line surveys are very important here in the determination of the rate at which the stars pass through this stage immediately before becoming simple white dwarfs. If indeed a significant fraction of all stars now becoming white dwarfs do so by means of the planetary nebula phenomenon, then the existence of faint nebular shells would greatly facilitate their identification, since this is a period of much faster internal cooling than is true in the most common white-dwarf objects.

The study of the planetary nebulae by themselves as a natural phenomenon is a justified and fascinating scientific activity; however, their study assumes a much greater significance when tied together with the study of the Galaxy with its stellar population and with the study of the evolution of stars. Fortunately, it is possible to say quite a bit about the nature of the ultimate source of these systems, but the question of the more immediate precursors is very open.

There are several qualitative considerations that lead one to inspect the rate of production of the planetary nebulae and their possible origin. The total number of identified planetaries now exceeds one thousand (Perek and Kohoutek, 1967), a number that, although small, maintains a particular significance when one considers that the lifetime of the phenomenon is only a few tens of thousands of years, indicating that the production rate of the planetary nebulae in the total Galaxy is quite high. The second remarkable general feature is the rather clear evidence that once a star has entered the state giving rise to the nebular phenomenon, that it is a one-way path towards the white-dwarf state – indicating that this arises at the end of the evolutionary sequence for at least some stars.

It is possible to discuss the rate of formation of possible precursors to the planetary nebula stage without knowing or assuming what the intermediate stages of stellar evolution are, a very great advantage for a general evaluation. The time-scale of the main-sequence stage of stellar evolution is now well known. This feature has been used by Schmidt (1963) to investigate the rate at which stars were formed in the past in our vicinity of the Galaxy in order to explain the present luminosity function and total stellar density. One can then take the best-fitting model from his work and calculate the rate at which stars of various masses are now leaving the main sequence.

Since these post-main-sequence states are short compared to the main-sequence state, this will also be the rate at which stars are being fed into the final stages of stellar evolution. It is advantageous to work not with a unit volume close to the Sun, but rather with a column of unit area perpendicular to and passing through the Galactic plane in the vicinity of the Sun since this frees one to a limited degree from some uncertainties of the Z-distribution of the objects considered, although for purposes of convenience we shall give both. We show in Figure 2 the results from calculations of

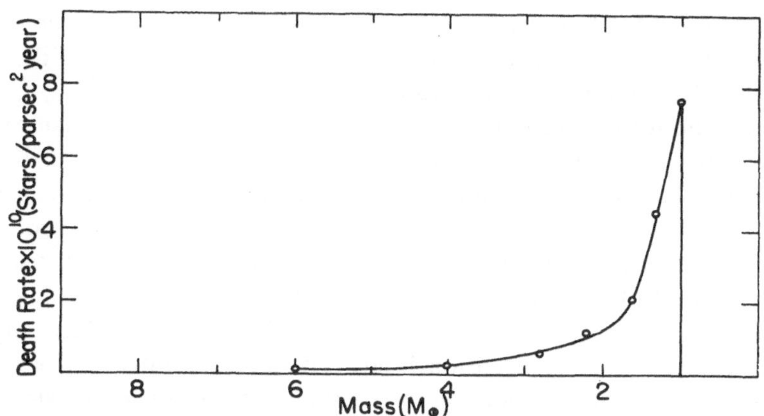

Fig. 2. *The death rate of all stars in a unit column perpendicular to the Galactic plane calculated from the Galaxy model of Schmidt (1963). It is seen that the greatest rate of production of stars possibly becoming planetary nebulae occurs at slightly greater than one solar mass. The hydrogen-burning lifetime adopted here is that where it is equal to the age of the Galaxy for a 1·0 $M_\odot$ star.*

the death rate of main-sequence stars from the best-fitting models of Schmidt – where the rate per unit mass is shown. The planetary nebula birth rates for Cahn's (1968) and O'Dell's distance scales are shown in Table 2.

It is quite obvious that upon O'Dell's distance scale, one would require most of the stars evolving from the main sequence – primarily those of close to solar mass due to the much greater rate there – to pass through the planetary-nebula stage. The space velocities of the planetary nebulae argue that most of them belong to a kinematically

Table 2

Density and Rate of Formation of Planetary Nebulae

Local Density	$= 1{\cdot}4 \times 10^{-8}$ planetary nebulae/parsec^3	(O'Dell)
Local Columnar Frequency	$= 8{\cdot}1 \times 10^{-6}$ planetary nebulae/parsec^2	(O'Dell)
Lifetime	$= 35000$ years	
Rate of Production	$= 3{\cdot}9 \times 10^{-13}$ planetary nebulae/parsec^3 year	(O'Dell)
	$4{\cdot}6 \times 10^{-13}$ planetary nebulae/parsec^3 year	(Cahn)
	$2{\cdot}3 \times 10^{-10}$ planetary nebulae/parsec^2 year	(O'Dell)

old component of the Galaxy, although not the very oldest, so that there are no inconsistencies with this argument of most low-mass stars passing through this state. The Harman and Seaton scale requires only a small fraction of all evolving stars to pass through this state, although even there the observed kinematics argue for an origin from a star originally not much greater than a solar mass. Neither of these rates of formation are sufficient to account for the observed total density of white dwarfs in the solar neighborhood if continued for 5×10^9 years. However, there is neither compelling reason to think that the rate of production has been constant over such a time interval nor to think that all white dwarfs have come through this phenomenon. It is more critical a test to compare rates of formation of the two classes of objects (cf. Weidemann's paper, below, p. 423).

Speculation about the nature of the *immediate progenitors* of the planetary nebulae has covered many years and many authors. So many, in fact, that I shall restrict my comments here to a brief condensation of the ideas of several workers in the last few years. Stellar evolution of solar mass stars is reasonably well understood through the red-giant phase and possibly the horizontal branch. It is certainly dangerous to assume *a priori* that planetaries are formed soon after this as there may exist several intermediate stages – so that linkage directly to these early evolutionary states is dangerous. In addition, there is so much left out of stellar interior theory (rotation, magnetic fields, etc.) that may become increasingly important in the later stages that additional caution about tying objects together is necessary. The safest procedure is probably to go forward from known stellar states and back in time from the planetary nebula state.

It is very interesting that we pick up the planetary nebulae when they have sizes of about 0·04 parsec radius. This poses a very interesting question of identification. Do their characteristics differ so much when smaller that they do not meet the usual criteria for classification as planetaries and are assigned to some other phenomenon or class of objects – or is the evolution of the nebulae so rapid that relatively few would be expected – or are the nebula essentially formed with this size (that is, perhaps the gas is not ionized before a time corresponding to about this size)? It is hard to believe that the very early evolution in the 2000 AU to 0·04 parsec range could proceed much faster than the later stages, because of the small importance of the gravitational field of the system. Abell and Goldreich (1966) have pointed out that the very earliest stages probably do go very rapidly, but this is earlier than we consider here. The possibility of the star heating up so that photo-ionization occurs at a radius of 0·04 parsec certainly exists, but is difficult to test and will not be further discussed here.

Projection backwards of the observed nebulae would indicate that the immediately preceding stage should be characterized by a very large optical depth in the Lyman continuum and the existence of a mostly neutral outer shell. The high densities expected would suppress the emission of certain forbidden lines (such as $\lambda 3727$) by collisional de-excitation and enhance emission from auroral with respect to nebular transitions, e.g., $\lambda 4363/\lambda 5007$. [Fe II] lines would also come up at the higher densities. The Balmer series would also be affected since the greater density would more than compensate for the shorter path length, so that an appreciable optical depth in the lower members of the Balmer series would exist. The nature of the expected continuum radiation is rather uncertain. If the hot star remains dominant, then it would be expected to be quite blue, corresponding to about a B spectral type. However, if a model such as that proposed by Sobolev is valid, then the nebular continuous emission should dominate, with its corresponding red color.

There are a few unclassified objects that generally satisfy these expected conditions for the proto-planetary. The most important two are Minkowski 1-2 and MHα 328-116. M 1-2 (VV8) was originally identified as a stellar planetary nebula and from its spectrum was inferred to have a very high nebular density by several investigators (Razmadze, 1960). Spectra of the continuum (O'Dell, 1966) revealed the surprising fact that the absorption-line spectrum corresponds approximately to that of a G-type supergiant star, while its continuum flux distribution does not fit that expected for a combination late and early-type binary even though there is a definite ultraviolet continuum present. MHα 328-116 is an object, originally found as a sixteenth-magnitude Me star, that has undergone an abrupt rise to eleventh magnitude (FitzGerald *et al.*, 1966; O'Dell, 1967). The spectrum indicates Balmer self-absorption (as does M 1-2), very high densities and an observed underlying continuum that is quite flat in wavelength interval – which becomes quite blue after correction for interstellar reddening effects. There do not seem to be any satisfactory spectra of the continua following the outburst although this is a very important feature. It is tempting to associate both of these objects with the proto-planetaries but there are many problems – one of the most important being the question of the origin of the G supergiant spectrum in M 1-2 and the previously found M-spectrum in MHα 328-116. More as a question than a solution, we should like to ask if these late spectra are formed in the neutral outer shell, following the mechanism roughly outlined by Sobolev (1960).

If one looks for a class of objects generally satisfying the prescribed conditions for a proto-planetary, one might look to the symbiotic stars. Although a very definite fraction of this class of stars is binary in nature, an appreciable fraction is probably very similar to the two objects described above and may be single stars with extended outer envelopes surrounded by neutral material. The emission-line shifts are comparable to those found in the planetary nebulae and the nature of the continuum is very uncertain, although an ultraviolet source is definitely within the system. There is an urgent need for continuous spectrographic and spectrophotometric observations of this very important class of objects along the lines of the pioneering efforts of Boyarchuk, Gershberg, and others (Belyakina *et al.*, 1963).

In summary, we can say that by careful selection of data and techniques, it is possible to bring together a rather comprehensive picture of the rapid evolution of the central stars of the planetary nebulae. Like many other areas, we have raised many additional questions to be answered, paramount among them being that of the linkage to other known states of stellar evolution.

References

Abell, G.O. (1966) *Astrophys. J.*, **144**, 259.
Abell, G.O., Goldreich, P. (1966) *Publ. astr. Soc. Pacific*, **78**, 232.
Aller, L.H. (1956) *Gaseous Nebulae*, Chapman and Hall, London, pp. 201–231.
Bautz, L.P. (1968) *Astr. J.*, **73**, 54.
Belyakina, T.S., Boyarchuk, A.A., Gershberg, R.E. (1963) *Publ. Crimean astrophys. Obs.*, **30**, 25.
Böhm, K.H., Deinzer, W. (1965) *Z. Astrophys.*, **61**, 1.
Cahn, J.H. (1968) in the present volume, p. 44.
Capriotti, E.R., Daub, C.T. (1960) *Astrophys. J.*, **132**, 677.
Collins, G.W. II, Daub, C.T., O'Dell, C.R. (1961) *Astrophys. J.*, **133**, 471.
Deinzer, W. (1967) *Z. Astrophys.* (in press).
FitzGerald, M.P., Houk, N., McCluskey, S.W., Hoffleit, D. (1966) *Astrophys. J.*, **144**, 1135.
Gebbie, K.B., Seaton, M.J. (1963) *Nature*, **199**, 580.
Greenstein, J.L., Minkowski, R.L. (1964) *Astrophys. J.*, **140**, 1601.
Harman, R.J., Seaton, M.J. (1966) *Mon. Not. R. astr. Soc.*, **132**, 15.
L'Ecuyer, Jean (1966) *Astrophys. J.*, **146**, 845.
Liller, W. (1955) *Astrophys. J.*, **122**, 240.
O'Dell, C.R. (1962) *Astrophys. J.*, **135**, 371.
O'Dell, C.R. (1963*a*) *Astrophys. J.*, **138**, 67.
O'Dell, C.R. (1963*b*) *Astrophys. J.*, **138**, 293.
O'Dell, C.R. (1966) *Astrophys. J.*, **145**, 487.
O'Dell, C.R. (1967) *Astrophys. J.*, **149**, 373.
O'Dell, C.R., Peimbert, M., Kinman, T.D. (1964) *Astrophys. J.*, **140**, 119.
Perek, L., Kohoutek, L. (1967) *Atlas of Planetary Nebulae.*
Razmadze, N.A. (1960) *Astr. J. Soviet Union*, **37**, 342.
Rose, W.K. (1966) *Astrophys. J.*, **144**, 1001.
Schmidt, M. (1963) *Astrophys. J.*, **137**, 758.
Seaton, M.J., Osterbrock, D.E. (1957) *Astrophys. J.*, **125**, 66.
Seaton, M.J. (1966) *Mon. Not. R. astr. Soc.*, **132**, 347.
Shaviv, G. (1967) *Astr. J.*, **72**, 319.
Shklovsky, I.S. (1956) *Astr. J. Soviet Union*, **33**, 315.
Sobolev, V.V. (1960) *Moving Envelopes of Stars*, S. Gaposchkin, translator, Harvard University Press, Cambridge, Mass., pp. 82–98.
Vila, S.C. (1966) *Astrophys. J.*, **146**, 437.
Würm, K., Singer, O. (1952) *Z. Astrophys.*, **30**, 387.
Webster, L.B. (1967) private communication.
Weidemann, V. (1968) in the present volume, p. 423.
Zanstra, H. (1926) *Phys. Rev.*, **27**, 644.

DISCUSSION

Böhm: Is it correct to say that stars having Wolf-Rayet spectral type lie always closest to the instability limit due to the action of radiation pressure in the atmosphere? I had the impression that W-R stars occur over a wide range of the evolutionary sequence up to $T_{eff} \simeq 1{\cdot}0 \times 10^5\,°K$, where the Harman-Seaton sequence is already relatively far from the instability line.

O'Dell: When one considers the uncertainties in the temperatures and luminosities as given here, the statement that W-R spectral types always lie close to the instability limit is more descriptive.

Seaton: Highly advanced, optically thick, planetaries may have approximately constant radii of the ionized region. Did you consider the possible use of this result in getting distances?

O'Dell: No.

Aller: The idea that combination or symbiotic variables are precursors of planetary nebulae has occurred to a number of people. At least it was the basis for the spectrophotometric program on these objects I initiated at Michigan in 1948. The Zanstra-Menzel temperatures of the hot sources in these stars fall at about the right level. The spectral changes appear very disorderly and complicated (as described e.g. by Merrill for BF Cygni). The object must settle down into a nice orderly planetary nebula. The connecting link between the symbiotic stars and the earliest recognized stage of a planetary nebula is yet to be established.

Van Horn: I'd like to ask O'Dell if there are any systematic concentrations of the continuous-spectrum stars along the evolutionary track such as he finds for the Wolf-Rayet type nuclei?

O'Dell: They are concentrated to the lower-luminosity region although not all are of the continuum type there.

Savedoff: What is the reason for difference between your quoted $L = 10^{4 \cdot 2}$ and my estimate of $L = 10^4 M/(1 + X)$ based on electron scattering?

Deinzer: The difference is probably due to chemical composition and mass. I used 0·4 $M_\odot$ and a chemical composition of population II. The limiting luminosity is then at about $10^{4 \cdot 2}$.

Westerlund: Miss Webster has observed a large number of galactic planetary nebulae in the Southern hemisphere. Several of those are in Norma with low reddening. She used narrow-band interference filters; in the continuum the wavelengths were 5300, 4200, and 3500. The nuclei are found near the hot end of the main-sequence relation in the two-color diagram and along the black-body line.

PROBABLE VARIABILITY OF NGC 6572

D. Koelbloed

(Astronomical Institute of the University of Amsterdam, The Netherlands)

1. Introduction

The suspected rapid rate of evolution of the central stars of planetary nebulae (Shklovsky, 1956) and the discovery by Seaton (1966) of their evolutionary track in the H-R diagram, have led to renewed interest in these stars. According to Seaton the evolutionary track starts at an atmospheric temperature of about 30000°K and a luminosity of 100 $L_{\odot}$. A maximum brightness is reached of 25000 $L_{\odot}$ at $T_* = 70000$°K. Then T_* increases to 100000°K, at approximately constant L, whereafter, at this temperature, L decreases finally below 100 $L_{\odot}$. This whole evolution is believed to take place in only 5×10^4 years. At such a rapid evolution, short-period fluctuations in brightness are expected and it should even be possible to detect secular variations in a lifetime. Aller and Liller (1957) discovered the variability of the spectrum of IC 4997 and Vorontsov-Velyaminov (1961) found variations in line intensities of NGC 6905. Khromov (1962) tried to explain the variations in both nebulae as a result of temperature changes of the central stars. Variations in the emission spectrum of IC 4997 were explained by Aller and Liller (1966) as a result of nebular expansion.

Changes, since the beginning of this century, of diameters of planetary nebulae were studied by Martha Liller *et al.* (1966) and of the magnitudes of central stars by Kohoutek (1966). In this note we compare A_ν-values of NGC 6572, derived in the interval 1927–66.

2. Comparison of A_ν-Values of NGC 6572 (1927–66)

A. Zanstra's measurements (1927–46)

Just 40 years ago (September 1927) Zanstra (1930, 1931) photographed the first standardized slitless spectra of NGC 6572 and NGC 6543 in order to derive the temperatures of the central stars, with his well-known A_ν-method. With this method one compares the total intensity of a monochromatic nebular image of an element with the intensity of the continuum of the central star at the same wavelength. The superiority, compared with other methods, is that only relative measurements are used, which need no interstellar-reddening corrections. When old and new observations are compared a change in response characteristics in the meantime has no effect on the results, an especially important consideration.

The author noticed that Zanstra's new determinations of the temperature-parameters T_H (Zanstra, 1960) 52000 °K in 1938 and 54000 °K in 1946, were considerably higher than the older ones, 41000°, found by him in 1927 (Zanstra, 1930, 1931). The old low values were confirmed by Berman (1930). Berman found $T_H = 43500$ °K from photographs taken by him in 1928.

For detecting variability, we cannot compare the published temperatures, because they strongly depend on the method used.* We have to compare the original A_ν-values. They are collected in Table 1. Berman used only $H\delta$, but did not publish the A_ν-value.

Table 1

Comparison of A_ν-values of NGC 6572

	1927		1938–46	1966
(1)	(2)	(3)	(4)	(5)
Hβ	0·19	0·20	–	0·265
Hγ	0·068	0·075	(0·118)	0·128 (0·110)
Hδ	0·039	–	(0·070)	0·067
He I, λ4471	0·009	0·008	0·019	0·0147
He I, λ4026			0·0066	

Column (2) are the values published by Zanstra (1930).

Column (3) are improved A_ν-values derived from the same negatives and published by Zanstra (1931).

As on the plates of 1938 and 1946, taken by Aller, the Balmer images were too strong to give reliable results, Zanstra had to use the A_ν-values of the He I lines 4471 Å and 4026 Å to find those for hydrogen. He therefore had to rely on the nebular line-intensity ratios Hδ/4026 = 10·5 and Hγ/4471 = 6·2 found in the literature (cf. *Bull. astr. Inst. Netherl.*, **15**, 244).

In column (4) the 1938–46 results are given. The values in parenthesis were derived from the He lines as described above.

Comparison of columns (2) and (3) with (4) shows that an increase of the A_ν-values with a factor of about 2 seems indicated. As it is improbable that such a large factor can be explained by observational errors one has to consider the possibility of a temperature increase of the central star in the short interval 1927–1938.

A systematic error between the old and new series is contradicted by the comparison

* For NGC 6572 see also the temperatures derived by Seaton in *Report Prog. Physics*, **23**, 1961, 313, and *Mon. Not. R. astr. Soc.*, **132**, 1966, 29.

of the results for NGC 6543 where only A(Hδ) can be compared. We find A(Hδ)= 0·039±0·006 (1927) from two films and 0·032±0·004 (1938–46) from three plates.

B. ALLER'S PHOTOELECTRIC TRACINGS (1966)

In connection with the suspected variability it seemed worthwhile to have new observations. At the request of the author Dr. Lawrence H. Aller, in cooperation with G. J. Stanley and S. J. Czyzak, secured observations of NGC 6572, on May 22, 1966, with the photoelectric scanner on the 60-inch Mount Wilson reflector. The tracings were kindly put at the author's disposal.

The continuum, on the tracings, contains contributions of both the star and the nebula. On spectra kindly loaned by Dr. Olin C. Wilson, no nebular continuum could be detected; also, Page (1942) was unable to measure it on plates, taken in 1939 and 1940, especially for the study of the nebular continuum. Neglecting this continuum, we find the A_ν-values given in column (5). For Hγ two values are given. The photoelectric profiles of Hβ, Hδ and He I 4471 are symmetric whereas the Hγ profile is disturbed at the red wing. Adopting a symmetrical profile for Hγ too, we find the value 0·128. The disturbance is caused by the Hg line 4359 Å of the sky background and by [O III] 4363 Å of the nebular spectrum. According to Aller and Kaler (1964) the ratio to the nebular lines Hγ/4363 is 4·3, derived from observations made in 1959–63. Correcting the asymmetrical line for both lines we find the value in parenthesis. The values for Hβ have low weight. On the photographic plates Hβ is disturbed by the strong N2-nebulium line, whereas on the photoelectric tracing the stellar continuum is too faint for deriving good results.

3. Conclusions

We conclude that the photoelectric results are in good agreement with Zanstra's 1938–46 results. Thus, if the change is real, it appeared to have occurred chiefly between 1927 and 1938. The results point out the desirability of further observations at regular intervals of planetary nebulae.

4. Further Evidences of Variability

In addition, two independent indications of variability of NGC 6572 can be mentioned. Strong evidence of an increase of the nebular diameter was given by Martha Liller *et al.* (1965). They compared photographs taken in 1916 and 1961 with the Crossley-reflector. Of the 14 nebulae investigated, the smallest probable error in the nebular growth was found for the minor axis of NGC 6572. The mean increase in angular size is 0.″81±0·10 in 100 years, based on a continuous increase during 1916–61. This corresponds to an increase of the mean diameter of a factor of 1·07 if the image of 1916 is compared with that of 1961.

The second indication of variability is the determination of the photographic magnitude of the central star. Kohoutek (1966) measured an AG K2 plate (1929) and an AG K3 plate (1961) and found an increase in luminosity of $0^m_.3$. He states that this difference is too large to be accounted for by observational errors. An increase of $0^m_.3$ in the photographic region corresponds roughly to an increase from $T_* = 43000\,°K$ to $53000\,°K$, if r_* remains constant.

The variations of A_ν, the diameter and m_{pg} may be explained by an increase of the temperature of the central star with a corresponding increase of the diameter of the surrounding H II region.

5. The Temperature Parameters T_H and T_{HeI} of NGC 6572

From the presence of [O I] lines in the nebular spectrum, it follows that the absorption by H I is complete in NGC 6572. Assuming with Zanstra an electron temperature of 15000 °K, we find, using Zanstra's method (1960), $T_H = 53300\,°K$ from the 1966 A (Hδ)-value. The same temperature is found from the A (Hγ) (symmetrical profile) and from the A (He I, 4471 Å). The A (Hγ) (corrected asymmetrical profile) gives 50800 °K. As the A (Hδ) has the larger weight we adopt $T_H = 53300\,°K$.

In Figure 1 T_{HeI} is plotted against T_H for Zanstra's group I (1960) (high-accuracy measurements). The 1966 value for NGC 6572 is added (black dot).

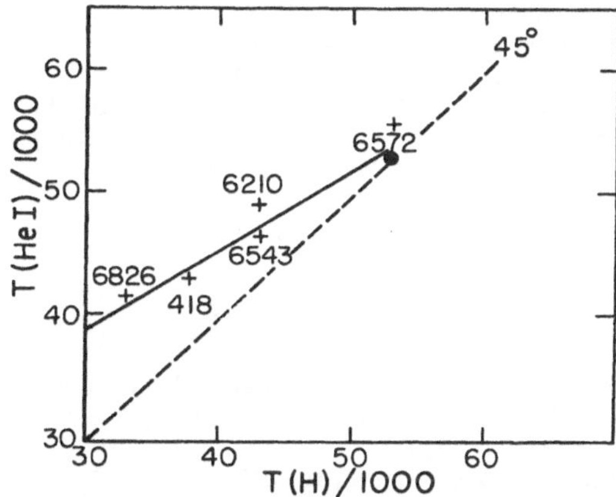

FIG. 1. *The* He I *Zanstra temperature plotted against the Hydrogen Zanstra temperature. The dashed line shows the one to one relation.*

According to Hummer and Seaton (1963) the helium images can only be used for temperature determinations if $T_* < 50000\,°K$. All nebulae of group I, with NGC 6572 as a probable boundary case, satisfy the condition.

An increase of the difference between the two temperatures for lower temperatures seems indicated.

From the approximate equality of T_H and T_{ee} (derived from forbidden lines) Zanstra concludes complete absorption by hydrogen for group I. Then the explanation of Figure 1 – if confirmed by more observations – must follow from atmospheric models of the central star (Gebbie and Seaton, 1963; Böhm and Deinzer, 1965*a*, *b*; Gebbie, 1967).

Seaton (1966, table I) on the other hand, arranges NGC 6572 and 6543 in class a_1 (thick He II, thick H I) and NGC 6826 in class a_2 (thick He II, thin H I). In this case the completeness of absorption in the nebulae has to be considered too in discussing Figure 1.

Acknowledgements

The author is very thankful to Dr. L. H. Aller and his collaborators for making the photoelectric scans of NGC 6572.

References

Aller, L. H., Kaler, J.B. (1964) *Astrophys. J.*, **140**, 621.
Aller, L. H., Liller, W. (1957) *Sky and Telescope*, **161**, 222.
Aller, L.H., Liller, W. (1966) *Mon. Not. R. astr. Soc.*, **132**, 337.
Aller, L. H., Zanstra, H. (1960) *Bull. astr. Inst. Netherl.*, **15**, 249.
Berman, L. (1930) *Lick Obs. Bull.*, **430**.
Böhm, K.H., Deinzer, W. (1965*a*) *Z. Astrophys.*, **61**, 1.
Böhm, K.H., Deinzer, W. (1965*b*) *Z. Astrophys.*, **63**, 177.
Gebbie, K.B., Seaton, M.J. (1963) *Nature*, **199**, 580.
Gebbie, K.B. (1967) *Mon. Not. R. astr. Soc.*, **135**, 181.
Harman, R.J., Seaton, M.J. (1966) *Mon. Not. R. astr. Soc.*, **132**, 15.
Hummer, D.G., Seaton, M.J. (1963) *Mon. Not. R. astr. Soc.*, **125**, 437.
Hummer, D. (1964) *Mon. Not. R. astr. Soc.*, **127**, 240.
Kohoutek, L. (1966) *Bull. astr. Inst. Csl.*, **17**, 318.
Khromov, G.S. (1962) *Sov. Astr. – A. J.*, **5**, 619.
Liller, Martha H., Welther, Barbara L., Liller, W. (1966) *Astrophys. J.*, **144**, 280.
Page, T. (1942) *Astrophys. J.*, **96**, 78.
Seaton, M.J. (1966) *Mon. Not. R. astr. Soc.*, **132**, 113.
Shklovsky, I.S. (1956) *Russian A. J.*, **33**, 315.
Vorontsov-Velyaminov, B. (1961) *Astr. Zu.*, **38**, 247. (Also in *Soviet Astr. – A. J.*, **5**, 186.)
Zanstra, H. (1930) *Publ. Dom. astrophys. Obs., Victoria*, **4**, 209.
Zanstra, H. (1931) *Z. Astrophys.*, **2**, 1.
Zanstra, H. (1960) *Bull. astr. Inst. Netherl.*, **15**, 237.

ON THE VARIABILITY OF THE NUCLEI OF THREE PLANETARY NEBULAE

M. A. KAZARIAN
(Burakan Astrophysical Observatory, U.S.S.R.)

The observations of nuclei of three planetary nebulae, NGC 6826, NGC 7662 and IC 4593, have been made as part of a more extensive program with the aim of obtaining their absolute spectrophotometric gradients (ϕ) and photographic magnitudes (from the spectrogram). The spectra were obtained with an 8-12″ Schmidt telescope combined with an objective prism (the dispersion is 420 Å/mm at Hγ). The preliminary data which were obtained at different times, particularly the spectrophotometric gradients of the nuclei of these nebulae, did not agree with each other.

In order to find the causes of this disagreement we have made a rather long series of observations of the nucleus of NGC 6826. Relatively shorter observational series have also been made of the nuclei of NGC 7662 and of IC 4593. The spectral interval of the investigation was λλ 4100–4700.

Twenty-two spectra of the nucleus of the NGC 6826 have been obtained on 15 different nights in the period from December, 1965 to October, 1966. The results show that the gradient changed within the limits 0·32 to 1·46, and the photographic magnitude between $9^{m}_{.}6$ and $10^{m}_{.}3$.

Eight spectra of the nucleus of NGC 7662 were obtained on eight different nights in the period from August to December, 1966. The results showed that the gradient changed within the limits 0·50 to 1·24. No changes of the photographic magnitude were noticed in the period of our observations.

Six spectra of the nucleus of the IC 4593 were obtained on four different nights in the period from March to August, 1966. The results show that the gradient remained almost unchanged, but the photographic magnitude changed within the limits $10^{m}_{.}2$ to $11^{m}_{.}1$.

Table 1

Summary of changes of nuclei of planetary nebulae

Nebula	n	Period of observation	Range of ϕ	Range of m_{pg}
NGC 6826	22	Dec., 1965 – Oct., 1966	0·32–1.42	$9^{m}_{.}6$–$10^{m}_{.}3$
NGC 7662	8	August–Dec., 1966	0·50–1.24	no changes
IC 4593	6	March–Aug., 1966	no changes	10·2–11·1

Osterbrock and O'Dell (eds.), Planetary Nebulae, 381–382. © *I.A.U.*

The quantity of the observational material for the nuclei of NGC 7662 and IC 4593 is not sufficient to make it possible to be certain about the variation of these two nuclei. Therefore we may suppose that the nuclei of these nebulae are possibly variable objects, while the nucleus of NGC 6826 is definitely variable.

ON THE VARIABILITY OF THE CENTRAL STAR OF THE PLANETARY NEBULA H_z 1-5

G. H. HERBIG and A. A. BOYARCHUK
(Lick Observatory, U.S.A.) (Crimean Astrophysical Observatory, U.S.S.R.)

During 1960–67 we have studied the variable star FG Sge = 377·1943 Sge, which is located exactly in the centre of an image of the planetary nebula

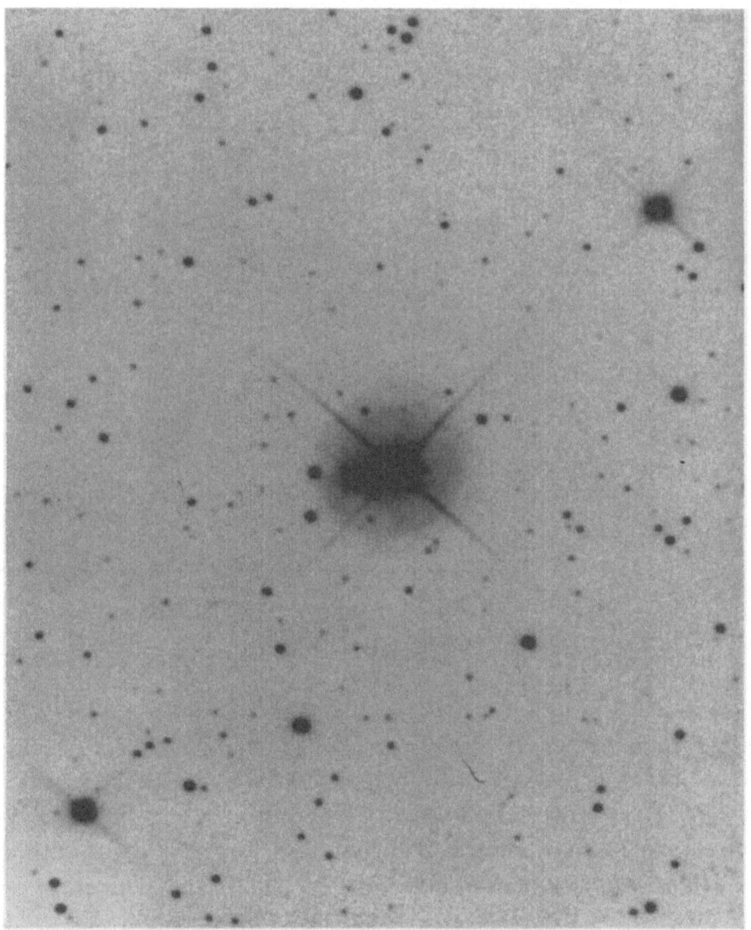

FIG. 1. *The nebulosity surrounding FG Sge.*

Osterbrock and O'Dell (eds.), Planetary Nebulae, 383–385. © *I.A.U.*

H_z 1-5, as shown on Figure 1. The most significant feature of the central star is the continuous increase of its brightness during the past 70 years, as one can see from Figure 2. During the same time, the temperature of the central star has apparently decreased, although direct information to this effect is available only since 1955. In 1955 the star had spectral type B4, while by 1967 it had become A5. The observational data can be explained by the continuous ejection of gas from the central star, which began at the end of the preceding century. The central star at that time must have had $T_* \approx 63300\,^\circ$K and $M_{pg} \approx +1{\cdot}3$, on this model.

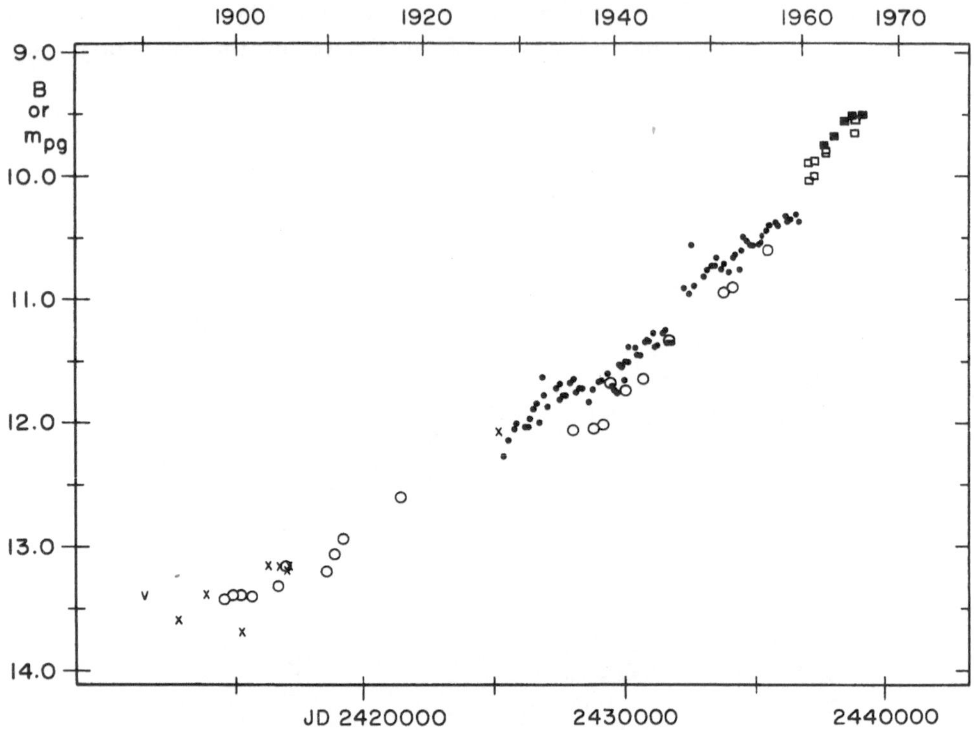

FIG. 2. *The blue magnitudes of FG Sge from the beginning of the 20th century.*

Those values agree very well with those of central stars of planetary nebulae. The expansion velocity of the new shell will, in about 3000 years, cause the new envelope to reach the size of the existing nebula H_z 1-5. Therefore, it is possible to produce more than one nebula by the same central star. Another possibility is that this central star is a double system, whose components have practically the same masses.

In any case, we believe that future investigations of this object will give us important information about the formation and evolution of planetary nebulae.

DISCUSSION

Münch: What kind of a spectrum does this most remarkable star have?

Boyarchuk: The spectral type in 1955 was B4 I with P Cyg features, but in 1967 it was A5 Ia without any P Cyg characteristics.

Kohoutek: I am able to confirm Boyarchuk and Herbig's result concerning the planetary H_z 1-5. Our photoelectric observations with the 65-cm reflector at Ondřejov between 1965–67 show an increase of the star brightness of about 0·2 mag. Also, the spectrum of the star in September 1965 from Asiago shows no trace of emission lines, so that perhaps the nebula no longer radiates.

LONG-PERIOD VARIABLES AND PLANETARY NEBULAE

DONALD H. MENZEL
(Harvard Observatory and Smithsonian Astrophysical Observatory, U.S.A.)

ABSTRACT

The occasional appearance of a red giant or long-period variable in planetary nebulae poses a problem for theoretical astrophysics. Such a cool nuclear star would not ordinarily provide a source of ultraviolet radiation necessary for the excitation of the spectrum of a gaseous nebula.

One possible solution of this problem postulates the existence of intense magnetic fields in the star. Second, the star itself has a structure resembling that of a miniature, highly compressed planetary, with a high-temperature nuclear star at the centre and a distended atmospheric shell enveloping chiefly the stellar equator.

The magnetic field induces a sort of pumping action that creates the tire-shaped envelope from matter ejected near the poles. As this shell grows denser, it radiates like a stellar photosphere at low temperature. Eventually the shell becomes unstable and disperses outward to form and maintain the nebula. A quasi-periodic situation occurs, which explains the variation of light. Ultraviolet light absorbed during the minima, when the shell has vanished, adequately accounts for the nebular excitation. A wide variety of such symbiotic stars occurs, including repeating novae as well as the long-period variables.

Planetary nebulae, in general, owe their luminosity to the presence of high-temperature nuclear stars, whose intense ultraviolet radiation excites the gaseous nebulosity by fluorescence. The occasional appearance of a red giant or long-period variable, in planetary nebulae, seems completely anomalous. Paul Merrill referred to such associations as 'symbiosis', a biological term signifying an intimate association (often mutually beneficial) of two dissimilar organisms.

Such objects, observed near minimum, often show spectra characteristic of hot, blue stars – a fact that has led some astronomers, including myself, to suggest that these objects were actually double stars, a cool, red variable with an O or W companion.

More than 20 years ago, however, I suggested (Menzel, 1946) an alternative model, that giant M stars, in general, possess a relatively condensed and stable core, which by itself would be classed as spectral type O or W. A distended, variable, dynamic atmosphere surrounds this core. The atmosphere probably does not cover the entire star. Like an enormous doughnut or tire, it covers the equatorial belt, leaving the polar regions relatively clear, so that the hot core may freely radiate its ultraviolet energy into distant space.

In an accompanying paper (Menzel, 1968), I gave the magnetohydrostatic solution for the distribution of density and temperature in the gaseous envelope of a star.

Osterbrock and O'Dell (eds.), Planetary Nebulae, 386–389. © *I.A.U.*

These formulae failed for planetary nebulae, but they seem adequate for a stellar envelope of relatively small radius.

The formulae are:

$$p = p_0(r) + \frac{15\mu^2 a^2 r^2 \sin^2 \theta}{4\pi (a^2 + r^2)^5} \tag{1}$$

$$\rho = \rho_0(r) + \frac{105\mu^2 a^2 r^2 \sin^2 \theta}{4\pi GM (a^2 + r^2)^6}, \tag{2}$$

where $p_0(r)$ and $\rho_0(r)$, functions of the radius r only, define the pressure and density along the polar axis; also, μ is the magnetic moment. The parameter a defines the radius of the currents producing the field. When $a=0$, the field becomes a simple dipole.

If we set these parameters equal to zero, we have a central star surrounded by an equatorial envelope, toroidal in shape.

The temperature distribution proves to be independent of the magnetic field. We have

$$T = pm/k\rho = GMm(a^2 + r^2)/7kr^3, \tag{3}$$

where m is the molecular mass and k Boltzmann's constant. For the Sun, with a set equal to the radius, we find the temperature at the inner boundary,

$$T_a \sim 4 \times 10^6, \tag{4}$$

a value suggestive of that for the inner corona. The number of atoms per cm^3, from (2), at the equator for $r=a$, with

$$\mu^2 = 2H^2 a^6, \tag{5}$$

where H is the field intensity at the pole, becomes

$$n = \rho/m = 1{\cdot}4 \times 10^8 H^2 \sim 3{\cdot}5 \times 10^9/\text{cm}^3, \tag{6}$$

a figure also very close to that of the base of the corona, with $H=5$ gauss.

I have not intended this demonstration to be a model of the solar corona; but the close agreement of the predicted temperature and density tend to confirm the validity of the model for its use in extended atmospheres. In summary, we may expect the temperature and density in a stellar envelope to follow laws like those of (3) and (2). However, if the atmospheres are very distended, we must adopt a value much larger than that of the stellar radius. The currents will be mainly in the distended atmosphere rather than in the central star. The steady state for which these distributions apply will almost certainly not exist.

A magnetohydrodynamic study of bipolar sunspots (Menzel and Shore, 1966) shows that solar flares probably result from a type of gaseous flow associated with strong magnetic fields. A search for magnetostatic or even steady-state solutions of the basic equations indicates that no such solutions exist for bipolar spots. In a steady-state solution, e.g., one seeks to find a gaseous flow along the magnetic lines of force. However, both spots of the pair attempt to 'pump' gas into the region between the spots. The pressure and temperature mount in this region until the magnetic field can no longer restrain the gas. The field expands, forming a shock wave that rushes over the solar surface.

Much the same thing can happen in a star containing a strong magnetic field. A static solution would require that regions near both magnetic poles be cooler and at lower pressure than the surroundings. Flow of heat into the regions, however, would prevent the static condition from ever occurring. Instead, each pole would begin to pump matter outwards, along magnetic lines of force, into the equatorial envelope of the star. The gradual accumulation of matter eventually leads to formation of a distended shell over the equator. As the opacity rises and radiative transfer takes control, the temperature falls. The shell converts the high-temperature radiation from the stellar nucleus into low-temperature radiation from the effective photosphere. The outer shell emits a spectrum characteristic of a long-period variable.

As for the sunspots, the pumping cannot continue indefinitely. Eventually, rising pressures cause the stellar magnetosphere to expand and suddenly – like an over-inflated balloon – to break and discharge its contents into space. Thus, the expanding shells gradually coalesce to form the gaseous envelope of the planetary. And the ultra-violet radiation of the core, absorbed during minimum light, when the shell is non-existent or incomplete, provides adequate energy for the nebula to shine steadily. The nebular gas has relaxation times far longer than the period of the associated variable.

Since the consecutive shells form and disperse over the equator, the ejecta over millenia conform at least roughly to the original toroid shape, distorted and expanded, but still toroidal. Viewed from the poles the nebula takes the form of a ring; from the equator it appears as two separate patches on either side of the star.

Fragments of the magnetic field may still persist in the expanding gas and serve to bind it together. Thus, many nebulae, such as the helical nebula in Aquarius, seem to consist of many small condensations rather than uniform continua. Each condensation could have resulted from a single outburst. (Could comets have arisen in some such manner?)

This model suggests that all planetary nebulae were once symbiotic objects. Conversely, it implies that all Mira-type and associated variables may eventually end up as standard planetaries. It further indicates a hitherto unsuspected kinship between giant M's, in general, long-period variables in particular, and the W-R stars that so often appear as planetary nuclei. Do any of these stars exhibit light variation?

Symbiotic stars include a variety of types. The repeating nova, RS Ophiuchi, prominently displays the red coronal line of Fe X at $\lambda 6374$ at a time when the associated star shows the Ti O bands characteristic of an M-type spectrum. Such a line requires for its excitation temperatures of the order of a million degrees. Z Andromedae,

AX Persei, RX Puppis, T Coronae Borealis, RW Hydrae, CI Cygni, and AG Pegasi are other well-known examples of late-type stars having nebular characteristics. R Aquarii appears to be a typical long-period variable, except for its occurrence as the nuclear star of a planetary nebula. Sargent's (1966) recent discovery of forbidden lines of N II in the spectrum of HR 8752, a G0 star, extends the relationship to stars other than M's.

References

Menzel, D.H. (1946) *Physica*, **12**, 768.
Menzel, D.H. (1968) in the present volume, p. 279.
Menzel, D.H., Shore, B.W. (1966) in *Comitato Nazionale per le Manifestazioni Celebrative del IV Centenario della Nascità di Galileo Galilei*, Ed. by G. Barbera.
Sargent, W.L.W. (1966) *Q. J. R. astr. Soc.*, **7**, 222.

DISCUSSION

Kippenhahn: Several years ago Schwarzschild and Stothers determined the eigenperiods of models for the ascending branch of globular clusters. The periods they derived agreed quite well with the observed periods of Mira stars. We have also computed eigenperiods for red-giant models and got periods of the order of that of normal Mira variables. I therefore have the feeling that the Mira stars are really pulsating stars as the Cepheids are. In your model a torus-like figure is pulsating. Have you estimated eigenperiods and got similar agreement?

Feast: The mutilation of the hydrogen lines in Mira variables which is generally taken to imply a fairly dense overlying cool atmosphere is explained by your model.

PULSATIONAL INSTABILITY AND THE ORIGIN OF PLANETARY NEBULAE

WILLIAM K. ROSE*

(Physics Dept, Massachusetts Institute of Technology, Cambridge, Mass., U.S.A.)

The calculations that will be discussed in this paper were undertaken to determine if plausible theoretical arguments could be found to indicate that planetary nebulae represent the stage in the evolution of a star immediately before it becomes a white dwarf. The model for the formation of a planetary nebula that will be described may also be applicable for more massive stars than have been studied. However, since it is known that some population-II stars become planetary nebulae and since there is evidence that evolved population-II stars are less massive than 1 $M_\odot$ (Christy, 1966; Hayashi *et al.*, 1962; Faulkner, 1966; Faulkner and Iben, 1966; Schwarzschild and Härm, 1966) only stars whose masses are less than 1 $M_\odot$ have been studied.

Recent investigations of models for horizontal branch stars (Hayashi *et al.*, 1962; Faulkner, 1966; Faulkner and Iben, 1966; Schwarzschild and Härm, 1966) indicate that the horizontal branch is an evolutionary stage such that the nuclear-energy generation results from helium burning at the centre of the star and hydrogen burning in a shell. In its subsequent evolution the population-II star will exhaust its supply of helium in the core. Helium shell-burning will result and the star will evolve up the red-giant branch for the second time. The model for the formation of planetary nebulae that is considered in this paper assumes that instabilities that arise during this helium shell-burning stage of evolution are responsible for the formation of planetary nebulae.

A number of investigators have computed stellar models that demonstrate the presence of thermal instability during the helium shell-burning stage of stellar evolution (Schwarzschild and Härm, 1965; Weigert, 1965; Rose, 1966). This instability has been found for models of stars that range in mass from 0·5 $M_\odot$ to 5 $M_\odot$. The physical basis for this instability lies in the high temperature sensitivity of the 3α process and the fact that the helium burning takes place in a thin shell. The calculations show that after a star has become thermally unstable it undergoes a series of relaxation oscillations. During each oscillation the rate of nuclear-energy generation is very high for a short period of time ($\approx$100 years for a typical oscillation). In general these periods of high nuclear-energy generation will not lead to the significant acceleration of mass that would be required to explain mass loss. However, under suitable conditions,

* Supported in part by the National Aeronautics and Space Administration (NsG-496).

high rates of nuclear-energy generation can lead to pulsational instability (Rose, 1967*a*). Although the calculations have so far only shown this to be true for blue stars, it is the assumption of this model for the formation of a planetary nebula that thermal instability can lead to pulsational instability in a red giant and that the presence of pulsational instability can lead to the formation of a planetary nebula.

There is an important distinction between pulsational instability in a very blue star and what we will call pulsational instability in a red giant. In the case of the blue star the injection of nuclear energy takes place in a time that is long as compared with the time for sound to traverse the star. Therefore, the adiabatic theory of pulsations is appropriate. On the other hand, in the case of the red giant the injection of nuclear energy can take place in a time that is comparable with or shorter than the time for sound to traverse the star. Therefore, the adiabatic theory of pulsations is not appropriate.

Although the low velocities of expansion observed for planetary nebulae ($\approx$20 km/sec) as well as independent arguments that are discussed in this paper indicate that planetary nebulae originate from red giants (a point that was first made by Shklovsky, 1956, and later amplified by Abell and Goldreich, 1966), the observed nuclei of planetary nebulae are very blue and must be almost completely hydrogen-deficient. This apparent paradox can be understood if it is recalled that for a helium shell-burning red giant almost all the gravitational and thermal energy of the configuration resides within a dense core. The core is composed of an interior region of mostly degenerate carbon and oxygen and an outer less massive region of semi-degenerate and non-degenerate matter that is mostly helium. If by some mechanism the outer hydrogen-rich envelope were removed from the helium shell-burning red giant, a blue star with a degenerate core that is surrounded by a less dense helium-rich region would remain. For this reason, it is of some interest to study helium shell-burning stars with pure helium envelopes as approximate representations for the nuclei of planetary nebulae.

Evolutionary sequences of models for helium shell-burning stars with pure helium envelopes have been computed up to the end of helium burning and into the white-dwarf state (Rose, 1966, 1967*a*; L'Ecuyer, 1966). Thermal stability analysis (Schwarzschild and Härm, 1965; Rose, 1966) has shown that some of these computed models are unstable. These unstable models undergo relaxation oscillations. During each of the relaxation oscillations the rate of nuclear-energy generation attains a high value for a short interval. Models with masses of 0·53 $M_\odot$, 0·75 $M_\odot$ and 0·95 $M_\odot$ have been studied. The results of these calculations show that the maximum rate of nuclear-energy generation is attained during the final relaxation oscillation, i.e. at the end of helium burning, and therefore just before the star contracts toward the white-dwarf state. The general behavior of the oscillations indicates a gradual increase in amplitude followed by a sharp increase for the final one or two oscillations. The maximum rate of nuclear-energy generation ($\approx 3 \times 10^7\ L_\odot$) was attained for the 0·75 $M_\odot$ star. This appears to be the case primarily because the helium-burning shell attained a higher density ($\approx 3 \times 10^4$ gm/cm^3) before the temperature of the helium shell went below that necessary to maintain helium burning for the 0·75 $M_\odot$ sequence as compared with the 0·53 $M_\odot$ and 0·95 $M_\odot$ sequences.

A sequence of models for a helium shell-burning star ($M = 0{\cdot}6\ M_\odot$) with an initial hydrogen envelope of 0·006 $M_\odot$ has been computed through a series of relaxation oscillations up to the end of helium burning. The calculations show that the star goes through a series of low-amplitude relaxation oscillations (peak nuclear-energy generation $\approx 2{\cdot}5 \times 10^4\ L_\odot$). The amplitude of these oscillations increase very slowly. Eventually the helium-burning shell, which is overtaking the hydrogen shell, stops burning because the temperature inside the helium shell falls below that necessary to maintain helium burning. At this point in the star's evolution the rate of hydrogen burning increases sharply until almost the entire luminosity of the star ($\approx 2200\ L_\odot$) is supplied by hydrogen burning. The helium layer beneath the hydrogen contracts and consequently the density inside the helium shell increases. As the hydrogen shell advances toward the surface the mass of the helium layer increases and eventually the temperature inside the helium shell begins to rise. Helium burning commences and after the helium-burning shell has reached about 100 $L_\odot$ it becomes unstable again.

At the onset of the instability the density within the helium burning shell is $\approx 3 \times 10^4$ gm/cm^3 as compared with $\approx 9 \times 10^3$ gm/cm^3 encountered during the earlier low-amplitude relaxation oscillations. The highest rate of nuclear-energy generation attained during the final oscillation is $\approx 3 \times 10^6\ L_\odot$ or approximately 10^2 times higher than is attained during the earlier relaxation oscillations. In addition, the helium convective zone associated with the final relaxation oscillation extended to a mass shell very close to the hydrogen shell so that mixing nearly took place. Hayashi *et al.* (1965) have discussed solutions where an extinguished helium shell is turned on again in a manner similar to what has been described above. In carrying out their solutions they assumed that the core is isothermal and that the shell must be turned on under degenerate conditions. In our case the helium shell goes out for the final time after one moderately-large amplitude oscillation. However, if the hydrogen-rich envelope of our star were more massive, the helium shell might have been turned on again under conditions more favorable for large amplitude oscillations.

The above discussion of unstable models may be summarized by saying that the calculations indicate that a thermally unstable star undergoes a series of relaxation oscillations that are of insufficient amplitude to cause any significant acceleration of mass. These initial relaxation oscillations are followed by a much smaller number (perhaps only one) of much more violent relaxation oscillations that may become important hydrodynamically.

O'Dell (1963) and Seaton (1966) have estimated the luminosities and effective temperatures for a number of planetary nuclei (see also Osterbrock, 1966). From the observed expansion velocities and estimated diameters for the nebulae it is possible to obtain a time-scale for their evolution if it is assumed that the nuclei represent a reasonably well-defined evolutionary track. The observations indicate that this assumption is justified. In a previous paper (Rose, 1967*a*) the final relaxation oscillation of a 0·75 $M_\odot$ star has been compared in the H-R diagram with the results of Seaton for the estimated tracks of observed central stars. These calculations indicate that it is possible to understand the high luminosities and short lifetimes estimated for these central stars if a large amplitude thermal instability has taken place at the time of the formation of a planetary nebula.

The computed models with pure helium envelopes are bluer than most of the observed nuclei of planetary nebulae. However, models of the final relaxation oscillation of the 0·6 $M_\odot$ sequence, which contains a 0·003 $M_\odot$ hydrogen-rich envelope at this stage of its evolution, are much redder than these nuclei. This is true because of the low molecular weight of hydrogen as compared with helium. The effect of the hydrogen on the radii is made more pronounced because of the high luminosities of the models. On the basis of these calculations it can be concluded that the mass of hydrogen-rich matter in typical central stars is $<0{\cdot}003\ M_\odot$. For some nuclei it must be considerably less than this value. Therefore, the very blue colors of central stars suggest that whatever the cause of planetary nebulae the energy source is beneath the hydrogen shell.

The observations suggest that the maximum luminosity of the central star does not occur immediately following the formation of the nebula but perhaps several thousand years later. It has been shown by computed models (e.g. Rose, 1966) that a decrease in surface luminosity occurs during the interval of high nuclear-energy generation if the thermal instability is sufficiently violent. More recent calculations of 0·6 $M_\odot$ models described above show that during the final relaxation oscillation the surface luminosity decreases from $\approx 2200\ L_\odot$ to $\approx 200\ L_\odot$ before increasing to higher luminosity ($\approx 4500\ L_\odot$) after the photons that were emitted during the thermal instability have had time to diffuse to the surface. The time between the model with the highest nuclear-energy generation ($\approx 3 \times 10^6\ L_\odot$ for this case) and the model with the highest surface luminosity is $\approx 10^3$ years. The observational evidence for relatively low luminosity for central stars during the very early stages in the formation of planetary nebulae is not very strong. However, if future observations confirm the effect it would provide evidence for the turning on of a nuclear source at approximately the time a planetary nebula is formed.

Eddington (1926) first pointed out that nuclear-energy generation can lead to pulsational instability in a star. However, most stars are known to be pulsationally stable with respect to excitation by nuclear sources. This is true primarily because the modes of oscillation are usually non-homologous i.e. the relative amplitudes of all the fundamental modes of oscillation are much higher near the surface layers where radiative dissipation usually dominates than in the interior region of the star where nuclear sources are situated. As a consequence, the stabilizing influence of radiative dissipation is usually sufficient to overcome the destabilizing influence of nuclear-energy generation. However, it has been known for some time that very massive stars

(Ledoux, 1941; Schwarzschild and Härm, 1959) and certain white dwarfs (Ledoux and Sauvenier-Goffin, 1950; Lee, 1950) ought to become pulsationally unstable. More recently it has been shown that thermal instability can lead to pulsational instability in a blue star (Rose, 1967*a*, *b*). It has been suggested that the occurrence of novae and certain X-ray sources may result because a thermal instability has caused a blue star to become pulsationally unstable. On the other hand, it has been argued that the progenitors of planetary nebulae are most probably red giants. For stars with extended envelopes the relative amplitude of the adiabatic modes of oscillation will be much smaller in the region of helium burning than in the extended convective envelope and therefore it is extremely difficult to excite adiabatic modes. However, for the very high rates of nuclear-energy generation, such as result from very large amplitude relaxation oscillations, the *e*-folding time for the development of the thermal instability will be less than the time for sound to traverse a red giant and so an injection of nuclear energy of $\approx 10^{49}$ ergs is expected in a time-scale that is less than the red-giant fundamental pulsation period ($\approx 10^7$ sec). It follows that for this case, which appears to be the case of interest as far as the formation of planetary nebulae is concerned, the results of *adiabatic* pulsation theory are not adequate. It remains to be shown that the amplitude of a relaxation oscillation can become sufficient to lead to non-adiabatic pulsational instability and mass loss.

The calculations that have been carried out so far do not make it appear probable that the input of nuclear energy will become sufficiently rapid to lead directly to the formation of a shock wave. A shock wave will be formed in the burning region only if the pressure varies significantly in less than 1 second.

It should be pointed out that a thermal instability associated with a helium-burning shell may lead to the mixing of hydrogen from the envelope into the hotter interior regions (Schwarzschild and Härm, 1967). In the calculation of Schwarzschild and Härm only a very small amount of hydrogen was mixed into the hot interior. However, if it could be shown that as much as $\approx 10^{-3}\ M_\odot$ of hydrogen were mixed into the helium zone then as much as 10^{49} ergs might be released by means of the CN cycle in several convective mixing times scales ($\approx 3 \times 10^5$ sec). It may turn out that this is a more favorable mode of releasing nuclear energy rapidly than by means of helium burning directly.

We summarize the arguments concerning the origin of planetary nebulae that have been presented in this paper:

(1) The lifetimes and luminosities observed for the nuclei of planetary nebulae can be explained even for nuclei with low mass.

(2) The observed very blue colors of central stars indicate that typical planetary nuclei have $<\frac{1}{2}\%$ hydrogen by mass. Some nuclei must have negligible hydrogen. This suggests that the energy source for the formation of planetary nebulae is below the hydrogen-burning shell.

(3) The maximum luminosity for planetary nuclei appears to come sometime after

the formation of the nebula. This can be understood as due to the turning on of a nuclear source.

(4) Instabilities associated with helium burning appear to arise as a direct consequence of stellar evolution.

References

Abell, G., Goldreich, P. (1966) *Proc. astr. Soc. Pacific*, **78**, 232.
Christy, R. (1966) *Astrophys. J.*, **144**, 108.
Eddington, A. (1926) *The Internal Constitution of the Stars*, Cambridge University Press, Cambridge.
Faulkner, J. (1966) *Astrophys. J.*, **144**, 978.
Faulkner, J., Iben, I., Jr. (1966) *Astrophys. J.*, **144**, 995.
Hayashi, C., Hoshi, R., Sugimoto, D. (1962) *Progress. theor. Phys.*, Suppl. **22**.
Hayashi, C., Hoshi, R., Sugimoto, D. (1965) *Progress. theor. Phys.*, **34**, No. 6.
L'Ecuyer, J. (1966) *Astrophys. J.*, **146**, 845.
Ledoux, P. (1941) *Astrophys. J.*, **94**, 537.
Ledoux, P., Sauvenier-Goffin, E. (1950) *Astrophys. J.*, **111**, 611.
Lee, T.D. (1950) *Astrophys. J.*, **111**, 625.
O'Dell, C.R. (1963) *Astrophys. J.*, **138**, 67.
Osterbrock, D.E. (1966) in *Stellar Evolution*, Ed. by L.F. Stein and A.G.W. Cameron, Plenum Press, N.Y., p. 381.
Rose, W.K. (1966) *Astrophys. J.*, **146**, 838.
Rose, W.K. (1967*a*) *Astrophys. J.*, **150**, 193.
Rose, W.K. (1967*b*) *Astrophys. J.* (in press).
Schwarzschild, M., Härm, R. (1959) *Astrophys. J.*, **129**, 637.
Schwarzschild, M., Härm, R. (1965) *Astrophys. J.*, **142**, 855.
Schwarzschild, M., Härm, R. (1966) Private Communication.
Schwarzschild, M., Härm, R. (1967) *Astrophys. J.*, **150**, 961.
Seaton, M. (1966) *Mon. Not. R. astr. Soc.*, **132**, 113.
Shklovsky, I. (1956) *Astr. Zu.*, **33**, 315.
Weigert, A. (1965) *Mitt. astr. Ges.*, p. 53.

DISCUSSION

Kippenhahn: In his models Rose has a carbon core and helium envelope. This would mean that his star must have already undergone mass loss. If, on the other hand, you want to get your pulsational instability in the red-giant region, you must put on top of your models a hydrogen envelope which is geometrically very thick. This would mean that the pulsational amplitudes will decrease rapidly inwards and will be fairly small at the region of the shells where the thermal pulses occur. Then the damping of the envelope might overcome the destabilizing effect of the shells. How will you overcome this difficulty?

Savedoff: Consider two cases: (a) mass loss by radiation pressure, (b) mass loss by limiting amplitude pulsations. When the radiation-pressure gradient is balanced gravitationally the gas expands into vacuum essentially at the speed of sound (Burger's expansion). Presumably release pressure will limit nuclear-energy rate. When pulsations build up to the escape velocity, the rapid energy loss will again maintain amplitude near the limiting value because of the tremendous energy drawn.

Rose: For the mechanism for the formation of a planetary nebula that I have discussed the energy input may take place in a time comparable with the time required for sound to traverse the red-giant envelope. Therefore, it may be possible to eject the envelope in a single shell.

In the evolution of the unstable 0·75 $M_{\odot}$ star that I have discussed only the final relaxation oscillation was sufficiently violent to approximately explain the short time-scale and high luminosity that are observed for the nuclei of planetary nebulae.

DYNAMICAL INSTABILITY OF THE ENVELOPES OF RED SUPERGIANTS AND THE ORIGIN OF PLANETARY NEBULAE

B. PACZYŃSKI and J. ZIÓŁKOWSKI
(Institute of Astronomy, Polish Academy of Sciences, Poland)

ABSTRACT

Deep convective envelopes of the red supergiants with masses smaller than 4 $M_\odot$ are found to be dynamically unstable for the luminosities in the range of 4×10^3–$5 \times 10^4 L_\odot$. The total energy of the envelopes is positive when the recombination energy of the ionized hydrogen and helium is taken into account, and is sufficient to expel a typical envelope up to infinite distance at a speed of 30 km/sec. We suggest that planetary nebulae are formed in this way.

Substantial progress has been recently achieved in the observational approach to the evolution of the planetary nebulae and their nuclei (O'Dell, 1963; Seaton, 1966). A number of theoretical papers have been devoted to the interpretation of the central stars of planetary nebulae in terms of stellar models composed of helium and heavier elements (L'Ecuyer, 1966; Rose, 1966, 1967; Vila, 1966). The evolutionary tracks and the time-scales obtained by Vila (1966) and Rose (1967) show the closest similarity to the empirical track of Seaton (1966). These theoretical sequences were obtained for stellar models of 0·75–1·2 $M_\odot$, cooling down towards the white-dwarf region. There is a widespread opinion that planetary nebulae originate from red giants. This idea was suggested by Shklovsky (1956), and recently Abell and Goldreich (1966) have summarized a series of arguments that tend to support this point of view. According to Osterbrock (1964) there was no satisfactory identification of the physical processes involved in the formation of the planetary nebulae. Recently, Rose (1966, 1967) suggested that thermal and pulsational instabilities found by him in the models of helium shell-burning stars may be connected with the origin of the planetary nebulae.

We found that red supergiants of sufficiently high luminosity have extended convective envelopes that are dynamically unstable. This phenomenon is due to a low value of the adiabatic exponent Γ_1 in the thick zones of partial ionization of hydrogen and helium. This dynamical instability is of the same character as the one encountered in the pre-main-sequence stars just before they reach the Hayashi track. There are two main differences between an old supergiant and a young proto-star: a red supergiant has a very dense core composed of helium and heavier elements and it evolves upwards on the H-R diagram towards the border of dynamical stability. It is not clear whether or not a helium shell-burning star may lose the hydrogen-rich envelope as a result of thermal or pulsational instabilities associated with the shell sources. If not,

Osterbrock and O'Dell (eds.), Planetary Nebulae, 396–399. © *I.A.U.*

then it seems that nuclear evolution of a single star may be terminated either as a result of the dynamical instability of the core, or because of the dynamical instability of the envelope. The first case was discussed by many authors as a likely cause of supernova explosions (Hoyle and Fowler, 1960; Colgate and White, 1966). The second will be discussed in this paper.

To construct static envelopes the computer program written by Ziółkowski (1967) was used. All the calculations were carried out with the GIER computer of the Warsaw University. Our program was almost identical with the one described by Baker and Kippenhahn (1962) and Hofmeister *et al.* (1964). There were only two minor differences: the dependence of the temperature on the optical thickness in the atmosphere was, in the present study, the same as that adopted by Henyey *et al.* (1965, eq. 49); and the opacities published by Cox and Stewart (1965) were used in our program. Model envelopes were integrated until the temperature increased to 5×10^5 °K. We adopted the ratio of mixing length to pressure scale height, $l/H_p = 1$. The luminosity L, and the chemical composition were assumed to be constant in a given model. Because of the high luminosities and the low effective temperatures which were of interest here the convection was strongly superadiabatic and the density changed very little in a typical envelope. The transition between the extended envelope and a small and dense core was very sharp, and in most cases occurred at fractions of the radius (0·0005–0·01).

The eigenvalue σ_0^2 corresponding to the fundamental mode of the adiabatic radial oscillations was computed for every static envelope. For a given total mass M of a star, and an effective temperature T_e, we searched for the luminosity L at which $\sigma_0^2 = 0$. As a by-product the mass fraction contained in a dense core was obtained for such a model. In this way it was possible to find a line on the M_{bol}–$\log T_e$ diagram which separated the stable and the unstable static envelopes. The unstable envelopes were, of course, found on the high-luminosity side of that line.

The total energy of an envelope that became unstable was positive when the recombination energy of the ionized hydrogen and helium was taken into account. Therefore, it is likely that instability may cause an expansion of the envelope to infinity. If the envelope expanded adiabatically it would attain a velocity of the order of 30 km/sec in a typical case. This is remarkably close to the expansion velocities observed in planetary nebulae.

The borderlines on the M_{bol}–$\log T_e$ diagram were obtained for dynamically unstable envelopes of the population-I stars of 1, 2, and 3 $M_\odot$, and for the population-II star of 1 $M_\odot$. The Kippenhahn I and II mixtures (Cox and Stewart, 1965) were used for the population I and II, respectively. The borderlines are shown in Figure 1 for the population-I stars. The dynamical instability appears for the population-II star at a bolometric magnitude 1^m above the line for 1 $M_\odot$ star of population I. At the highest luminosities allowed by the assumptions that stellar models are in hydrostatic and thermal equilibrium the envelopes of stars of 5 and 15 $M_\odot$ were dynamically stable.

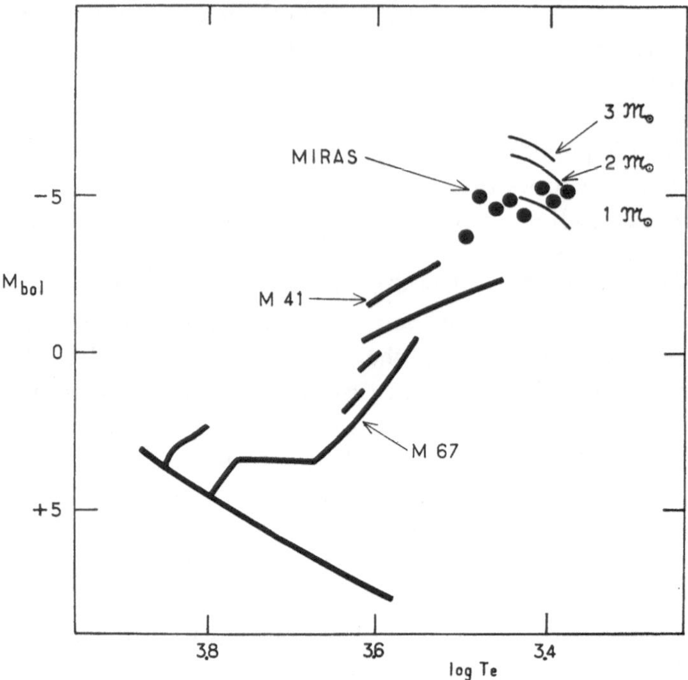

FIG. 1. *The borderlines for dynamically unstable envelopes of population-I stars of 1, 2, and 3 $M_\odot$ are shown on the M_{bol}–$\log T_e$ diagram. The mass fraction contained in the dense cores of the stars varies from* 0·2 *to* 0·8 *between the low- and high-temperature ends of each line. Positions of some galactic clusters are adopted from Sandage (1957) and of the Mira variables from Smak (1966).*

We propose the following scheme for the late phases of stellar evolution. After the exhaustion of helium in the core the star evolves into the region of red supergiants and moves up on the H-R diagram very close to the Hayashi border (see e.g. Kippenhahn *et al.*, 1965; Weigert, 1966; Sugimoto and Yamamoto, 1966). The mass of the helium and carbon core and the luminosity due to the helium and hydrogen-shell sources increase. A star with total mass smaller than about 4 $M_\odot$ will terminate this type of evolution with an outflow of hydrogen-rich matter as a result of the dynamical instability of the extended envelope. We suggest that planetary nebulae are formed in this way. This process seems to be impossible for a star with larger mass. In the latter case the mass of the core in which all the nuclear-energy sources are exhausted will finally exceed the Chandrasekhar limit for degenerate configurations. Dynamical instability of the core, followed by the supernova explosion, may be expected.

It transpires from Figure 1 that the region in the H-R diagram in which stellar envelopes might be dynamically unstable is occupied by Mira-type stars at maximum light (Smak, 1966). This may suggest that there is some connection between our instabilities and the light changes in the long-period variables, and also that those stars might be the direct ancestors of the planetary nebulae.

We would like to point out that stellar models with the highest luminosity obtained by Sugimoto and Yamamoto (1966) for 1 $M_{\odot}$ should have dynamically unstable envelopes. The dense core of 0·8 $M_{\odot}$ obtained by Weigert (1966) at the end of his evolutionary track for 5 $M_{\odot}$ had a luminosity sufficiently high to induce instability of a hydrogen-rich envelope of 0·8 $M_{\odot}$. Therefore, some evolutionary tracks already available approach the phase in which a star may lose its outer layers and form a planetary nebula.

The numerical values of the luminosities and masses of stars with dynamically unstable envelopes are highly uncertain because of the poorly known theory of convection, opacities at low temperatures, and the boundary conditions for the stability tests. More detailed description of our work and more complete discussions of the results obtained will be published soon in *Acta Astronomica*.

References

Abell, G., Goldreich, P. (1966) *Publ. astr. Soc. Pacific*, **78**, 232.
Baker, N., Kippenhahn, R. (1962) *Z. Astrophys.*, **54**, 114.
Colgate, S.A., White, R.H. (1966) *Astrophys. J.*, **143**, 626.
Cox, A.N., Stewart, J.N. (1965) *Astrophys. J. Suppl.*, **11**, 22.
Henyey, L., Vardya, M.S., Bodenheimer, P. (1965) *Astrophys. J.*, **142**, 841.
Hofmeister, E., Kippenhahn, R., Weigert, A. (1964) *Z. Astrophys.*, **59**, 215.
Hoyle, F., Fowler, W. (1960) *Astrophys. J.*, **132**, 565.
Kippenhahn, R., Thomas, H.-C., Weigert, A. (1965) *Z. Astrophys.*, **61**, 241.
L'Ecuyer, J. (1966) *Astrophys. J.*, **146**, 845.
O'Dell, C. R. (1963) *Astrophys. J.*, **138**, 67.
Osterbrock, D.E. (1964) *A. Rev. Astr. Astrophys.*, **2**, 95.
Rose, W.K. (1966) *Astrophys. J.*, **146**, 838.
Rose, W.K. (1967) *Astrophys. J.*, **150**, 193.
Sandage, A.R. (1957) *Stellar Populations*, ed. by D.J.K. O'Connell, Pontifical Acad. of Science, Rome, 41.
Seaton, M. (1966) *Mon. Not. R. astr. Soc.*, **132**, 113.
Shklovsky, I. (1956) *Astr. Zu.*, **33**, 315.
Smak, J. (1966) *Acta astr.*, **16**, 1.
Sugimoto, D., Yamamoto, Y. (1966) *Progress Theor. Phys.*, **36**, 17.
Vila, S.C. (1966) *Z. Astrophys.*, **64**, 395.
Weigert, A. (1966) *Z. Astrophys.*, **64**, 395.
Ziółkowski, J. (1967) in preparation.

PRODUCTION OF PLANETARY NEBULAE

M. P. SAVEDOFF, G. S. KUTTER, and H. M. VAN HORN
(C. E. K. Mees Observatory, Dept. of Physics and Astronomy, University of Rochester, Rochester, N.Y., U.S.A.)

For various reasons, we have been studying evolution in the pre-white dwarf phase at Rochester. Our attention to the relevance of this work to planetary nebulae came as a result of calculations of the evolution of a one solar-mass iron star carried out at Rochester by S. Vila. Here the neutrino processes drive the peak luminosity to log $L/L_\odot = 4{\cdot}26$, at an effective temperature $\log T_{\rm eff} = 5{\cdot}53$. Although these models are brighter than 100 $L_\odot$ for 500000 years, they are brighter than 1000 $L_\odot$ for 4000 years, and exceed 10000 $L_\odot$ for 900 years. We are therefore near the luminosity of the planetary-nebula nuclei, but considerably hotter, for a period of the order of the planetary lifetimes. Except for the temperature discrepancy, these models are in rough agreement with the observationally determined evolutionary sequence found by O'Dell (1963) and by Harman and Seaton (1964) and Seaton (1966).

The space density of planetary nebulae and white dwarfs and their Galactic kinematics (to the extent that they are known) are consistent with this evolutionary connection, provided the planetary lifetime is approximately 10000 years.

Let us examine two observational properties of planetaries for guidance in discovering modes of planetary production. Clearly, the expansion velocities in the range of 30–100 km/sec are well separated from those of the novae and supernovae, which we may call explosive objects. This flow is characteristic of sound speeds of the order of 10^5 °K rather than 10^9 °K. We are led to consider models in which things proceed slowly compared to pulsational periods. This does not preclude Rose's results as being applicable. On the time-scale of 10^3 years, a mass of $0{\cdot}5\ M_\odot$ can be accelerated to 30 km/sec by a mean luminosity of 100 $L_\odot$ assuming total efficiency. Removing the material from a deep potential well and possible inefficiencies can easily require the available energy to be nearly a thousand-fold higher.

Secondly, there exist some indications that an object may have only one planetary nebula phase in contrast to the behavior of novae. However, if more than one phase does occur, we suggest a minimum recurrence time for episodes of 10000 years. This may be compared with the pulses found by Rose (1966) and Weigert (1966) of 10^5 and 4000 years.

We propose to put nuclear fuel into Vila's stars and watch what happens. These differ from earlier models in including neutrino effects; e.g. that the central temperatures remain below a peak temperature which propagates outwards, increasing in

Osterbrock and O'Dell (eds.), Planetary Nebulae, 400–406. © *I.A.U.*

amplitude until it reaches a maximum peak temperature $\log T = 8{\cdot}92$ at $M/M_* = 0{\cdot}73$, this peak decreases now in peak temperature as it propagates out to a fractional mass at least as large as 0·99. The time-scale of non-thermal equilibrium is related to the photon diffusion time,

$$t \simeq kM/Rc \simeq 10^4 \text{ years}.$$

Models of pure carbon are reported as the simplest system for which the temperature inversion is expected and for which a large temperature-sensitive energy source is available. This last condition causes us to omit models with helium and hydrogen as the energy sources responsible for the flash. The abundance of elements in white dwarfs is uncertain, although one of us has suggested evidence for a mean charge of 6–8 for one group (Van Horn, 1968).

We estimate that for masses less than 1·08 $M_\odot$, neutrino losses will dominate and hence that no mass loss will occur. In contrast, for M greater than 1·6 $M_\odot$, nuclear reactions begin before the temperature peak is formed, hence the energy-generation peak is in the centre. These models should resemble main-sequence models. Calculations have been carried out for 1·1 $M_\odot$ from initial conditions shown in Table 1,

Table 1

Initial conditions for 1·1 $M_\odot$ C^{12} Star

Polytrope of index 3
$R_* = 0{\cdot}4\ R_\odot$
$\rho_c = 10^3$ g/cm^3
$T_c = 80 \times 10^6$ °K
$L = 10^3\ L_\odot$
$T_{eff} = 46000$ °K

through the white dwarf stage, assuming hydrostatic equilibrium throughout. The initial model is derived distantly from a polytrope and has no degeneracy, no neutrino emission, etc.

Evolution in the L, T_e plane (see Figure 1) is similar to the $M = 1\ M_\odot$ iron model of Vila (1965) except for the second loop driven by the C^{12} burning. We find that all the nuclear energy is essentially converted into gravitational energy through the work done in expanding the star adiabatically. No real instability (in the sense of Mestel's thermal run-away) occurs.

In Figure 2, note the development of an extended convective shell driven by the extreme peak in the energy generation. The peak rate of surface expansion was only 30 cm/sec, vanishingly small compared to escape velocities exceeding 10^8 cm/sec. Temperature distributions for some of the models listed in Table 2 are shown in Figs. 3a and 3b.

Can these models be consistent with mass loss? We had originally expected an extended model, $R = 10^{12}$ cm with convection throttled in the surface layers, but the

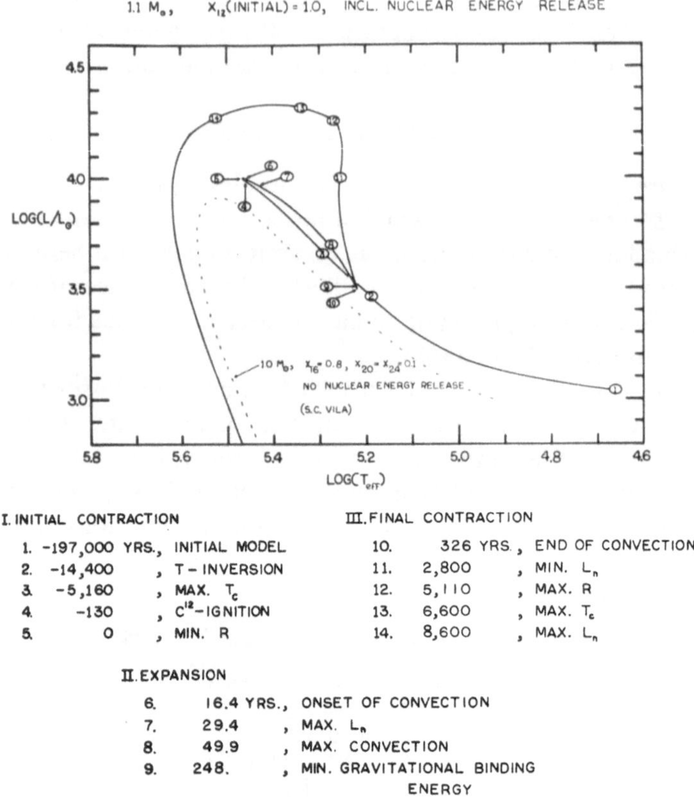

FIG. 1. *L, T_e diagram for* 1.10 $M_\odot$ C^{12} *star.*

model allows convection only out to $M/M_* = 0{\cdot}96$. Ledoux has suggested that these models should be vibrationally unstable, with pulsation amplitude increasing until damped by mass loss from the surface. We have been unable to follow this idea into the non-linear regions. Rough estimates show that the amplitude must be comparable to the stellar radius for the velocity to reach the escape velocity.

We had hoped for mass loss through a 'stellar wind' analogous to the Parker 'Solar Corona' Model. This 'corona' could be produced by sonic waves generated in the convective region. Lighthill (1952) and Proudman (1952) has shown that the rate of sonic noise generation by an isotropic turbulent medium is approximately

$$\frac{L_{\text{sonic}}}{4\pi R^2} = 30\rho v_s^3 (v/v_s)^8 .$$

Using conventional mixing length theory we estimate $v = 2$ km/sec and $v_s = 10^3$ km/sec, hence $L = 10^{26\cdot7}$ ergs/sec. We fall short of the needed energy by a factor of 10^{10}.

Seaton (1966) has noted that radiation pressure may dominate the stellar models in the region of the planetary nebula nuclei. Setting

$$\frac{GM\rho}{r^2} = \frac{L\kappa\rho}{4\pi r^2 c}$$

we find for electron scattering

$$\frac{L_c}{L_\odot} \sim \frac{10^{4\cdot 8} M/M_\odot}{(1+X)}.$$

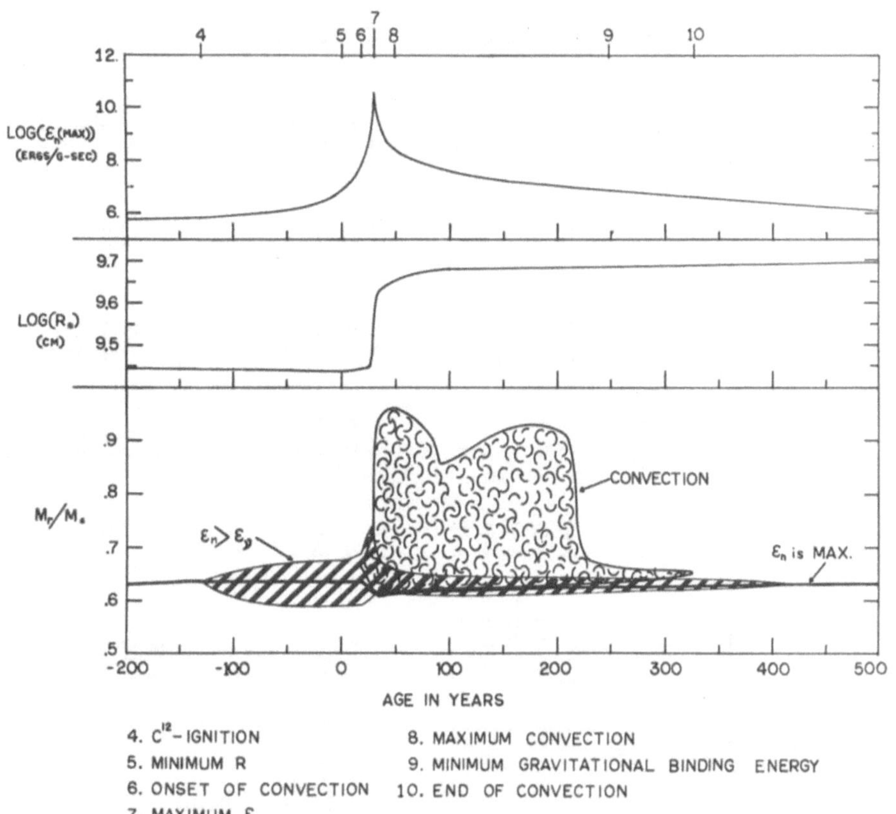

FIG. 2. *Development of* $\log \varepsilon_n$ *(max),* $\log R_*$*, and extent by mass fraction of nuclear burning* ($\varepsilon_n > \varepsilon_\nu$) *and convection vs. time during the* C^{12} *flash of the* 1·1 $M_\odot$ *star.*

At the second pass (see Figure 1) peak log luminosity reached 4·33 at that phase where the temperature peak was at $M/M_* = 0{\cdot}84$.

Thus an increase of the opacity or the luminosity by a factor of 3 would suffice to free the material gravitationally. We have come close. For $L > L_c$, as a first approxima-

Table 2

Summary of the evolution of the 1·1 $M_\odot$ C^{12} star

Description	Age (years)	$\log(T_c)$ (°K)	$\log(\rho_c)$ (g/cm³)	$\log(R)$ (cm)	$\log(T_{eff})$ (°K)	$\log(\frac{L}{L_\odot})$	$\log(\frac{L_\nu}{L_\odot})$	$\log(\frac{L_n}{L_\odot})$	$E_{gravitation}$ (ergs)
I. Initial Contraction									
1. Initial Model	−197000	7·92	3·00	10·56	4·66	3·04	–	–	$-0{\cdot}155 \times 10^{50}$
2. *T*-Inversion	−14400	8·61	5·74	9·73	5·19	3·46	3·87	−0·93	−1·16
3. Maximum T_c	−5160	8·64	6·30	9·62	5·29	3·66	4·61	1·90	−1·73
4. C^{12}-Ignition	−130	8·51	6·99	9·44	5·46	3·99	5·21	4·71	−3·27
5. Minimum *R*	0	8·50	7·01	9·44	5·47	4·00	5·24	5·41	−3·29
II. Expansion									
6. Onset of Convection	16·4	8·49	6·99	9·45	5·46	4·00	5·24	5·49	$-3{\cdot}20 \times 10^{50}$
7. Maximum L_n	29·4	8·44	6·88	9·50	5·43	3·96	5·73	8·45	−2·84
8. Maximum Convective Zone	49·9	8·35	6·69	9·66	5·28	3·68	5·16	6·08	−2·16
9. Min. Gravitational Binding Energy	248·	8·33	6·67	9·68	5·22	3·51	4·90	4·88	−2·06
III. Final Contraction									
10. End of Convection	326	8·33	6·67	9·68	5·22	3·50	4·87	4·69	$-2{\cdot}07 \times 10^{50}$
11. Minimum L_n	2800	8·36	6·80	9·89	5·26	4·06	4·71	3·13	−2·46
12. Maximum *R*	5100	8·37	6·96	9·97	5·26	4·26	4·88	3·46	−2·99
13. Maximum T_c	6600	8·38	7·10	9·85	5·34	4·32	5·00	3·87	−3·50
14. Maximum L_n	8600	8·36	7·30	9·46	5·53	4·27	5·01	4·21	−4·31
15. White-Dwarf Model	856000	7·89	7·71	8·72	4·79	−0·15	0·58	–	−6·27

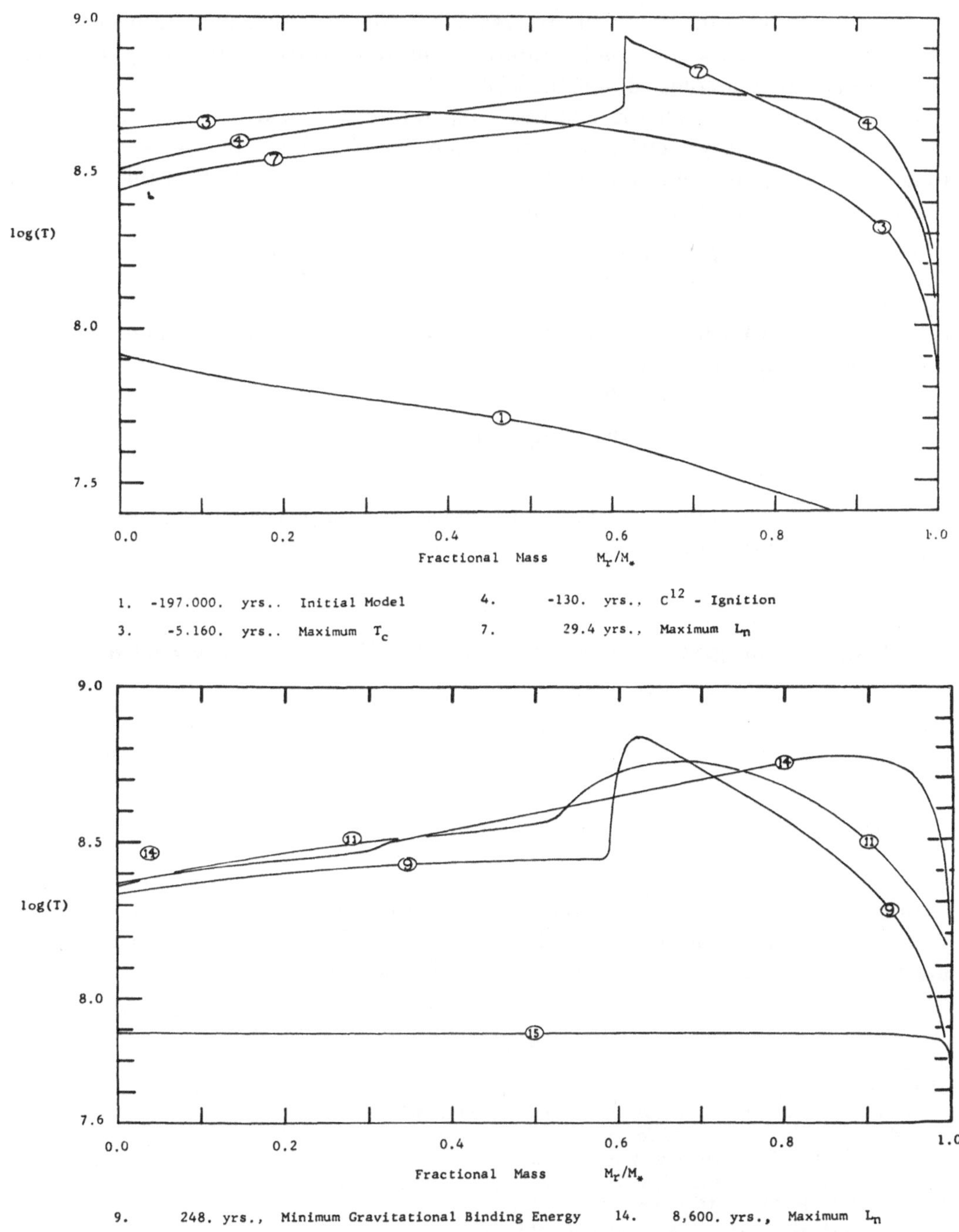

FIG. 3. *Distribution of log T vs. fractional mass for selected models of the* 1·1 $M_\odot$ C^{12} *star.*

tion, the behavior of the gas would resemble free flow into a vacuum until such time as its opacity decreased (recombination) or the luminosity decreased so that the gravitational force becomes dominant again.

The apparent discrepancy in effective temperature and hence radius between our models of the planetary nuclei may be removed by judicious additions of low-molecular weight material to the outer layers of the star. To estimate the effects of this inhomogeneity upon the calculations we require temperature continuity at radius $r_1 = r_2$, that

$$T = \frac{\beta}{4{\cdot}25} \frac{HGM}{kr_1} \mu_1 \left(1 - \frac{r_1}{R_1}\right) = \frac{\beta}{4{\cdot}25} \frac{HGM}{kr_1} \mu_2 \left(1 - \frac{r_1}{R_2}\right)$$

be the same both for the homogeneous composition (1) and for the boundary of the outer layer (composition (2)),

$$\frac{R_2}{R_1} = \frac{1}{\dfrac{\mu_1}{\mu_2} - \left(\dfrac{\mu_1}{\mu_2} - 1\right)\dfrac{R_1}{r_1}}.$$

In Table 3 we see that these radii can be appreciably enlarged by the substitution of a 1% helium envelope or a 10^{-5}% hydrogen envelope.

Table 3

Radii of inhomogeneous models based upon 1·1 $M_\odot$ C^{12} star, age 7000 years

$\log T$	R_1/r_1	R_2/R_1	$\log M/M_*$	$\log \rho$	κ
		Helium–Carbon			
8·28	3·6	3·9	−·004	3·11	0·14
8·14	3·0	2·3	−·002	2·62	·16
		Hydrogen–Carbon			
6·93	1·42	∞	-2×10^{-7}	−1·37	·26
6·78	1·38	12·	-5×10^{-8}	−1·92	0·30

In summary, we have made another pass through the region of the planetary nuclei; we have come close, but we did not find a nebula. Proposed extensions of this work are being investigated as they seem likely to provide possible planetaries.

References

Harman, R.J., Seaton, M.J. (1964) *Astrophys. J.*, **140**, 824.
Lighthill, M.J. (1952) *Proc. R. Soc. London*, **A211**, 564.
Proudman (1952) *Proc. R. Soc. London*, **A214**, 119.
O'Dell, C.R. (1963) *Astrophys. J.*, **138**, 67.
Rose, W.K. (1966) *Astrophys. J.*, **146**, 838.
Seaton, M.J. (1966) *Mon. Not. R. astr. Soc.*, **132**, 113.
Van Horn, H.M. (1968) *Astrophys. J.*, **151**, 227.
Vila, S. (1965) Thesis, University of Rochester.
Weigert, A. (1966) *Z. Astrophys.*, **64**, 395.

ON THE QUESTION OF THE ORIGIN OF PLANETARY NEBULAE

I. A. KLIMISHIN
(Astronomical Observatory of Lvov University, U.S.S.R.)

While the physics of some processes which are going in planetary nebulae grows to be one of the most elaborated parts of theoretical astrophysics, the origin of these objects is enigmatic. Ten years ago Shklovsky (1956) made a supposition that the formation of planetary nebulae occurs at the last evolutionary stage of a red giant as a result of the violation of mechanical equilibrium in the envelope of the star. In connection with this the possibility of the separation of an envelope of a red giant under the influence of a shock wave which moves with the velocity of the order of 100 km/sec was considered by Kaplan and Klimishin in 1959. Later on numerical calculations of the motion of a weak shock wave in the envelope of a red giant, were made which showed the basic possibility of formation of a planetary nebula with a mass of the order of 0·5 $M_{\odot}$ and an initial velocity about 95 km/sec (Sakashita and Tanaka, 1962). The ballistic character of the mechanism of the formation of the planetary nebulae has been recently discussed also by Abell and Goldreich (1966).

Usually the appearance of a shock wave on the surface of a star is more naturally associated with phenomena like novae or supernovae. An intrinsic peculiarity of such phenomena is the presence of an essential velocity gradient, owing to which the envelope quickly dissipates into space. That is why the possibility of formation under the influence of a shock wave of such compact envelopes as we have in the case of planetary nebulae could be considered a disputable one.

However, it is not difficult to show that the motion of an ejected envelope is defined by conditions such that either the undisturbed envelope was in thermal equilibrium or in convection. Such calculations have been done recently by us (Klimishin, 1967) with the help of the quasi-stationary method of Chisnell-Withem.

From general considerations it follows that the strength of the shock wave (the jump of pressure p/p_0 at the front of the wave) will grow in direct proportion to the value $m(r)=(\mathrm{d}\ln p_0)/(\mathrm{d}\ln r)$. Since at each non-dimensional distance in the convective envelope the value $m(r)$ is about half of its corresponding value for the thermal envelope, patterns of the velocity changes of the shock wave with the distance for envelopes of different structures are not the same. In the thermal envelope the velocity of the shock quickly grows with the distance, and the ejected envelope moves with a large gradient of velocity and quickly dissipates in the space. This is typical for nova outbursts.

As for the convection envelope, the velocity of the shock wave is practically constant, the velocity gradient in the envelope is 0, and the ejected envelope moves as a whole. This is probably the most attractive peculiarity of the shock-wave mechanism for the formation of planetary nebulae.

These calculations enable us to conclude that both phenomena (the flashes of nova and the formation of planetary nebulae) can be interpreted in terms of the same mechanism, namely shock waves. Nevertheless, the shock-wave mechanism of formation of planetary nebulae meets some serious difficulties.

Let us note in particular that the radius of a remnant of the star after the envelope has been ejected (that is, the possible nucleus of the nebula) according to calculations by Sakashita and Tanaka, is of the order of 50 $R_{\odot}$. At the same time it is known (Sobolev, 1966) that the observed radii of nuclei of planetary nebulae are about 10^{10} cm or about 0·1 R. This means that the escape velocity on the surface of such a star is 2000 km/sec and the energy of separation of the envelope with mass 0·10 $M_{\odot}$ is 10^{49} ergs. The velocity of the shock wave producing the separation of the envelope with the following formation of a planetary nebula is 2600 km/sec. As the velocity of the shock will increase during its way out of the chromosphere, the formation of each new planetary nebula must be accompanied by a rather intensive explosion. In other words, the combination of the small radii of nuclei with the ideas of the small velocity of the shock in the envelope of red giants seems to be impossible.

In conclusion, with regard to the hypothesis of Shklovsky, we should remark that it has not proven possible to build models of red giants and supergiants at low stellar temperatures. Numerical calculations inevitably lead to a rapid increase of density at small distances from the surface of the star, that is, to the appearance of a spherical layer with nothing inside (Nadezhin, 1967). In conclusion, one should note that in the atmospheres of red supergiants the assumption of hydrostatic equilibrium does not hold and it is very likely that it is there that we should search for the answer of the question of the formation of planetary nebulae.

References

Abell, G., Goldreich, P. (1966) *Publ. astr. Soc. Pacific*, **78**, 232.
Kaplan, S.A., Klimishin, I.A. (1959) *Soviet Astr. J.*, **36**, 410.
Klimishin, I.A. (1967) *Circ. astr. Obs. Lvov Univ.*, **43** (in press).
Nadezhin, D.K. (1967) private communication.
Sakashita, S., Tanaka, Y. (1962) *Progress. theor. Phys.*, **27**, 127.
Shklovsky, J.S. (1956) *Soviet Astr. J.*, **33**, 315.
Sobolev, V.V. (1967) *Kurs teoreticheskoy astrofiziki*, Moskva.

EVOLUTION OF CENTRAL STARS OF PLANETARY NEBULAE THEORY*

EDWIN E. SALPETER
(Cornell University, Ithaca, N.Y., U.S.A.)

1. Elementary Evolution Theory

Before discussing observational inputs and actual model calculations, I want to give a very elementary review of the relevant parts of stellar evolution theory. We shall only be dealing with stellar masses M below the Chandrasekhar limiting mass $M_{ch}(\sim 1{\cdot}2\text{–}1{\cdot}45\ M_{\odot}$, depending on chemical composition). The inequality $M < M_{ch}$ implies that relativistic effects are not of overriding importance and I will not mention them further (however, all quantitative model calculations which I will mention later include all the relativistic corrections to the equation of state, including radiation pressure which is also not very important). Let us try to estimate how the central temperature T_c and (total bolometric) luminosity L varies with central density ρ_c or radius $R(\rho_c \sim MR^{-3})$ for a star of fixed mass.

At low density we can surely use the classical perfect gas law for the equation of state and the mean thermal energy per particle ($\sim kT_c$) is comparable with the mean gravitational potential energy per particle ($\sim GM/R$), according to the Virial Theorem. T_c thus *increases* with increasing density ($T_c \sim MR^{-1} \sim M^{2/3}\ \rho_c^{1/3}$) for small T_c. However, the (non-relativistic) Fermi energy increases more rapidly ($\sim \rho_c^{2/3}$) with density, electron degeneracy becomes more important (when $\rho_c \sim \text{const}\ T_c^{3/2}$) and the electrons finally become fully degenerate at *zero* temperature when ρ_c approaches a critical value $\rho_{c,\max}$. The evolutionary phase of *decreasing* temperature as the density increases slightly towards its final value represents the white-dwarf phase, of course, but here we are interested in the phase just before.

Since T_c as a function of ρ_c first increases and then decreases it must reach a maximum value $T_{c,\max}$ at an intermediate value of density (a few times smaller than $\rho_{c,\max}$) when the electrons are partially degenerate. From the arguments above one can easily derive that

$$T_{c,\max} \sim M^{4/3};\ \rho_{c,\max} \sim M^2 .$$

Conversely, for a star to be able to heat up beyond a certain central temperature, its mass must exceed a certain minimum value. Thus only stars more massive than

* This work supported in part by the National Science Foundation under NSF grant GP-6928.

some M_{min} can burn some nuclear fuel (we shall be interested in carbon) in its interior (Deinzer and Salpeter, 1964). Other things being equal, the luminosity L of a star increases with its interior temperature, and L will have a maximum value at an evolutionary stage fairly close to (and slightly before) the one at which T_c reaches its maximum. Further, the numerical value of L_{max} is larger for larger masses, as illustrated schematically in Figure 1. In this figure we also see a schematic color-magnitude diagram which illustrates that the effective *surface* temperature T_e also reaches a maximum value soon after the luminosity does, since $T_e^4 \sim LR^{-2}$ and the radius R is already fairly close to its final (smallest) white dwarf value.

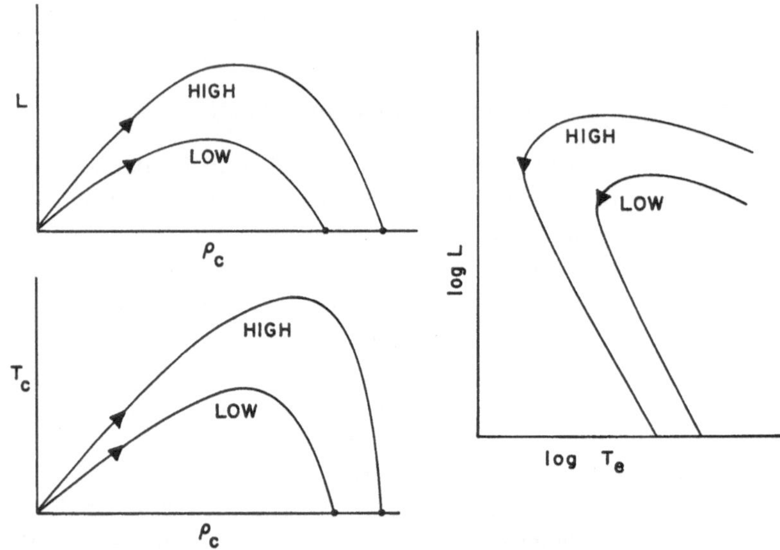

FIG. 1. *Schematic evolutionary tracks of partially degenerate stars.*

2. Mass and Composition

Figure 2 is a schematic color-magnitude diagram, from a review by Stothers (1966), showing the observed positions of the central stars of planetary nebulae as well as the white-dwarf region and the evolutionary track for globular cluster stars. The high values of L and T_e for the central stars and a comparison with Figure 1 make it very plausible that these stars are at an evolutionary phase near their maximum luminosity and interior temperature. This qualitative statement is agreed upon by all, but the quantitative features are still highly controversial. The uncertainties are in a large part due to the fact that we do not have quantitative information on the mass and chemical composition of these stars.

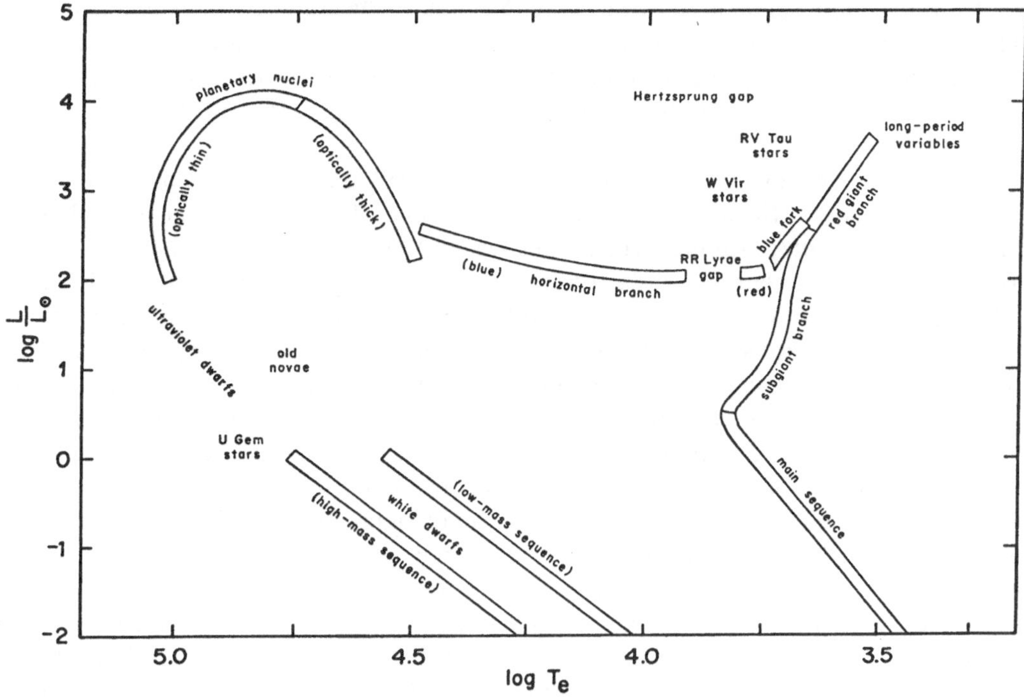

FIG. 2. *Schematic color-magnitude diagram (from Stothers, 1966).*

Central stars of planetary nebulae are fairly highly evolved and some mass loss during or after the red-giant phase (but *before* the mass ejection of the planetary nebula itself) is at least a possibility, but not certain. The amount of this previous mass loss is important (even if the present mass were known) in determining the present composition since carbon, oxygen and neon, as well as helium, are built up in the star's interior whereas the mass ejection is mainly from the hydrogen-rich outer layers. If the total age and original chemical composition of the star were known, its original mass could be determined (since the main sequence age depends on mass and the subsequent evolution is very rapid).

Previous speakers have shown that planetary nebulae belong to stellar population II (including one known member of a globular cluster) but not in its extreme form. This implies some spread in ages and hence a spread in original masses upwards from that of globular cluster red giants M_{gl}. The degree of spread is not known, as previous speakers emphasized, and the 'typical' (or mean) original mass could be only very slightly larger than M_{gl} or it could easily be $1{\cdot}5\ M_{gl}$, say. The numerical value of M_{gl} is itself in doubt at the moment (Faulkner and Iben, 1966; Hartwick *et al.*, 1967): M_{gl} could be as large as $1{\cdot}2\ M_{\odot}$ if the original composition was helium-poor or as low as $0{\cdot}7\ M_\odot$ if the original stars were as helium-rich ($\sim 35\%$ by mass) as population I. The present masses of central stars of planetary nebulae are also not known. As Dr. Weidemann will discuss in more detail, there is also a spread in the masses of white dwarfs with masses up to $\sim 0{\cdot}6\ M_\odot$ 'typical' and (since it is not known whether further mass loss occurs between the planetary nebula and white-dwarf stages) we can only be fairly certain that the central stars have $M \gtrsim 0{\cdot}5\ M_\odot$. The upshot of these uncertainties is that the amount of pre-planetary mass loss could either be negligible or very considerable; original 'typical' masses could be as low as $0{\cdot}7\ M_\odot$ and as high as $2\ M_\odot$ and the present mass of a 'typical' central star in the range of 0·5 to $1{\cdot}2\ M_\odot$.

As mentioned above, there is considerable uncertainty in the present chemical composition of the central stars, but a few things are clear: (i) During the previous evolutionary phases considerable nuclear burning must have taken place and the central regions must be rich in C^{12} (and O^{16}, Ne^{20}) and intermediate regions in He^4. (ii) Since only the *extreme* population-II stars are metal-deficient, the original composition of most (but not necessarily all) of the central stars must have involved normal abundances of the metals and oxygen. (iii) The planetary nebulae themselves are not anomalously rich in helium and were ejected from the central stars. Thus, while a large fraction of the outermost hydrogen-rich (original composition) layers may have been ejected, the ejection could *not* have penetrated to the helium-rich layers beneath.

Evolutionary model sequences through helium shell burning are so far only available for a few cases (Iben, 1966; Kippenhahn *et al.*, 1966; Hofmeister, 1967). For a star of mass $5\ M_\odot$, e.g., there is one evolutionary state in which a central core of mass $0{\cdot}9\ M_\odot$ consists mainly of C^{12} (and O^{16}, Ne^{20}), followed by a *thin* intermediate layer of helium. If mass ejection happened at this stage down to the boundary between the helium layer and the hydrogen envelope, one would have an almost pure carbon star of about $1\ M_\odot$. As we have discussed, typical masses of the original stars of planetary nebulae must be considerably less than $5\ M_\odot$, and it is not yet clear whether almost pure carbon stars are possible candidates for the central stars of planetary nebulae. Nevertheless, I will discuss such homogeneous stellar models next, if only for their simplicity.

3. Homogeneous Models

Before discussing actual models, let us consider Figure 3 which illustrates one general point contrasting models with a concentrated vs. those with an extended energy source. With a concentrated energy source, such as nuclear burning in the core, the heat flux is already close to the full luminosity even in the core and the temperature gradient is appreciable in the core. As a consequence, the temperature in the intermediate regions of the star (for given central temperature T_c and density) is lower and so is the temperature gradient from there on out which determines the value of the luminosity. Conversely, with an extended energy source such as gravitational energy release (or even more so with nuclear shell-source burning), the temperature in the inter-

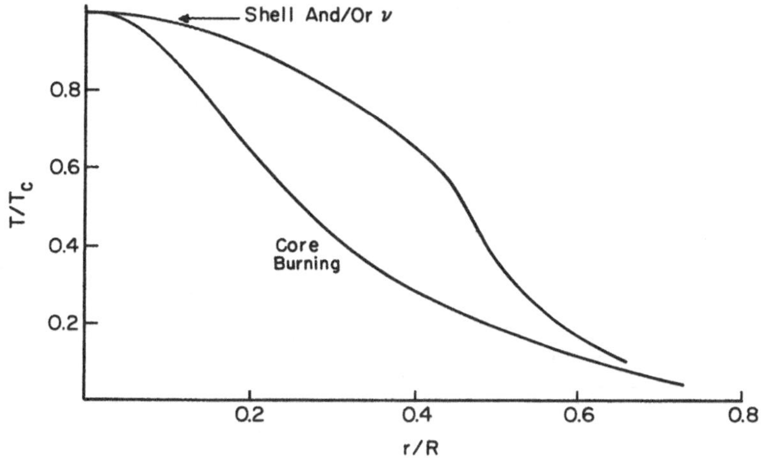

FIG. 3. *Schematic temperature distributions in different types of models.*

mediate regions is close to T_c and the temperature gradient further out, and hence the luminosity is larger. Neutrino energy *loss* (which is fairly concentrated) accentuates the effects of an extended energy source. We shall indeed find that the onset of nuclear core burning in a gravitationally contracting star *lowers* the luminosity L at first (with a slight increase in T_c), whereas the onset of shell burning *increases* L (and models with neutrino energy loss generally have larger optical L).

Explicit evolutionary sequences of models of homogeneous stars consisting mainly

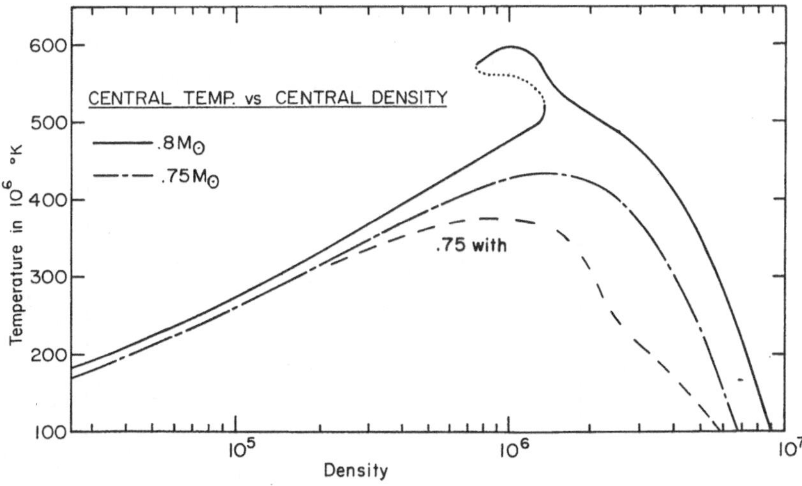

FIG. 4. *Central temperature plotted against central density for pure carbon models ('with' denotes inclusion of neutrino rates).*

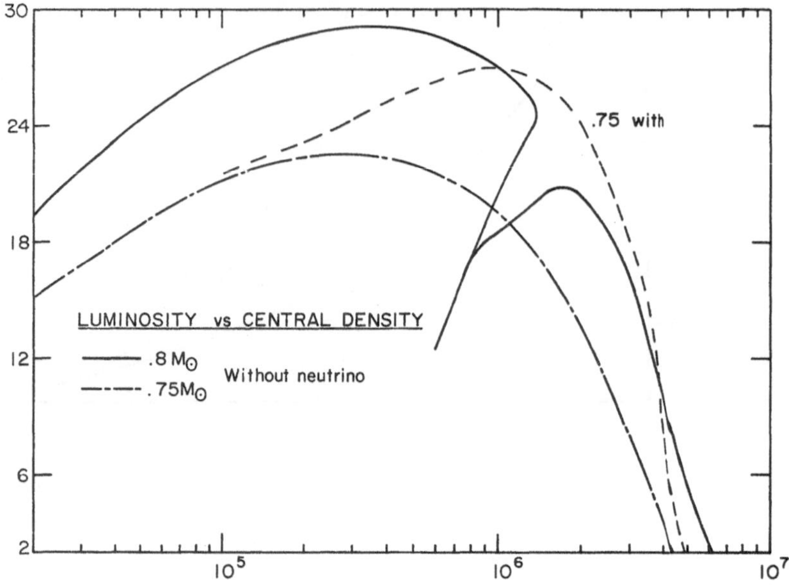

FIG. 5. *Bolometric luminosity plotted against central density for pure-carbon models.*

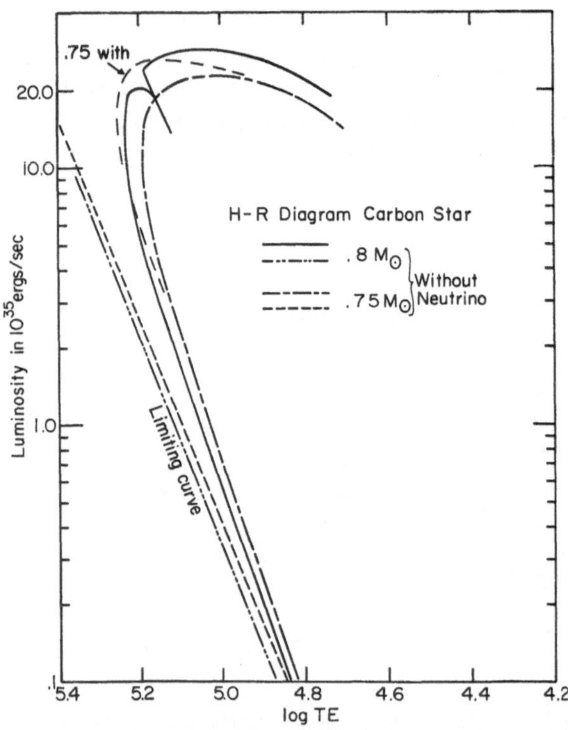

FIG. 6. *Color-magnitude diagram including constant-radius 'limiting curves'.*

of carbon and oxygen (50% by mass of C^{12}) have now been carried out by a number of authors (L'Ecuyer, 1966; Beaudet, 1967; Beaudet and Salpeter, 1968; Rakavy and Shaviv, 1967). Figures 4–6 are the counterpart of Figure 1, adopted from some actual models for 0·75 $M_{\odot}$ and 0·8 $M_{\odot}$ without neutrino reactions and for 0·75 $M_{\odot}$ with neutrinos. Only the 0·8 $M_{\odot}$ model burns carbon in its core and its onset is marked by the sudden dip in luminosity. The 'limiting curves' marked on Figure 6 are the color-magnitude tracks at constant radius corresponding to the final equilibrium radius for 0·75 and 0·8 $M_{\odot}$.

A word about the remaining uncertainties in the models (for a homogeneous star of given mass): The relevant opacity is mainly electron scattering plus a little free-free opacity from the abundant carbon and oxygen atoms. The opacity is therefore fairly well known and errors in the luminosity due to this cause exceeding a factor of 2, are not likely. The various neutrino energy-loss processes have by now been studied fairly carefully. All these processes are predicted on the presence of the Universal Fermi Interaction, which is likely to be correct (but has not yet been rigorously proven from laboratory experiments). If these neutrino reactions *are* present at all, their total combined rate as a function of temperature and density is now known quite accurately ($\pm 20\%$; Beaudet *et al.*, 1967). In all the models published so far, Reeves' (1963) estimate for the energy production rate of the ($C^{12}+C^{12}$)-reaction was used. More recent Cal. Tech. experiments indicate that these rate estimates will have to be lowered, possibly by as much as a factor of about 40.

The minimum mass $M_{\min}$ for which a homogeneous carbon star can initiate appreciable carbon burning in its interior is of interest. As a comparison of the curves for 0·75 and 0·8 $M_{\odot}$ in Figure 4 shows, the onset of carbon burning is a sudden matter: for the lower mass there is too little burning to affect the model at all. For the slightly larger mass there is just enough carbon-burning just before T_c should reach a maximum for it to raise the value of T_c slightly which in turn makes the carbon-burning rate more appreciable, etc., so that this star burns an appreciable amount of carbon before T_c starts decreasing towards zero again. Without any neutrino processes included $M_{\min} \approx 0{\cdot}8\ M_{\odot}$ and with neutrinos included $M_{\min} \approx 1{\cdot}00\ M_{\odot}$, if the old carbon-burning rate is used. If this rate is to be lowered by a factor ~ 40 (as seems likely) these two masses will be raised to $M_{\min} \approx 0{\cdot}95\ M_{\odot}$ and 1·10 $M_{\odot}$, respectively (Beaudet, 1967).

Figure 7 gives a comparison of the three illustrative models discussed above with the region in the color-magnitude diagram in which the central stars of planetary nebulae are observed. Note that these models have too low a luminosity L and too high a surface temperature. For models with neutrino rates included (which raises L and inhibits carbon burning) slightly higher masses (say 0·9 $M_{\odot}$ – 1·1 $M_{\odot}$) will have total bolometric luminosity comparable with the observed ones. However, these homogeneous models with higher mass and luminosity have very much larger surface temperatures than those observed. It behooves us now to look at more realistic inhomogeneous models.

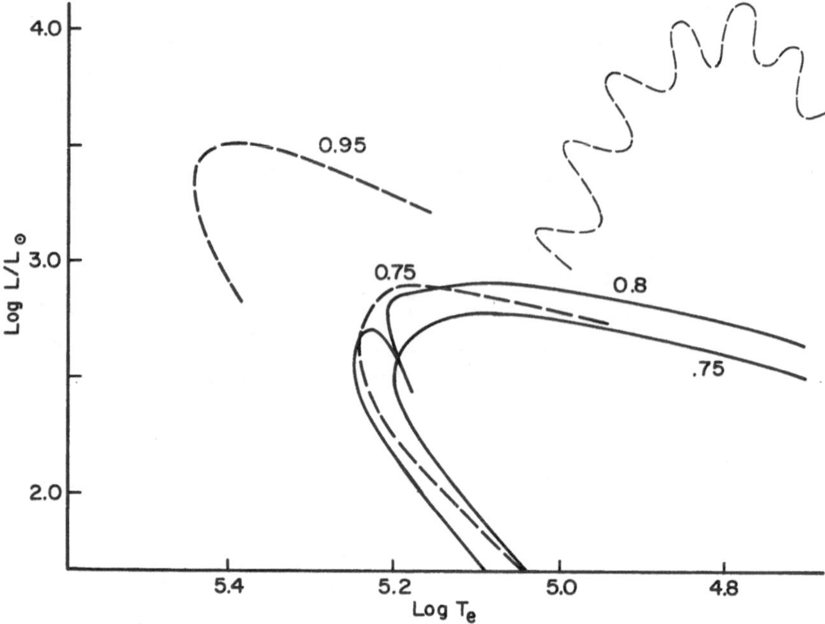

FIG. 7. *Color-magnitude diagram for pure-carbon models (with the region of observed central stars shown schematically in top right-hand corner).*

4. Inhomogeneous Models and Discussion

One kind of chemical inhomogeneity whose effect is easy to estimate is a hydrogen-rich envelope on a star consisting mainly of heavier elements (Cox and Salpeter, 1961). Extremely small amounts of hydrogen are sufficient to appreciably increase the radius and hence decrease the surface temperature T_e, even when the amount of hydrogen is too small for the bottom layer of the envelope to be undergoing any hydrogen burning. In such cases the luminosity is hardly affected at all and only T_e is decreased, but by a smaller and smaller fractional temperature shift as the central density increases. This effect is shown (semi-schematically) in Figure 8 in the curve labelled '10^{-4}H' which represents the 0·75 $M_{\odot}$ model with neutrinos (dashed curve to the left) but with an envelope of pure hydrogen of mass 10^{-4} $M_{\odot}$.

As one can see, this evolutionary track in Figure 8 is sufficiently 'reddened' to agree fairly well with the observed positions of central stars of planetary nebulae as regards surface temperature, although not in luminosity L. However, similar arguments apply to more massive carbon stars (0·9–1·2 $M_{\odot}$ say) which can have reasonable values of L as well as of T_e if they have a small amount of hydrogen in an envelope. Evolutionary time-scales (see also Figure 9) for the stars near the upper end of this mass-range are also at least of the right order of magnitude, $\lesssim 10^5$ years. On purely observational

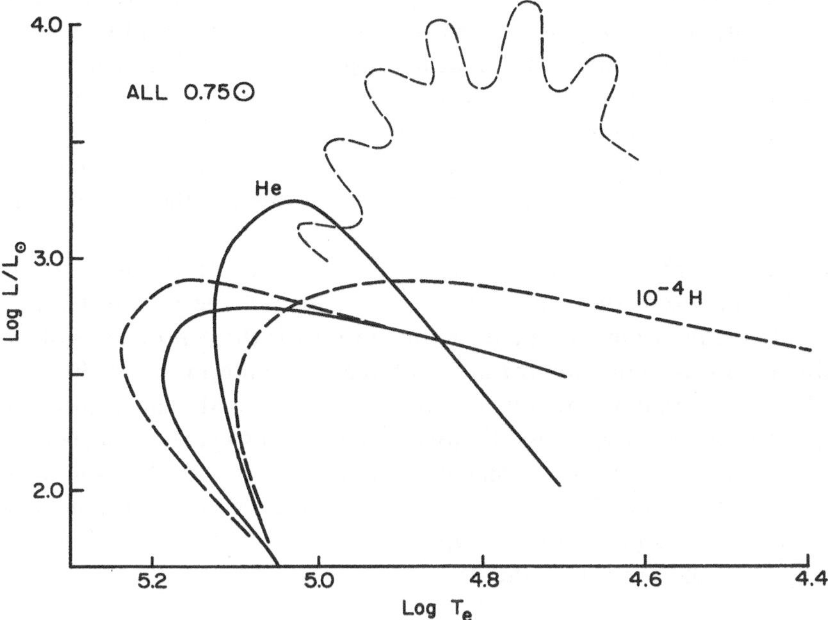

FIG. 8. *Color-magnitude diagram for various models (*He *indicates a helium-burning shell,* 10^{-4}H *a hydrogen envelope).*

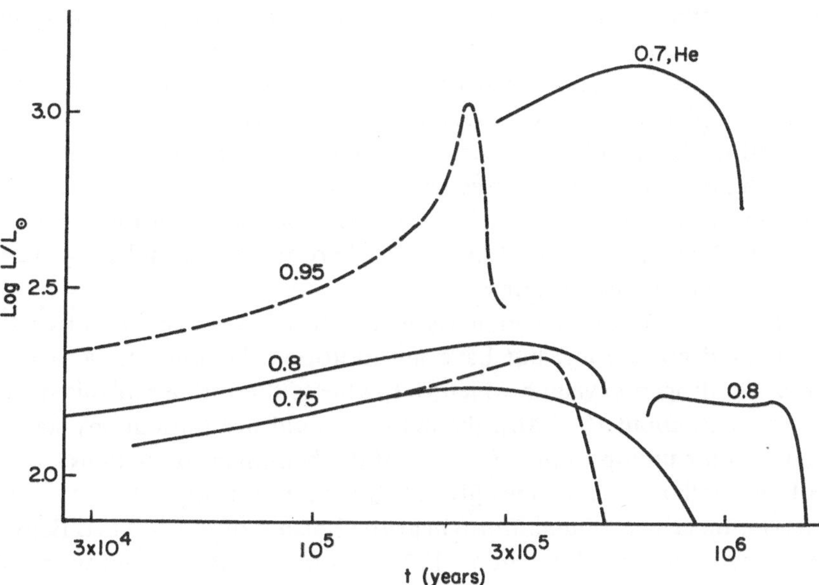

FIG. 9. *Bolometric luminosity plotted against evolutionary time for various models (starting at a stage slightly before maximum luminosity).*

grounds such stars of moderately large mass (but well below the Chandrasekhar limit) consisting mainly of carbon and oxygen (but not burning carbon) with a 'puffed up' envelope consisting of a small amount of hydrogen are possible candidates for central stars of planetary nebulae. On this picture the fairly rapid evolution ($\sim 10^5$ years) implied by the observations would be simply the rather rapid first stages of cooling down from the maximum-luminosity stage toward the white-dwarf stage, which are speeded up by the rather large neutrino luminosities at these high central temperatures.

This picture of carbon stars near 1 $M_\odot$ (a) is predicated on the assumption that the Universal Fermi interaction, and hence our neutrino energy-loss mechanisms, is correct and (b) may or may not be the explanation for the central stars of *some* planetary nebulae, but certainly *not* for all. To have such an appreciable mass consisting largely of carbon (and oxygen) now, the mass of the original main-sequence star must have been considerably larger. As discussed by previous speakers, planetary nebulae may belong to a somewhat mixed stellar population and some fraction of them may be young and have come from original stars of large mass, but presumably a large fraction of them are about as old as 'typical stellar population II' and therefore have come from original stars of moderate mass (0·7–1·4 $M_\odot$ as discussed for globular clusters).

Before turning to stars of smaller mass, a side remark which is likely to apply to all kinds of models: A hydrogen-rich envelope is required to 'redden' highly-evolved models, but the mass in this envelope is likely to be quite small (or else the models would be much redder than the observed central stars). The previous 'blowing-off' of the planetary nebula itself must therefore have removed a large fraction of the originally much more extensive hydrogen-rich envelope of the star. On the other hand we know that the planetary nebula gas is *not* overabundant in helium or oxygen; thus the blowing-off of the nebula could not have removed much of the layers below the envelope which are rich in helium or heavier elements. We therefore have the interesting (but not unexpected) indication that the lower boundary of the ejected material was close to the hydrogen-'heavier' interface where the mean molecular weight (and other quantities) had a discontinuity.

Let us return to stars of smaller mass which still consist mainly of carbon (and oxygen) in their deep interior, but have an additional helium-rich layer (as well as the outermost hydrogen envelope which only affects the radius). If this helium layer contained less than about 10^{-2} $M_\odot$, the helium would not burn at any stage and the layer would be quite unimportant. However, if the helium layer contains a few percent or more of the stellar mass, it does give shell-source burning (at least in the stages near the maximum central temperature) near the bottom of this layer (L'Ecuyer, 1966; Rose, 1966; Sugimoto and Yamamoto, 1966). As mentioned previously, shell-source burning can increase the bolometric luminosity of a star considerably. The curve labelled He in Figure 8 (and in Figure 9) represents (in a semi-schematic way) an evolutionary model whose carbon core is gravitationally contracting but with some helium burning in an outer shell.

As Figure 8 shows, shell-source burning increases the luminosity of a low-mass star somewhat, but not sufficiently so, and (as Figure 9 shows) gives much too long an evolutionary time-scale. However, this gloomy picture applies only to the time-averaged model and does not take account of 'thermal instabilities' (Schwarzschild and Härm, 1965). As Rose (1967) showed recently, a star with a helium-burning shell can lead to thermal instabilities which in turn can lead to even more violent events (such as pulsational instability). It will be a while before all the possible ramifications of thermal instabilities have been explored and explicit models involving them are proposed for central stars of planetary nebulae. At the moment I only want to point out optimistically that thermal instabilities can represent extreme forms of relaxation oscillations where peak luminosities are very much higher for a much shorter period (than on steadily evolving models), so that models for low-mass stars ($\lesssim 0{\cdot}7\ M_{\odot}$) may become compatible with the observations.

References

Beaudet, G. (1967) Ph.D. Thesis, Cornell University.
Beaudet, G., Salpeter, E. (1968) CRSR 299, Cornell University.
Beaudet, G., Petrosian, V., Salpeter, E. (1967) *Astrophys. J.*, **150**, 979.
Cox, J.P., Salpeter, E. (1961) *Astrophys. J.*, **133**, 764.
Deinzer, W., Salpeter, E. (1964) *Astrophys. J.*, **140**, 499.
Faulkner, J., Iben, I. (1966) *Astrophys. J.*, **144**, 995.
Hartwick, F., Härm, R., Schwarzschild, M. (1967) *Astrophys. J.*, (in press).
Hofmeister, E. (1967) *Z. Astroph.*, **65**, 164.
Iben, I. (1966) *Astrophys. J.*, **143**, 483.
Kippenhahn, R., Thomas H., Weigert, A. (1966) *Z. Astroph.*, **64**, 373, 395.
L'Ecuyer, J. (1966) *Astrophys. J.*, **146**, 845.
Rakavy, G., Shaviv, G. (1967) *Astrophys. J.*, **150**, 131.
Reeves, H. (1963) in *Stellar Structure*, ed. by L.H. Aller, University of Chicago Press, Chicago, p. 113.
Rose, W.K. (1966) *Astrophys. J.*, **146**, 838.
Rose, W.K. (1967) *Astrophys. J.*, **150**, 193.
Schwarzschild, M., Härm, R. (1965) *Astrophys. J.*, **142**, 88.
Stothers, R. (1966) *Astron. J.*, **71**, 943.
Sugimoto, D., Yamamoto, Y. (1966) *Progress Theor. Phys.*, **36**, 17.
Vila, S. (1966) *Astrophys. J.*, **146**, 437.
Vila, S. (1967) *Astrophys. J.*, **149**, 613.

DISCUSSION

Reeves: The one numerical value that we have obtained in Dr. Aller's talk is that the ratio of neon to oxygen in planetary nebulae is about ten times smaller than in ordinary stars. This observation does not seem to be compatible with the views presented in your talk.

Salpeter: I consider the abundance ratios in general an embarrassment: We were also told that the ratio of C + O + Ne abundances to H are comparable with or smaller than 'cosmic'. Therefore, these elements could not have been contributed to the nebula by its central star (or else the sum of these abundances would be larger); therefore, one cannot explain anomalous ratios of C to O to Ne in terms of reactions in the central star itself.

Böhm-Vitense: Why must the time-scale be an evolutionary one? Could it be that a small mass-loss determines the time-scale if a small change in a very thin hydrogen shell can move the star appreciably in the HR diagram? Since we estimated yesterday that in this region of the HR diagram a small mass-loss requires a large amount of energy, or continuously decreasing mass-loss might perhaps give an increase of luminosity as observed. Would you think this to be possible?

Savedoff: Not only burning hydrogen, but merely mixing it will change radius and hence color.

Salpeter: We are certainly all agreed that hydrogen envelopes of various kinds can change the color of a star greatly. I personally doubt whether very small amounts of hydrogen can change the luminosity greatly, but it might – at least indirectly, by triggering some instability or dynamic effect.

EVOLUTION TOWARDS THE WHITE-DWARF STAGE

S. C. VILA
(Indiana University, U.S.A.)

We have followed the evolution of stellar models of 1 and 1·2 $M_{\odot}$ from a non-degenerate configuration to the stages where degeneracy becomes important and eventually causes the star to become a white dwarf. By detailed computer calculation using an evolution program kindly lent to us by Schwarzschild and Härm, we have obtained evolution sequences of stars of the mentioned masses with a core composition of 80% O, 10% Ne and 10% Mg and a thin outer envelope of He (10^{-6} of total mass) under the assumption of energy loss due to photons alone, and also under the assumption of losses due to photons plus neutrinos.

These models were intended to represent the remnant cores of more massive stars that have lost their outer layers. Further evolution reduces to contraction and release of the gravitational and thermal energies but some interesting features appear where the neutrino losses are included that we think might be related to the evolution of the stellar nuclei of planetary nebulae.

In the calculation of the models matter has been assumed to be completely ionized, its pressure and entropy being the sum of the ion, electron and radiation components. The opacities were kindly calculated by A. N. Cox using the Los Alamos opacity program and were extended to the electron conduction region using the work of Mestel (1950).

The neutrino reactions included were: pair annihilation of electron pairs into neutrinos, photoneutrinos and plasma neutrinos. All these stem from the hypothesis of universality of weak interactions for which there is no direct experimental confirmation, but in whose existence there is a confident belief among physicists.

Evolution with no neutrinos proceeds for a time almost as a homologous contraction with constant luminosity. When degeneracy becomes important the electron pressure stops the contraction and electron conduction causes the core to become isothermal. Further evolution asymptotically approaches a line of cooling at constant radius. The evolution with neutrino losses included (started from the same stellar model) proceeds in a similar way until the central temperature is high enough for an appreciable neutrino production. In that case, neutrinos are produced at the center and their escape and subsequent energy loss causes the core to evolve much more rapidly towards the degenerate state. The central temperature falls faster than the envelope temperature because as the equation of state of a degenerate gas is almost insensitive to the temperature, the depletion of energy by neutrino emission is not compensated

Osterbrock and O'Dell (eds.), Planetary Nebulae, 421–422. © *I.A.U.*

by contraction and subsequent release of gravitational energy. This causes energy to flow from the envelope to the core to supply the neutrino losses. This sudden cooling and contraction of the core causes the outer layers to collapse and release gravitational energy at a fast rate. This causes an increased luminosity and effective temperature as reflected in the HR diagram.

The evolution line obtained for the neutrino-star models invites comparison with the evolution of the central stars of planetary nebulae recently proposed by Harman and Seaton (1964) and Seaton (1966). Though the luminosities are of the same order of magnitude, the effective temperatures in the calculated models are systematically higher. Presumably a more realistic model with an extended envelope of H or He could reduce the temperatures while keeping the same luminosities.

The most suggestive feature in this comparison is that the lifetime estimated by Seaton (1966) for the nuclei of planetary nebulae is on the order of 10^4 years while the lifetime obtained by gravitational collapse alone is hundreds of times greater.

On the negative side, we must point out that in our scheme there is no explanation for the ejection of the planetary shell, but perhaps such a phenomenon can result from a not too deep shell of H or He that is ignited by the rise of envelope temperature in the neutrino-collapse phase. This is at present being investigated.

Although we have presented no conclusive proof, we would like to suggest that the late stages of evolution of solar-mass stars with neutrino losses might provide an explanation for the evolution, of central stars of planetary nebulae and at the same time might provide evidence in favor of the universality of weak interactions between Fermi particles.

Further details of the models, as well as a figure showing tracks in the HR diagrams are given in Vila (1966).

References

Harman, R.J., Seaton, M.J. (1964) *Astrophys. J.*, **140**, 824.
Mestel, L. (1950) *Proc. Camb. Phil. Soc.*, **46**, 331.
Seaton, M.J. (1966) *Mon. Not. R. Astr. Soc.*, **132**, 113.
Vila, S.C. (1966) *Astrophys. J.*, **146**, 437.

CHEMICAL COMPOSITION AND DISTRIBUTION OF WHITE DWARFS

VOLKER WEIDEMANN
(Universität Kiel, Germany)

This contribution makes a few points which are relevant to the question of the fraction of white dwarfs formed via the planetary-nebula phenomenon.

There is spectroscopic evidence for at least two kinds of white dwarfs which differ in atmospheric composition. Those belonging to spectral type DA (60–80%) having hydrogen-determined spectra, and those of spectral types DB, DC λ4670, which are hydrogen-free. DA stars form a well-defined sequence in the two-color and HR diagrams, while the scatter for the other types seems to be larger. Masses for DA's are confined to $0{\cdot}6 \pm 0{\cdot}3$ solar masses. The average mass for hydrogen-free stars may be slightly higher, but this is not significant in view of the uncertainty in parallaxes as well as in blanketing and bolometric corrections. In particular, there is no evidence for the existence of two separate white-dwarf sequences of smaller and larger mass.

Evolutionary considerations imply that the interiors of white dwarfs should be H- and He-free, i.e. consist of C, O or heavier elements. Exceptions are white dwarfs formed by mass exchange in binaries according to Kippenhahn *et al.* (1967).

Since white dwarfs follow closely the theoretical luminosity function derived from the cooling law of Mestel and Ruderman (1967), an estimate can be given of the total number of white dwarfs as a function of age or lower limit of luminosity (Weidemann, 1967). By fitting the luminosity function to empirical data (from Luyten, Eggen, Greenstein and Sandage) and extrapolation to $M_{bol} = 15{\cdot}5$, corresponding to a cooling time of $5 \times 10^9 a$, the space density of white dwarfs younger than $5 \times 10^9 a$ turns out to be most probably 1×10^{-2} pc^{-3}, corresponding to a birth rate of $\chi_{WD} = 2 \times 10^{-12}$ pc$^{-3}a^{-1}$. Comparison with corresponding figures for planetary nebulae, $\chi_{PN} = 1 \times 10^{-13}$ or 11×10^{-13} pc$^{-3}a^{-1}$ for Seaton's or Abell's distance scales respectively, show that 5–50% of all white dwarfs should be formed via planetary nebulae. These are expected to be of type DA since planetary nebulae are not hydrogen-deficient.

Extension of the luminosity function to higher temperatures gives no indication of a 'gap' at $0 < \log L/L_\odot < 2$ that could be taken as evidence for neutrino cooling.

Although the majority of white dwarfs have moderate space velocities (<100 km sec^{-1}) there are some, at first thought to be pygmies, which have velocities up to 250 km sec^{-1} and thus show extreme population-II characteristics (Eggen and Sandage, 1967).

Osterbrock and O'Dell (eds.), Planetary Nebulae, 423–424. © *I.A.U.*

References

Eggen, O.J., Greenstein, J.L. (1965) *Astrophys. J.*, **141**, 83.
Eggen, O.J., Sandage, A.R. (1967) *Astrophys. J.*, **148**, 911.
Kippenhahn, R., Kohl, K., Weigert, A. (1967) *Z. Astrophys.*, **66**, 58.
Mestel, L., Ruderman, M.A. (1967) *Mon. Not. R. Astr. Soc.*, **136**, 27.
Weidemann, V. (1967) *Z. Astrophys.*, **67**, 286.

IMPLICATIONS OF WHITE-DWARF CRYSTALLIZATION FOR THE CHEMICAL COMPOSITION OF THE PLANETARY NUCLEI

H. M. VAN HORN
(Dept. of Physics and Astronomy, University of Rochester, U.S.A.)

It now seems to be reasonably well-established that the central stars of planetary nebulae evolve directly into white dwarfs. Evidently a knowledge of the chemical composition of the white dwarfs would therefore be of considerable importance in helping to identify the point in the evolution at which the mechanism responsible for expulsion of the nebular shell becomes operative. For this reason I would like to present some evidence which provides a direct suggestion for the internal composition of some of the white dwarfs and to examine briefly the implications of this suggestion for the relation between the planetary nuclei and the white dwarfs.

It was first pointed out by Kirzhnits (1960) and independently by Salpeter (1961) that matter in the interior of a cold 'white' dwarf should be in a crystalline rather than a gaseous state. More recently Mestel and Ruderman (1967) have shown that crystallization will begin in a white dwarf of central density ρ when the central temperature T falls to a value of

$$T \sim 3 \times 10^5 \left(\frac{\rho}{10^6}\right)^{1/3} Z^{5/3}\,{}^\circ\mathrm{K}\,, \tag{1}$$

where ρ is in g cm^{-3}, and Z is the mean nuclear charge. In this phase transition there is a release of energy in the form of the heat of crystallization, $\Delta Q \sim kT$, where T is the crystallizing temperature. The release of this energy, which is roughly comparable to the total thermal energy of the crystallizing white dwarf at this late phase of the evolution, markedly reduces the rate of cooling at this stage and leads to a temporary still-stand along the cooling track (Van Horn, 1968). For a given chemical composition the locus of points in the Hertzsprung-Russell diagram which mark this still-stand for stars of different masses, thus defines a 'crystallizing sequence' of white dwarfs, which is more densely populated than adjacent regions of the diagram. As is evident from Equation (1), stars consisting of elements of higher atomic number will crystallize at higher central temperatures, which are reached earlier on the cooling tracks and hence at higher luminosities. It is important to note that the rather strong Z-dependence of the crystallizing temperature in Equation (1) results in a significant separation in luminosity between the crystallizing sequences for different elements,

Osterbrock and O'Dell (eds.), Planetary Nebulae, 425–427. © *I.A.U.*

thus providing the possibility of discriminating observationally among white dwarfs of different compositions.

When the crystallizing sequences predicted theoretically for different assumed compositions are compared with the observations by Eggen and Greenstein (1965), it is found that the fainter of the two sequences which they observed agrees quite well with the crystallizing sequence for stars composed principally of carbon and oxygen (Van Horn, 1968). This fainter, observed sequence contains virtually all of the hydrogen-deficient objects – mainly the DC stars, which have continuous spectra – that have been identified by Weidemann (1968) as being intrinsically different in atmospheric composition from the bulk of the white dwarfs. It is therefore strongly suggestive that these stars are in fact also different in internal composition from the majority of the white dwarfs and consist almost exclusively of carbon and oxygen. This conclusion, together with (1) the observation by Greenstein and Minkowski (1964) that the central stars of several old planetaries appear to be rich in α-particle nuclei and deficient in hydrogen, (2) the fact that there appears to be a clean separation between the hydrogen-rich nebular shell and the hydrogen-poor central star – as has been pointed out by Osterbrock (1966) and by Salpeter (1968) – and (3) the observation by O'Dell (1968) that many of the fainter planetary nuclei possess continuous spectra, as do the DC white dwarfs, therefore strongly suggests that the hydrogen-deficient white dwarfs are the descendants of the central stars of planetary nebulae. The statistics of these stars are also consistent with Weidemann's (1968) result, that only 5–50% of the white dwarfs can have been formed in a planetary nebula phase, since the DC stars comprise only about 14% of the total number of white dwarfs.

The situation with respect to the brighter group of white dwarfs is unfortunately less clear, however. If Weidemann is correct and there is no *luminosity* gap between the two sequences that Eggen and Greenstein find to be separated in absolute magnitude, then there is no evidence for a crystallizing sequence lying above the carbon-oxygen sequence, and one would therefore expect these brighter stars to consist of carbon-oxygen cores with extensive, helium-rich envelopes. On the other hand, if the luminosity gap is real, then the evidence suggests that the brighter group of white dwarfs corresponds to a crystallizing sequence which is bracketed by the crystallizing sequences for stars composed entirely of Mg^{24} or of Fe^{56}. In view of the present uncertainties in the observations, however, it does not seem to be possible at present to draw any satisfactory conclusions about the composition of these stars.

Finally, it must be noted that even if the composition identification suggested for the fainter group of white dwarfs is correct, this does not constitute a *proof* that these stars actually are the descendants of the planetaries. On the other hand, such an hypothesis does seem to be consistent both with the observational evidence and with the inferences which can be drawn from evolutionary model calculations (Salpeter, 1968) and thus seems to indicate an evolutionary connection between the planetary nuclei and the (apparently carbon-oxygen-rich) DC white dwarfs.

References

Eggen, O.J., Greenstein, J.L. (1965) *Astrophys. J.*, **141**, 83.
Greenstein, J.L., Minkowski, R. (1964) *Astrophys. J.*, **140**, 1601.
Kirzhnits, D.A. (1960) *Soviet Physics – JETP*, **11**, 365.
Mestel, L., Ruderman, M.A. (1967) *Mon. Not. R. astr. Soc.*, **136**, 27.
O'Dell, C.R. (1968) in the present volume, p. 361.
Osterbrock, D.E. (1966) in *Stellar Evolution*, Ed. by R.F. Stein and A.G.W. Cameron, Plenum Press, New York, p. 381.
Salpeter, E.E. (1961) *Astrophys. J.*, **134**, 669.
Salpeter, E.E. (1968) in the present volume, p. 409.
Van Horn, H.M. (1968) *Astrophys. J.*, **151**, 227.
Weidemann, V. (1968) in the present volume, p. 423.

ON THE EVOLUTION OF THE CENTRAL STARS

WILLI DEINZER
(University of Heidelberg, Germany)

A most puzzling feature of the evolution of the central stars of planetary nebulae is their rapid contraction (O'Dell, 1963; Seaton, 1966); assuming homologous contraction, an enormous amount of gravitational energy is released therefore. For reasons of hydrostatic equilibrium a large part of this energy must be radiated away. As this does not show up in the luminosity of the central stars it must either leave them by some other means – e.g. neutrinos (Vila, 1966) – or it does not originate from the outset; in other words, the actual contraction must deviate strongly from the homologous way. I should like to discuss briefly the second possibility.

For this purpose the following type of stellar model is considered: The central stars should consist of a core, which contains almost all of the stellar mass and the radius of which has almost decreased to the limit set by complete degeneracy (Schwarzschild, 1958, § 26). They should further consist of an envelope containing a negligible amount of mass but of very extended radius. If at the end of the mass-loss phase producing the planetary nebula such a star has no nuclear-energy sources available anymore, it will start contracting. This contraction will now proceed in a strongly non-homologous way. The envelope will change its radius by a large amount, which is accompanied by a negligible amount of gravitational-energy release because of the negligible amount of mass taking part in this contraction. The core can change its radius only very little; but in view of the large amount of mass taking part in this contraction an appreciable amount of gravitational energy could be released. Besides, thermal energy could be released due to the effect that contraction of a degenerate configuration will lead to cooling.

Is it possible in this way to obtain evolutionary time-scales and evolutionary tracks comparable to those observed for the central stars? To answer this question, calculations were carried out for a simplified model. Its partially degenerate core is assumed to be isothermal (Schwarzschild, 1958, § 13); the prevailing opacity of its radiative envelope should be due to Thomson scattering only. The solutions for the envelope can be obtained analytically and hence can easily be fitted to the core solutions. Results obtained are shown in Figure 1. They belong to stars of $0{\cdot}4 M_{\odot}$ consisting of a helium core and of a population-II envelope. The tracks differ in the mass content of the envelope (10^{-2}, 10^{-3}, 10^{-4} of the total mass). They are limited towards high luminosities by the restriction that radiation pressure in the envelope can at the most equal the total pressure; and they are limited towards the left by the white-dwarf

Osterbrock and O'Dell (eds.), Planetary Nebulae, 428–430. © *I.A.U.*

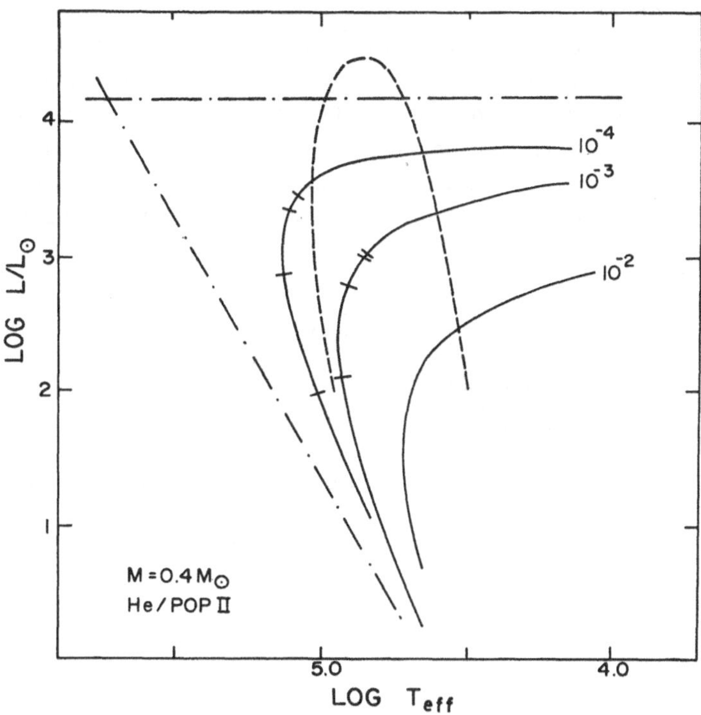

FIG. 1. *Evolutionary paths for theoretical stars whose atmospheres contain the designated fraction of the total mass of the star. The dashed line indicates the Harman-Seaton sequence: the upper dash-dot line designates the radiation-pressure boundary, while that on the left is the complete degeneracy boundary. The lower three bars on the evolutionary paths indicate the stellar positions at times of 10^2, 10^3 and 10^4 years.*

radius corresponding to 0·4 $M_\odot$. Along each track the energy release from the core was calculated; together with the luminosity a time-scale was obtained. Bars on the tracks indicate how far a star has evolved after 10^2, 10^3 and 10^4 years; a star evolves indeed from high luminosities and large radii to the white-dwarf region. Similar results were obtained for a star of 0·5 $M_\odot$ consisting of a carbon core and a helium envelope.

Finally let me summarize the simplifications of the present models. There is first the assumption of an isothermal core. It could be justified, if the nebula is produced in a shell-source burning stage of stellar evolution; for this reason also the chemical discontinuity was introduced. Since, in the degenerate state of matter, pressure and density are very insensitive to the temperature, the hydrostatics does not depend very strongly on the assumed temperature distribution (Hayashi *et al.*, 1962, § 6E.1). There is next the assumption that all the energy released from the core is transported immediately to the outside of the star. As Baglin (1966) pointed out, the transport time-scales are of the order of 10^6 years, even in the case of heat conduction by degenerate electrons. Hence, only a small part of the energy from the core will become available during the central-star stage and the evolution of the discussed model stars would proceed faster. Making the envelope a little thicker in mass and taking its gravitational-energy release into account would probably help to extend the time-scale. Thus stars starting at the top of the Harman-Seaton sequence with a very extended envelope and a highly contracted core could in principle account for the short evolutionary time-scales during which the central stars move through the falling branch of this sequence (for further details see Deinzer, 1967).

References

Baglin, A. (1966) private communication.
Deinzer, W. (1967) *Z. Astrophys.*, **67**, 342.
Hayashi, C., Hoshi, R., Sugimoto, D. (1962) *Progress. theor. Phys.*, Suppl. **22**.
O'Dell, C.R. (1963) *Astrophys. J.*, **138**, 67.
Schwarzschild, M. (1958) *Structure and Evolution of the Stars*, Princeton University Press, Princeton.
Seaton, M.J. (1966) *Mon. Not. R. astr. Soc.*, **132**, 113.
Vila, S.C. (1966) *Astrophys. J.*, **146**, 437.

THE EVOLUTION OF HELIUM SHELL-BURNING STARS

D. J. Faulkner
(Mount Stromlo Observatory, Australia)

Evolutionary sequences of stellar models have been obtained, using the Henyey method, for stars of mass $0{\cdot}8\ M_{\odot}$, $0{\cdot}9\ M_{\odot}$ and $1{\cdot}0\ M_{\odot}$. The starting-point of each sequence is a star with a core of carbon and oxygen, comprising 25% of the star's mass, surrounded by helium-rich material. It is assumed that any remaining hydrogen-rich material has been ejected in the form of the planetary shell. The star contracts gravitationally until helium ignites in a shell at the composition discontinuity, and the subsequent evolution of the shell burning has been followed. No neutrino-loss processes have been considered in this investigation.

As the evolution proceeds, the shell becomes thinner, but an increase in its temperature outweighs this effect and the resultant energy production, and hence the luminosity of the star, increases sharply. The radial distance to the shell decreases slightly, but the exterior layers of the star expand at first, and the total radius increases. In the case of the $1{\cdot}0\ M_{\odot}$ sequence this expansion is quite considerable. The radius of the star passes through a maximum and decreases again. Eventually, the contraction of the region interior to the shell source becomes insufficient to maintain the rate of energy production in the source, and the luminosity, too, passes through a maximum. The temperature in the core becomes high enough for carbon burning to commence, and when this occurs, the layers immediately surrounding the core expand and the helium-burning shell dies out. The star then settles down to a phase of carbon burning in a convective core. The present evolutionary sequences have been terminated at this point.

The latter parts of the computed tracks in the Hertzsprung-Russell diagram match the observed track for the nuclei of planetary nebulae; the computed times of evolution are longer than the observed time by a factor of 5–10. They are, however, considerably shorter than the evolution times obtained (in the absence of neutrino processes) for stars without shell burning, which are too long by two orders of magnitude (Vila, *Astrophys. J.*, **146**, 1966, 437). It thus appears that the inclusion of shell burning will render less sharp any test of the existence of the universal Fermi interaction based on a comparison of the observed and computed evolution times of the nuclei of planetary nebulae.

The present results suggest that the planetary nebula phase of a star's evolution may occur between helium burning and carbon burning, rather than directly before the white-dwarf stage. A more detailed description of this work will appear elsewhere (*Mon. Not. R. astr. Soc.*, in press).

GENERAL DISCUSSION – SIXTH SESSION

Minkowski: The absence of small planetaries from the size distribution might not be more than a selection effect if planetaries in an early stage of their evolution, where they are small, are not recognized as planetary nebulae. There is, as a matter of fact, a group of objects that have been classified as Bep by Merrill with a low Balmer gradient, but no, or in a few cases, very faint, forbidden lines. One of these objects was found to be a nebula of about 3″ diameter. It seems quite possible that these objects are not stars, but show an early phase of planetary nebulae with small diameters and, of course, very high densities.

Westerlund: Miss Webster found in the study of unresolved or nearly unresolved planetary nebulae in Henize's and my catalogue, several low-excitation objects which must have very small (radii 0·04 parsec) and small hydrogen mass. They represent most likely the youngest evolutionary stage so far observed for planetaries. Their spectra contain forbidden lines of [O II], [N II] and [S II], but not as strong as would be expected. No [O III] lines have been seen.

Abell: I am inclined to agree with the evolutionary sequence described by O'Dell, but am bothered about one circumstance. Some or all of the large faint nebulae studied by O'Dell and me are probably optically thin. Thus, the temperatures we derive for the stars are lower limits. Even these lower limits, however, show that the bulk of the stellar energy lies shortward of the Lyman limit. Thus, our computed lower limits to the stellar luminosities are too small by roughly the fourth power of the factor by which our temperatures are too small. Now since the energy required to keep a given spherical mass of gas ionized is proportional to the reciprocal of the volume of that gas mass, our underestimate of the stellar luminosities would go as the inverse cube of the nebular diameters, if all nebulae had the same mass and all of their stars the same luminosities. In fact, a plot of $\log L$ vs. $\log D_{\mathrm{neb}}$ for O'Dell's and my objects exhibits exactly this relation expected, due entirely to the nebulae enlarging and becoming more and more optically thin. We should make certain therefore, that the apparent evolution of the central stars is not partly due to our method of determining the properties of those stars. In fact, the central stars may not drop much in luminosity during the relatively brief times during which the nebulae can be recognized.

Osterbrock: Many theoretical ideas seem to be converging on long-period variables as the progenitors of planetary nebulae. I wonder what statements can be made about the galactic distribution of long-period variables with respect to the galactic distribution of planetaries; also, with respect to the relative numbers of long-period variables and of planetaries.

Osterbrock and O'Dell (eds.), Planetary Nebulae, 432–438. © *I.A.U.*

Feast: The distribution and kinematics of the Mira variables are remarkably similar to that of the planetaries, so that a connection between these two classes of objects would seem quite possible on these grounds – though, of course, it does not prove a connection. The Mira variables are much more frequent in the galaxy than the planetary nebulae so that the planetary phase would have to be much shorter than the Mira phase.

Westerlund: The emission-line spectra of symbiotic stars are of high excitation. Is it possible to explain how the low-excitation young planetary nebulae can follow in evolution immediately after the symbiotic stars?

O'Dell: The dilution is so much less in the symbiotic stars that this might account for the higher degree of ionization.

Boyarchuk: The search for related objects is very important for understanding of the evolution of planetary nebulae and their central stars. I think that the symbiotic stars are such objects. The nature of symbiotic stars is not understood completely thus far. However, one can explain most of the existing observational evidence on a supposition of their being binary stars.

According to this hypothesis symbiotic stars consist of a cool giant and a hot dwarf, both enveloped by a rather small dense gaseous nebula.

Investigations of continuous spectra of symbiotic stars offer a possibility to determine the relative brightness of the three sources mentioned. Consideration of absorption spectra as well as data on the galactic distribution of giants enable us to suppose that the luminosity of the cool component is equal to that of normal cool giants.

On the basis of this supposition one can determine the position of the hot component on the temperature-luminosity diagram. In Figure 1 it is seen that their position coincides fairly well with that of the central stars of planetary nebulae.

I think that symbiotic stars give us an example of such double-nuclei planetary nebulae. A hot component of a larger initial mass has reached the evolutionary stage of the planetary nebula nuclei, while another, initially less massive, component is still in the stage of a cool giant.

Abell: We can put a lower limit on the luminous energy required to remove 0·2 $M_\odot$ from the surface of a star of radius R to infinity. The momentum required to remove a unit mass from the star to infinity is the velocity of escape. Thus the total momentum is

$$0{\cdot}2\ M_\odot \frac{2Gm^{1/2}}{R^{1/2}} \text{ gm cm sec}^{-1}.$$

For $M = 1\ M_\odot$, we find this momentum to be $2{\cdot}5 \times 10^{40}/R^{1/2}$ with R in solar units. If this momentum is to be supplied by radiation pressure, we must have

$$\frac{Lt}{c} = 2{\cdot}5 \times 10^{40}/R^{1/2},$$

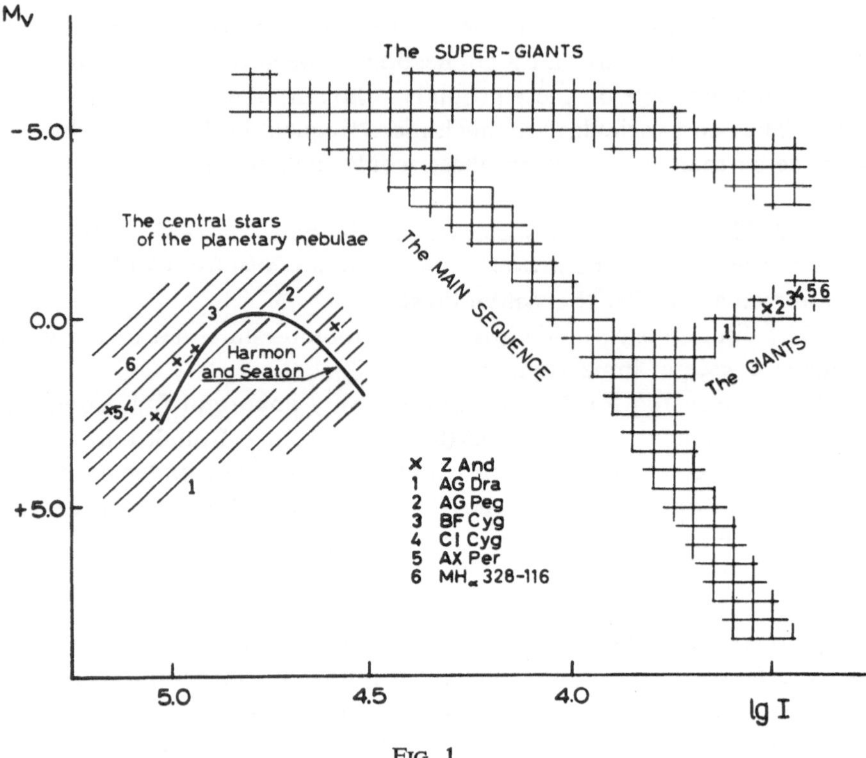

FIG. 1.

where L is the luminosity of the star and t is the time taken to transfer the momentum. If $t = 10^4$ years, and $R = 1$, we have $L = 6 \times 10^5$, solar luminosities, assuming complete opacity of the ejected material.

O'Dell: However, if the radius is as large as in late-type supergiants, a condition indicated several times in this meeting, then the luminosity requirements are much lower for "reasonable" timescales.

Abell: I should like to point out one other elementary fact. According to the vis-viva equation, the energy E with which the gas shell is ejected is

$$E = E_{es} + E_{\infty},$$

where E_{es} is the escape energy, and E_{∞} is the energy the material has at infinity. The nebulae now have, essentially, E_{∞}. Assuming 1 $M_{\odot}$ for the star and 0·2 $M_{\odot}$ for the nebula, we find

$$E_{es} = 7{\cdot}7 \times 10^{47}/R \text{ ergs},$$

with R in solar units. The ratio E_{es}/E_{∞} is thus about $400/R$. For extreme giants, $R \sim 10^2$–10^3, and E_{es} and E_{∞} are comparable; but if the matter were ejected from small stars, it would be most remarkable that E was equal to E_{es} within such narrow limits – one part in 400 for $R=1$, one part in nearly 10^5 for radii like those of the smallest central stars observed. Thus the evidence favors ejection from very large stars.

Kohoutek: Some years ago only one case of the binary nature of central stars was known: the optical binary NGC 246. At the present time:

(1) We have the following objects for which there exists evidence, more or less guaranteed, that they could be binary stars: NGC 1514, H_z 1-5 (V 377·1943 Sge), M 1-2 (VV8), K 1-2, IC 4406.

(2) There are also some stars, which are or could be variable: Hb 6=AS Sgr, H 3-29=S 5337, M 3-18, 12 central stars from Abell's list: A 14, 20, 30, 36, 41, 46, 51, 61, 63, 74, 78, and M 1-67, II 2149, NGC 2346, NGC 6891, II 4997, which show on AGK2 and AGK3 astrographic plates differences in brightness larger than observational errors.

(3) Some stars could be expected as not responsible for radiation of the nebula: NGC 3132, VV 68, NGC 2346; or some objects: II 2149, NGC 2392, NGC 6826, NGC 6572 show differences in temperature.

One can say that there are about 20 central stars (10% of all central stars known) which we could be suspicious of as being binaries.

Is this stage final, or could we expect – when collecting more precise photoelectric and spectrographic material – additional evidence for variability or the binary nature of planetary nuclei?

I believe that we have now a similar situation as in the problem of the binary nature of novae some years ago, and that the next years (and further observations) may show, that perhaps most of planetary nuclei are binary stars.

It is, of course, only an assumption now that planetary central stars are binaries, but if this assumption were correct, then it would bring a new idea to the problem of their evolution. Then we could explain the origin of planetary nebulae as a process connected with the evolution of binary stars.

Hekela and I suppose that the assumption of being binary may explain the origin of planetary nebulae much more easily than the assumption of a single star.

Let us assume moreover, that planetary nuclei are *close* binaries. Then, when the mass of primary components reaches the Roche critical surface, mass exchange will take place, the mass flows out to the second component (as calculated by Kippenhahn and Weigert, or for the larger mass stars by Plavec). Such a rapid process could cause a rapid change in hydrodynamic equilibrium of these stars, or an increase of radiation pressure. Then it is not very difficult to imagine that a part of the exchanged mass (0·1 $M_{\odot}$ is not more than about ten percent of the exchanged mass) could leave the central star and create the planetary nebula.

Reeves: Qualitatively, the $N^{14}+He^4$ reaction could well play an important role in amplifying thermal instabilities, because of its large temperature sensitivity. May I urge model builders to include this reaction in their models.

Rose: On the basis of the calculations of models for the nuclei of planetary nebulae that have been discussed it would appear that it is difficult to explain the short life-times and high luminosities observed for the nuclei of planetary nebulae unless the masses of the nuclei of planetary nebulae are greater than 1 $M_{\odot}$ or unless high rates of nuclear-energy generation have taken place in these stars immediately before the mass ejection. In addition, any mechanism for the formation of planetary nebulae must prevent the continuation of helium burning after the mass ejection if it is to explain the rapid evolution of the observed nuclei.

Weidemann: I was inclined to assume that white dwarfs of spectral type DA are the descendants of PN, since planetary nebulae and the atmospheres of the central stars are not hydrogen-free. However, arguments given by Salpeter and others favor models in which there is nearly no unburned material left on top of the C/O-PN nucleus star, opening the possibility that we have to tie the non-DA white dwarfs to PN. This would be all right as far as space density and mass is concerned (masses somewhat higher than the average DA mass of 0·6 $M_{\odot}$, say 0·8) but it then leaves us with the task to explain the majority of WD's in the DA sequence in another way. The masses of the DA sequence are almost certainly not higher than 0·6 $M_{\odot}$, as model calculations indicate, and as shown the best-determined mass of 40 Eri B is about 0·4 $M_{\odot}$.

Böhm-Vitense: It then seems to me that there is a serious discrepancy with respect to the mass because the theoreticians need masses of the order of 1 $M_{\odot}$, while the white dwarfs, which are supposed to be the descendants of the planetary nuclei do not have masses larger than 0·5 $M_{\odot}$. Where does the excess mass go?

Van Horn: If you believe the story about the crystalline nature of the interior of the white dwarfs, then there is a natural explanation for the apparent absence of the white dwarfs more massive than about 1·0 or 1·2 $M_{\odot}$. In such a star, the central temperature will fall below the Debye temperature of the crystalline material while the star is still quite bright, and the star will then cool very rapidly to invisibility because the specific heat quickly falls below the classical value of $3k/2$ per ion. Thus the apparent absence of such stars is not necessarily inconsistent with masses of the order of 1·0 to 1·2 $M_{\odot}$ for the central stars of the planetary nebulae.

Savedoff: On the problem of low-mass planetary nebulae it is difficult (as pointed out by Salpeter) to get any nuclear source for energy supply. A model of an iron star $M = 0{\cdot}631\ M_{\odot}$ computed by Savedoff, Van Horn, and Vila had a maximum peak $\log T = 8{\cdot}560$ at $M(r)/M = 0{\cdot}805$. A shell of helium could easily be ignited under these conditions. It is not suggested that this structure of iron core and helium envelope is very probable.

Reeves: The $C^{12}(d, \nu)O^{16}$ rate has been determined recently at Cal. Tech. The rate is such that the ratio of C/O at the end of helium burning is about 50–50 more or less independent of the stellar mass. May I suggest that such a mixture be used in models.

Rose: Abundance differences can significantly influence the evolution of a star.

For example, the existence of a horizontal branch appears to be associated with metal abundance. Therefore, it is possible that if mass loss should take place on the horizontal branch (e.g. during the RR Lyrae phase) the ultimate mass of the star might be a function of initial chemical abundance and could be important in determining whether or not a star with a given initial mass evolves through a planetary nebula phase.

Aller: The brighter nuclei of planetaries that show absorption lines always show hydrogen. A possible example of a hydrogen-deficient atmosphere is that of the NGC 246 nucleus. The Wolf-Rayet stars may be hydrogen-deficient – we have no quantitative information on this problem.

Faulkner: Concerning the similarity of the compositions observed in planetary nebulae shells and in the interstellar medium, are we sure that the mass loss from planetary nebulae in the Galaxy is not sufficient to actually produce the current interstellar medium, and so determine its composition?

Salpeter: I believe that the statistics is such that an appreciable fraction but *not* a majority of star deaths go through a PN stage, so I feel the current interstellar composition is not strongly dominated by mass loss from PN.

Seaton: What can be said about the number of planetaries at earlier times in the history of the galaxy?

Salpeter: O'Dell in his calculations has taken into account ideas of an increased birthrate function at earlier stages of the Galaxy as estimated a few years ago by Schmidt and by myself.

Aller: To what extent can we always fix the conditions of the outburst that *only* the H-rich envelope is ejected, as Salpeter emphasized? A small residue is left. We might expect to find eventually some planetary nebula with an enriched heavy-element composition.

Mathews: The faint outer shells surrounding planetary may represent mass ejection during the next-to-last pulsation of an unstable star. Whether this feature represents a double ejection or it is a natural consequence of the nebular evolution remains to be seen.

Kippenhahn: I would like to know how the different groups which have computed evolution near the white-dwarf stage have dealt with the electron conductivity in the relativistic degenerate regions. We have been quite careless and have just extrapolated Mestel's formulae into the relativistic region. I wonder whether other people have been more careful.

Salpeter: I am not aware of actual calculations of electron conductivity with relativity included, but I would guess that the effects are not enormous for stars of relatively low mass.

Seaton: What happens to the models of Faulkner and Rose when neutrino processes are included?

Faulkner: I would expect the carbon burning to remain unimportant until higher masses, and the times of evolution would be shorter.

Rose: The presence of neutrino emission is of little importance for the evolution of the nuclei of planetary nebulae if the masses of these stars are less than or equal to about 0·7 $M_{\odot}$.

Salpeter: From the fact that only one PN has been observed among the globular clusters which have been looked at, one should be able to see if this fact is compatible with *all* stars in extreme population-II moving through a PN phase for 10^4 years, using methods of Sandage for 'semi-empirical evolutionary tracks'. My guess is that it will *not* be compatible, i.e. that too few PN are found in globular clusters.

Underhill: An expanding envelope similar to that in Population-I Wolf-Rayet stars might be detected in WR nuclei of planetary nebulae by looking carefully at the profile of C IV 5801·12 for shortward displaced absorption components.

Aller: We have not observed absorption lines of 5801, 5804 C IV in the central stars. The observations are very difficult. In the NGC 6543 nucleus, Lick coudé plates suggest that Hγ and Hδ may have P Cygni profiles. The Wolf-Rayet nuclei probably are ejecting material.

Menzel: Barnard, early in this century observing with the 40-inch refractor of the Yerkes Observatory, once reported variability and complete disappearance of the nucleus of NGC 7662. I took occasional plates of this planetary between the years 1926 and 1932, with the Crossley Reflector, but found no evidence of variability.

O'Dell: This reported variability is probably due to the low contrast between the star and the bright nebula, in the sense that under poor seeing conditions the stellar image blends into that of the larger nebula. This is described in detail elsewhere (*Sky and Telescope*, **29**, 1965, 85).

Perek: It is proposed that new discoveries of planetary nebulae be communicated to L. Kohoutek and their finding charts and coordinates published in the *Bull. astr. Inst. Csl.* in order to keep the list of planetary nebulae up to date.

Session VII

CONCLUDING DISCUSSION

PHYSICAL PROCESSES

D. E. OSTERBROCK
(Washburn Observatory, University of Wisconsin, U.S.A.)

When Dr. Seaton asked me to chair the epilogue session of our conference I could not help but wonder if it were to be considered an example of type-casting. He told you that when an Englishman is asked to give a prologue, he immediately begins thinking of quotations from Shakespeare, so I must tell you that when an American is asked to deliver an epilogue, his thoughts turn naturally to Ernest Hemingway. I therefore suggest to you as the theme for our symposium the closing words of the main character of *To Have and Have Not*, which may for publication here be bowdlerized to read "*One* man *alone* ain't no —, — good at all." Over and over again in the last week we have seen that discussions between astronomers have provided new insights, have opened up new fields for investigation, have revealed unsuspected connections between apparently unrelated problems. I believe this has been a very fruitful conference because there have been so many fruitful discussions among the participants. I shall therefore try to summarize briefly my impressions of the things we have learned and the things we may reasonably hope to learn in the next few years about physical processes in planetary nebulae, and then five of our colleagues will summarize the other sessions of our conference. We invite discussion from all of you after each of these summaries, and also during the general discussion at the close.

First of all I think I can say that we have the theoretical tools we need to investigate planetary nebulae. The collision strengths as described by Seaton and Czyzak, the transition probabilities as summarized by Garstang, and the recombination coefficients and photo-ionization cross-sections as summarized by Flower, all seem to be reasonably accurate. Using them we can calculate models of spherically symmetric nebulae. This first approximation agrees reasonably well with the observed data, and we can make some statements about the properties of the central stars. Some particularly quite specific processes are important in determining the structure of the nebulae, e.g., as Hummer has told us, the He II Ly-α photons are important in the ionization of H, particularly because they 'create' high energy electrons in the nebula. A good deal of work has been done on radiative-transfer methods for treating the ultraviolet ionizing radiation. We have heard of the importance of diffuse radiation field, including specific resonance lines such as those I described just a moment ago. The interaction of the diffuse radiation field of one ion (such as the He II Balmer continuum) with another ion (the ionization of H in the present case) is often important. Looking into the future it seems to me that with large computers we will be able to make giant

Osterbrock and O'Dell (eds.), Planetary Nebulae, 441–444. © *I.A.U.*

strides on this problem. Probably Monte Carlo calculations will become more and more important; also, however, numerical solutions of the relevant transfer equations will become less time-consuming.

We have discussed the importance of fluorescence, not only due to nebular emission lines (as in the Bowen mechanism), but also due to radiation in the stellar continuum (which might be described as case C), as Seaton has pointed out. A good deal of work will undoubtedly be done in the near future to follow up this idea of Seaton and to compare its predictions in a detailed way with observational data. A very hurried and superficial comparison I have been able to make is that in agreement with the predictions of this idea, permitted lines of the same multiplicity as the ground level (such as O I $3^3S–3^3P$ $\lambda 8446$) tend to be strong in planetary nebulae as compared with lines of different multiplicity than the ground level (such as O I $3^5S–3^5P$ $\lambda 7774$).

We still do not understand completely the H-line spectrum. Here is one problem which should be calculable to high accuracy, and the results should not depend critically on the physical conditions. The observations seem (by comparison of results of different observers taken with different instruments using different procedures) to be accurate to within 10%. Nevertheless, in some nebulae they disagree with the theoretical ratios by amounts larger than this probable error, and we do not understand why. Possibly the answer may be found in fluorescence in the higher Lyman lines. I think we must be sure we understand the H problem before we can say with certainty that some physical process has not been overlooked.

Once we subject almost any kind of observations of planetary nebulae to detailed scrutiny, we find that they do not agree in a detailed numerical way with calculations based on theoretical models. For instance, the values of T_e, N_e, and T_* calculated from observations of lines of different ions vary widely within a single nebula. The reason must be that different parts of the nebula have different temperatures. Some regions in planetaries cannot be observationally analyzed at all; for instance, at the present time we cannot measure spectroscopically the electron temperature in the most highly ionized regions of planetary nebulae where there is no [O III] or [N II]. This points out the need for ultraviolet observations, e.g. in this example, observations of [Ne III] and [Ne V] lines which can only be made from above the earth's atmosphere.

The problem of the interpretation of the helium emission-line spectrum still exists. The question is, What physical process destroys metastable helium (He I 2^3S) so effectively in planetary nebulae? The $\lambda 10830$ line is much weaker than we would predict for objects at the known electron densities of observed planetary nebulae, which must mean that in some way or other He I atoms are removed from the 2^3S level. Interpretations based on a high density of H I Ly-α radiation are faced with the difficulty that other expected consequences of this radiation are not observed.

Many of the papers at this conference have emphasized the importance of fine-scale structure in planetary nebulae. There are differences from point to point in a nebula of N_e, T_e, ionization, etc. Discrepancies in the 'observed' T_e, as measured e.g. by the

[O III] method and by the radio-frequency method, undoubtedly are due to these temperature fluctuations. Comparison of different kinds of observations (as in the example I have just cited) will give information on the scale and amplitude of these fluctuations. I think we are coming to realize more and more the importance of this small-scale structure. Our observational measurements must be aimed, as much as possible, at analyzing individually the small condensations within nebulae. I think we will see more concentration (with specialized instruments that combine large scale and high luminosity) on individual features within a relatively few planetary nebulae. We will have to be selective, and try to understand well a few typical examples, from which we may then hope to generalize to the class.

Actually of course, we cannot discuss the physical processes in a nebula independently of its dynamics; stationary, equilibrium planetaries do not exist. The nebula is expanding, the central star is evolving, and these non-stationary physical effects have to be included in our attempt at understanding planetary nebulae.

Several theoretical papers have described the possible importance of non-thermal particles in planetary nebulae – e.g., a stellar wind was mentioned by several speakers. We should make observational searches for these non-thermal particles, that is, we should see if some features of the planetary-nebulae spectra point to the presence (or absence) of fast protons or of fast electrons in nebulae.

The abundance determinations show that there are no large composition differences between nebulae and common stars, nor between one nebula and another. However, all the abundance determinations have a relatively large margin of error at the present time, because of our uncertainty about specific numerical values of T_e, N_e, etc. With better knowledge of these quantities, and with observations of all stages of ionization (made in the satellite ultraviolet and infrared regions, as well as in the photographic and visible regions accessible from the ground) we can possibly begin to look for subtle abundance changes (smaller than factors of 3) and try to understand the nuclear evolution of planetary nebulae and their progenitors.

DISCUSSION

Gurzadian: Can you explain the observations of Aller, Minkowski and Kaler, that the high Balmer lines appear to be stronger than calculated?

Seaton: I do not have an explanation, but I would like to make a suggestion for future observational work. The intensities of the high Balmer lines should be measured relative to the intensity of the Balmer continuum. This would enable us to deduce b_n factors from observation. If we assume that the recombination theory is correct for the low Balmer lines, the intensities of the high lines lead to b_n factors greater than unity. It would be very valuable to check this by measuring the high lines relative to the continuum.

Aller: It is indeed possible to measure the intensities of the high Balmer lines with respect to the continuum, but there is trouble with the Hg (mercury) lines near 3650. Kaler noted that the deviations occurred not only for H, but also for He I and He II, the deviation depending on the number of j-states involved!

Capriotti: The classical Case C of Aller, Menzel and Baker assumes that there is no absorption of the Lyman-line radiation that is produced in the nebula and that radiation from the central star in the spectral region right below the Lyman limit is absorbed in 1s–*n*p transitions. The predicted relative Balmer-line intensities from Case C deviate from the observed intensities when *n* is large as do the predicted intensities from Cases A and B of Menzel and Baker. If one includes the effect of the absorption radiation that is produced in the nebula in a modified Case C, one can produce the observed Balmer-line intensities with the proper combination of optical thickness at the Lyman limit in the continuum, stellar black-body temperature and fraction of stellar Lyman continuum radiation that is converted to Balmer-line radiation by the nebula.

Seaton: I have also considered this possibility, but find that it is capable of explaining the observations only if one assumes a large excess of the stellar radiation in the region of the high Lyman lines.

Aller: Kaler tried to correlate abnormal H-line intensities with density, electron temperature, properties of central star, and filamentary structure. He found no clear-cut correlation, save some suggestion that the more inhomogeneous the nebula the greater the effect.

Capriotti: How did Kaler consider the correlation between the central star characteristics and the deviations of the observed relative Balmer-line intensities with respect to the predicted intensities?

Aller: Kaler tried to see if the Zanstra-Menzel temperature of the central star – or the brightness of the central star – influenced the intensities of the high Balmer lines. He found no clear-cut correlation.

Van Horn: In the study by Kaler, was the other obvious correlation tried of the strength of this effect relative to position along the Harman-Seaton sequence?

Aller: Kaler did not have enough observational data to allow him to correlate effects with position of the central star on the Harman-Seaton sequence. You need high dispersion spectra which restricts consideration to bright nebulae.

GALACTIC DYNAMICS AND GALACTIC DISTRIBUTION

M. W. Feast
(Radcliffe Observatory, South Africa)

The main papers at this symposium dealing with the galactic dynamics and distribution of the planetary nebulae were those of Perek and Cahn, and I shall not attempt to summarize their work again. Rather, I want to draw attention to what I believe are some of the main problems we have to solve in order to make further progress in this field.

Aside from the problem of deciding whether a particular object is a planetary or not, a far from trivial problem, there seem to be two basic problems to be solved. First, there is the problem of distance determinations, and second, the problem of the heterogeneous nature of the planetary population in the Galaxy.

Taking the second point first, it seems to me that there are good reasons for believing that the planetaries have a wide range of ages, and that these ages span the period of the initial collapse of the Galaxy to a disk. For instance, the presence of a planetary nebula in the globular cluster M 15 which is metal deficient by a factor of the order of 100, indicates that some planetaries belong to the Halo population. Nevertheless, the overall distribution of planetaries, unlike that of globular clusters, shows a marked concentration to the galactic plane, indicating a substantial fraction of disk objects. Furthermore, the velocity dispersion of the planetary nebulae, radial to the galactic centre, increases steadily as one goes from the Sun in towards the galactic centre. This was shown very clearly by Perek using the velocities of Mayall and Minkowski. This result is paralleled in a remarkable way by the Mira variables. Studies of the Miras have the great advantage that although we know that they cover a rather wide range of ages, they may be separated into groups of relatively small age dispersion by grouping according to the period. Thus, in the case of the Miras, we are able to see that the increase in the velocity dispersion towards the galactic centre is simply a reflection of decreasing mean period; that is, of increasing mean age, as the galactic centre is approached. For Miras of a given period (i.e. of a given age) the velocity dispersion seems to be independent of distance from the centre; a result which, incidentally, is in accord with the predictions of classical ellipsoidal theory. Evidently, therefore, the change in the velocity dispersion of the planetaries with distance from the centre, indicates a range in ages for these objects. A further point is the fact that in the solar neighbourhood the asymmetrical drift for the planetaries (-16 ± 7 km/sec) matches that of the younger (that is, the longer-period) Miras.

Evidently, therefore, before much further progress can be made in discussing the

Osterbrock and O'Dell (eds.), Planetary Nebulae, 445–447. © *I.A.U.*

galactic distribution and dynamics of the planetaries there is an urgent requirement that we find some physical characteristic of the planetaries which is an age index. Very little seems to be known on this point. Some long while ago, H.M. Johnson (*Astrophys. J.*, **120**, 1954, 182) suggested that there was a statistical correlation between the distance of the planetaries from the galactic plane and their excitation class (N2/Hβ ratio). A much more extensive investigation of this and other possible age-dependent properties of the planetaries is needed. Both Dr Westerlund and I mentioned the different frequency distributions of excitation classes in the two Magellanic Clouds, and in the LMC there is some suggestion of a correlation between absolute magnitude and velocity dispersion for the planetaries. These may perhaps be clues to the type of age index for which we are looking. Clearly, wholesale determinations of chemical abundances would be of very great importance in subdividing the planetaries into coeval groups, but we have heard from Aller of the difficulties of this work.

While we can be fairly certain that there are planetary nebulae ranging in age from the oldest Halo objects down to the younger disk objects, we are quite uncertain as to how old the youngest planetaries can be. Westerlund and Miss Webster have suggested that some of the planetaries in the SMC may have come from stars with masses as large as 1·8 $M_{\odot}$. This would make them relatively young objects, and of course there is the interesting case of NGC 1514 discussed by Kohoutek, where the planetary nucleus appears to have an early A-type companion. On the other hand, it seems rather unlikely on kinematic grounds that there can be any very high proportion of extremely young planetaries in the Galaxy. This seems to be shown fairly clearly in the Magellanic Clouds also. In both Clouds, although the planetaries seem to have the same overall kinematic flow pattern, as is shown by the very young supergiants and the gas, there is a lack of detailed agreement of positions and velocities for the planetaries and the very young objects.

It must be emphasized that until we solve this problem of dividing the planetaries into homogeneous groups, we can only study the galactic distribution and dynamics of these objects in a rather limited way.

The other major problem in discussing the distribution and kinematics of the planetaries is the problem of distance determination. A great deal of work has been, and must still be, put into the solution of this problem. It does, however, seem necessary to stress one point which appears to have received little attention up till now. We require not only a means of distance modulus estimation which is correct on the average, but also an estimate of the standard error of the distance modulus. Straightforward statistical arguments show that if the standard error of a modulus determination is high, then large corrections to the mean distances may be necessary when the distribution or kinematics of the planetaries are being studied. This is, of course, a classical problem in statistical astronomy; but since one or two people apparently misunderstood me when I raised the point briefly in the general discussion, I should perhaps give a simple example. If we look in a direction in which the space density of the planetaries is increasing with distance, and select all the planetaries to which we have assigned the distance *r*, then this group obviously contains some planetaries whose true distance is greater than *r* and some whose true distance is less. Because of the density gradient there are more stars at greater *r* so that a group selected in this way contains more underestimated distances than overestimated distances. The resulting systematic error in the mean value of *r* may be quite large and of considerable importance in dynamical studies, especially for distant objects in the direction of the galactic centre where the density gradient is large. Evidently here again the Magellanic Clouds should prove invaluable both in helping fix a distance scale as well as in giving some indication of the errors of an individual modulus.

To conclude let me mention three observational problems. Firstly, and obviously, the extension of radial velocity work in the Southern Hemisphere based on the many new Southern planetaries discovered by Henize is very desirable. Secondly, it must now be clear to all of us just how important a place in our ideas is occupied by the planetary in M 15. It would be a great reassurance if we could find at least a couple of field planetaries that also have low O/H and Ne/H ratios. Finally, there has been an indication this week that more planetaries are expected in globular clusters than are actually found. Further searches for such objects are evidently desirable, and it is important that investigators whose searches have proved negative be encouraged to publish full details of their work.

DISCUSSION

O'Dell: The suggestion by H. M. Johnson that there exist variations in the excitation of planetaries correlated with galactic latitude is not confirmed by photoelectric measurements. The apparent effect seen by Johnson was probably due to uncertainties in the photographic spectrophotometry.

I believe that the expected number of planetaries that should exist in globular clusters is about 5 in the entire Galaxy, which indicates that we are at about the correct level of discovery since only one nebula has been seen.

Minkowski: One should consider not only planetary nebulae in globular clusters, but also planetaries at large distances in the halo. The only known object of this kind is a planetary of about magnitude 15 close to the galactic pole discovered by Haro. A search for more objects of this kind might be of interest.

Feast: While I think that a search for planetaries far from the galactic plane would be very valuable, I would guess from the distribution of the much more frequent Mira variables that one would not perhaps expect to find many such planetaries, so that the search would have to be very extensive.

RADIO EMISSION

J. G. Davies
(Nuffield Radio Astronomy Laboratories, England)

I understand that my task is to summarize the present state of radio observation of planetary nebulae, and to give us radio observers our marching orders for the next few years. Where, then, do we stand at the end of this Symposium? Well over 100 planetary nebulae have been looked at by one or more groups at one or more frequencies, and nearly that number have been detected. Most nebulae which have been observed at several frequencies show optically thin spectra, about eight to a wavelength of 40 cm, and three to 70 cm. The fluxes of these nebulae agree well with the values predicted from the observed Hβ radiation, after correction for interstellar extinction. Some are clearly optically thick at long wavelengths, but none have been shown to remain so at the shortest wavelengths. NGC 6853 becomes optically thin near 75 cm, and NGC 6857 may become thin near 5 cm. Eight further nebulae become thin at intermediate wavelengths.

Non-thermal spectra have been reported, but the low-frequency observations on which these are based are not free from the effects of confusion, and are open to considerable doubt.

One spectral line, the 109α recombination line, has been reported in one nebula, NGC 7027. There has been no report of any detectable polarization.

Angular detail is now available on a few nebulae, and shows general agreement with optical size; but a detailed correlation of radio features with optical data will yield valuable information.

Almost all of this information has been accumulated since the IAU met in Hamburg in 1964. What must we do between now and the next symposium on Planetary Nebulae?

It is doubtful if observations of more nebulae selected rather arbitrarily from the catalogues is of much value, although it would be worthwhile to observe at 2 or 4 cm a few which might be optically thick at short wavelengths. In view of the anomalous infrared spectra of NGC 7027, this object should be observed in the millimetre waveband.

More details of the angular distribution of radio emission at more than one wavelength, preferably concentrating on a few of the larger and brighter nebulae, are needed, since all that we can do at the moment is to assume uniform emission from the optical object.

A fairly brief interferometric survey of the reported non-thermal nebulae should

Osterbrock and O'Dell (eds.), Planetary Nebulae, 448–449. © *I.A.U.*

indicate whether the radiation is really coming from the nebula or from a nearby confusing source.

The spectra of a selected list of nebulae should be studied in detail, and with the highest accuracy, since both the wavelength at which the nebula becomes optically thin and the shape of the spectrum in that region will yield information on the distribution of emission within the nebula. When sufficiently accurate spectra covering several octaves about this point are available, they can be compared with theoretical curves derived from various models of electron temperature and density distribution.

In order to do this, more accurate low-frequency observations are required, and this must be a task for the interferometers, since pencil-beam instruments do not have the resolving power required to avoid the effects of confusion.

Two final points: the report of the 109α recombination line in NGC 7027 should be followed up in other nebulae, and in other lines; and since magnetic fields are proposed in some models of planetary nebulae, a search for polarization should be made.

I am sure that the radio observers here this afternoon have taken notes of the points for which their instruments are best suited, and will catch the night train back to their telescopes, so that it need not be too long before we can meet together again in a place as delightful as Tatranská Lomnica.

DISCUSSION

Seaton: I am worried about some of the results for T_e from radio observations. The earlier work of Terzian and Menon gave results in good agreement with those from [O III], but the recent radio work of Thompson for IC 418, using isophotal contours, gives a value of T_e which seems to be surprisingly low. I hope that further careful work of this sort will be made, in order to establish whether there is any real evidence for discrepancies between temperatures from the radio surface brightness and from forbidden-line intensity ratios.

Thompson: The disagreement between the temperatures deduced from the radio spectra by myself and by Menon and Terzian results essentially from the different solid angles of the models in the two cases. Menon and Terzian assumed that only the material in a shell is optically thick, but this is not what one would expect from the Balmer line isophotes of IC 418 and NGC 6572. NGC 6572 shows a central concentration of brightness with very little evidence of shell structure.

Aller: A slitless coudé plate of NGC 6572 secured by Wilson was analyzed to obtain the spatial distribution of emitting gas. The nebula consists of a ring surrounded by an amorphous outer envelope, suggesting a very inhomogeneous structure.

Minkowski: The appearance of NGC 6572 on direct photographs matches that found on the slitless spectrograms by Wilson. NGC 6572 is definitely not a nebula without structure.

Menon: The electron temperatures computed in the paper by Menon and Terzian are based on the fluxes in the black-body region and an assumed solid angle based on optical appearance. In the case of IC 418 the new interferometer results support the angular size used in that paper. Hence, I feel that the electron temperatures computed in our paper are not overestimates, but, if anything, underestimates.

KINEMATICS AND DYNAMICS

G. A. GURZADIAN
(Branch of Byurakan Astrophysical Observatory, U.S.S.R.)

I would probably not be mistaken in saying that for the first time in the practice of solid scientific meetings the problem of the dynamics of planetary nebulae has become a subject of broad and sufficiently many-sided discussion. At the same time, I am inclined to estimate the successes achieved at the present symposium as more than modest. However, there have been no sensations. To the contrary of diminishing the importance and value of the reports made here, I only want to underline the fact that despite its attractiveness, the problem of the dynamics of planetary nebulae remains as before one of the difficult and complicated fields of theoretical and practical astrophysics. It cannot be solved without the cooperation of contiguous branches of science, particularly gas dynamics, magnetohydrodynamics and others. These are new components without which the physics of planetary nebulae has, nevertheless, managed. At the same time, almost all of the physics of planetary nebulae is comprised in the problem of the dynamics. Finally, the problems of the dynamics of planetary nebulae cannot be solved in an approach detached from the problem of the origin of planetary nebulae themselves and also from the origin and development of their nuclei. In this sense, the dynamics of planetary nebulae is becoming to a certain degree a cosmogonic science.

After these remarks of a general character, permit me to dwell on some concrete questions of the dynamics of planetary nebulae. To my mind, the following problems are of particular interest.

(1) *The problem of internal motions of planetary nebulae.* Taking into consideration the small magnitudes of the velocities of these motions, it is quite obvious that great care should be taken in applying delicate and complex methods of observation. We have seen successful examples of the application of such methods in the reports of Osterbrock, Münch, Courtès and Sheglov. It is expected that the field of application will expand. It is especially necessary to raise the sensitivity of these methods. However, the absence of small-scale (chaotic) motions inside nebulae should not be considered finally proven.

(2) *The problem of planetary nebulae having two envelopes.* As far as I know, only two works have been devoted to the problem of double-enveloped planetary nebulae – that of the Gurzadian (1962) and of Minin (1958). In both works an attempt was made to show that the origin of the second (external) envelope is an unavoidable consequence of the peculiar activity of the pressure of Ly-α radiation in the nebulae. The

Osterbrock and O'Dell (eds.), Planetary Nebulae, 450–455. © *I.A.U.*

peculiarity lies in the fact that at a definite stage of development, i.e. expansion of the planetary nebula, Ly-α-radiation pressure strongly and spasmodically increases, which leads to the breaking off of parts of the external layers of the nebula from its primary mass. The process of breaking off may take place only once in the life of a nebula as its optical thickness in the frequencies of Lyman continuum radiation can only once be of the order of 5–10, a necessary condition for breaking off.

The rather trivial assumption that double envelopes may appear as the result of repeated outbursts of gaseous matter from the central star of a nebula seems unreal (see details in Gurzadian, 1962).

A curious picture of the evolution of planetary nebulae is obtained: every nebula comes into existence as single-enveloped, then becomes double-enveloped, and finally, with the expansion of the nebula, the outer envelope disappears earlier while the basic (first) one obtains a ring-like form in its projection in the sky.

The number of double-enveloped planetary nebulae is rather large, about 30. It will grow with the accomplishment of the technique of photographing faint (but not star-like) planetary nebulae. The universality, for all planetary nebulae, of the phenomenon leading to the birth of second envelopes is becoming obvious.

Under such conditions of frequency of occurrence it is extremely desirable to carry out special experiments to test the basic positions of the above-mentioned theory regarding the formation of double-enveloped nebulae. According to this theory the expansion of the outer envelope must take place with a velocity of about 10 km/sec greater than the expansion of the first (inner) envelope.

The measurement of such a difference in the velocities of expansion does not present in itself a major problem, but in an overwhelming majority of cases the brightness of the external envelope is very small compared with the brightness of the basic envelope, wherein lies the difficulty. Nevertheless, I hope that Osterbrock and Münch, with the help of large telescopes, will be able to carry out such work, if only for a few nebulae the outer envelopes of which are sufficiently bright. I have in mind particularly NGC 2392 (which has the greatest known velocity of expansion), NGC 7009, NGC 7662, and others.

(3) *The problem of magnetic fields in planetary nebulae.* Bipolarity is a frequent property of the structure of planetary nebulae. It is present in the tendency of a nebula to form two regions with heightened brightness ('cap') set symmetrically in relation to the nucleus. Examples of bipolar nebulae are NGC 2474-5, NGC 7026, anon. $16^h10^m.5$, A 70, NGC 3587, NGC 7293, and many others.

The circumstance of the bipolar structure being very well defined (always two 'caps' set strictly at opposite directions relative to the nucleus) motivated the present author to suggest ten years ago the hypothesis of the electromagnetic nature of that property. In this manner there appeared the first suggestions of the existence of magnetic fields in planetary nebulae. It is surprising that the supposition of a bipolar magnetic field, having nothing in common with the bipolar magnetic field of the central star, explains very well the diversity of structure, external shapes and forms. It appears that the elongation or extension in form of the planetary nebulae is primarily a consequence of the activity of their own bipolar magnetic fields.

I cannot say in what manner the bipolar magnetic fields arise and remain in planetary nebulae. It is indisputable that the bipolar structure cannot exist for long. Their continual existence would require the activity of some continual force. In the basic envelope a magnetic field of the non-pointed dipole-like type (when the dimensions of the dipole are comparable to the dimensions of the nebula) may explain the existence of 'caps'. However, the existence of a chaotic, irregular magnetic field within the central rarefied region of a nebula is not excluded.

These arguments do not, of course, prove the existence of magnetic fields in planetary nebulae of precisely the bipolar type. But we have independently obtained sufficiently convincing evidence of the existence in general, of magnetic fields in planetary nebulae. I have in mind the fact of the radio emission of non-thermal (synchrotron) nature of some planetary nebulae. However, radio observations do not yet say anything about the nature of the magnetic field itself. New ways must be explored in this direction. I do not know in what manner to obtain relevant data on the character of the magnetic fields in planetary nebulae. Perhaps it can be done by radio-polarimetric observations carried out on the relation of planetary nebulae with non-thermal radio emission.

The number of bipolar nebulae is great, constituting almost half of all the planetary nebulae photographed sufficiently well to permit an understanding of their structure. This fact alone is sufficient ground for attracting special attention to the bipolar phenomenon.

Perhaps one of the most characteristic aspects of our symposium is that for the first time we have begun to speak directly about the magnetic fields in planetary nebulae. In this connection I particularly like the report by Menzel, who underlined the necessity of the existence of magnetic fields connected with the nebula, for understanding the peculiar features of its structure. I hope that in the future both theoretical and observational investigations in this direction will make our understanding of the magnetic fields in planetary nebulae more precise.

(4) *The problems of gas dynamics.* At present we do not have many results on the problems of gas dynamics of planetary nebulae. In this respect we must welcome the initiatives of Kahn, Mathews and others in formulating these interesting problems. It is necessary to continue these investigations, however, using more refined physical assumptions. Perhaps we should be wary of repeating the errors of theoreticians working on the problems of the evolution and internal constitution of stars. We should have in mind that the possibility of checking theory with observations is more likely in the case of gaseous nebulae than in the case of stellar interiors; in any event, we may see the nebulae throughout.

I think that it is necessary to continue theoretical investigations regarding the stability of gaseous envelopes in general and the stability of the forms of planetary nebulae in particular, taking into account the action of magnetic fields. Gaseous envelopes, originating around stars by diverse ways (expulsion, outflow, breaking off), may possess various degrees of stability of shape. Considerations about the stability of forms of envelopes may in the end show us the way the nebulae originate.

(5) *The problem of condensations in NGC 7293.* Two concepts of the formation of condensations in the nebula NGC 7293 – static and dynamic – have existed until recently. The static concept was suggested by Zanstra (1955) and may be regarded as arising in the following manner. Condensations are regions of decreased temperature and consequently increased density of matter in a nebula in pressure equilibrium with the substrate. The condensations themselves have come into being as a result of the energy expended by free electrons in the excitation of the forbidden lines. The dynamic concept, suggested by Gurzadian (1962), connects the origin of condensations wholly with processes of non-stationary character, occurring in the central star of a nebula.

When Daub (1963), in exchange for the strongly idealized model of Zanstra, examined a more realistic model of the nebula with due regard for its chemical composition and the changes of the state of ionization, it appeared that the formation of condensations was simply impossible. The same problem, but from the viewpoint of pure thermal stability of a planetary nebula, was examined also by Field (1965) and Sofia (1966). They have come to the conclusion that within the broad limits of the physical conditions (the dilution factor of the radiation, chemical composition and density in a nebula, the temperature of the nucleus, etc.) the thermal equilibrium is stable, and consequently, the formation of condensations as a result of the disturbance of the thermal stability is impossible.

Thus, Daub, Field and Sofia have shown convincingly the inconsistency of the static concept of the condensations. We have to consider such a conclusion as progress in the problem discussed, for in this way we substantially narrow the field of possibilities. The dynamic nature of the condensations becomes more attractive.

No concrete mechanism of the formation of condensations within the framework of dynamic processes has been suggested. However, the qualitative picture follows. First, the appearance of condensations is to be connected only with the activity of the nucleus of the planetary nebula. They are clots or filaments of gaseous matter ejected or emitted from the nucleus. The velocity of their expulsion considerably increases the velocity of the expansion of the nebula itself.

The very ejection of clots is one of the results of the spontaneous rise in the activity of the nucleus taking place some time, probably some hundreds or thousands of years, previously. Then there was a long period of comparative calm, during which the process of the small-scale corpuscular outflow of charged particles does not stop. When overtaking condensations ejected earlier with great velocity, these particles bring about the phenomenon which we observe. This is similar to the interaction of the corpuscular stream of the Sun with the heads of comets. That is how, apparently, one may explain the rather clearly revealed polar forms of those condensations in the photographs of NGC 7293. No calculations have been made to quantitatively test the above hypothesis, it being more advisable to first have some confirmation or refutation of this hypothesis through direct observations. In particular, an attempt should be made to measure the proper motions of these condensations. Mrs Liller has informed us that according to her measurements, the velocity of radial motions of these condensations is about 10 km/sec, if the distance of this nebula is 200 parsec. But this value of the distance is overestimated at least by a factor of 2 or 3; the new estimates, obtained by astrophysical methods, give for the distance of NGC 7293 a value of less than 100 parsec, equal to 75 parsec (Gurzadian, 1963). Therefore, it is desirable to re-examine this problem, especially if we have in mind the preliminary character of these measurements, as I have learned from W. Liller.

It is desirable to obtain photographs in various wavelength regions of the filaments with the purpose of obtaining some data about the distribution of energy in their spectrum. Decreases in the brightness in the course of time should be expected for the condensations themselves. It is clear that any result obtained through such observations would clarify our initial assumptions. I would like to emphasize that until now, as far as I know, no special observations have been carried out on these condensations, and it would be very well if the present representatives of observatories having powerful telescopes showed initiative in this work.

Although we have spoken here about the condensations in the nebula NGC 7293 only, it is quite obvious that we are concerned with a general phenomenon. It is possible that such a generalization is premature, in fact, except for NGC 7293 and NGC 6270, since we do not possess any data about the presence of condensations in other planetary nebulae. Moreover, we occasionally come across statements that they do not exist in, e.g., the nebula IC 418 or NGC 7662. This may be the case, for the position of the dynamic concept of the origin of condensations indicates that during certain periods of the life of a nebula they may not exist at all. However, and here one should not hurry, one must bear in mind that the conditions for discovery of the condensations in the case of NGC 7293 are exceptionally favorable primarily due to its being one of the nearest, if not the nearest nebula. The author feels that condensations in planetary nebulae may serve as important and independent indicators of the processes and phenomena taking place in the nuclei of planetary nebulae.

References

Daub, C.T. (1963) *Astrophys. J.*, **137**, 185.
Field, G.B. (1965) *Astrophys. J.*, **142**, 531.
Gurzadian, G. (1962) *Planetary Nebulae*, Moscow.
Gurzadian, G. (1963) *Commun. Burakan Obs.*, **34**, 59.
Minin, I. (1958) *Problems of Cosmogony*, Moscow, p. 211.

Sofia, S. (1966) *Astrophys. J.*, **145**, 84.
Zanstra, H. (1955) in *Vistas in Astronomy*, Vol. I, Pergamon Press, London and New York, p. 256.

DISCUSSION

Kahn: I feel much less optimistic than Gurzadian about the possibility of invoking magnetic-field effects to explain the structures observed in planetary nebulae. In the study of problems involving the structure of the Galaxy or the behaviour of the interstellar gas, people have often tried to give explanations in terms of a magnetic field; but the magnetic field has usually proved rather ineffective. It seems likely that students of planetary nebulae will have a similar experience.

Woyk: As to whether or not it is meaningless to consider magnetic field models in planetaries there is the question of other types of energies which perhaps should be taken into account when drawing limits of the magnetic energy.

Capriotti: J.P. Harrington has made a study of the origin of the condensations as a thermal instability phenomenon. He has taken into account the changing behavior of the nebular radiation field with distance from the central star while in previous studies the assumption was always made that the nebular radiation field had a blackbody spectral distribution. Negative results were obtained in that seemingly unrealistically high stellar-effective temperatures were needed in order to produce thermal instabilities.

In nebulae that are optically thick in the Lyman continuum, the Ly-α-radiation pressure may exceed the gas pressure. A realistic study of the origin of the condensations in these objects should include the effects of the work done by Ly-α radiation.

Kahn: In essence, the restriction put on the magnetic field by Woltjer's theorem relates the positive energy of the magnetic field laced through the nebula to the negative energy of the nebula in the gravitational field of the central star.

Osterbrock: What do you think the distance of NGC 7293 is?

Gurzadian: I think that the distance of NGC 7293 is nearly 60 parsec, or, in any case, less than 100 parsec.

Liller: I took the distance to NGC 7293 to be 200 parsec simply because I was commenting on Vorontsov-Velyaminov's paper and this is roughly the distance derived by him. I certainly did not intend to imply that this value was sacred.

STRUCTURE

R. MINKOWSKI
(Radio Astronomy Laboratory, University of California, Berkeley, Calif., U.S.A.)

My first task is to give summaries of the sessions on observations of the spectra of planetary nebulae and on the spectra of the central stars.

About the spectra of planetary nebulae there is little more to say than that the observations of line intensities in the range from the near ultraviolet to the near infrared have made impressive progress. Most important is the extension of the observations into the infrared which has led to the discovery by Gillet, Low and Stein of unexpectedly high intensities in the continuous spectrum of NGC 7027 between 4μ and 14μ (75 to 22 THz). In this respect NGC 7027 and the Seyfert galaxy NGC 1068 resemble each other. Similarity of the compositions of the emission-line spectra of these two objects has been noted long ago; some lines of low ionization – [O II], [S II] –, however, are stronger in the Seyfert galaxies. The physical significance of the similarity of the infrared continua is not clear at this time, but I see no reason to reject the classification of NGC 7027 as a planetary nebula. It has a very irregular brightness distribution with much structure, but an outline which is roughly elliptical and the usual expansion pattern with a velocity of expansion of 21 km/sec.

Theoretical predictions of the line-emission spectra in the far ultraviolet and infrared have been made. These regions require observations from above the atmosphere. A predicted [S IV] line at $\lambda = 10{\cdot}53\,\mu$ may have been observed by Gillet, Low and Stein.

The far ultraviolet spectrum of the central stars is of fundamental importance for the understanding of the physical processes in planetary nebulae. We may hope that space observations will permit us in the not too distant future to check and to replace the computations of model atmospheres on which we must rely now to avoid the unrealistic assumption that the central stars have black-body spectra. The most important part of the spectrum, below the Lyman limit, may remain hidden forever by the interstellar absorption of hydrogen.

Few of the central stars are bright enough to be investigated spectroscopically in great detail. For the fainter stars, the superposition of the nebular emission lines of hydrogen and helium on the corresponding stellar absorption features adds to the difficulties of observations. New information becomes available only at a slow rate. Contradictions between stellar temperatures derived from Zanstra methods and from spectral information are not infrequent. Until such difficulties are understood, the HR diagram is not well established and not a safe basis for discussions of the evolution of the central stars.

Osterbrock and O'Dell (eds.), Planetary Nebulae, 456–462. © *I.A.U.*

Photometry of the central stars was a badly neglected field that now seems to find more attention. Variability of central stars has been established in some cases, and Kohoutek has succeeded in showing that, as has often been suggested, the A0 type central star in NGC 1514 is a binary whose unobserved companion is the real nucleus of the nebula.

Turning to the general aspects of our problems, I want to emphasize a point that concerns both observers and theoreticians and that has been disregarded almost completely, except for a remark by Abell and the report on chemical abundances by Aller, whose plea for increased attention to the complicated structure of planetary nebulae I want to join and to amplify. I believe that one of our most urgent problems is that of trying to assess the consequences of picturing a planetary nebula as an envelope or shell with constant density, constant ionization, and constant electron temperature. I would be foolish to ask for attempts to analyze even one planetary nebula in full detail. But we should try to estimate at least the order of magnitude of the errors that are now being committed by total disregard of the properties of real planetaries.

The problem concerns the observer. The photoelectric observations which are our most reliable source of information on the flux densities in the emission lines, and many photographic determinations, give us values integrated over the whole nebula. But, if we are to be able to assess the effect of filaments and condensations on the analysis of the results, we must know the conditions in these features. We need reliable spectrophotometry of small details and local areas. Even such observations give no more than an integral over the line of sight through the nebula. Differential observations of condensations and their surroundings are necessary to ascertain the conditions in condensations. Here is perhaps the moment to point out that differences between photoelectric and photographic results do not necessarily demonstrate the superiority of the photoelectric method, but often are likely to show differences between observations that integrate over the whole nebula and localized observations with a slit spectrograph. Good photographic photometry is difficult, but not necessarily much inferior, and possibly the method of choice for detailed studies.

The theoretician must remember that observed intensities are not proportional to volume emissivities, but to integrals over all or part of the nebula. Even crude models of inhomogeneities may be useful for estimating the errors that may arise from the use of unrealistic and oversimplified descriptions of a planetary nebula. It is easy to see that such errors can be very large. The errors concern all quantities that interest us. How much can we really trust, for instance, the evolutionary track derived for the central stars?

I will not try to show here how dangerous these errors easily can be. But I hope it will be instructive to show you a few photographs of planetary nebulae which demonstrate the problems.

NGC 1501, photographed in the red (Hα and [N II]) lines (see frontispiece), has a well-defined nearly elliptical outline. The appearance in the green [O III] is very

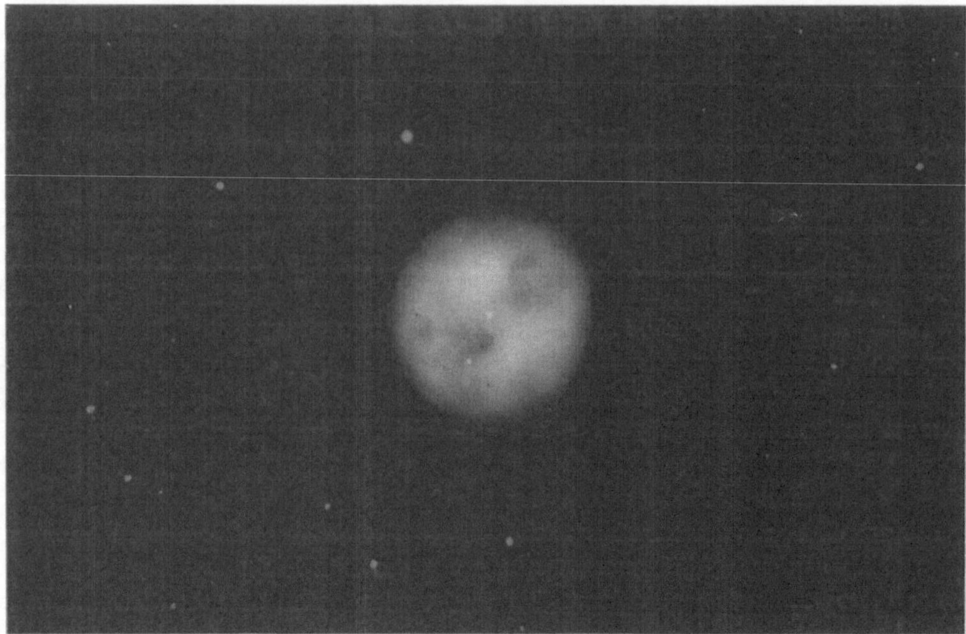

FIG. 1. *NGC 3587 photographed with the 200-inch telescope in* [O III]. *Scale 6·5″/mm.*

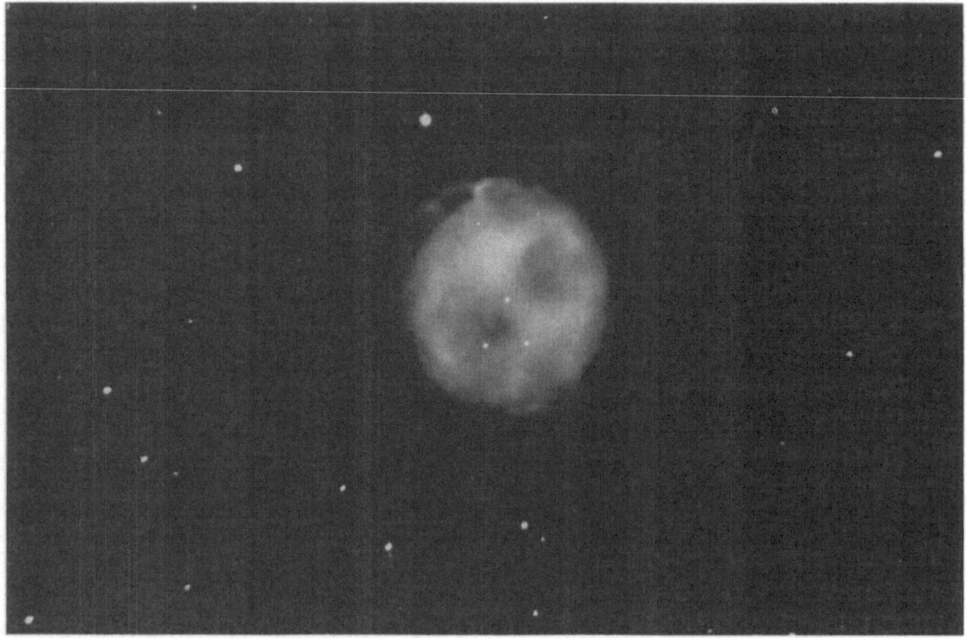

FIG. 2. *NGC 3587 photographed with the 200-inch telescope in Hα and* [N II]. *Scale 6·5″/mm.*

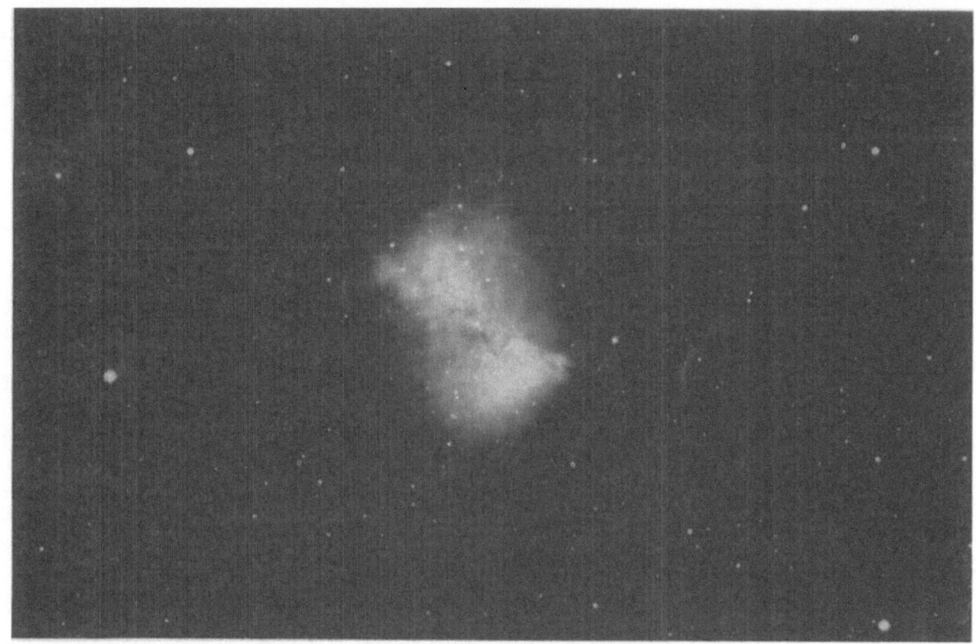

FIG. 3. *NGC 6853 photographed with the 200-inch telescope in* [O III]. *Scale 8·6″/mm.*

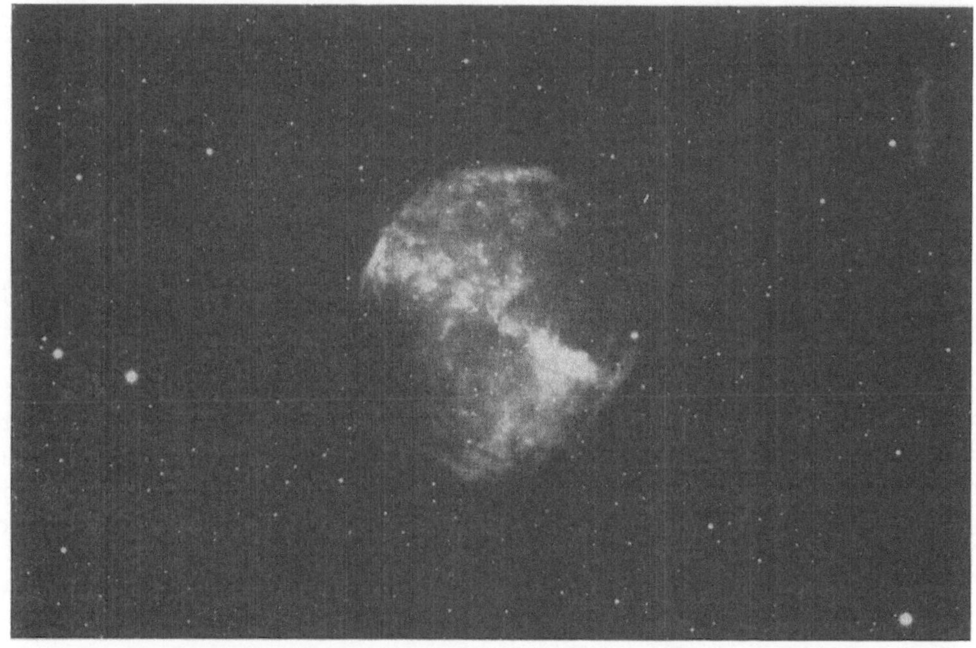

FIG. 4. *NGC 6853 photographed with the 200-inch telescope in* $H\alpha$ *and* [N II]. *Scale 8·6″/mm.*

similar. The size obviously can be stated without difficulty. But the nebula is a mass of small irregular condensations, with some big holes. What is the filling factor? Which fraction of the stellar radiation escapes through the holes, and how much too low is the temperature of the star derived on the assumption that the star is completely surrounded by the nebula and that there is no loss of radiation?

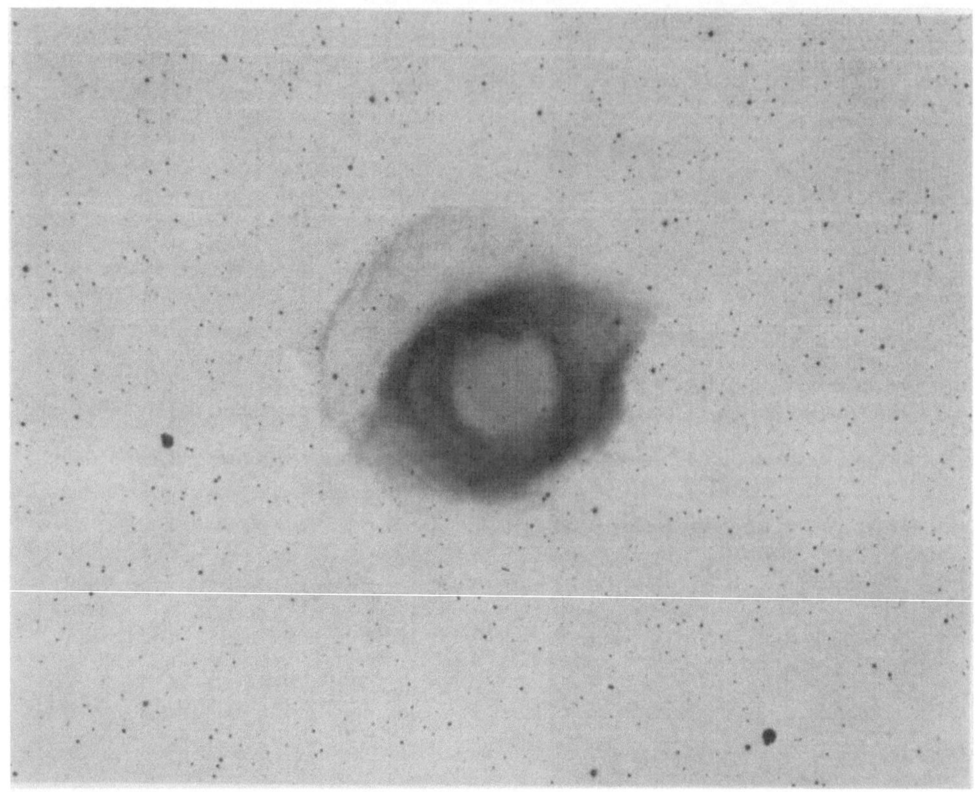

FIG. 5. *NGC 7293 photographed with the 48-inch Schmidt telescope in Hα and* [N II]. *Scale 28″/mm.*

NGC 3587 (Owl nebula) is shown in Figure 1 photographed in [O III], in Figure 2 photographed in Hα and [N II]. The main difference is the enhanced brightness of the filamentary structure near the Northern edge which is most likely caused by [N II]. The outline is fairly regular, the size not too uncertain. But, what is the filling factor, and how misleading is it to take the observed ratios of integral line intensities as ratios of volume emissivities?

NGC 6853 (Dumb-bell nebula) is shown in Figure 3 photographed in [O III], and in Figure 4 photographed in Hα and [N II]. The [O III] picture shows the structure which gave the name to this nebula, but the red picture looks quite different, with numerous small condensations which probably show mainly [N II]. [O II] probably would look quite similar. Stronger exposures show outer loops in position angle 115° where the diameter is 470″. There is an exceedingly faint approximately circular envelope with a diameter of about 840″. What is the size, what is the filling factor, how completely does the nebula surround the star? Slit spectra might show a wide variety of relative intensities, different from integral values, depending on where the slit is placed.

NGC 6781 appears as a ring in Hα and [N II], as a disk in [O III] (for illustrations see Minkowski, 1964). Broad-band photographs in the blue look similar to the red picture because [O II] is strong on the outside. The difference in appearance shows,

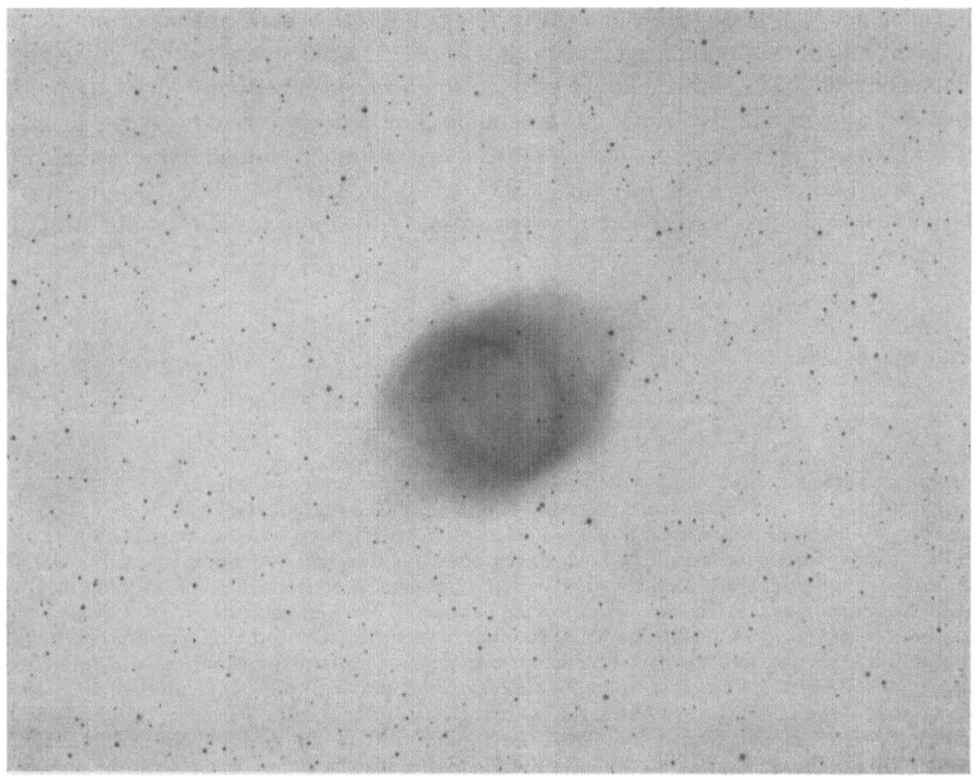

FIG. 6. *NGC 7293 photographed with the 48-inch Schmidt telescope in* [O III]. *Scale 28″/mm.*

of course, the ionization structure. Nebulae such as this might permit detailed tests of computations of the ionization structure such as those reported here by Williams who has compared for the first time computed integral values of line intensities with observed intensities without, however, considering the possible effects of structure and density variations in the nebula.

NGC 7293 is shown from photographs with the Palomar 48-inch Schmidt telescope in Hα and [N II] in Figure 5, in [O III] in Figure 6. The red picture shows a faint outer arch in the North. Here, you may begin to wonder how to define the size. The difference between the red and the [O III] picture again shows the ionization structure. If the nebula were bright enough to be photographed in He II λ4686 Å, it probably would appear as a small disk. Photographs with the 200-inch telescope in Figure 1 of Seaton's introduction (p. 1) and Vorontsov-Velyaminov's paper (p. 256) (an enlargement of an area in the Northern part) reveal the full complexity of the structure which is beyond the resolving power of the 48-inch Schmidt telescope. The heads of the well-known 'comets' range down in size to unresolved features smaller than 1″ of arc. Such structures might be present in many nebulae which appear quite smooth, but are at a distance much larger than NGC 7293, the planetary nebula closest to us.

The examples that I have shown were selected because they are large enough for detailed studies and fairly simple. Such nebulae should permit us to explore the difficulties that stand in the way of a more realistic discussion. In smaller nebulae these difficulties will be less obvious, but this does not mean that they are not present. To worry at the present stage about the worst examples of complex forms and structures would be premature.

Reference

Minkowski, R. (1964) *Publ. Astr. Soc. Pacific*, **76**, 197.

DISCUSSION

Menon: Is there any evidence for the presence of dust in planetary nebulae?

Kahn: What upper limit can you set on the amount of dust in a planetary nebulae?

Minkowski: There is no indication of the presence of dust. The violet and the red components of the lines – showing the front and the back of the nebula – show no systematic intensity difference. The central stars, which are seen through half the nebula, show no reddening in addition to the interstellar reddening. Galaxies are often seen in nebulae of large angular diameter.

Miller: Are any direct photographs being taken of planetaries, say in the light of one spectral line, which would serve as good first epoch plates for measuring nebular expansions at some future time?

Liller: We are not taking first epoch monochromatic photographs. One must emphasize the uncertainties in the establishment of the zero point of the distance scale; therefore, I sincerely believe that when Mrs Liller and I have completed the reduction of the Mount Wilson photographs and have derived good angular motion rates for the *filaments* (not the edges), then it will be possible to evaluate the distance scale zero point to an accuracy not yet attainable.

Mathews: While I do not want to detract from the importance of proper motion measurements in planetary nebula, dynamic models indicate that the combination of proper motion data and measured radial velocities to determine nebular distances will be very difficult indeed. The same problems of interpretation may arise even when proper motions of smaller features are measured.

Liller: As we have pointed out, we are trying very hard to use well-defined filaments instead of edges for our expansion measures. We feel that by doing this, we are measuring true motion of material.

ORIGIN AND EVOLUTION

MALCOLM P. SAVEDOFF
(The University of Rochester, Rochester, N.Y., U.S.A.)

"It is said: He who is content with his opinion runs into danger. Blessing to the owner."*

Discussants have been asked to synthesize and prophecy rather than to summarize. The variety of models and views expressed here warn that any report is only another opinion. Blessing to the reader.

The major impression from these sessions is a general discontent with simple qualitative answers while encouraging more realistic exploration of both theoretical and observational questions. Let us examine some of the major questions raised.

What is the ancestor of the planetary nebula? Abell, Menzel, Paczyński, and Rose answer: Red giants – though for different theoretical reasons. It remains probable from the presentations that this is a phase differing from those currently discussed in evolution from the main sequence in having a very dense core and a large envelope. It is suggested that these cores lie between $0{\cdot}5\ M_{\odot}$ and $1{\cdot}5\ M_{\odot}$ and have envelopes comparable to the resultant planetary both in mass and composition. The Rochester group emphasizes that the major mass loss may occur prior to planetary formation and attempts to produce planetaries in relatively simple objects of uncertain parentage.

What is the energy source for planetary production? The energy is bounded by the observed kinetic energy, 10^{46} ergs, and the estimated 10^{49} ergs required to lift this much material from a degenerate core. Most agree that nuclear energy is involved, except Paczyński who seems to be depending upon the ionization energy of H and He.

Why are planetaries characterized by a low velocity of expansion compared to the velocity of escape of the central star? In Paczyński's view, this is a natural consequence of the positive energy of the extended envelopes, the velocity of expansion being equivalent to the energy available to a recombining proton and electron. What is expected from the pulsational instabilities discussed by Rose? It was suggested to me by Ledoux in May 1965 that pulsations in these models may build up until the surface layers at maximum expansion velocity exceed the velocity of escape. The added energy sink (mass loss) may limit the amplitude of the pulsation provided the input is not too large. Thus, if the energy in pulsation grows on a time-scale comparable to the

* Inscription on Afrasiyab-ware plate from Nishapur, Iran, 10th century A.D. Freer Gallery No. 57.24, Washington, D.C.

Osterbrock and O'Dell (eds.), Planetary Nebulae, 463–466. © *I.A.U.*

period, an equilibrium would be established. Perhaps novae reflect still larger amplification factors.

The only form of mass loss for which we have quantitative models is the solar wind. A complete story would need a theory of how the 'corona' was formed (from convection?) and the prediction of densities and velocities. Scaling from solar parameters, one finds excessive velocities if the critical potential is measured by the observed radius of central stars, and one is again led to a 'red giant model' in which the critical potential is that of a hundredfold larger radius than observed.

K.H. Böhm has called our attention to a new class of atmosphere problems in which the luminosity satisfies the inequality

$$L > 10^{4 \cdot 8} M/(1 + X),$$

all quantities being in solar units and specialized to electron-scattering opacity. Here radiation pressure dominates gravitation, and no purely static or even convective solutions are expected. I have not encountered discussions of this problem previously and would like to speculate on its dimensions. Do we need to study the 'real' problem or is there much to be learned from the 'grey' atmosphere approximation? Is the plane parallel approximation at all useful? How sensitive are the hydrodynamic solutions to starting-conditions, and to the little approximations that must be made?

Our previous discussion is concerned with the removal of the gas shell from the star. The dynamics of the gas shell itself has been the subject of papers here by Menzel, Gurzadian, Khromov, Kahn, and Mathews. Although there has been some isolated support for dominant magnetic fields and non-thermal processes, these seem not required by the observations reported at this conference. One should recall that, in the absence of an adequate description of how mass-loss occurs, the initial conditions for the gas-shell dynamics are the pressure, temperature, velocity of the gas at some initial time within some volume, the incident radiation and corpuscular fluxes (which are not only a function of time but may also be dependent upon frequency and position on the surface of the volume). Models fitting observed planetaries are obviously sufficient but not necessary in the presence of such vast unexplored degrees of freedom. Thus corpuscular fluxes and magnetic fields appear to be interesting deductions but, pending considerable investigation, unproven.

Some topics of more specialized interest in the area principally of the physics of white dwarfs require comment. In principal, the high electron conduction damps thermal memories of the star's earlier history so that we need only guess the composition. Even the composition is obscured by the low dependence of the equation of state of degenerate matter on composition ($\mu_e \sim 2$). The simplicity of evolution below $0 \cdot 1 \; L_\odot$ permits interpretation of the rather weak observational statistics available. I will not discuss the mass-radius relation as it is difficult to assess possible systematic errors in the data for Sirius B.

Evolutionary models of white dwarfs permit prediction of the luminosity or temperature distribution of the white dwarfs. This has led Weidemann to estimate a production rate of 2×10^{-12} yr^{-1} pc^{-3}, a rate which is perhaps 2–20 times that of the planetary nebulae. This seems based on the present density of white dwarfs between $M_v = 11{\cdot}5 \pm 0{\cdot}5$, and the assumption of uniform production in time over the past 5×10^9 years. If we choose to base our estimates on all white dwarfs brighter than $V = 14$, in a particular band of color, we would eliminate the need for parallaxes. The evolutionary curves would provide an estimate of the volume occupied by the sample and the transit time. For this magnitude limit, even the brightest planetary nucleus lies within a scale-height of the galactic plane permitting the assumption of constant density. A color-dependence of the apparent production rate need not be interpreted as galactic evolution but would be related to details of the evolutionary tracks not observable in other ways. Aside from observational selection, four physical phenomena would be revealed in this way.

An absence of the bluest stars may result from the blue limit to the surface temperature of a contracting non-degenerate star which results from the onset of core degeneracy. This can be interpreted as evidence of too high a mass in the models. Savedoff, Rose and others have emphasized that the radius of these brightest models is strongly affected by small hydrogen envelopes. Thus the same empirical evidence might imply the mass of the hydrogen envelope has been underestimated. Again, neutrino losses have been shown to speed evolution in the range of colors bluer than U-B$= -1{\cdot}0$ for masses exceeding perhaps $0{\cdot}5\ M_{\odot}$. Alternatively, evolution in the range U-B$=0{\cdot}0$ may be slowed by the heat of crystalization discussed by Van Horn. I have received a private communication from Ostriker noting that the specific heat of a crystal star below the Debye temperature is sufficiently small so that the evolution rate is increased and a deficiency of very red stars would be expected. This low specific heat allows a star to reach zero temperature in a finite time. It is not clear how these effects can be separated, but it would be useful if the observers could find the color distribution so that an attempt can be made to interpret the shape of the distribution in these terms.

Lastly, we must admit that the physics of matter and the contents of our models leave the theorists with some unfinished business. The calculations of the rate of the carbon-carbon reaction have been questioned by Reeves on the basis of recent measures by Fowler's group. The neutrino rates have been estimated by Salpeter to be correct to within 10 %, provided that the elementary interaction

$$e^+ + e^- \to \nu + \tilde{\nu}$$

exists and is correctly given by the theory of the universal Fermi interaction. These hypotheses appear untestable in the laboratory. It must be noted that the conductivity of the electrons in the region of relativistic degeneracy has not been treated, and although the conductivity is so high in this region that its effect should be unobservable, the present situation is a bit untidy. Model calculations are plagued by the expense resulting from relaxation oscillations and uncertainties arising from the physics of mixing. As always effects of rotation and magnetic fields remain on the edge of the calculable, although there are better ideas of the magnitude of the effects which may be expected.

In conclusion, we have many opinions of the explanation of the planetary phenomena and the evolution from the central stars into the white-dwarf regions, but one has little cause to be content with his own opinions.

DISCUSSION

Salpeter: Appreciable mass-loss by radiation pressure from stars of small radius (like central stars of PN) would require a lot of energy drain from the optical luminosity (via red-shift of photon). Can anyone give an upper limit to such mass-loss which would not contradict observed spectral data?

GENERAL DISCUSSION - SEVENTH SESSION

Seaton: I think it would be useful to summarize the estimates which have been made for the calibration of Shklovsky's distance scale for optically thin nebulae. Table 1 gives: the calibration constant, relative to O'Dell's value; the nebular mass, including helium; the mean distance, $|\bar{Z}|$, from the plane. (The table was completed by contributions from other participants.)

Table 1

Author	Calibration Constant	M_{neb}	$\|\bar{Z}\|$	Remarks
O'Dell	1·00	0·2	280	
Seaton	1·45	0·6	360	method valid for $0{\cdot}06 \leqslant R \leqslant 0{\cdot}6$, where R is the nebular radius in parsecs
Šklovsky	0·63	–	–	
Abell	0·8	–	–	
Webster	–	0·2	–	
Oort	–	–	340–450	From $\|\dot{Z}\| = 18$–24 km sec^{-1}

Perek: From what year is the value by Oort of $|\dot{Z}| = 18$ km/sec?

Seaton (to Perek)*:* It is the value quoted at the Vatican Conference in 1957.

Perek: Then it was before the radial velocities by Minkowski and Mayall were known. I get from planetaries above 20° latitude a value of 24 km/sec.

Seaton: I would like to report on the conclusions reached in a private discussion between O'Dell, Perek, Cahn and myself, concerning space densities. Let χ be the number of planetaries evolving per cubic parsec per year in the local region. The following results have been obtained (Table 2), assuming in all cases an expansion velocity of 20 km/sec^{-1}.

Table 2

Author	$\chi \times 10^{13}$	Adopted distance scale
O'Dell	3·9	O'Dell
Cahn	4·6	Seaton
Cahn	15	O'Dell

When the same distance scale is used, Cahn's value of χ is several times larger than O'Dell's. Cahn fits the data to an assumed galactic distribution while O'Dell counts the number in a certain finite volume. Perek has deduced the density by counting numbers in spheres of finite radius and extrapolating to zero radius. His results are in general agreement with those of Cahn.

Weidemann: I want to bring up the question of the planetary nebula mass again; how do we feel about it at the end of this meeting?

Abell: Most of the planetaries that I have investigated are at distances between 2 and 4 K_{pc}, because they are large and faint and would be hidden by extinction at larger distances. In general, due to the uncertainties in observational data and the uncertainties of the assumptions employed, I would not take seriously these differences in the distance scales.

As for the masses of typical shells, the very few independent determinations we have are obtained from measuring the flux from objects whose distances are assumed known from other kinds of observations. Because we suspect there to be unresolved filamentary structure, however, we really do not necessarily obtain the correct mass, but a function of mass and filling factor. The actual masses of the shells, in other words, could be much less than the values we generally assume.

O'Dell: The method used for optically thin nebulae shown in the review paper figure uses an average value for ε determined by the calibration independent of eye estimates and hence should not be greatly sensitive to unresolved filaments.

Weidemann (to Seaton):

(1) How would your mass change if you take the distance scale of O'Dell?

(2) How certain is your value in view of the fact that the best fit given in your publication was drawn in a rather compressed ordinate scale, 1 cm corresponding to a factor of 10?

Williams: I would like to point out that, assuming the validity of Seaton's mass vs. radius curve, the nebular mass one obtains for an assumed electron temperature of 10000 °K is 0·4 $M_{\odot}$, instead of 0·6 $M_{\odot}$.

Gurzadian: I have a comment in connection with the evolution of the nuclei. From the point of view of the evolution of central stars of the planetary nebulae, perhaps there should be some interest in the search for and investigation of stars which are former nuclei. Particularly, it is not unexpected that some part of the Humason-Zwicky blue objects in high galactic latitudes may be former nuclei of planetary nebulae.

Salpeter: I would like to give an oversimplified summary of the similarities and differences of some of the models we have heard about:

(1) The three sets of models by Savedoff and Van Horn, by Vila and by Beaudet and Salpeter essentially form one class of models. The essential feature of these models is an almost homogeneous star consisting mainly of C^{12} and slightly heavier nuclei. The exact abundance ratios of C-O-Ne-Mg, the numerical values used for reaction rates and opacity and the range of stellar masses considered varied somewhat, but the main conclusions are at least semiquantitatively the same: to obtain luminosities comparable with the observed ones, rather large masses are required for these homogeneous models, $\sim$1·0 to 1·2 $M_{\odot}$ (slightly larger if neutrino reactions are absent).

With neutrino reactions included, the evolutionary time-scales of the large-mass models are also of the right order of magnitude.

(2) The two sets of models by Rose and by Faulkner also form one class, a more realistic class of models which also contain a core of carbon (and oxygen, etc.) but are allowed a substantial helium-envelope in which helium burning can proceed. Again a range of masses was considered ($\sim$0·6 to 1·0 $M_{\odot}$) and here slightly lower masses are sufficient to give the observed luminosities ($\lesssim$0·8 $M_{\odot}$). The models of this class carried out so far show that thermal instabilities are likely to be important, but the precise kind and phase of instability to be encountered in central stars of planetary nebulae is not yet clear.

(3) Realistic models are not yet available for stars that have recently lost their outer layers dynamically (suddenly) and are in the process of relaxing from this mass ejection, but Deinzer's models may give some qualitative hints.